Optical Communications
Components and Systems

Optical Communications

Components and Systems

Analysis • Design • Optimization • Application

(with 326 Figures)

J.H. Franz
V.K. Jain

CRC Press

Boca Raton   London   New York   Washington, D.C.

Narosa Publishing House

New Delhi   Chennai   Mumbai   Calcutta

Prof. Dr.-Ing. Jürgen H. Franz
Fachhochschule Düsseldorf
Faculty of Electrical Engineering
Josef-Gockeln-Straße 9
40474 Düsseldorf, Germany

Prof. Dr. Virander K. Jain
Department of Electrical Engineering
Indian Institute of Technology, Delhi
Hauz Khas, New Delhi-110 016, India

Library of Congress Cataloging-in-Publication Data

Franz, Jürgen.
 Optical communications components and systems/Jurgen H. Franz, Virander K. Jain,
 p. cm.
 Includes bibliographical references and index.
 ISBN 0-8493-0935-2 (alk. paper)
 1. Optical communications. 2. Optoelectronic devices. I. Jain, Virander K. II. Title.

TK5103.59.F72 2000
621.382'7-dc21   00-057982

Exclusive distribution in North America only by CRC Press LLC

Direct all inquiries to CRC Press LLC, 2000 N.W. Corporate Blvd., Boca Raton,
Florida 33431. E-mail: orders@crcpress.com

PHYSICAL CONSTANTS AND CONVERSION FORMULAS

PHYSICAL CONSTANTS

Constant	Physical meaning	Quantity
c_0	Velocity of light	$2.998 \cdot 10^8$ m/s
e	Electron charge	$1.601 \cdot 10^{-19}$ As
h	Planck's constant	$6.624 \cdot 10^{-34}$ Ws^2
k_B	Boltzmann's constant	$1.379 \cdot 10^{-23}$ Ws/K
ϵ_0	Permittivity constant	$8.854 \cdot 10^{-12}$ As/(Vm)
μ_0	Permeability constant	$1.256 \cdot 10^{-6}$ Vs/(Am)

CONVERSION FORMULAS

$1 \ \mu m = 10^{-6} \ m = 10^{-4} \ cm$

$1 \ \text{Å} = 10^{-4} \ \mu m = 10^{-10} \ m$

$1 \ Np = 8.686 \ dB$

Frequency f in Hz $\approx 3 \cdot 10^{14}$/wavelength λ in μm

Bandwidth Δf (at centre wavelength λ) $\approx (c/\lambda^2) \cdot$ bandwidth $\Delta \lambda$

PREFACE

For several years now, optical fiber communication systems are being extensively used all over the world for telecommunication, video and data transmission purposes. Optical communications offer advantages of ultrahigh speed and highly reliable information transmission. Further, light intensity modulation and direct detection transmission links are cost effective as well. Optical communication links have now been preferred over the other high bit rate point-to-point communication links. In communication engineering, optical fiber and free-space communications have become more and more important. Besides microelectronic and software, optical communications represent a key technology of modern telecommunication systems.

Since the middle of the nineties, optical fiber technology has progressed from fiber links to optical networks. Besides transparency, these networks offer flexible optical routing based on optical crossconnects and wavelength division multiplexing. Such networks provide digital broadband accesses to end users by means of fiber-to-the-building and fiber-to-the-home. The optical fiber systems represent a key component of information superhighways which are required for high-quality, interactive multimedia services.

With advanced techniques such as coherent detection, optical communications can reach a new horizon, characterized by a number of new applications. Coherent optical communication systems offer significant advantages based on improved receiver selectivity and increased sensitivity. The second advantage is partially reduced in comparison to direct detection system with an optical preamplifier, but the first advantage still remains. It is quite important in coherent multichannel communication systems which offer the possibility of exploiting fully the large optical bandwidth available with the fibers.

This book has been written with the aim of providing basic material required for advanced study in theory and applications of optical fiber and space communication systems with and without optical amplifiers. The background required to study the book is only that of typical engineering students. Specifically, it is presumed that the reader has been introduced to the principles of electromagnetic theory and communication engineering. It would be helpful if the reader has some exposure to spectral analysis and statistics. Some relevant topics are briefly reviewed in this book also to maintain continuity.

The book is recommended to those, who have interest in optical communications. It can be used for an introductory level course as well as for a senior level course to engineering students. The practising engineers and physicists will also find it useful to update their knowledge in the field. In addition, this book will also be useful as a working reference in the selection and design of optical fiber and free-space communication systems.

We are pleased to thank our colleagues, students and friends who made many valuable suggestions and skilful services in the preparation of the manuscript. Last but not least, we would like to thank our wives and children for their patience during the time we devoted to write this book.

The authors wish all readers a successful study.

J. H. Franz

V. K. Jain

CONTENTS

Optical Communications
Components and Systems

1   INTRODUCTION

To begin with, this Chapter presents a short review of the history of optical communications which started more than 2000 years ago (Section 1.1). Thereafter, an introduction to modern optical communications, its background and some discussion on related technical and nontechnical aspects are presented in Section 1.2. Finally, Section 1.3 gives an overview of contents and organization of the book.

1.1   HISTORY

When someone talks about optical communications, people normally think of lasers, optical fibers and high bit rate data transmission. This basically represents the modern, high-tech part of optical communications. On the other hand, there is a part of optical communications which was used more than 2000 years ago. It has been verified that around 800 BC, *fire signals* were used to transmit information out of a limited number of selected messages known to the operators. Fire signals are optical signals, since their spectrum is located in the optical frequency domain just as the spectrum of a laser or a light emitting diode (LED). Information transmission with fire signals represents an optical free-space communication system as described in Chapter fifteen. Two hundreds years later, message of final conquest of Trojan was transmitted by fire signals via eight intermediate relay stations to Argos which is about 500 km away. This historical relay link is known as the *torchpost of Agamemnon*.

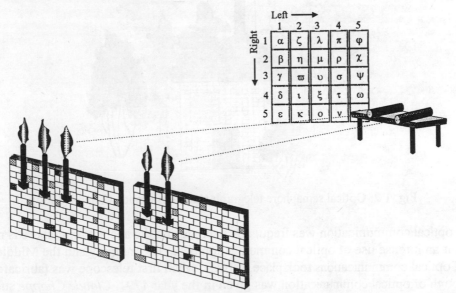

Fig. 1.1: Coding-based optical communication system and coding table of Polybios (200 BC)

In the ancient time, many other optical relay links were realized. However, they all show one main disadvantage. Instead of possibility to transmit every kind of information, only a limited number of selected messages were available which were a result of a previous agreement of the users. Around 200 BC, the Greek *Polybios* developed a torch-based transmission system which was able to transmit single letters instead of fixed messages. The basic idea of this system was to use a *coding table* as shown in Fig. 1.1. Depending on the number of torches behind the left and right wall, each letter out of 24 letters could be coded, transmitted, received and finally decoded. If, for example, two torches were behind the right wall and three torches behind the left one, the Greek letter μ was sent. With trained and educated operators, about eight letters were sent in one minute using this *coding-based optical communication system*. By taking into account an information capacity of five bits per letter, it results in a bit rate of 0.67 bit/s in contrast to some Gbit/s at present.

Fig. 1.2: Optical semaphore telegraph station of Claude Chappe (1800)

While optical communication was frequently used during the ancient time, there is no knowledge about an intense use of optical communication between Antiquity and the Middle Ages. A rebirth of optical communications took place in 1600, when first telescope was fabricated. A first breakthrough of optical communication was given in the year 1791. *Claude Chappe* successfully experimented an optical message transmission system based on semaphores shown in Fig. 1.2.

Step by step, his *optical semaphore telegraph* kept on extending, becoming a real star network fanning out from Paris. It marked the beginning of a telecommunication network spanning the French territory. In 1844, the network included no less than 5000 kilometres of lines with over 500 stations. In the first few years of the 19th century, the optical-relay semaphore moved into other countries such as Belgium, Italy, Germany, Egypt, Algeria and many more. The semaphore telegraph included a mast of four to five metres high, in the middle of which was mounted a mobile beam called the regulator. At each end of the beam was hinged an indicator, consisting of an arm of one metre length swivelling about a shaft. Each position of the indicators was corresponding to a digit or number which was observed by a neighbouring station operator using a telescope and simply repeated to the next station. Although this type of optical communication was very successful during that time, it soon found its end due to the discovery of electricity. In 1876, Graham Bell demonstrated his telephone and in 1887, Heinrich Hertz discovered the existence of electromagnetic waves by practical experiments. Some years later in 1895, first radio was developed by Guglielmo Marchese Marconi.

In 1917, Albert Einstein presented his theoretical work about stimulated emission of radiation which is the basic concept for realizing a laser. However, it took more than forty years until first laser, a gas laser, was fabricated in 1960. Only two years later in 1962, first *semiconductor laser* was realized. However, no *low-loss fiber* was existent at that time. A first experiment on *coherent optical communication* was performed in 1967 and second in 1974 by using gas lasers and free-space medium [1, 3]. However, atmospheric distortion and small transmission range limited interest in such systems at that time. In 1970, optical fibers showed an attenuation of about 20 dB per kilometre and lifetime of semiconductor laser was about one thousands hours. Experiments on coherent optical communications were restarted in 1980, but now by employing optical fibers and semiconductor laser diodes [6]. Improvements in fiber fabrication finally yielded fibers with an attenuation of 0.2 dB per kilometre in 1982. From this date on, era of modern optical communications has started.

1.2 MODERN OPTICAL COMMUNICATIONS

Optical communications make use of very high carrier frequencies located in the optical frequency domain as shown in Fig. 1.3. Wavelengths λ and frequencies f given in this figure are related by the well known equation

$$\lambda \cdot f = c \qquad (1.1)$$

where c represents the velocity of light. In vacuum, c equals $c_0 = 3 \cdot 10^8$ m/s called as free-space velocity whereas in material velocity of light is slower depending on the refraction index n which is about 1.5 in an optical fiber. Thus,

$$c = \frac{c_0}{n} \le c_0 \qquad\qquad (1.2)$$

Except some special applications, where visible green ($\lambda \approx 500$ nm) or red ($\lambda \approx 670$ nm) light is used, modern optical communications normally operates with carrier frequencies located in the near infrared (IR). Typical and wavelengths used are 1300 nm and 1550 nm which correspond to a frequency of 230 THz and 193 THz, respectively.

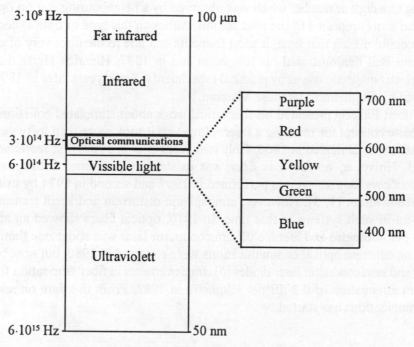

Fig. 1.3: Optical frequency spectrum

Over the past decades, the growth of *optical-fiber technology* in undersea, terrestrial long haul interoffice trunklines and cable television (cable TV) systems has been explosive. In the last few years, the speed of commercially available systems has increased exponentially doubling after every 2.4 years. In many areas of the world, long-distance conversation might already travel part of its way with some 40,000 other conversations on a 2.5 Gbit/s fiber optic line. While more and more national network operators now provide fiber-based SDH (synchronous digital hierarchy) overlay networks with a bit rate of 2.5 Gbit/s (STM-16, OC-48). First 10 Gbit/s systems are tested in various field trials and pilot projects all over the world. Besides fiber optics, innovations in micro-electronics and in software have also been most important to reach this remarkable success. As an example, Table 1.1 shows progress in transmission capacity expressed in number of voice channels per link given in selected *trans-Atlantic-transmission* (TAT) links. The complete table is given in the Appendix A1. Similar results are given in trans-Pacific links and other trans-oceanic links. Since 1988, fiber is exclusively used in TAT [4].

Table 1.1: Progress in trans-Atlantic-transmission (TAT) capacity

Year	1956	1963	1970	1976	1988	1997	2001
Medium	Coaxial	Coaxial	Coaxial	Coaxial	Fiber (1.3 μm)	Fiber (1.55 μm)	Fiber (1.55 μm)
Voice channels	84	128	720	4000	40,000	60,000 (5 Gbit/s)	≈120,000 (50·STM1)

In the middle of the eighties, optical fiber communication was used for long-haul intercity links transmitting a data rate of 140 Mbit/s. Since the 1980s, fiber has been replacing copper in telephone feeder systems and cable television trunks step by step. In many countries, fiber optics is exclusively used in telephone long-haul links since the middle of the eighties e.g., since 1987 in Germany. At present, single-mode fiber is the preferred transmission medium for *long-distance, point-to-point links* such as telephone company intercity trunks. Typically these links operate at data rates between 45 Mbit/s and 565 Mbit/s, but now also 1.6 Gbit/s to 1.7 Gbit/s (US) and 2.4 Gbit/s to 2.5 Gbit/s (Europe) systems are available at many places. The 2.5 Gbit/s long-haul fiber optic transmission link consists of a number of regenerator spans, each about 40 km long. It should be noted that only about 50% to 60% of the installed fiber is actually in use at present. The remaining fiber is "dark" fiber which will be used for future traffic. Current point-to-point transmission links only use between 1% to 5% of the fiber's potential.

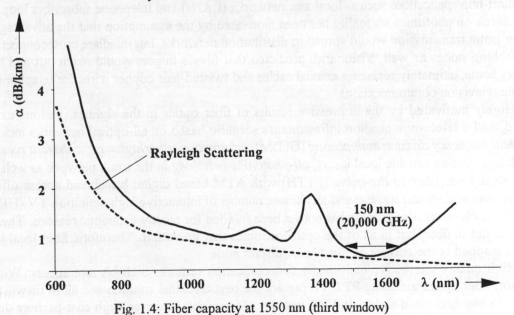

Fig. 1.4: Fiber capacity at 1550 nm (third window)

As shown in Fig. 1.4 above, in the 1.55 μm wavelength range, which is known as the third window, available fiber bandwidth is approximately 20,000 GHz or 150 nm. This very large bandwidth equals a potential i.e., *fiber capacity* of more than 300 million 64 kbit/s-ISDN channels or about four million analog TV channels! In addition, in third window fiber attenuation coefficient α is only about 0.18 dB/km. Summarizing discussion and results above we can conclude [6]:

> Fiber optics is on the road to become the key technology of information superhighways offering ultrahigh bit rate transmission of up to tens of terabit per second i.e., 10^{12} bit/s to some 10^{13} bit/s.

Now-a-days, technology required to realize *ultrahigh speed transmission links* is available, but cost barriers primarily limit the speed of progress. Looking back to the eighties and nineties, a remarkable progress in fiber optics can be recognized, but also increasing problems in the introduction of new technologies as discussed below. In the eighties, digital transmission was based on a fixed number of 64 kbit/s ISDN channels, whereas the nineties opened the way to multi-Mbit/s access based on *asynchronous transfer mode* (ATM) which offers flexible data transmission independent of bit rate.

The explosive advancement of light wave technology over the past decades has revolutionized long distance communications, whereas it is still unclear what role *photonics technology* will play in short-hop applications such as local area networks (LANs) and telephone subscriber loops [2]. Research on photonics switching has been motivated by the assumption that the advantages of fiber optic transmission would spread to distribution networks, intermediate crossconnects and switching nodes as well. Visionaries predicted that fiber's fingers would reach out and touch every home, ultimately replacing coaxial cables and twisted-pair copper wires for telephone and cable television communications.

Highly motivated by the impressive results of fiber optics in the eighties, various experts predicted a telecommunication infrastructure scenario based on all-optical networks including *optical frequency division multiplexing* (OFDM), *information superhighways, coherent receivers* with high-quality tunable local lasers, *all-photonics switching* in the trunk network as well as in the local loop, *fiber-to-the-home* (FTTH) with ATM-based digital broadband access offering *interactive multimedia services* and an ultimate number of interactive high definition TV (HDTV) channels. However, this promise has not yet been fulfilled for mostly economic reasons. The high cost is not in fiber, but in all of the optoelectronics that go with it. Therefore, fiber local loops have stopped at the curb or an even-more distant node.

High quality interactive multimedia communication requires high bit rate access. Without doubt, fiber optics including FTTH represent the best technical solution and all of us will have fiber in our home and at our desktop one day. At present, however, high cost-barriers do not allow a general global implementation of FTTH. In addition, various other parameters limit the speed of realizing FTTH as shown in Chapter fourteen. On the other hand, there is a continuously increasing demand of higher transmission capacity, new services (in particularly interactive services) and multimedia communication [7]. Hence, there is a need for advanced optical networks. Introduction of video on demand (VoD) will additionally influence the demand for transmission capacity in the near future. But high replacement costs for users require continued use of copper infrastructure. Thus systems and strategies for transport and access networks heavily depend on installed infrastructure and range of services permitted by regulation. In the nineties, most telecommunication network operators worked for the development of low expensive techniques which allow to upgrade the existent copper infrastructure and offer the user a first generation broadband access. This trend will also continue in the next decade. Upgraded networks will act as a bridge between current narrowband and future broadband telecommunication networks. However, it is not clear which technology and which network configuration will match the unknown users demands most efficiently.

Nevertheless, there is a continuous progress in fiber optics independent of different types of problems existent at present and most likely in the near future. It is quite obvious that

> Fiber optics is progressing from point-to-point links to optical networks!

In conventional optical communication systems, information is carried only by modulating the optical *power* or intensity of a laser or an LED. In the receiver, demodulation is simply performed by converting the modulated light to an equivalent electrical current by means of a photodiode. This detection scheme is termed as *direct detection*. At present, *intensity modulation* and direct detection (IM/DD) are exclusively used in all commercially available optical communication systems. However, a direct detection receiver is the optical analogue of an old-fashioned crystal radio, whereas a coherent receiver is that of a modern superheterodyne receiver used in radio and television systems. Just as heterodyning has revolutionized radio communications, coherent optical detection can open the door to more sophisticated advanced optical communications.

In *coherent optical communication systems*, information is contained either in amplitude, frequency or phase of an optical carrier signal. The receiver analyses the incoming electric field with respect to modulation scheme applied to the optical carrier. This is done by mixing the electric fields of received optical signal and local laser, which operates as a local oscillator (LO). Thereby, an intermediate frequency (IF) equal to the difference between LO and received carrier frequencies is generated. The IF signal is simply a frequency-translated replica of the original incoming optical signal. As the IF is usually in the GHz range, well-established microwave signal processing techniques can be employed to process this signal. Besides a gain in sensitivity of up to 20 dB, coherent receivers show a significant improvement in selectivity. This offers the possibility to realize a real OFDM system with a large number of broadband channels at different closely-spaced optical carriers, which is termed as *coherent multichannel communication* (CMC). In the receiver, optical channels are separated in the electrical domain by means of a tunable local laser and sharp microwave filters. Theoretically, CMC systems with more than hundred optical high-bandwidth channels are possible. Thus, CMC systems with coherent detection can make use fully the large optical bandwidth available with the optical fiber. However, all advantages of coherent optical communications are at the cost of an increased complexity. Additional components such as the local laser, optical coupler and additional control and tracking circuits like the polarization control unit are required. Moreover, some advantages mentioned above have partially been reduced by the advent of optical amplifiers and dispersion equalization in the optical domain.

To summarize this Section, let us review the most important steps of progress of telecommunication. Since many years, telecommunication market is world wide growing very rapidly and the pace of growth is still increasing. The enormous progress in communication technology during the last decades was primarily based on various milestones such as digitalization, microprocessor-controlled switching, optical fiber transmission, LAN, ISDN, cellular mobile radio and ATM. Now-a-days, modern communications is fundamentally characterized by three key technologies: *microelectronics*, *software-controlling* and last but not least, *optical communications*. The next generation of telecommunication will primarily be characterized by highly sophisticated, high bit rate *information superhighways*, interactive broadband *multimedia services* and intelligent multiple-access networks combining ISDN with broadband services which results in broadband ISDN (BISDN) and combining the flexible ATM transmission scheme with the standardization of SDH. Furthermore, photonics may more and more replace electronics. In advanced optical communications, coherent systems can represent an alternative technology of importance.

1.3   ORGANIZATION OF THE BOOK

As indicated in the title, this book is divided in two parts. First part i.e., Chapter two to eight is dedicated to components, whereas part two i.e., Chapter nine to fifteen is focussed on systems. In the component-based Chapters all relevant passive and active optical and optoelectronic components used in point-to-point links as well as in networks are discussed. The system-based Chapters contain analysis and optimization of incoherent and coherent systems. In addition, they give an introduction to fiber optic link design, discuss physical limits and present applications such as optical networks and optical space communications.

Analysis, optimization and design of optical communication systems require a basic knowledge of the physics of optical sources. Therefore, *Chapter two* gives a brief review of the basic principle of the important optical sources which are light emitting diode (LED), semiconductor laser and solid-state laser. This Chapter concludes with a brief discussion of optical transmitters which include the optical source module as well as all the electronics required to control temperature and stabilize optical output power.

The performance of high bit rate, long-range optical communication systems as well as coherent optical transmission systems can seriously be deteriorated by intensity and phase noise of the optical sources. In particular, laser phase noise is a dominant source of noise in coherent optical communication systems. In *Chapter three*, statistical properties of laser noise by means of a simple model are derived which especially allow to determine its degrading effect on the communication system. This Chapter also briefly examines the effects of laser relaxation oscillations and presents some interesting and generally valid aspects on filtering signals perturbed by phase noise. Here, a remarkable difference between phase noise and additive Gaussian receiver noise (i.e., shot noise and thermal noise) has been illustrated. Finally, some important technical solutions to reduce the effects of laser noise are suggested.

Chapter four is dedicated to optical fibers. It describes the theory of optical propagation and explains the physical background of relevant fiber parameters such as attenuation and dispersion. Chapter four also describes fiber design and selection issues and discusses the effects of fiber non-linearities. Finally, fiber gratings are introduced as excellent tools to compensate fiber dispersion and to fabricate components such as optical filters, add-drop multiplexers and more.

Besides laser noise, polarization fluctuations in the fiber also seriously deteriorate the performance of optical communication systems, in particularly long-range links and coherent systems. *Chapter five* is focussed on reasons, effects and technologies of handling the problems arising from polarization fluctuations.

Chapter six is focussed on optical detectors and receivers. It presents the basic principle of main detectors such as PIN photodiode and avalanche photodiode (APD). Noise in photodiodes, minimum detectable power and different receiver configurations have been discussed.

One of the most exciting development in the field of optics has been the discovery of optical amplifier. It is probably the most significant development since the discovery of single-mode fiber. *Chapter seven* describes different types of optical amplifiers and presents their comparative study. Further, various applications of optical amplifiers in optical communications are discussed.

Fiber optics is progressing from point-to-point links to networks. *Chapter eight* describes all relevant components required to realize optical networks. The following components are discussed: couplers and splitters, wavelength multiplexers and demultiplexers, isolators, modulators, optical crossconnects (OXC) and add-drop multiplexer (ADM).

Chapter nine provides the background required for study of optical communication systems. This Chapter describes the principle of optical receivers and gives an introduction to the basic laws. Performance gain of coherent over direct detection receiver is analysed in terms of sensitivity and selectivity. In the frame of selectivity, wavelength division multiplexing (WDM), dense WDM (DWDM) and optical frequency division multiplexing (OFDM) are discussed. Chapter nine also gives an overview of the various components and signals in an optical link. Finally, it reviews some important aspects of digital communication systems which are required to analyse coherent optical communication systems. The eye pattern technique and probability of error, also referred as BER, are refreshed as two important criteria for systems comparison.

Based on the fundamentals of Chapter nine, *Chapter ten* gives the analysis of incoherent optical communication systems i.e., IM/DD systems which are based on light intensity modulation and direct detection. In the analysis, all relevant sources of noise and system imperfections have been considered. In this Chapter, system optimization is always accomplished with the goal to minimize the probability of bit error.

Chapter eleven discusses the analysis and optimization of various coherent optical communication systems. Modulation and demodulation schemes are used as a characteristic feature to distinguish and to group the different systems. Similar to Chapter nine, all relevant sources of noise and system imperfections have been considered. For this purpose, various methods of calculation and approximation are explained and compared. In addition, this Chapter also presents some interesting results obtained by computer simulation.

Main objective of *Chapter twelve* is to compare optical communication systems based on the analysis given in Chapters ten and eleven. For comparison, different criteria are taken into account such as bit rate, laser line width requirements, eye pattern and applications. The second goal of this Chapter is to highlight the physical limits of optical communications. For this purpose, quantum limit, shot noise limit, Shannon limit and dispersion limit are discussed.

Over the past several years, optical fiber communication systems are extensively used all over the world for telecommunication and data transmission purposes. Therefore, design of optical links to meet a given requirements has become an important issue. In *Chapter thirteen*, design procedure of a digital point-to-point link has been given.

Chapter fourteen first presents an introduction to the principle of network topologies and design valid for both electrical and optical networks. Some fundamental technical and non-technical backgrounds of optical networks are discussed. A main part of this Chapter is dedicated to fiber-supported networks which include optical as well as electronic components and all-optical networks where all network functions are exclusively realized by optical components including the switching function as well as the access. Finally, this Chapter highlights some important aspects of fiber-based digital broadband networks and information superhighways as a fundamental precondition for high-quality multimedia communication.

In *Chapter fifteen*, first of all, a very promising application of optical communications for space links i.e., optical free-space communications has been discussed. This last Chapter serves to provide an overall view to applications and advantages of optical space communications relative to other transmission media such as fiber and microwave communications.

1.4 REFERENCES

[1]   Goodwin, F. E.: A 3.39-micron infrared optical heterodyne communication system. IEEE J. QE-3(1967)11, 524-531.
[2]   Henry, P. S.: High-capacity light wave local area networks. IEEE Commun. Magazine, 27(1989)10, 20-26.
[3]   Nussmeier, T. A.; Goodwin, F. E.; Zavin, J. E.: A 10.6-μm terrestrial communication link. IEEE J. QE-10(1974)2, 230-235.
[4]   Runge, P. K. : Future directions of undersea systems. ECOC (1994), 927-932.
[5]   Saito S.; Yamamoto, Y.; Kimura, T.: Optical heterodyne detection of directly frequency modulated semiconductor laser signals. Electron. Lett. 16(1980)22, 826-827.
[6]   Smith, D. W. : The road to superhighways. ECOC (1994), 903-906.
[7]   Stordahl, K.; Murphy, E.: Forecasting long-term demand for services in the residential market. IEEE Commun. Magazine, 33(1995)2, 44-49.

2   OPTICAL SOURCES AND TRANSMITTERS

This Chapter is focussed on optical sources and transmitters used in optical communication systems. After a brief introduction (Section 2.1) which presents an overview of both the most important optical sources LED and laser, Section 2.2 describes the physical background of optical sources including absorption, spontaneous emission, stimulated emission and light amplification. In Sections 2.3 and 2.4 design, realization and characteristic features of LED and laser are discussed. These Sections also present, as an example, some typical data from commercially available sources. Some realization aspects of optical transmitter are finally described in Section 2.5. The phenomenon of laser noise which is a fundamental source of noise in most advanced optical communication systems such as high-speed long-range data transmission systems and coherent optical systems is discussed in the next Chapter.

2.1   INTRODUCTION

In optical communication systems two components are at least required to transmit and to receive information: (i) the optical source and (ii) the optical detector discussed in Chapter six. The two most important optical sources are

- Light emitting diode (LED) and
- Laser (light amplification by stimulated emission of radiation).

Semiconductor lasers are normally used in fiber-based systems, whereas in free-space systems solid-state lasers are preferred. The characteristics, communication-based features of LED and laser are given in Table 2.1 and discussed in the following Sections.

Table 2.1 Typical features of LED and laser

LED	Laser
Incoherent	Coherent
Without optical resonator	With optical resonator (laser cavity)
For multimode fibers only	For multi- and single-mode fibers
Up to some 100 Mbit/s	Up to some 10 Gbit/s
Relative large beam divergence	Relative low beam divergence

In contrast to an LED, a laser is primarily characterized by communication-based features and advantages.

> Advantages of the laser are:
>
> · Higher modulation rate (data rates)
> · Narrower spectral width
> · Less dispersion-induced signal distortion
> · Higher fiber-coupling efficiency
> · Greater transmission distance

The LED also exhibits some very attractive advantages which makes it a well-suited device for low-cost optical communication systems.

> Advantages of the LED are:
>
> · Easier fabrication
> · Lower cost
> · Simpler transmitter circuit design
> · Lower temperature dependence

Optical sources such as laser and LED are characterized by a number of different parameters, for example, optical output power, wavelength, bandwidth of emission spectrum et. al.. All these data can normally be obtained from a data sheet. As an example, Fig. 2.1 shows emission spectra of an LED and a laser diode taken from a typical commercially available data sheet. Both sources operate at a centre wavelength of 1.55 μm.

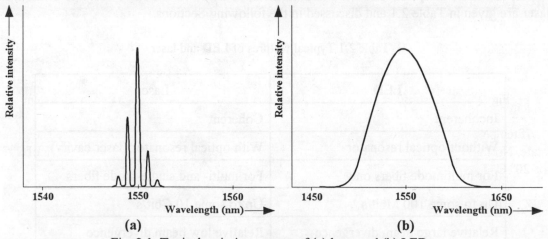

Fig. 2.1: Typical emission spectra of (a) laser and (b) LED

2.2 PRINCIPLE OF OPTICAL SOURCES

Designing optical communication links with simple on-off-keying (OOK), moderate bit rate and moderate transmission length, a special knowledge of laser physics is normally not required and a simple black-box consideration of laser or LED source is quite adequate for such applications. However, when advanced optical communication systems such as coherent systems and high-speed long-range links have to be analysed and optimized, this simplified black-box consideration is not sufficient. The analysis of advanced optical communication systems requires a much more detailed knowledge of the physics of optical sources, in particular a fundamental knowledge of the statistical properties of noise. For this reason, a brief review of the physical principle of laser and LED operation is given in the following Subsections.

2.2.1 ABSORPTION, SPONTANEOUS AND STIMULATED EMISSION

Each atom of any material is characterized by a perfectly specified number of electrons which are distributed on a fixed number of orbits. As an example, Fig. 2.2 shows two orbits with energy W_1 and W_2. The orbit with lower energy W_1 is called the *ground state*, *equilibrium state* or *conduction band* and with higher energy W_2 the *excited state* or *valence band*.

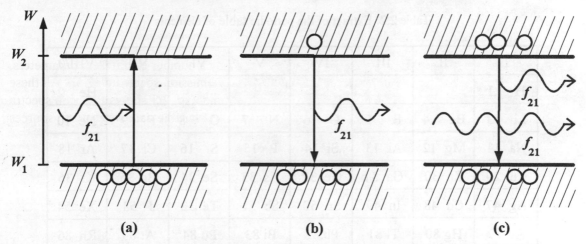

Fig. 2.2: (a) absorption, (b) spontaneous and (c) stimulated emission in a two-level system

If temperature is absolutely zero (0 K), then all atoms of a molecule are located in the ground state. Normally, an atomic system is also in the ground state for an indoor temperature of 293 K or 20° C. The difference W_{21} in energy follows the well-known physical relation

$$W_{21} = W_2 - W_1 = hf_{21} \qquad (2.1)$$

Here, $h = 6.6625 \cdot 10^{-34}$ Ws^2 = $4.135 \cdot 10^{-15}$ eVs is the Planck's constant and f_{21} the frequency of light. If semiconductor material is used, then the energy difference W_{21} ranges from approximately 0.1 eV to 2 eV. It is seen from Fig. 2.2 that three different physical processes must be distinguished in an atomic system:

- Absorption,

- Spontaneous emission and

- Stimulated emission.

Spontaneous emission is the basic process in an LED, whereas laser operation is based on stimulated emission. An electron moving from the upper energy level to the ground state can either produce thermal radiation (heat) or a photon i.e., a light wave of frequency f_{21}. The first case is known as an *indirect band-gap*, the second as a *direct band-gap*. Considering semiconductor materials, none of the normal single-element semiconductors such as Silicon (Si) are direct-gap materials and emission of light is not possible. Emission of light is only possible in direct band-gap semiconductor compounds. These are made from elements out of the third and fifth group of the periodic table (Table 2.2). They are known as *III-V compounds* or AIII-BV compounds.

Table 2.2: Extract of the periodic table of elements

I	II	III	IV	V	VI	VII	VIII/0
H 1							He 2
Li 3	Be 4	B 5	C 6	N 7	O 8	F 9	Ne 10
Na 11	Mg 12	Al 13	Si 14	P 15	S 16	Cl 17	Ar 18
Cu 29	Zn 30	Ga 31	Ge 32	As 33	Se 34	Br 35	Kr 36
Ag 47	Cd 48	In 49	Sn 50	Sb 51	Te 52	J 53	Xe 54
Au 79	Hg 80	Tl 81	Pb 82	Bi 83	Po 84	At 85	Rn 86

Examples of III-V semiconductor compounds are GaAs and GaP referred as binary compounds, $Ga_{1-x}Al_xAs$ and $GaAs_{1-x}P_x$ known as ternary compounds and $In_{1-x}Ga_xAs_yP_{1-y}$ as quaternary compounds. The ratio of mixtures x and y define the energy difference of the band-gap and, hence, the frequency of the peak emitted radiation. At present, GaAlAs laser diodes operating in the wavelength range of 800 nm to 900 nm and InGaAsP laser diodes which ranges from 1100 nm to 1600 nm are the most important optical sources in optical communication systems. Some more aspects of semiconductor lasers are considered in Section 2.4.

(i) ABSORPTION

An atom in the ground state is only able to pick up or absorb energy. If this energy equals the energy difference W_{21}, then this atom is transferred from the ground state to the excited state as shown in Fig. 2.2a. Usually, absorption is accomplished within a radiation field. In that case, only photons with frequency f_{21} can be absorbed by the atom. If, for example, a large number of atoms are located within the radiation field, then the intensity of radiation decreases exponentially with the absorption [5].

In an atom, absorption process strongly depends on the energy density ρ of external radiation field and number of electrons N_1 in the ground state W_1. Absorption decreases the number of ground state electrons with rate

$$R_a = -\frac{dN_1}{dt} = B_{12}\rho N_1 \tag{2.2}$$

Here, B_{12} represents the Einstein's absorption coefficient.

(ii) SPONTANEOUS EMISSION

As the excited state is an unstable state, an atom which has been excited by absorption only remains in this state for a certain time called the lifetime τ_{21}. Afterwards, this atom returns to the ground state by itself spontaneously as shown in Fig. 2.2b. Thereby, a photon of energy W_{21} and frequency f_{21} is emitted. As this process occurs without any external stimulation, it is called spontaneous emission.

Normally, an atom contains of several different energy levels. Therefore, spontaneous emission yield light radiation of many different frequencies. Considering a light wave at single frequency, a random and absolutely uncorrelated phase can be observed. This imply that spontaneous emission generate incoherent light. In an LED, spontaneous emission is the result of recombination of excited electrons in the conduction band (excited energy level) and holes in the valence band (ground level). Since the light of an LED is incoherent, LEDs are exclusively used in direct detection system. In contrast, they are absolutely unusable in coherent optical communication systems.

Spontaneous emission is not influenced by an external radiation field. As an statistical process, spontaneous emission only depend on the population N_2 of excited electrons in state W_2. By using the Einstein's coefficient A, emission rate R_{sp} of spontaneous emission is given by

$$R_{sp} = -\left(\frac{dN_2}{dt}\right)_{sp} = AN_2 = \frac{N_2}{\tau_{21}} \tag{2.3}$$

The spontaneous emission rate R_{sp} describes the temporal removal of electrons in the excited state W_2 and, hence, the temporal increase of spontaneously generated photons. This rate as seen from Eq. (2.3) is directly proportional to the population N_2 and inversely proportional to the average lifetime $\tau_{21} = 1/A$ of excited electrons in state W_2. If lifetime is high, excited electrons only return to ground state very rarely and the spontaneous emission rate R_{sp} is low. In contrast to this, emission rate R_{sp} becomes high when lifetime is short.

(iii) STIMULATED EMISSION

In contrast to absorption, there are two different mechanisms for emission: one is the spontaneous emission as described above and the other is stimulated or induced emission as shown in Fig. 2.2c. Unlike spontaneous emission which is a random process, stimulated emission is a deterministic process providing the supply of external energy. Stimulated emission only occur if a stimulating photon of energy W_{21} and frequency f_{21} impinges upon the system. Thereby, a new photon of same frequency, phase, polarization and direction of propagation as the stimulating photon is generated. Thus, the incoming photon triggers the generation of an additional photon. Stimulating photon itself is not influenced by this process which means that the optical waves of both photons are coherent. Thus, stimulated emission of radiation offer the possibility to amplify a light wave true in phase and frequency.

The physical process of amplification of a light wave by stimulated emission is the reason for the well-known abbreviation *LASER* which stands for *Light Amplification by Stimulated Emission of Radiation*. In an LED, light generation is primarily performed by spontaneous emission, whereas in a laser light is predominantly generated by stimulated emission. Unlike spontaneous emission, stimulated emission is directly proportional to the energy density ρ of external radiation field. In addition, emission rate R_{st} of stimulated emission increases, of course, with the increase in number N_2 of excited electrons. Hence,

$$R_{st} = -\left(\frac{dN_2}{dt}\right)_{st} = B_{21}\rho N_2 \tag{2.4}$$

Here, B_{21} is the Einstein's stimulated coefficient.

(iv) BUDGET OF EMISSION AND ABSORPTION

Indeed, absorption and stimulated emission are mutual processes. Moreover, they are absolutely homogeneous that is,

$$B_{12} = B_{21} = B \tag{2.5}$$

The relationship between coefficients A and B is given by

$$A = 8\pi h \lambda_{21}^{-3} n^3 B = 8\pi h \left(\frac{f_{21}}{c_0} \right)^3 n^3 B \tag{2.6}$$

which was given by Einstein [5]. Here, λ_{21} is the wavelength of emitted light wave, n the refraction index and $c_0 = 3 \cdot 10^8$ m/s the velocity of light in vacuum i.e., free-space.

Considering a system to be composed of a large number of atoms, some atoms are in the ground state and the others in the excited state. In this atomic system, numerous absorption and emission processes occur (Fig. 2.3). Each absorption process attenuates the incident light wave, whereas each stimulated emission process amplify it. In addition, a large number of spontaneous emission processes are superimposed randomly. Unlike light waves generated by stimulated emission, light waves caused by spontaneous emission exhibit a random phase and are likewise transmitted in all directions. The small number of spontaneously emitted photons which accidentally have the same direction as the photons generated by stimulated emission, is the fundamental physical reason of laser phase, laser amplitude and laser intensity noise.

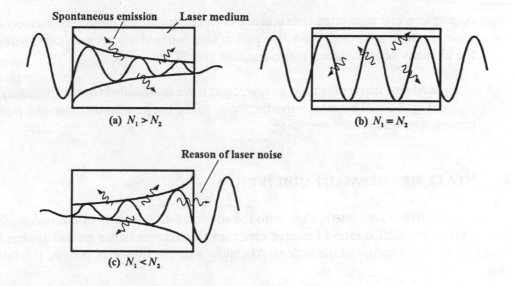

Fig. 2.3: Influence of number of electrons in ground state and excited state on incident light wave

In a closed system, absorption and emission processes are in state of thermal equilibrium. This means that

$$R_a = R_{st} + R_{sp} \tag{2.7}$$

Using Eqs. (2.2) to (2.4), it becomes

$$B_{12}\rho N_1 = B_{21}\rho N_2 + AN_2 \tag{2.8}$$

Laser operation i.e., amplification of incident light wave provides that stimulated emission is dominant compared with absorption. To determine which of these two processes is actually dominant, the ratio

$$\frac{R_{st}}{R_a} = \frac{B_{21}\rho N_2}{B_{12}\rho N_1} = \frac{N_2}{N_1} \tag{2.9}$$

of stimulated emission rate to absorption rate is a well-suited parameter. It becomes clear that this ratio depends only on the population ratio of electrons in the excited state and ground state. On the basis of above ratio, following classification can be made:

$N_1 > N_2$:   Absorption is dominant. Hence, the intensity of incident light wave is exponentially decreased (Fig 2.3a).

$N_1 = N_2$:   Absorption and stimulated emission are in equilibrium. This imply that incident light wave is travelling through the active part of laser without any attenuation. Therefore, the intensity of light wave remains constant (Fig. 2.3b).

$N_1 < N_2$:   Stimulated emission is dominant. Incident light wave is amplified true in frequency and phase (Fig. 2.3c). This most important case, called the population inversion, must be provided to realize laser operation.

2.2.2   STATE OF THERMAL EQUILIBRIUM

A closed system without any interrelation with the environment is in state of thermal equilibrium. In this state, population ratio of excited electrons to electrons in the ground state is only influenced by the temperature of the system. Assuming a simple two-level system, this ratio is given by

$$\frac{N_2}{N_1} = \exp\left(-\frac{W_{21}}{k_B T}\right) = \exp\left(-\frac{hf_{21}}{k_B T}\right) \tag{2.10}$$

Here, $k_B = 1.38 \cdot 10^{-23}$ Ws/K $= 8.62 \cdot 10^{-5}$ eV/K is the Boltzmann's constant and T the temperature measured in Kelvin.

Example 2.1

With a difference in energy $W_{21} = 0.827$ eV (i.e., $f_{21} = 200$ THz, $\lambda_{21} = 1.5$ µm) at an indoor temperature of $T = 293$ K (i.e., 20° C), electron-population ratio given in Eq. (2.10) yields $N_2/N_1 = \exp(-32.74) = 6.02 \cdot 10^{-15}$. If temperature is very high, for example $T = 1273$ K i.e., 1000° C, then we obtain $N_2/N_1 = \exp(-7.54) = 5.33 \cdot 10^{-4}$.

It becomes clear from the above example that even for a very high and unrealistic temperature of 1000° C, electrons are almost in the ground state. Thus, a *population inversion* which is required for amplification of a light wave can never be achieved within a closed system in state of thermal equilibrium. For this, state of equilibrium must be disturbed first. This, for example, can easily be accomplished by supplying an external energy which is usually called *pumping*. In a semiconductor laser diode, inversion is simply achieved by the injection current. In contrast to that, a population inversion and, hence, pumping is not required in an LED.

2.3 LIGHT EMITTING DIODES

In contrast to a laser which is discussed in the next Section, light in a light emitting diode (LED) is exclusively produced by incoherent spontaneous emission processes but not by stimulated emission. In an LED, emission of light starts with a very low injection current since a population inversion $N_1 < N_2$ and, hence, an injection current above a threshold I_{th} is not required. There is also no need for an optical resonator which results in a continuous emission spectrum in contrast to a laser where emission spectrum is composed of several modes. The linewidth Δf of an LED is some order of magnitudes broader (typical 50 to 100) than the linewidth of a laser because no light amplification is achieved in an LED. Speed of modulation ranges from few MHz to some hundred MHz. A typical emission spectrum and a power-injection current curve which is also known as P/I-curve are given in Fig. 2.4.

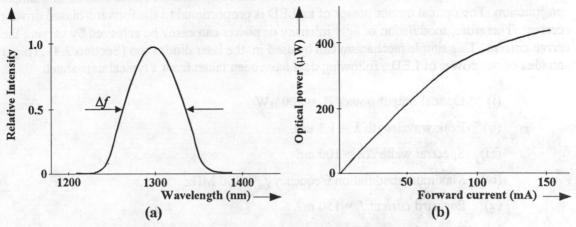

Fig. 2.4: (a) Emission spectrum and (b) P/I-curve of an LED

Although technical features of an LED are not as good as of a laser, this device is the most suitable light source in communication systems with moderate bit rates up to a few hundred Mbit/s together with multimode fibers. In comparison to a laser, an LED needs a much less complex drive circuit in the optical transmitter since no thermal stabilization is required. In addition, LEDs can easily be fabricated with less cost. The incoherent optical light from an LED can only be coupled into multimode fibers with a sufficient efficiency.

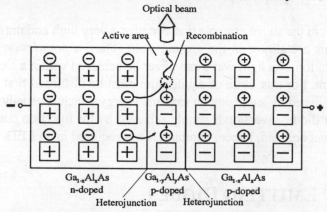

Fig. 2.5: Electrons and holes in a forward biased pn-junction of an LED

LEDs are made from semiconductor compounds of a group III element such as Al, Ga or In and a group V element such as P, As or Sb since only these compounds show a direct band-gap suitable for light transmission. Similar to conventional semiconductor diodes, an LED consists of a *pn junction* which is responsible for the characteristics of an LED. When LED is reverse biased, only an insignificant small current flows through the diode and LED works as an isolator. In order to generate light radiation, an LED must be *forward biased*. In this case, electrons from n-side and holes from p-side diffuse into the *depletion region*, also referred as *space charge region*, where both electrons and holes recombine (Fig. 2.5). While recombining, optical radiation is produced. Now-a-days, the double-heterostructure is normally used for LEDs instead of a simple pn-junction. The optical output power of an LED is proportional to the forward-biased driving current. Therefore, modulation of light intensity or power can easily be achieved by varying the drive current. This simple mechanism can be used in the laser diodes too (Section 2.4). To get an idea of the power of LEDs, following data have been taken from a typical data sheet.

(i)   Optical output power P_o = 200 µW

(ii) Peak wavelength λ = 1.3 µm

(iii) Spectral width $\Delta\lambda$ = 100 nm

(iv) Maximum modulation frequency f_m = 150 MHz

(v)   Forward current I_f = 150 mA

(vi) Forward voltage U_f = 2 V

2.4 LASERS

Considering lasers, several different types have to be distinguished such as gas lasers, solid state or crystal lasers and semiconductor lasers. In particular, heterojunction-structured semiconductor laser diodes, also known as injection laser diodes are indispensable key components of advanced optical fiber communication systems. Solid state and gas lasers are used in atmospheric and free-space links. Despite the difference in their size, material and pumping scheme, principle of operation is same for all types of lasers. As mentioned in the previous Section, *population inversion* is required for laser operation. In contrast, population inversion is not required in an LED (Section 2.3)

2.4.1 POPULATION INVERSION

Population inversion means to disturb the state of thermal equilibrium of closed system by an external energy supply to achieve an electron population ratio $N_2/N_1 > 1$. For this, a three- or four-level instead of a simple two-level structure and an appropriate pump energy have to be considered. Fig. 2.6 shows two typical examples.

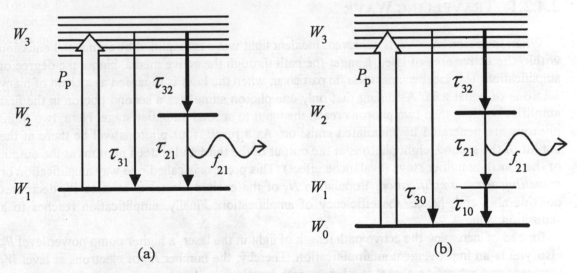

Fig. 2.6: Realization of population inversion by means of (a) three- and (b) four-level lasers

Besides the ground state W_1, two excited states W_2 and W_3 are used in a three-level structure. Regarding the lifetime of electrons in the excited states, special conditions must be fulfilled: First, the lifetime τ_{31} of electrons crossing from the excited energy level W_3 to the ground level W_1 should be very large, whereas the lifetime τ_{32} for downward transition from level W_3 to W_2 should be very small ($\tau_{32} \ll \tau_{31}$). Thus, a transition from state W_3 to W_1 is very rare, while a transition from W_3 to W_2 is very probable. In addition, a pass over from level W_3 to W_2 should be without any radiation. Second, to collect a sufficient large number of electrons at the excited energy level W_2 and, consequently, to achieve the desired population inversion, level W_2 must be a so-called meta-stable energy level with a high lifetime τ_{21}.

With an external energy source, electrons at level W_1 are pumped to the higher energy state W_3 as shown in Fig. 2.6a. For this purpose, a pump power P_P is required. To reach a high efficiency, level W_3 should be a rather broad energy level. Due to very low lifetime $\tau_{32} \ll \tau_{31}$, most of the pumped electrons pass over to the meta-stable energy level W_2 within a very short time. If pump power P_P is appropriately high enough, then required population inversion $N_1 < N_2$ is achieved.

In a three-level laser, inversion can be obtained only when pump power exceeds a relative high threshold power P_{th}. This is a serious drawback of a three-level system. Consequently, four- or multilevel lasers are used. As an example, Fig. 2.6b shows the simplified structure of a four-level laser. Here, close to the primary level W_0 an additional fourth energy level W_1 is used. The lifetime τ_{10} for a downward transition from energy level W_1 to level W_0 is very low i.e., $\tau_{10} \ll \tau_{31}$. Thus, the energy level W_1 nearly contains no electrons and population inversion can be achieved with a very low pump power.

2.4.2   LIGHT AMPLIFICATION

2.4.2.1   TRAVELLING WAVE

When inversion $N_1 < N_2$ is achieved, incident light wave is amplified by stimulated emission within the active area of laser. Longer the path through the active media, higher the degree of amplification. This fact becomes clear, in particular, when the laser is regarded as a chain of many sections of equal gain. Assuming that only one photon stimulates a second photon in the first amplifier section, then two photons reach the input to second amplifier stage. Here, two more photons are generated by stimulated emission. As a result, four photons will be there at the output of the second, eight photons at the output of the third and sixteen photons at the output of the fourth amplifier stage (avalanche effect). This process is called one way amplification or *travelling wave amplification*. Population N_2 of the excited electrons at state W_2 decreases considerably and, hence, the efficiency of amplification. Finally, amplification reaches to a saturation value.

Instead of increasing the active path length of light in the laser, a higher pump power level P_P also yields an improvement in amplification. Thereby, the number N_2 of electrons at level W_2 increases and saturation occurs at a longer path length.

2.4.2.2   OPTICAL CAVITY

In order to improve light amplification by increasing the length of optical path in the active area, this area is normally embedded inside an *optical cavity* also called an optical resonator. By means of high reflecting plates at both sides of the resonator (e.g., two mirrors), length of the actual optical path becomes a multiple of cavity length. Due to a very large number of reflections (optical feedback), light amplification becomes very efficient.

In the practical realization of an optical cavity, plates with a reflectivity of somewhat less than 100% are used to extract and transmit a certain part of the laser light. The laser process usually starts with spontaneously emitted photons and reaches real laser operation within a very short time provided the pump energy is sufficiently large.

2.4.2.3 LASER EFFICIENCY

The optical output power P_o of a laser, which is extracted from the laser resonator, is used to transmit information through an optical fiber or open space. To generate this optical power, electrical pump power P_P is required as mentioned above. The efficiency of a laser, also known as *quantum efficiency*, is defined as

$$\eta = \frac{P_o}{P_P} \quad \text{or} \quad \eta = \frac{P_o}{P_{el}} \tag{2.11}$$

where P_{el} represents the electrical input power required to produce the pump power P_P. Laser efficiency η of lasers suitable for optical communication varies from 0.01% to 0.1% for He-Ne lasers, about 0.1% to 1% for solid state lasers, 1% to 10% for semiconductor lasers and about 20% for CO_2 lasers.

2.4.3 SPECTRAL FEATURES OF LASERS

2.4.3.1 EMISSION SPECTRUM OF RADIATING ELECTRON TRANSITION

Up-to-now, we have assumed that an atom is characterized by discrete energy levels. Because of this, each downward transition from higher to lower energy level always yields a photon of well-defined frequency (Figs. 2.2 and 2.6). However, this is actually not true since atomic energy states are continuous. As a consequence, each radiating transition from the exited state W_2 to the ground state W_1 exhibits a characteristic emission power spectral density $G_{21}(f)$. As shown in Fig. 2.7, this spectrum is primarily characterized by its centre frequency f_{21} and 3 dB bandwidth Δf_{21}. At the centre frequency f_{21}, radiated power reaches maximum since most of the electrons are located in the centre of continuous energy level W_2 with very high probability. In contrast, the probability to find an electron in some distance from this centre decreases rapidly with the increase in the distance (Fermi distribution). Therefore, optical power spectral density decreases for frequencies $f > f_{21}$ and $f < f_{21}$.

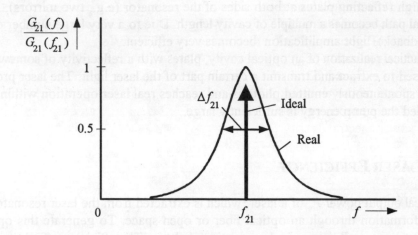

Fig. 2.7: Emission spectrum $G_{21}(f)$ of radiating energy transition under ideal and real conditions

The physical reason for the variations in the energy level is given by various interrelations between the different atoms of an atomic system, caused by pressure, pushes, temperature fluctuations and other internal mechanisms. In a gas laser, for example, fast moving radiating gas atoms give rise to optical Doppler effect which also broaden the energy levels. Line broadening caused by the above effects is called the *inhomogeneous line broadening*. In this case, shape of the emission spectrum $G_{21}(f)$ is Gaussian.

Electrons which are located at the excited energy level W_2 return to the ground level W_1 after an average lifetime τ_{21} as already explained in the previous Section. Due to variations in lifetime, an uncertainty $\Delta W_{21} \approx h/\tau_{21}$ in the transition energy in accordance with the well-known uncertainty relation can be observed. This uncertainty gives rise to a broaden emission spectrum as shown in Fig. 2.7. This type of broadening due to inherent uncertainty is frequently called the *natural* or *homogeneous line broadening* [5, 6]. In this case, shape of the emission spectrum follows the equation

$$G_{21}(f) = \hat{E}_{21}^2 \, \frac{2}{\pi \Delta f_{21}} \, \frac{1}{1 + \left(\dfrac{f - f_{21}}{\Delta f_{21}/2}\right)^2} \tag{2.12}$$

which is called the *Lorentzian line shape* [5, 6]. In the above equation

$$\Delta f_{21} = \frac{1}{\tau_{21}} \tag{2.13}$$

represents the *3 dB linewidth* or *spectral half-width* and $\hat{E}_{21}$ the amplitude of radiated electric field. It should be noted that $G_{21}(f)$ is given in unit of $(V/m)^2/Hz$. Hence, $G_{21}(f)$ is the power spectral density of electric field. The area under $G_{21}(f)$ is independent of linewidth Δf_{21} and yields the constant quantity $(\hat{E}_{21})^2$. Thus, this area equals the square of the absolute of complex electric field $\underline{E}(t) = \hat{E}_{21} \exp(j2\pi f_{21} t)$ and is, therefore, a direct measure of the optical power.

The real shape of the emission spectrum $G_{21}(f)$ i.e., Gaussian or Lorentzian depends on whether the homogeneous or inhomogeneous line broadening is dominant in the system. The normalized power spectral density

$$g_f(f) = \frac{G_{21}(f)}{\hat{E}_{21}^2} = \frac{2}{\pi \Delta f_{21}} \frac{1}{1 + \left(\dfrac{f - f_{21}}{\Delta f_{21}/2}\right)^2} \tag{2.14}$$

corresponds to the probability density function (pdf) of the random frequency f. Hence, the function $g_f(f)$ given in unit of per Hz represents a direct measure of determining the occurrence probabilities of all possible frequencies within the broaden energy transition W_{21}. The probability of a transmitted photon with frequency f in the range of df will be $g_f(f)df$. Thus, photons with frequency close to the centre frequency f_{21} are transmitted very frequently, whereas photons with higher or lower frequencies are transmitted very rarely. The normalized power spectral density $g_f(f)$ is called the *normalized line shape*. As a pdf, this function is characterized by

$$\int_0^{+\infty} g_f(f)\,df = 1 \tag{2.15}$$

provided that $\Delta f_{21}/f_{21} \ll 1$.

2.4.3.2 SPECTRAL FEATURES OF AMPLIFIED LIGHT WAVE

In principle, amplification of an incident light wave by means of stimulated emission is only possible within the emission spectrum $G_{21}(f)$ of a radiating energy transition. Light waves with frequencies outside this spectrum are not amplified and, consequently, they are insignificant as compare to the transmitted laser light wave. According to occurrence probability $g_f(f)df$, light waves with frequency close to the centre frequency f_{21} are amplified much more efficiently than those light waves of frequency not close to f_{21}. This fact becomes clear particularly when the optical amplifier is considered in multiple stages as in Section 2.4.2.1. Assuming that only one photon of frequency f_{21} and only one photon of double frequency $2f_{21}$ enter the input to the first amplifier stage. We also assume that each amplifier section amplifies a light wave of frequency f_{21} by a factor of four and a light wave of frequency $2f_{21}$ by a factor of two only. Hence, we observe four photons of frequency f_{21} and two photons of frequency $2f_{21}$ at the input to the second amplifier stage. At the inputs to the third and fourth amplifier stages, we obtain 16 photons and 64 photons of frequency f_{21} and only 4 photons and 8 photons of frequency $2f_{21}$ respectively.

It is clear from this that the optical amplification strongly depends on the frequency of incident light wave. Further, bandwidth of emitted light wave at the output of last amplifier stage decreases rapidly with the increase in number of stages or reflections inside the laser cavity. Hence, spectral bandwidth Δf_g of amplified light wave is usually some order of magnitudes smaller than the bandwidth Δf_{21} of normal light wave caused by a simple energy transition ($\Delta f_g \ll \Delta f_{21}$).

2.4.3.3 MODES OF LASER CAVITY

Owing to the discrete size of the cavity which equals size of active area of laser, only a limited number of discrete and specific light waves called *laser modes* are able to propagate. These modes are perfectly specified by standing waves inside the cavity. All other waves can not exist. Due to three dimensions of a cavity, three orthogonal directions for transmitting light will be there in principle.

As an example, the principle physical structure of a *gain-guided semiconductor laser* realization is shown in Fig. 2.8. The thickness d of the optical cavity is usually in the order of some one tenth of μm and cavity length L is in the order of some 100 μm. The average length is about 300 μm. In x-direction, active area of this laser diode is confined by an electrode. This is used for the injection current to provide required pump energy and, hence, achieve required population inversion. The width w of this electric stripe guide is approximately in the order of some μm. In the y-direction, active area of this laser in confined by n-doped emitter layer and p-doped buried substrate, both characterized by a refraction index less than the index of the active layer.

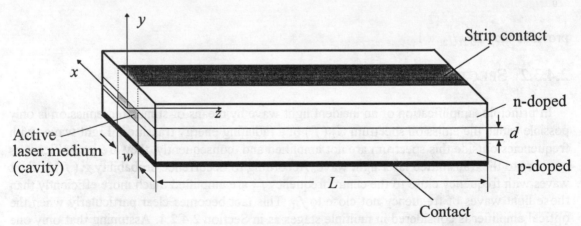

Fig. 2.8: Principle physical structure of gain-guided semiconductor laser

An alternate solution to confine the active area of laser is to realize the cavity as passive optical waveguide offering the advantage of much stronger optical guiding. Here, active area is confined by means of a change in the refraction index also in x-direction. This type of laser is called *index-guided laser*. In both types of lasers mentioned above, optical power is emitted from the facets only which can be regarded as semi-transparent mirrors. Since the principle structure of the above laser diode is similar to a Fabry-Perot resonator, this special kind of laser is frequently called Fabry-Perot lasers.

The modes which are able to travel in x- and y-directions are called the *lateral* and the *transverse modes* respectively. In optical communication systems, the longitudinal or the axial modes in z direction are of practical importance only. For an active media of refraction index n, axial mode a and wavelength λ_a for axial modes, following simple relationship can be derived:

$$2L = a\frac{\lambda_a}{n} = a\frac{c_0}{nf_a} \tag{2.16}$$

It should be noted that the wavelength λ_a is related to vacuum. In the above equation, mode numbers $a \in \{1, 2, 3, \cdots\}$ are in accordance with the number of possible half waves inside the confinement of laser in the z-direction. It becomes clear that frequency f_a or wavelength λ_a of a laser strongly depend on the mode number a. Considering the frequency separation δf_a between two neighbouring axial modes, we obtain

$$\delta f_a = f_{a+1} - f_a = \frac{c_0}{2nL} \tag{2.17}$$

Thus, the frequency separation δf_a is constant within the emission spectrum of a laser. As seen from Eq. (2.17), this separation is small if the length of laser cavity is large and vice versa. In contrast to constant frequency separation, spectral wavelength separation

$$\delta\lambda_a = \lambda_a - \lambda_{a+1} = \lambda_a \lambda_{a+1}\frac{1}{2nL} \approx \frac{\lambda_a^2}{2nL} \tag{2.18}$$

depends on the wavelength itself. In brief, spectral separation $\delta\lambda_a$ is small for modes which are close to the central wavelength λ_a and large for modes which are not as close.

Example 2.2

Consider a laser cavity of length $L = 200$ μm and refraction index $n = 3.6$. At the central wavelength $\lambda_a = 1.5$ μm, spectral separation from Eq. (2.18) is $\delta\lambda_a = 1.56$ nm. It corresponds to a frequency separation of $\delta f_a = 208$ GHz. In comparison to the bandwidth of typical information signals (for example, 5 MHz for an analog TV signal), this is really a very high frequency separation.

2.4.3.4 SPECTRAL FEATURES OF TRANSMITTED LASER LIGHT WAVE

In Fig. 2.9, power spectral density $G_{21}(f)$ of light wave caused by a simple transition from the excited energy level W_2 to the ground energy level W_1 and emission spectrum $G_g(f)$ of an amplified (gained) light wave due to stimulated emission are shown.

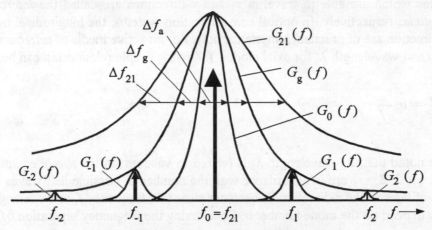

Fig. 2.9: Power spectral density of light wave

In addition, this figure also shows the modes which are determined by the size of laser cavity as mentioned above. However, modes are given different numbers as compared to Eq. (2.16). The emission spectrum $G_{21}(f)$ of the energy transition exhibits a broad Lorentzian line shape with a linewidth Δf_{21} as given in Eq. (2.12). Due to selective amplification in the active area, 3 dB bandwidth Δf_g of amplified emission spectrum $G_g(f)$ is much smaller than the 3 dB bandwidth of emission spectrum $G_{21}(f)$ i.e., $\Delta f_g \ll \Delta f_{21}$. Within the emission spectrum $G_g(f)$, only a certain and limited number of waves called the laser modes are able to exist. The laser modes are charac-terized by standing waves within the discrete size of the laser cavity. In absence of special steps of realization, a laser generally oscillates in many different axial and transverse modes. In Fig. 2.9, axial modes are represented by Dirac delta functions. It becomes clear that this type of laser is, of course, not a monochromatic laser. As it oscillates in many different modes, we call this type of laser a *multimode laser*. The stimulated and amplified light waves (the modes) are accompa-nied by spontaneously emitted light waves. Thus, the stimulated waves are disturbed by random light waves called the *laser noise*. This most fundamental noise in coherent optical communica-tion systems arises from spontaneous emission and results in a serious broadening of linewidth of each mode. In Fig. 2.9, this linewidth is represented by Δf_a. Laser noise i.e., laser phase and amplitude or intensity noise will be discussed in Chapter three.

Similar to the line shape of spectrum $G_{21}(f)$ of an energy transition, shape of power spectral density $G_a(f)$ of the modes are also Lorentzian (3.2.5). From Fig. 2.9, following relationship can be given:

$$\Delta \lambda_a \ll \Delta \lambda_g \ll \Delta \lambda_{21}$$

$$(2.19)$$

For commercially available semiconductor diode lasers without any special technological steps to reduce the emission bandwidth, 3 dB bandwidth mentioned above are in the order of

$$\Delta f_a \approx 10^8 \text{ Hz}, \Delta f_g \approx 10^{11} \text{ Hz and } \Delta f_{21} \approx 10^{13} \text{ Hz}.$$

The 3 dB bandwidth is also termed as *Full Width at Half Maximum* (FWHM) or as mentioned above *spectral half-width*.

The power spectral density of the emitted laser light wave in Fig. 2.9 can be expressed mathematically by

$$G(f) = \sum_{a} G_a(f) \qquad (2.20)$$

where the spectrum $G_a(f)$ of each single mode is again defined by the Lorentzian line shape which is

$$G_a(f) = \hat{E}_a^2 \frac{2}{\pi \Delta f_a} \frac{1}{1 + \left(\dfrac{f - f_a}{\Delta f_a/2}\right)^2} \qquad (2.21)$$

Here, $\hat{E}_a$ represents the amplitude of electric field.

2.4.3.5 SINGLE-MODE OPERATION

The transmission capacity of optical communication systems strongly depends on the spectral emission bandwidth of the lasers i.e., transmitter laser and in addition local laser when coherent detection is applied. The smaller the bandwidth, the higher the possible bit rate and the larger the possible repeaterless transmission span. In the ideal case, laser should be a real monochromatic source with one single frequency and zero linewidth i.e., $G(f) = \delta(f - f_0)$ is a Dirac delta function at frequency f_0.

In order to get such a laser, first step is to realize a single-mode laser and second step is to reduce its residual linewidth. In principle, single-mode operation can be accomplished either by decreasing the length L of the laser cavity or decreasing the pump power level P_P as shown in Fig. 2.10. It is observed from this figure that if pump power is low enough, only one single mode is able to exist. In case of all other waves, pump power is below the threshold power. If pump power is increased, laser will start operating again in multimode. With an appropriate short laser cavity, frequency separation of the modes can be made sufficiently large (Eq. 2.17). This again results in single-mode operation. In addition, Fig. 2.10 illustrates that the centre frequency f_0 of a single-mode laser may be different from the frequency f_{21} of energy transition.

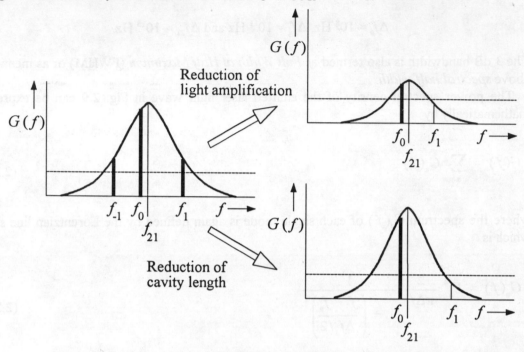

Fig. 2.10: Influence of laser cavity length and amplification on laser modes

In order to achieve highly reliable single-mode operation, more efficient solutions are generally used. DFB (Distributed Feedback) lasers and DBR (Distributed Bragg Reflection) lasers, principally based on change in laser structure (next Subsection), offer a very stable and highly efficient single-mode operation. In high bit rate direct detection system, DFB lasers are particularly most popular. Even if stable single-mode operation is achieved, a single-mode laser is still not a real monochromatic laser since the linewidth $\Delta \lambda_a$ of the mode is still not zero. To enhance system performance, further reduction of the remaining laser linewidth is required particularly when a coherent optical PSK homodyne system is needed (Chapter eleven). By using appropriate technologies, linewidths in the kHz range can, for example, be achieved with semiconductor laser diodes (Chapter three).

2.4.4   TYPES OF LASERS

For optical communications, there are different types of lasers available which can be used to transmit the information. The types of lasers can be principally distinguished or classified on the basis of either the material used for the active area or by the pumping process. The most important active materials are *semiconductors*, *gasses* and *crystals*; the most important pumping schemes are *optical pumping* by means of a second optical source and *injection current pumping* employed in semiconductor lasers. Despite of differences in their pumping mechanisms, active material and dimension, basic principle is the same for all lasers. The following graphic shows some of the relevant types of lasers used in optical communications as well as in other applications.

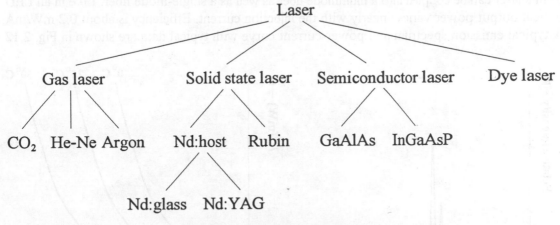

Fig. 2.11: Classification of lasers

2.4.4.1 SEMICONDUCTOR LASER DIODES

One of the most commonly used laser in advanced optical communications, in particularly in fiber optic systems, is the *heterojunction-structured semiconductor laser diode* also referred as *injection laser diode*. Like in an LED (Section 2.3), light in a laser diode is produced by the recombination of charge carriers i.e., electrons and holes as explained in Section 2.1 and 2.2. Due to small laser cavity all single emission processes are strongly correlated, whereas in an LED these emission processes are completely uncorrelated. As a result, light wave of an LED is incoherent with a spectral half-width of some 10 nm, whereas light wave of a laser diode is coherent with a spectral half-width in the order of typical 0.1 nm. With DFB single-mode laser diodes described below, spectral half-width about 0.0001 nm is possible. Main advantage of small spectral half-width is that dispersion induced signal distortion decreases and, therefore, maximum transmission distance increases (Section 12.4.4). In addition to the small linewidth, light beam of a laser diode is much more focussed in comparison to the light beam of an LED i.e., beam divergence is less. This results in a higher source-to-fiber coupling efficiency.

Similar to an LED, a semiconductor laser diode consists of a pn-junction based on direct-band-gap III-V semiconductor compounds (Section 2.2). When laser diode is *forward biased*, electrons from the n-side and holes from the p-side diffuse into the *depletion region* where both electrons and holes recombine (see Fig. 2.5). While recombining, optical radiation are produced. Frequently, depletion region which is much smaller in a laser as compared to an LED is also referred as *space charge region*, *active region*, *active area* or *recombination region*.

Under forward biased condition when the drive or injection current I_f is below the threshold current I_{th}, a lasers diode generates incoherent optical radiation similar to an LED. Above threshold, population inversion and, hence, spatial and temporal coherent radiation is achieved. Now, optical output beam is highly monochromatic and very directional. However, spatial coherence is less compared to gas and solid state lasers described below. Optical output power from a laser can be coupled into a multimode fiber as well as a single-mode fiber. Like in an LED, optical output power varies linearly with the injection current. Efficiency is about 0.2 mW/mA. A typical emission spectrum and power-current curve with typical data are shown in Fig. 2.12.

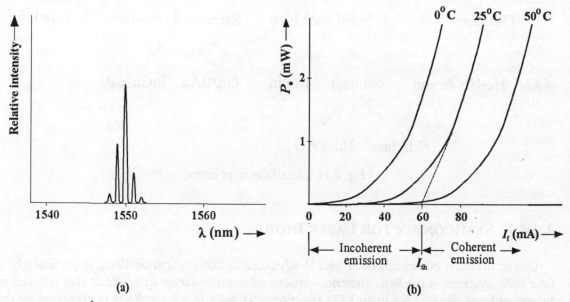

(a)   (b)

Fig. 2.12: (a) Emission spectrum and (b) P/I curve of a multimode semiconductor laser diode

Semiconductor laser diodes show a strong temperature dependence of the optical output power and threshold current. Temperature coefficients are typically in the order of 100 μW/K or about 1 mA/K, respectively. The peak emission wavelength varies with temperature with a coefficient of about 1 nm/K. This represents a main disadvantage as compared to LEDs, since complex stabilisation circuits are required in laser diode transmitters (Section 2.5).

The spectral half-width of the laser normally given in the data sheet of the laser is valid only for continuous wave operation. When laser is modulated by varying the injection current, then *chirping* occurs which effectively broadens the linewidth of the laser diode. As a result chromatic dispersion and, hence, signal distortion is increased. On the other hand, chirping is utilized for frequency modulation of laser diodes as described below. Physically, chirping is produced by a change of the refraction index of the active semiconductor material due to a change in the driving current. This is explained in more detail in Chapter three.

Semiconductor laser diodes are realized by employing the epitaxy process. Here, the required III-V layers are vapoured step-by-step on a GaAs- or InP-substrate. As a result, multilayer devices are obtained where each layer exhibits its own characteristic inner composition and doping. The transition between two of such layers is known as a *heterojunction or a heterotransition*. As the active layer is surrounded on *both sides* by different semiconductor materials which are differently doped, it is called a *double heterojunction*. In such devices, the active layer acts as a waveguide directing the light out to the edge since these materials have different refraction indexes.

The realization principle of a GaAlAs laser diode with a double heterostructure is shown in Fig. 2.13. This structure limits the recombination process on a very small area with a thickness of only some tenth μm (typical 0.1 μm) which confines the active area in the vertical direction. Because of this small thickness, laser efficiency is high. A narrow metallic electrode stripe running along the whole length of the diode limits the active area to about 5 μm to 15 μm in the lateral direction. In the longitudinal direction, active region is confined by a pair of flat, partially reflecting mirrors which are directed towards each other to enclose the cavity. These mirrors are made by cleaved facets constructed by parallel cleaves along the natural cleavage planes of the semiconductor crystal.

Within the active area which serves as a resonator known as *Fabry-Perot resonator*, light waves are generated by stimulated emission processes and guided through the well-confined active region as in a waveguide. Emission is from both the facets with similar radiation characteristics. Normally one facet is coupled to the fiber or another external optical device such as optical isolator or external optical modulator, whereas the other facet is used to monitor the output to control the laser operation.

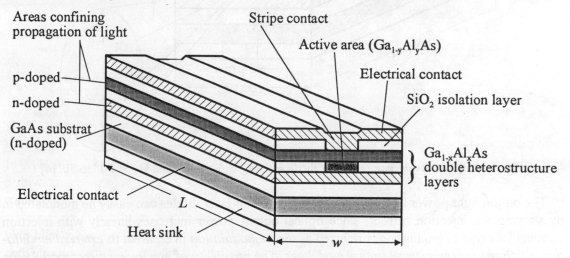

Fig. 2.13: GaAlAs double heterojunction structure

Both layers next to the active area are made of $Ga_{1-x}Al_xAs$ and have a thickness of about 1 μm. Total length L and width w of the laser diode are approximately a few 100 μm (typical 300 μm) and about 10 μm, respectively. Thus, dimensional characteristics is compatible with those of the optical fiber. Laser employing an electrode stripe to confine the active region are commonly called as a *gain-guided laser*.

As mentioned in the previous Section, dielectric waveguide structures fabricated in the lateral direction can be used as an alternative. This type of laser is known as an *index-guided laser*. Here, optical guiding is provided by the lower refraction index of the material that surrounds the active zone. This confinement results in a much stronger optical guiding compared to the first confinement method. Although gain-guided lasers can emit powers exceeding 100 mW, it has strong instabilities in light transmission (quality of spectrum) compared to the index-guided one.

The transmitted light at the output of a semiconductor laser diode is deflected, since spatial dimensions of the active emitting area are in the order of the wavelength. Therefore, beam divergence of a laser diode is much higher compared to a gas or solid state laser where these dimensions are some order of magnitudes larger than the wavelength of the light. Moreover, beam divergence is different in the vertical and horizontal axes, since about half the light travels in the confining layers and emitting area is unequal in width and thickness as shown in Fig. 2.14.

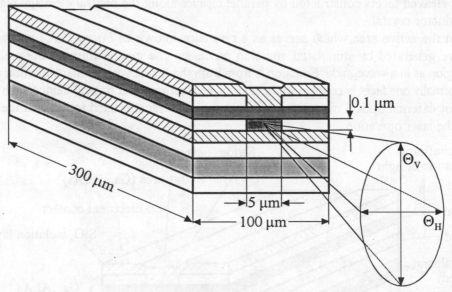

Fig. 2.14: Beam divergence of a laser diode with $\Theta_H \approx 5°$ to $10°$ and $\Theta_V \approx 30°$ to $50°$ [4]

The output light power or intensity of semiconductor laser diodes can easily be modulated in by varying the injection current, since optical output power increases linearly with injection current. This type of modulation is referred as *direct modulation* in contrast to *external modulation* which requires an external optical modulator. The possibility of employing direct modulation is a most important advantage of semiconductor lasers compared to other lasers such as gas or solid state laser. In addition to intensity modulation (IM), frequency modulation (FM or FSK) can also be easily achieved by varying the driving current. Laser diodes can be modulated by digital as well as analog signals, because they exhibit a very good linearity. As an example, Fig. 2.15 illustrates the analog intensity modulation of a laser diode. Intensity modulation is the preferred modulation scheme. Analog intensity modulation is frequently used for TV broadcasting and digital intensity modulation, also referred as on-off keying (OOK), is normally employed for data and digital voice transmission e.g., in an ISDN telephone network.

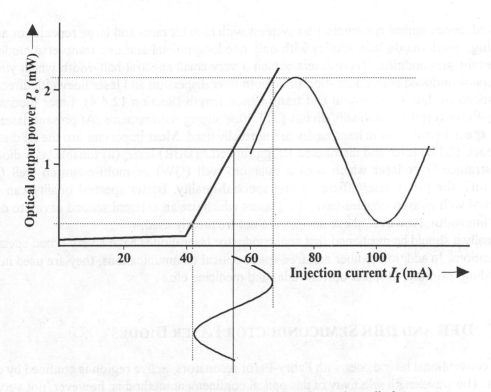

Fig. 2.15: Intensity modulation of semiconductor laser diode

The chirping effect of a semiconductor laser diode which is responsible for an undesired broadening of the spectral half-width, is also the basis of *laser frequency modulation* (FM). By varying the injection current, optical output power as well as laser peak wavelength change simultaneously (Chapter three). Efficiency of modulation is described by the FM coefficient which is in the order of 1 GHz/mA. FM can be employed either in coherent or incoherent optical communication systems using IM/DD. In the latter case, FM is normally used for TV broadcasting. Here, multiple AM-VSB video signals provided for example, in conventional cable TV are simultaneously transformed onto the optical carrier.

In summary, semiconductor laser diodes are the preferred optical sources in fiber optic transmission links and fiber-based optical networks. They are used in systems with simple analog or digital intensity modulation and direct detection, but also in coherent optical communication systems with advanced modulation schemes such as FSK, DPSK and PSK. In the latter case, strong requirements with respect to spectral purity (single-mode operation, small linewidth) and frequency stability have to be fulfilled. In PSK homodyne systems (Chapter eleven), these requirements are in particularly very high. Therefore, such systems could only be realized with gas lasers for a long time which exhibit an extremely narrow linewidth (Section 2.4.4.3). Now-a-days, optical homodyne systems can easily be realized by using solid state lasers as explained in Section 2.4.4.2. A first realization of a semiconductor laser diode-based homodyne systems was performed in 1988. However, external resonators were used to decrease linewidth of both transmitter and local lasers.

In advanced optical communication systems with high bit rates and large repeater or amplifier spacing, single-mode laser diodes with only one longitudinal and one transverse mode are an important precondition. These lasers exhibit a very small spectral half-width which yields less dispersion-induced signal distortion since both fiber dispersion and laser linewidth directly limit the speed of data transmission and transmission length (Section 12.4.4). Laser diodes with a Fabry-Perot resonator normally do not fulfil these strong requirements. At present, laser diodes with special structures in laser cavity are normally used. Most important are the (i) distributed feedback (DFB) laser and distributed Bragg reflector (DBR) laser, (ii) tunable laser diodes and (iii) strained-layer laser which uses a quantum well (QW) or multi-quantum well (MQW) structure for better laser efficiency and spectral quality. Better spectral quality can also be achieved with cleaved-coupled-cavity (C^3) lasers which use an external second cavity to decrease laser linewidth. However, these devices are not frequently used.

Finally it should be mentioned that semiconductor laser diodes have a very broad spectrum of applications. In addition to fiber and free-space optical communications, they are used in optical recording, printing, Doppler optical radar and medicine etc..

(i)   DFB AND DBR SEMICONDUCTOR LASER DIODES

In conventional laser diodes with Fabry-Perot resonators, active region is confined by cleaved facets. The wavelength selectivity of this optical confinement method is, however, not very good. This means that light reflection at the facets is broadband which does not allow to realize stable single-mode operation with small linewidths.

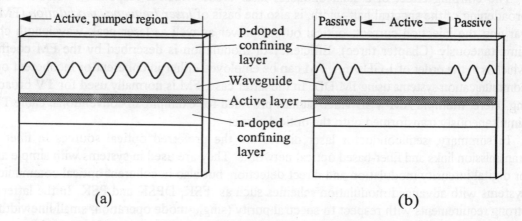

Fig. 2.16: Principle structure of (a) DFB and (b) DBR lasers

In a *distributed feedback* (DFB) laser, the cleaved facets are replaced by a periodic variation of the refraction index of the semiconductor material along the length of the active area i.e., the laser cavity. Thus, the grating for wavelength selection is formed over the entire laser cavity. In a *distributed Bragg reflector* (DBR) laser, cleaved ends of a Fabry-Perot resonator are replaced by an optical grid with Bragg reflection located at both the ends of the normal cavity. The schematic structures of both DFB and DBR lasers are shown in Fig. 2.16. DFB and DBR lasers make use of a so-called built-in frequency selective resonator. As a result of the very good wavelength selectivity of both cavity structures, DFB and DBR lasers exhibit an extemely stable single-mode operation. This important advantage is also available at very high modulation speeds of 10 Gbit/s and more.

(ii) TUNABLE LASER DIODES

In advanced optical communication systems, *continuously-tunable single-mode laser diodes* are key components. These can be employed in the transmitter of multichannel systems based on wavelength division multiplexing (WDM) and in coherent optical receivers performing the channel selection. Wavelength tuning in laser diodes is achieved by changing the refraction index of the active material which is inversely proportional to the wavelength. In principle, this can be achieved by temperature and pressure changes which, however, only allow a very slow modulation speed. Fast tuning can be obtained by electronically changing the carrier density. Different structures have been developed to achieve a wide continuous tuning range while maintaining a narrow spectral half-width. Most important are the multisection DFB- and DBR-devices and the tunable twin-guide (TTG) laser [1]. In general, tuning causes an additional broadening of the linewidth. So, a trade-off between tuning range and linewidth is required.

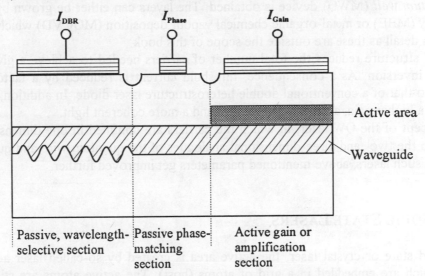

Fig. 2.17: Principle structure of a tunable three section DBR laser

The basic idea of a tunable laser, is to divide the conventional laser structure in three subsections which are optically coupled via a low-loss waveguide as shown in Fig. 2.17. Each subsection has its own driving current, whereas in the conventional laser structure only one injection current is required. The first region, called the gain section, is an active region. It is required to generate the light wave of high power, similar to a conventional laser. Second section is a passive section, which serves as a wavelength-selective reflector e.g., a DBR section. Between both these sections, a passive phase shifting section is located. Its function is to adjust the effective optical length of gain and phase shifting sections together to the wavelength selected in the DBR section. Tuning is achieved by controlling the injection currents of all three sections. The disadvantage of simultaneously controlling three injection currents, is avoided in a TTG laser where gain and wavelength selection regions are integrated transversely. Instead, of a separate DBR section, a DFB grating is used which extends over the whole cavity. The essential advantage of TTG lasers is that only two control currents instead of three are required.

Tunable laser diodes typically provide a tuning range of some nanometre with a linewidth of a few megahertz.

(iii) MQW LASER DIODES

In order to improve the operation of semiconductor laser diodes in terms of linewidth, output power, threshold current and tunability, *quantum well* (QW) structures are quite advantageous. When a layer in a laser diode is very thin, typically 10 nm or less, then quantum effects become important [2]. We talk about a quantum well structure, when location of the carriers i.e., electrons and holes is restricted to one dimension. A QW structure consists of one or more very thin layers with a narrow band-gap semiconductor material interleaved with layers of a wide band-gap semiconductor material. If many thin layers are arranged e.g., about hundred, then a *multi-quantum well* (MWQ) device is obtained. The layers can either be grown by molecular beam epitaxy (MBE) or metal-organic chemical vapour deposition (MOCVD) which can not be discussed in detail as these are outside the scope of this book.

The QW structure reduces the total number of carriers needed to achieve a given level of population inversion. As a consequence, threshold current is reduced by a factor of 10 as compared to that of a conventional double heterostructure laser diode. In addition, QW-lasers give rise to a higher gain, a narrower linewidth and a more coherent light.

The concept of the QW laser, in which the carriers are confined in one dimension, can be extended to the two- and three-dimensional analogues of the quantum wire and quantum box laser [2]. In such lasers, above mentioned parameters get improved further.

2.4.4.2 SOLID STATE LASERS

In a solid state or crystal laser, the active area is created by so-called laser active atoms (guests) which are embedded in a grid of atoms (host). The active atoms are stimulated by injection of a powerful light which is called *optical pumping*. Suitable pump sources are flash lamps or lasers, in particularly semiconductor lasers. By employing optical pumping, electrons located in the ground level are pumped to an upper level with higher energy. Laser operation is finally achieved by stimulated emission, whereby the electrons of the upper energy level returns to the ground level as described in detail in Section 2.2.

In optical communication systems, especially in atmospheric and free-space communication links, the Nd:YAG laser is the most important one. In this laser, a Yttrium Aluminium Garnet (YAG) crystal represents the host grid and Neodym (Nd) the embedded active guest atoms. Determined by the Nd atoms, a Nd:YAG laser operates at a peak wavelength of 1064 nm. By employing appropriate birefringent crystals, wavelength can be divided in half i.e., 532 nm which represents a green visible light. The advantage of using 532 nm instead of 1064 nm is due to the fact that laser light beam is much more focussed at the shorter wavelength. Thus, beam divergence is less. This advantage is in particularly of interest in long-range, free-space communication links e.g., from the planet Mars to Earth (78 million kilometres) as shown in Chapter fifteen. However, current birefringent crystals exhibit high internal loss which cancels out the advantage of low beam divergence.

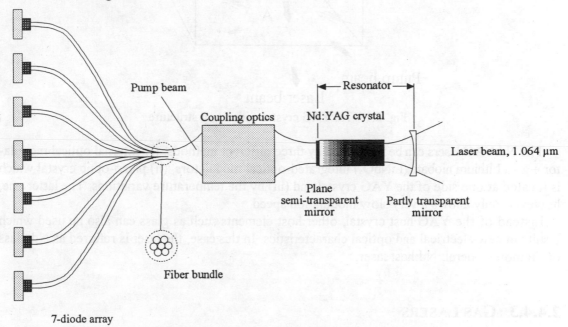

Fig. 2.18: Schematic structure of a diode-pumped Nd:YAG Laser

Solid state lasers are characterized by an extremely stable transmission frequency and an excellent spectral quality. Linewidths less than 50 kHz can be achieved without any practical problem. The aperture of the laser beam is in the order of 0.1 degree. Thus, beam divergence is also very small.

The principle structure of a Nd:YAG solid state laser employing a laser diode array as a pump is shown in Fig. 2.18. This structure can practically be realized either in a planar or in a ring configuration. As an example, Fig. 2.19 shows the path of a Nd:YAG laser light beam in case of a ring structure. In this structure, point A represents a partly transparent mirrored surface with a reflection of about 90 % and points B, C and D mirrored surfaces with 100 % reflection. The ring geometry eliminates the problem of spatial hole burning present in conventional linear or standing wave lasers which severely limits their single frequency output power. To ensure operation in only one direction, the ring must include an unidirectional device normally constructed by employing a polarizer, a Faraday rotator and a half-wave plate.

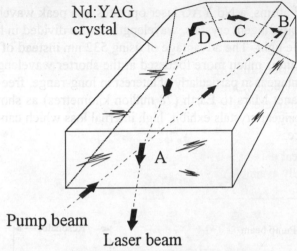

Fig. 2.19: Nd:YAG crystal with a ring structure

The Nd:YAG lasers can be modulated by three different methods: (i) external optical modulator e.g., a Lithium niobate ($LiNbO_3$) integrated-optical modulators, (ii) piezo-optic crystal which is located at one side of the YAG crystal and (iii) by the temperature variations. The latter one, however, only allows a very slow modulation speed.

Instead of the YAG host crystal, other host elements such as glass can also be used which results in new electrical and optical characteristics. In this case, the laser is referred as Nd:glass or, in more general, Nd:host laser.

2.4.4.3 GAS LASERS

As far as spatial and spectral quality of the laser light beam are concerned, the gas laser provides the best results compared to all other types of lasers. For this reason, coherent optical communication systems with PSK modulation and homodyne detection were first realized by employing gas lasers in the beginning of the eighties. In a gas laser, gas atoms or gas molecules are employed to produce the active material. The spectral linewidth of gas lasers is in the order of some Hz only. However, this linewidth only considers the laser phase and frequency noise due to spontaneous emission process as the only source of linewidth broadening (Chapter three). This noise is also known as quantum phase or quantum frequency noise. The real linewidth of a gas laser is, however, broader (in the order of 10 kHz), since there is normally an additional frequency noise as a result of external acoustical disturbances. This noise is an extremely low frequency noise, whereas the quantum frequency noise is approximately a white noise (Chapter three). Therefore, this additional frequency noise can be drastically reduced by employing an optical phase looked loop as described in detail in Chapter eleven. Although gas laser exhibit excellent spectral features, they are normally not used in advanced optical communication systems due to their large size.

2.4.5 LASER DATA

This Subsection describes important laser data usually presented in a data sheet. The values given below, represent typical values for a semiconductor laser diode and a solid state Nd:YAG laser. The following parameters are of interest when realizing an optical communication system (Table 2.3 and Fig. 2.20).

Table 2.3: Typical data of a DFB semiconductor laser diode and Nd:YAG laser taken from commercially available data sheets

	Parameter		Example	
			Laser diode	Nd:YAG
(i)	Wavelength λ		1.55 μm	1.064 μm
(ii)	Multimode or single mode		single mode	single mode
(iii)	Optical output power P_o		4 mW	40 mW
(iv)	Spectral width $\Delta\lambda$ or Δf		0.1 nm	40 kHz
(v)	Efficiency $\eta = P_o/P_{el} \cdot 100\%$		5 %	0.1 %
(vi)	Beam divergence	horizontal Q_H	25°	0.15°
		vertical Q_V	35°	0.15°

In order to compare spectral half-width, it is required to calculate either the frequency-related bandwidth of the laser diode which is $\Delta f = c_0/\lambda^2 \cdot \Delta\lambda \approx 12.5$ GHz or the wavelength-related bandwidth of the Nd:YAG laser which is $\Delta\lambda = \lambda^2/c_0 \cdot \Delta f \approx 1.5 \cdot 10^{-7}$ nm. Here, c_0 represents the velocity of light. It becomes clear that difference in spectral half-width is quite large. When laser diodes are considered, following additional parameters are also of main importance:

(vii) Threshold current I_{th} = 40 mA

(viii) Voltage supply U_f = 2 V

(ix) Maximum modulation speed f_M = 4 GHz

(x) Relative intensity noise RIN = -160 dB/Hz

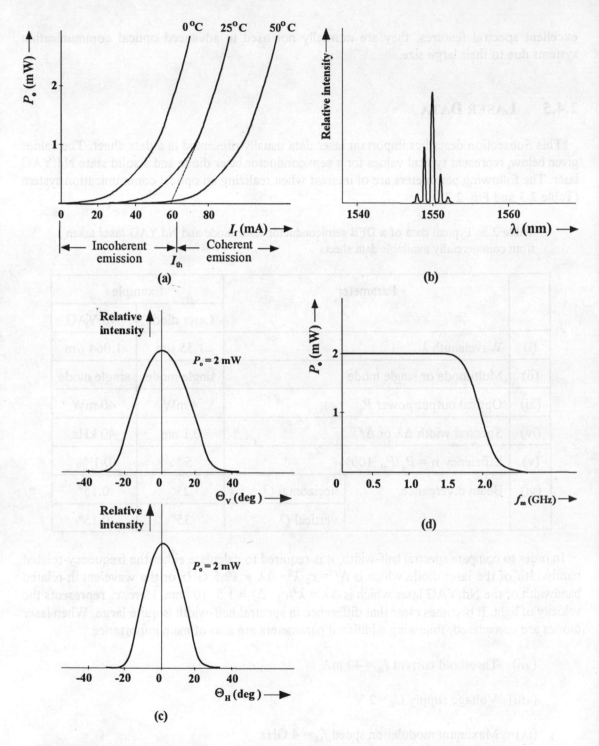

Fig. 2.20: Typical measurement results given in a laser diode data sheet (a) optical output vs. driving current, (b) emission spectrum, (c) far-field pattern and (d) frequency response

The parameters (i) to (ix) given above have already been discussed in the previous Subsections. Last parameter given in (x) describes the influence of laser intensity noise which can seriously deteriorate quality of signal transmission in analog communication systems. This so-called *RIN* parameter is described in more detail in Chapter three. When a data sheet of a laser is well prepared, all required information needed to employ the laser are included. In such a case, data sheet also consists of measurement results presented in the form of appropriate curves. Typical examples are given in Fig. 2.20. In addition to the parameters and measurement results given above, a data sheet also contains information about type of casing including the number of pins, operation temperature e.g., from -40° C to +70° C, fiber pigtail and type of connector.

2.5 OPTICAL TRANSMITTER

Besides the optical source described in the previous Sections, an optical transmitter also consists of a lot of electronics to stabilize temperature, driving current and input signal amplitude. A typical transmitter configuration is shown in Fig. 2.21.

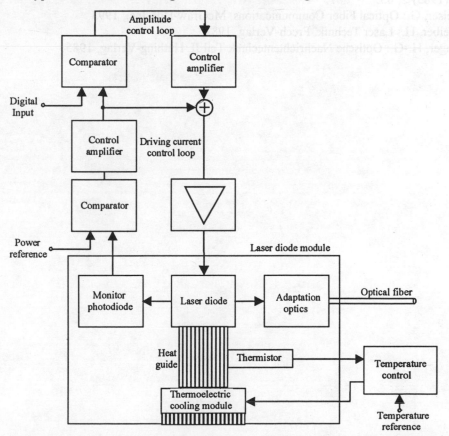

Fig. 2.21: Block diagram of an optical transmitter with a laser diode module

In this transmitter, the optical output power at the monitor output of the laser diode is lunched into a photodiode. The photodiode current, which is proportional to the laser diode output power is taken to control the laser diode driving current in two control loops. In the driving current control loop, optical output power is stabilized by comparing the photodiode current with a reference. The amplitude control loop is needed to stabilize the extinction ratio. Temperature control is achieved by measuring the temperature with a thermistor e.g., a NTC resistor and comparing with a reference.

2.6 REFERENCES

[1]   Amann, W.-C.; Thulke, W.: Continuous tunable laser diodes: Longitudinal versus transversal tuning scheme. IEEE J. SAC-8(1990)6, 1169-1177.
[2]   Bouley, J.-C.; Destefanis, G.: Multi-Qantum Well Lasers for Telecommunications. IEEE Commun. Magazine 32(1994)7, 54-60.
[3]   Kane, T.G.; Byer, R.L.: Monolithic, unidirectional single-mode Nd:YAG ring laser. Opt. Lett. 10(1985) 2, 65.
[4]   Keiser, G.: Optical Fiber Communications. McGraw-Hill, Inc., 1993.
[5]   Treiber, H.: Laser Technik. Frech-Verlag, 1982.
[6]   Unger, H.-G.: Optische Nachrichtentechnik Teil II. Hüthing-Verlag, 1985.

3   LASER NOISE

The gain of advanced optical communication systems is reduced by the noise of optical source. Coherent optical communication systems are in particular very sensitive to the noise of transmitter and local laser. In optical communication links employing an LED, transmitter noise is normally negligible in comparison to the other sources of noise in the system such as shot noise of photodiodes and thermal noise of resistors and electrical amplifiers (also called the circuit noise). Laser noise i.e., *laser intensity noise* and *laser phase noise*, is only negligible in communication systems with moderate bit rate and moderate transmission length. However, laser intensity noise can seriously degrade the quality of transmission in high bit rate links and laser phase noise is a dominant source of noise in coherent optical receivers with optical or electrical synchronous detection; for example, PSK homodyne and heterodyne receivers which require a phase-locked loop (PLL). Unlike coherent detection, conventional optical communication systems with intensity modulation of light and direct detection (IM/DD) are absolutely insensitive to laser phase noise. This, of course, is a significant advantage of using direct detection systems instead of coherent detection systems.

Modulation in direct detection system is easily performed by turning the laser light on and off as per the digital binary information signal to be transmitted. Hence, this kind of modulation is usually called on-off keying (OOK) or in a more devaluated way "smoke-sign modulation". A special knowledge of laser physics is not required and a simple black-box consideration of laser source is quite adequate for such applications.

In order to analyse and optimize advanced optical communication systems, this simplified black-box consideration is not adequate. It requires a much more detailed knowledge of the laser physics, in particular a fundamental knowledge of the statistical properties of laser noise. A brief review of the basic principle of laser operation has already been given in Chapter two. Physical reasons for laser noise is discussed in Section 3.1 below. With growing interest in realizing high-quality coherent communication links in the early eighties, physical reasons of laser noise had been examined very intensively. The results were primarily focussed on physical than on communication aspects and published in many technical papers, reviews and books e.g., [4, 14-16, 23, 24, 26-29, 31]. In this book, communication aspects of laser noise have been primarily considered.

Section 3.2 derives and discusses most important statistical properties of laser noise by means of a simple model that especially allows to highlight the communication aspects of laser noise. Laser phase noise, laser frequency noise and so-called phase-change noise are considered in detail. In addition, harmonic oscillations disturbed by phase noise are also discussed.

Section 3.3 briefly examines the effects of laser relaxation oscillations and Section 3.4 presents some interesting and generally valid aspects of filtering signals perturbed by phase noise. The concluding Section 3.5 presents some important technical solutions to reduce the effects of laser phase noise, especially if semiconductor lasers are used in coherent communication systems.

3.1   REASONS AND FORMATION OF LASER NOISE

In an ideal laser, transmitted optical wave is composed of stimulated emission only i.e., spontaneous emission processes does not occur. Therefore, an ideal single-mode laser transmits a real monochromatic light wave with a single frequency $f_a \to f_0$ only and, in addition, a time-invariant constant phase $\phi_a \to \phi_0$. In this book, we exclusively concentrate our interest on single-mode lasers. Hence, the mode number a as a special index is no longer required. Instead, we use the new index "0" to represent a laser that operates without any laser noise (zero noise). In the ideal case, power spectral density or emission spectrum $G_a(f) \to G_0(f)$ of a single-mode laser is completely defined by a Dirac delta function as shown Fig. 3.1a.

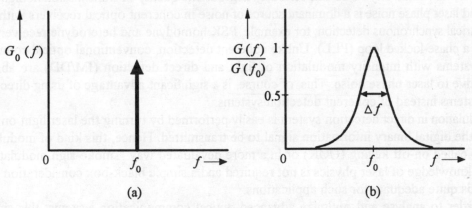

(a)   (b)

Fig. 3.1: Emission spectrum of (a) ideal and (b) real single-mode laser

Using the complex representation, time dependence of the electric field at the output of ideal single-mode laser can be described by the equation

$$E_0(t) = \hat{E}_0 \, e^{j2\pi f_0 t} = \hat{E}_0 \, e^{j\phi_0} \, e^{j2\pi f_0 t} \qquad (3.1)$$

It simply represents a harmonic oscillation. The electric field in addition of time as given by Eq. (3.1) is also a function of position. In the study of laser noise, parameter of location is insignificant. However, it has to be considered in Chapter five when polarization effects are discussed.

3.1.1   REASONS OF LASER NOISE

As mentioned in the previous Section, laser noise arises from spontaneous emission processes which are unavoidable in a laser. In gas lasers, physical reason of spontaneous emission is primarily the local fluctuations of laser mirrors. These fluctuations are in turn caused by the changes in temperature and external mechanical disturbances.

Spontaneous emission yield a time-varying amplitude $|\underline{E}(t)| = \hat{E}(t)$ and, in addition, a time-varying phase $\phi(t)$. As a result, laser emission spectrum is substantially broaden as shown in Fig. 3.1b. The random phase $\phi(t)$ is called as *laser phase noise* and the random amplitude $|\underline{E}(t)|$ as *laser amplitude* or *laser intensity noise*. The electric field of the transmitted laser wave, which is disturbed by laser noise, can be expressed as

$$\underline{E}(t) = \hat{E}(t) \; e^{j(\phi_0 + \phi(t))} \; e^{j2\pi f_0 t} \tag{3.2}$$

The fundamental measure to assess the strength of laser noise is given by the spectral laser linewidth $\Delta f_a - \Delta f$, which is termed as *3 dB linewidth*, *spectral half-width* or *linewidth at FWHM*. This characteristic linewidth is defined by

$$G\left(f_0 \pm \frac{\Delta f}{2}\right) = 0.5 \, G(f_0) \tag{3.3}$$

Formation of laser noise i.e., the conversion of single spontaneous emission process into time varying random amplitude $|\underline{E}(t)|$ and phase $\phi(t)$ is discussed in the following Subsection.

3.1.2 FORMATION OF LASER NOISE

By making use of a simple model shown in Fig. 3.2, formation of laser noise will now be discussed in more detail [14-16].

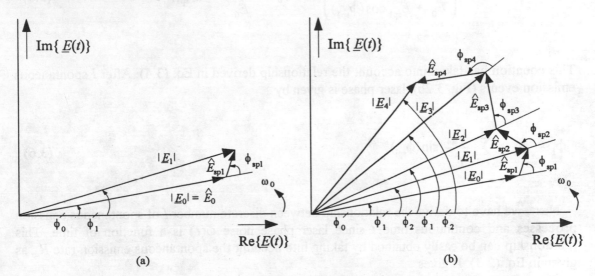

Fig. 3.2: Model of distortion of stimulated optical wave by (a) one and (b) four spontaneous emission events

In a single-mode laser with discrete energy states, spontaneous emission as well as stimulated emission only generate optical waves at frequency f_0. Phases of the stimulated waves are all synchronous, whereas they are random and absolutely uncorrelated in case of spontaneously emitted light waves. Each spontaneous emission event yields a spontaneous wave $E_{spi}(t) = \hat{E}_{spi} \exp[j(2\pi f_0 t + \phi_{spi})]$ which is superimposed onto the stimulated wave (see Eq. 3.2).

As shown in Fig. 3.2a, stimulated optical wave is disturbed by a single spontaneous emission event only. This figure presents the complex electric field vectors of both stimulated optical wave $E_0(t)$ and spontaneously emitted light wave $E_{sp1}(t)$. The spontaneous phase ϕ_{sp1} is absolutely random and each value lies in the range $-\pi$ to π and can occur with same probability. Fig. 3.2b shows in successive stages the formation of optical laser wave $E(t)$ now disturbed by four spontaneous emission events. It becomes evident from Figs. 3.2a and 3.2b that the resulting optical wave at the output is disturbed in phase (laser phase noise) as well as in amplitude (laser amplitude or intensity noise). Since spontaneously generated waves are not amplified to the same extent as stimulated waves, random amplitudes of spontaneous waves are always some order of magnitudes lower than the amplitude of the stimulated wave. Hence, following relationship is valid:

$$\hat{E}_{spi} \ll \hat{E}_0 \qquad i \in N \tag{3.4}$$

The resulting laser phase ϕ_1 after one spontaneous emission event from Fig. 3.2 is

$$\phi_1 = \phi_0 + \arctan\left(\frac{\hat{E}_{sp1}\sin(\phi_{sp1})}{\hat{E}_0 + \hat{E}_{sp1}\cos(\phi_{sp1})}\right) \approx \phi_0 + \frac{\hat{E}_{sp1}}{\hat{E}_0}\sin(\phi_{sp1}) \tag{3.5}$$

This equation has taken into account the relationship derived in Eq. (3.4). After I spontaneous emission events (Fig. 3.2b), laser phase is given by

$$\phi_I \approx \phi_0 + \sum_{i=1}^{I} \frac{\hat{E}_{spi}}{\hat{E}_0}\sin(\phi_{spi}) \tag{3.6}$$

Next, we have to derive the relationship between discrete number I of spontaneous emission processes and continuous time t since laser phase noise $\phi(t)$ is a function of time. This relationship can be easily obtained by taking into account the spontaneous emission rate R_{sp} as given in Eq. (2.3). It gives

$$I = I(t) = N_2(0) - N_2(t) = R_{sp}t \qquad t \geq 0 \tag{3.7}$$

It means, on an average I spontaneous emission processes occur or I spontaneous photons are generated in time t. Further, disturbance to the stimulated laser light wave due to spontaneous emission starts at $t = 0$. The electric field of optical wave at the output of laser can now be described by

$$E(t) = \hat{E}_0 \, e^{j\Phi_0} \, e^{j\phi(t)} \, e^{j2\pi f_0 t} \qquad\qquad (3.8)$$

with

$$\phi(t) \approx \sum_{i=1}^{I(t)} \frac{\hat{E}_{spi}}{\hat{E}_0} \sin(\phi_{spi}) \qquad I(t) = R_{sp} t \qquad\qquad (3.9)$$

In order to make it more clear, results of computer simulation experiment based on $I(t) = 10^5$ spontaneous emission events are shown in Fig. 3.3.

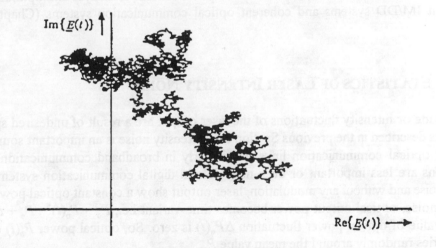

Fig. 3.3: Simulation results on the disturbance of stimulated light wave by spontaneous emission

It shows the electric field of a laser light wave disturbed by laser noise. In the simulation, the ratio $\hat{E}_{spi}/\hat{E}_0$ has been chosen to be constant (10^{-3}). In practice, this ratio is actually lower. For instant, numerical values in the order of 10^{-5} are usually obtained for semiconductor laser diodes. With an amplitude ratio of 10^{-3}, much lower number of simulated spontaneous emission events are required. Thus, much lower simulation run time is needed to obtain recognizable effects and achieve representative results. Moreover, the results obtained would be very similar and in principle be the same as with 10^{-5}.

3.2   STATISTICAL PROPERTIES OF LASER NOISE

In the first part of this Section, fundamental statistical features of laser intensity or power noise $P(t)$ (Section 3.2.1), laser phase noise $\phi(t)$ (Section 3.2.2) and other important laser random processes are derived and discussed in detail. Statistical properties of the *phase noise change* $\Delta\phi(t, \Delta T)$ and temporal derivation $d\phi(t)/dt$ which is called the *laser frequency noise* are discussed in Sections 3.2.3 and 3.2.4 respectively.

The most important statistical quantities which have to be derived are expected value (or mean value), variance (i.e., square of standard deviation), probability density function (pdf), autocorrelation function (acf) and power spectral density (psd). In coherent optical communication systems, "harmonic" random processes $\sin(\phi(t))$, $\cos(\phi(t))$ and $\exp(j(\phi(t)))$ which include phase noise $\phi(t)$ as argument are of fundamental importance in particular. They are considered in detail in Section 3.2.5.

In order to determine the statistical properties of laser noise, communication-based considerations have been kept in view. Indeed, the more profound physical consideration of laser noise is useful to study laser as an optoelectronic component. However, this consideration is not well-suited in the frame of this book. The statistical results obtained in the following Sections are summarized at the end of each Section excepted Section 3.2.1. It gives a small collection of important formulas, which will turn out to be a powerful tool to analyse and optimize advanced incoherent IM/DD systems and coherent optical communication systems (Chapters nine to eleven).

3.2.1   STATISTICS OF LASER INTENSITY NOISE

Amplitude or intensity fluctuations of the laser output are a result of undesired spontaneous emission as described in the previous Section 3.1. Intensity noise is an important source of noise in analog optical communication links, particularly in broadband communications. Intensity fluctuations are less important or even negligible in digital communication systems. Without intensity noise and without any modulation, laser output show a constant optical power P_o. With intensity noise, optical output power becomes time-variant i.e., $P_o \to P_o(t) = P_o + \Delta P_o(t)$. The average value of optical power fluctuation $\Delta P_o(t)$ is zero. So, optical power $P_o(t)$ of the laser output varies randomly around the mean value P_o.

In a laser data sheet, intensity or optical power fluctuations are expressed in terms of *relative intensity noise* (RIN) per unit bandwidth e.g., $RIN = 10^{-15}/\text{Hz}$ or $RIN = -150$ dB/Hz. The definition of this characteristic laser parameter is

$$RIN = \frac{1}{B}\left(\frac{\text{E}\{\Delta P_o^2(t)\}}{P_o^2} \right) = \frac{1}{B}\left(\frac{\text{E}\{[P_o(t) - P_o]^2\}}{P_o^2} \right) \tag{3.10}$$

where B is the noise-equivalent bandwidth and $E\{\Delta P_o^2(t)\}$ the mean square of the optical power fluctuations. The *RIN*, as given in the above equation is usually expressed in unit per Hz or % per Hz or in dB/Hz and ranges from about -120 dB/Hz to -180 dB/Hz. If *RIN* equals 10^{-12}/Hz, then *RIN* in dB/Hz is -120 dB/Hz. Thus, logarithm is taken of the numerical value only without considering the physical unit Hz. This must be taken into account to avoid confusion which frequently occurs when dealing with the *RIN* parameter. Considering, for example, the total bandwidth of a 30 channel cable TV signal, which is about 400 MHz (Chapter fourteen) and a *RIN* = -120 dB/Hz (worst case), then ratio of intensity noise $E\{\Delta P_o^2(t)\}$ to mean square value P_o^2 is $4 \cdot 10^{-4}$ or -34 dB.

The *RIN* of a laser depends on the injection current as well as on the modulation frequency. Intensity noise increases as the injection current decreases. Below some hundred megahertz, intensity noise is independent of modulation frequency, whereas it increases with modulation frequency when the modulation frequency is in the gigahertz range. At about 10 GHz, intensity noise shows a peak. Intensity noise is very sensitive to optical reflections back to the laser. Reflected light can increase the *RIN* by up to 20 dB.

In an optical receiver, laser intensity noise is directly transferred to fluctuations in the photodiode current which results in deterioration of receiver and, hence, system performance as shown in Chapter nine. In addition, receiver performance is seriously deteriorated by the shot noise of the photodiode and the thermal noise of the electrical preamplifier. The decision whether laser intensity noise can be neglected or not (compared to the other sources of noise in the receiver), strongly depends on the optical power P_r at the receiver input. Influence of intensity noise on system performance decreases when the optical input power decreases. As derived in Chapter nine, intensity noise can normally be neglected in comparison to the shot noise of the photodiode, if $RIN << -110$ dB/Hz - 10 $\log_{10}(P_r/\text{mW})$ where second term represents the optical input power in dBm.

3.2.2   STATISTICS OF LASER PHASE NOISE

(i)   EXPECTED VALUE

The expected value (mean value, average) of random laser phase $\phi(t)$ from Eq. (3.9) is given by

$$E\{\phi(t)\} = \frac{1}{\hat{E}_0} \sum_{i=1}^{I(t)} E\{\hat{E}_{spi} \sin(\phi_{spi})\} = \frac{1}{\hat{E}_0} \sum_{i=1}^{I(t)} E\{\hat{E}_{spi}\} E\{\sin(\phi_{spi})\} = 0 \qquad (3.11)$$

As spontaneous amplitudes $\hat{E}_{spi}$ and phases ϕ_{spi} are statistically independent random variables, expected value of the product in the first part of Eq. (3.11) can be separated in two as given in the second part. The phases ϕ_{spi} of a spontaneous emission event are uniformly distributed in the range $-\pi$ to $+\pi$ and the random variables $\sin(\phi_{spi})$ are symmetrical about zero. The expected value of the random variables $\sin(\phi_{spi})$ and, therefore, $\phi(t)$ will be zero.

(ii) VARIANCE

By using the fundamental statistical relation

$$\sigma_\phi^2(t) = E\{\phi^2(t)\} - E^2\{\phi(t)\} \tag{3.12a}$$

variance of $\phi(t)$ can be determined as follows:

$$\sigma_\phi^2(t) = \frac{1}{\hat{E}_0^{\,2}} E\left\{\left[\sum_{i=1}^{I(t)} \hat{E}_{spi} \sin(\phi_{spi})\right]^2\right\} = \frac{1}{\hat{E}_0^{\,2}} \sum_{i=1}^{I(t)} \sum_{j=1}^{I(t)} E\{\hat{E}_{spi}\hat{E}_{spj}\} E\{\sin(\phi_{spi})\sin(\phi_{spj})\}$$

$$\tag{3.12b}$$

$$= \frac{1}{\hat{E}_0^{\,2}} \sum_{i=1}^{I(t)} E\{\hat{E}_{spi}^2\} E\{\sin^2(\phi_{spi})\} = \frac{1}{2} \frac{E\{\hat{E}_{spi}^2\}}{\hat{E}_0^{\,2}} R_{sp}t = \frac{1}{2} \frac{P_{sp}}{P_0} R_{sp}t = K_\phi t$$

In the above equation, transformation from a single summation to double is based on the consideration that the expected values $E\{\sin(\phi_{spi}) \sin(\phi_{spj})\}$ are zero for $i \neq j$. The next change in Eq. (3.12b), which finally yields the result, takes into account Eq. (3.7) and the fact that the expected values of $\sin^2(\phi_{spi})$ are 0.5. In the last step, P_{sp} and P_0 represent the average light power of the spontaneous and stimulated emission processes respectively ($P_{sp} \ll P_0$). The constant of proportionality

$$K_\phi = 2\pi\Delta f \tag{3.13}$$

is related to the laser linewidth Δf as given in Fig. 3.1b and defined in Eq. (3.3). This simple relationship will be derived in Section 3.2.5. It becomes clear from Eq. (3.12) that the variance of laser phase noise is time-varying. Hence, it is called a *non-stationary random process*. This typical statistical feature of laser phase noise is illustrated in Fig. 3.4 by means of three different sample processes simulated by using a computer. Non-stationarity of laser phase noise becomes clear immediately when these processes are considered at different point of times e.g., t_1 and $t_2 > t_1$.

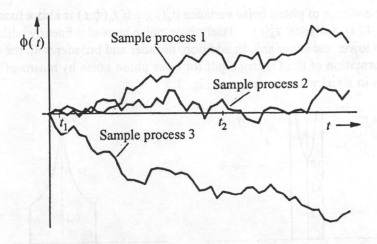

Fig. 3.4: Typical sample processes of laser phase noise $\phi(t)$

(iii) PROBABILITY DENSITY FUNCTION (pdf)

As seen from Eq. (3.9), laser phase noise is formally generated by a sum of sine functions with spontaneous emission phases ϕ_{spi} as arguments which are statistically independent and uniformly distributed in the range of $-\pi$ to $+\pi$. Due to inherent non-linearity of the sine function, pdf of the random variables $\sin(\phi_{spi})$ are not uniformly distributed as shown in Section 3.2.5. The sum given in Eq. (3.9) is characterized by two special properties: First, this sum contains a very large number of terms. Considering a typical spontaneous emission rate $R_{sp} = 10^{12}$ s^{-1} for instant, then this sum includes 1000 terms after a very short time interval of $t = 1$ ns only. It should be remembered that this time interval corresponds to the point of time when the laser source starts operation. The second characteristic feature is the statistical independence of the individual sum terms due to independence of phases ϕ_{spi} as mentioned earlier.

Both features permit the use of well-known central limit theorem of statistics [22]. This theorem states that if the random variables are independent (here $\sin(\phi_{spi})$), then the probability density function of their sum tends to follow a normal curve as the number of terms is sufficiently large. With this theorem, we obtain that laser phase noise is a *Gaussian random process* or a normal process.

Using Eqs. (3.11), (3.12b) and (3.13), pdf of laser phase noise is given by

$$f_{\phi}(\phi, t) = \frac{1}{\sqrt{2\pi}\sigma_{\phi}} \exp\left(-\frac{\phi^2}{2\sigma_{\phi}^2}\right) = \frac{1}{2\pi\sqrt{\Delta f t}} \exp\left(-\frac{\phi^2}{4\pi\Delta f t}\right) \qquad t \geq 0 \qquad (3.14)$$

Due to time dependence of phase noise variance $\sigma_\phi^2(t)$, pdf $f_\phi(\phi,t)$ is also a function of time. As given in Eq. (3.12a), variance $\sigma_\phi^2(t)$ of laser phase noise increases linearly with time. Thus, pdf $f_\phi(\phi,t)$ becomes lower and lower and, in addition, broader and broader with the increase in time.

Temporal formation of the Gaussian pdf for laser phase noise by means of six spontaneous emission events in step by step is shown in Fig. 3.5.

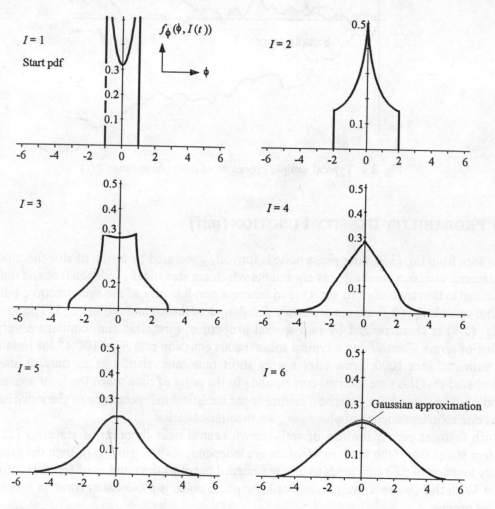

Fig. 3.5: Probability density function $f_\phi(\phi, I(t))$ of laser phase noise ϕ after $I = R_{sp}t$ spontaneous emission events. The Gaussian approximation in case of $I = 6$ follows Eq. (3.35) with $\sigma_\phi^2 = 2\pi\Delta ft = 0.5\ R_{sp}t = I/2 = 3$. In this figure, pdf is normalized to a constant field ratio $\hat{E}_{spi}/\hat{E}_0$.

After the first spontaneous emission process i.e., $I = 1$, pdf $f_\phi(\phi, I(t))$ equals the pdf of the random variable $\sin(\phi_{sp1})$. Remember that the laser output phase ϕ equals $\sin(\phi_{sp1})$ after the first spontaneous emission event (Fig. 3.2 and Eq. 3.5). Due to non-linearity of the sine function, this pdf called the "start-pdf" is non-Gaussian and, in addition, restricted to $|\phi_1| \le 1$. All other pdfs shown in Fig. 3.5 (i.e., $I = 2$ to $I = 6$) are obtained in step by step by taking convolution with the "start-pdf" ($I = 1$) in successive stages. Since all sum terms are statistically independent and additive, convolution with a single pdf is allowed. A detailed analytical derivation of the "start-pdf" is given in Section 3.2.5. It becomes clear from Fig. 3.5 that $f_\phi(\phi, I(t))$ is nearly Gaussian after six spontaneous emission events only. Thus, Gaussian pdf given in Eq. (3.14) is a well-suited approximation to study influence of phase noise in advanced communication systems.

(iv) AUTOCORRELATION FUNCTION (acf)

Due to non-stationarity of laser phase noise, it cannot be described by a simple one-dimensional autocorrelation function with $\tau = |t_2 - t_1|$ as argument. In such a case, two-dimensional acf which is a function of t_1 and t_2 has to be taken into consideration. We maintain that

$$R_\phi(t_1, t_2) = E\{\phi(t_1)\,\phi(t_2)\} = 2\pi\Delta f \min(t_1, t_2) \qquad t_1, t_2 \geq 0 \qquad (3.15)$$

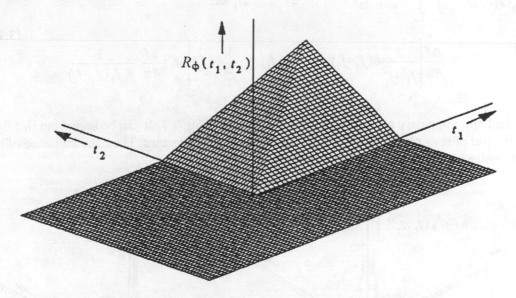

Fig. 3.6: Autocorrelation function $R_\phi(t_1, t_2)$ of non-stationary laser phase noise $\phi(t)$

Proof:

With $t_1 \leq t_2$, expected value in Eq. (3.15) can be written as follows:

$$E\{\phi(t_1)\,\phi(t_2)\} = E\{\phi(t_1)\,[\phi(t_2) - \phi(t_1)] + \phi^2(t_1)\}$$

$$= E\{\phi(t_1)\,[\phi(t_2) - \phi(t_1)]\} + E\{\phi^2(t_1)\} \qquad (3.16a)$$

The phase noise difference $\phi(t_2) - \phi(t_1)$ is independent of the previous temporal phase $\phi(t_1)$ since spontaneous emission events are statistically independent. For this reason, first expected value in the second line of Eq. (3.16a) is zero. Thus,

$$E\left\{\phi(t_1)\ \phi(t_2)\right\} = E\left\{\phi^2(t_1)\right\} = \sigma_\phi^2(t_1) = 2\pi\Delta f t_1 \qquad t_1 \leq t_2 \tag{3.16b}$$

By changing t_1 and t_2, that is $t_1 \rightarrow t_2$ and $t_2 \rightarrow t_1$, reader may prove Eq. (3.15) for $t_2 \leq t_1$.

(v) POWER SPECTRAL DENSITY (psd)

The power spectral density $\underline{G}_\phi(f_1, f_2)$ of the laser phase noise $\phi(t)$ is a complex quantity. It is obtained from the real acf $R_\phi(t_1, t_2)$ by taking two-dimensional Fourier transform. It is given by

$$\underline{G}_\phi(f_1, f_2) = \int_{-\infty}^{+\infty} \int_{-\infty}^{+\infty} R_\phi(t_1, t_2)\ e^{-j2\pi f_1 t_1}\ e^{-j2\pi f_2 t_2}\ dt_1\ dt_2$$

$$= \frac{\Delta f}{4\pi}\left[\frac{2}{f_1^2 + f_2^2}\delta(f_1 + f_2) - \frac{1}{f_2^2}\delta(f_1) - \frac{1}{f_1^2}\delta(f_2)\right] + j\frac{\Delta f}{4\pi}\ \frac{1}{f_1 f_2 (f_1 + f_2)} \tag{3.17}$$

The real and imaginary parts of $\underline{G}_\phi(f_1, f_2)$ are shown in Fig. 3.7. It can be seen from this figure that the psd of laser phase noise decreases rapidly at high frequencies. Thus, the influence of laser phase noise is much more in the low frequency range.

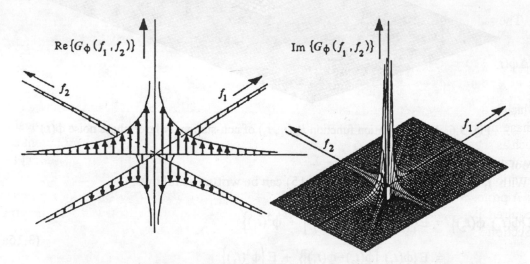

Fig. 3.7: Complex power spectral density $\underline{G}_\phi(f_1, f_2)$ of non-stationary laser phase noise $\phi(t)$

Usually, a psd is primarily characterized by its 3 dB bandwidth or FWHM bandwidth. However, in case of non-stationary phase noise, it is impossible to define a noise-equivalent bandwidth neither by defining a 3 dB cut-off frequency nor by determining an equivalent cube or cylinder with a volume in accordance with the volume of two-dimensional function $G_\phi(f_1, f_2)$. The reasons for this problem are $1/f^2$ shape of the psd and poles at the origin $f = 0$.

Statistical properties of laser phase noise $\phi(t)$ are summarized as follows:

(i)   non-stationary

(ii)   expected value: $E\{\phi(t)\} = 0$

(iii) variance: $\sigma_\phi^2(t) = 2\pi\Delta f\, t \qquad t \geq 0$

(iv)   pdf: $f_\phi(\phi,\, t) \;=\; \dfrac{1}{2\pi\sqrt{\Delta f t}}\; \exp\!\left(-\dfrac{\phi^2}{4\pi\Delta f t}\right)$

(v)   acf: $R_\phi(t_1, t_2) = 2\pi\Delta f \min(t_1, t_2) \quad t_1, t_2 \geq 0$

(vi)   psd: $G_\phi(f_1, f_2) \;\bullet\!\!-\!\!\circ\; R_\phi(t_1, t_2)$

Finally, it may be mentioned that laser phase noise exhibits same statistical features as the *Brownian motion* of molecular particles.

3.2.3   STATISTICS OF PHASE NOISE DIFFERENCE

The random process

$$\Delta\phi(t,\, \Delta T) = \phi(t+\Delta T) - \phi(t) \tag{3.18}$$

which is defined by a temporal change in laser phase during the time interval ΔT will be called a phase noise difference or a phase noise change in this book. This noise is not a physical noise such as shot noise of photodiode or phase noise of laser, but to be considered in coherent optical communication systems. The time interval ΔT is in general arbitrary, but usually equal the bit duration.

A profound knowledge of the statistical features of phase noise change $\Delta\phi(t, \Delta T)$ is especially required if the statistical features of laser frequency noise $d\phi(t)/dt$ have to be determined (Section 3.2.4) or optical DPSK system has to be analysed (Section 11.3.3).

(i) EXPECTED VALUE

Like laser phase noise $\phi(t)$, phase noise difference $\Delta\phi(t, \Delta T)$ is also a random process of zero mean. Thus, its expected value is also zero i.e.,

$$E\{\Delta\phi(t, \Delta T)\} = E\{\phi(t+\Delta T)\} - E\{\phi(t)\} = 0 \tag{3.19}$$

(ii) VARIANCE

Using Eqs. (3.12b) and (3.15), its variance can be determined as

$$\sigma_{\Delta\phi}^2 = E\{[\phi(t+\Delta T) - \phi(t)]^2\}$$
$$= E\{\phi^2(t+\Delta T)\} - 2E\{\phi(t+\Delta T)\,\phi(t)\} + E\{\phi^2(t)\} \tag{3.20}$$
$$= 2\pi\Delta f\,[(t+\Delta T) - 2\,\min(t+\Delta T,\,t) + t] = 2\pi\Delta f|\Delta T|$$

Unlike the variance of $\phi(t)$ given in Eq. (3.12b), variance of $\Delta\phi(t, \Delta T)$ is not a function of time. In fact, it is a function of constant time difference ΔT only. It may be mentioned that each phase is actually a phase change. Thus, laser phase itself is also a phase change from a fixed reference phase $\phi(0)$ at $t = 0$ when the phase noise process has started. Hence,

$$\phi(t) = \Delta\phi(0,\, t) \tag{3.21}$$

This means that $\Delta T \rightarrow t$ has to be considered as a variable of time in phase noise, whereas ΔT is a constant in case of phase noise difference.

As mentioned at the beginning of this Section, phase noise difference is important in the analysis of coherent optical DPSK transmission system. In Section 11.3.3, it has been derived that the standard deviation $\sigma_{\Delta\phi}$ of $\Delta\phi(t, \Delta T)$ must be less than 0.24 to achieve a bit error rate of less than 10^{-10}. For example, at a bit rate of 155.52 Mbit/s (STM1 of SDH) i.e., $\Delta T = T = 1/155.52$ µs, Eq. (3.20) gives a maximum allowed laser linewidth of about 1.43 MHz.

(iii) PROBABILITY DENSITY FUNCTION (pdf)

In Section 3.2.2, a Gaussian pdf was derived for laser phase $\phi(t)$. The phase noise change $\Delta\phi(t, \Delta T)$ is also a *Gaussian random process* since $\Delta\phi(t, \Delta T)$ is a linear combination of two random laser phases as defined in Eq. (3.18). Therefore, we have

$$f_{\Delta\phi}(\Delta\phi) = \frac{1}{2\pi\sqrt{\Delta f|\Delta T|}} \exp\left(-\frac{\Delta\phi^2}{4\pi\Delta f|\Delta T|}\right) \qquad (3.22)$$

By comparing this result with Eq. (3.14), a close relationship between $\Delta\phi(t, \Delta T)$ and $\phi(t)$ becomes evident again. As laser linewidth Δf increases, $f_{\Delta\phi}(\Delta\phi)$ becomes broader. It approaches to Dirac delta function i.e., $f_{\Delta\phi}(\Delta\phi) \rightarrow \delta(\Delta\phi)$ when laser linewidth Δf equals zero. In this unrealistic ideal case, laser phase remains constant and no difference between $\phi(t)$ and $\phi(t+\Delta T)$ can be observed. Thus, $\Delta\phi(t, \Delta T) = 0$.

(iv) AUTOCORRELATION FUNCTION (acf)

The acf of $\Delta\phi(t, \Delta T)$ can be calculated by

$$
\begin{aligned}
R_{\Delta\phi}(t_1, t_2) &= E\{\Delta\phi(t_1, \Delta T)\,\Delta\phi(t_2, \Delta T)\} \\
&= 2\pi\Delta f \begin{cases} |\Delta T| - |\tau| & \text{if } |\tau| = |t_2 - t_1| \le |\Delta T| \\ 0 & \text{otherwise} \end{cases} \qquad (3.23) \\
&= R_{\Delta\phi}(\tau) = R_{\Delta\phi}(-\tau)
\end{aligned}
$$

Proof:
Using Eqs. (3.15), (3.18) and, in addition, the linear property of expected values, above equation can be expressed as

$$
\begin{aligned}
R_{\Delta\phi}(t_1, t_2) &= E\{\,[\phi(t_1+\Delta T) - \phi(t_1)]\,[\phi(t_2+\Delta T) - \phi(t_2)]\,\} \\
&= 2\pi\Delta f\,\big[\min(t_1+\Delta T, t_2+\Delta T) - \min(t_1+\Delta T, t_2) \qquad (3.24) \\
&\qquad - \min(t_2+\Delta T, t_1) + \min(t_1, t_2)\big]
\end{aligned}
$$

Now we have to consider all possible interrelations between the points of time t_1, $t_1+\Delta T$, t_2 and $t_2+\Delta T$. Evaluation of all these cases and their substitution in Eq. (3.24) yield Eq. (3.23).

In contrast to the autocorrelation function $R_\phi(t_1, t_2)$ of non-stationary phase noise $\phi(t)$, auto-correlation function $R_{\Delta\phi}(t_1, t_2) = R_{\Delta\phi}(t_2-t_1) = R_{\Delta\phi}(\tau)$ of $\Delta\phi(t, \Delta T)$ only depends on the time difference $\tau = |t_2-t_1|$. For this reason, $\Delta\phi(t, \Delta T)$ is called a *stationary random process*. Thus, its autocorrelation function is simply a one-dimensional function in variable τ. The typical triangular shape of $R_{\Delta\phi}(\tau)$ is shown in Fig. 3.8a.

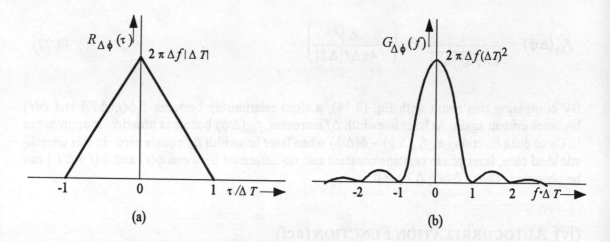

Fig. 3.8: (a) Autocorrelation function $R_{\Delta\phi}(\tau)$ and (b) power spectral density $G_{\Delta\phi}(f)$ of the phase noise difference $\Delta\phi(t, \Delta T)$

(v) POWER SPECTRAL DENSITY (psd)

Due to stationarity of $\Delta\phi(t, \Delta T)$, psd $G_{\Delta\phi}(f)$ can easily be obtained by taking Fourier transform of $R_{\Delta\phi}(\tau)$. We obtain

$$G_{\Delta\phi}(f) \bullet\!\!-\!\!\circ R_{\Delta\phi}(\tau)$$

$$G_{\Delta\phi}(f) = 2\pi\Delta f(\Delta T)^2 \, si^2(\pi f\Delta T) \tag{3.25a}$$

where

$$si(x) = \frac{\sin(x)}{x} \tag{3.25b}$$

The power spectral density $G_{\Delta\phi}(f)$ is shown in Fig. 3.8b. Like phase noise, phase noise difference is also dominant at low frequencies.

Statistics of phase noise difference $\Delta\phi(t, \Delta T) = \phi(t+\Delta T) - \phi(t)$ are summarized as follows:

(i)   stationary

(ii)   expected value: $E\{\Delta\phi\} = 0$

(iii)   variance: $\sigma_{\Delta\phi}^2 = 2\pi\Delta f \, |\Delta T|$

(iv) pdf:
$$f_{\Delta\phi}(\Delta\phi) = \frac{1}{2\pi\sqrt{\Delta f |\Delta T|}} \exp\left(-\frac{\Delta\phi^2}{4\pi\Delta f |\Delta T|}\right)$$

(v) acf:
$$R_{\Delta\phi}(\tau) = 2\pi\Delta f \begin{cases} |\Delta T| - |\tau| & \text{if } |\tau| \leq |\Delta T| \\ 0 & \text{otherwise} \end{cases}$$

(vi) psd:
$$G_{\Delta\phi}(f) = 2\pi\,\Delta f\,(\Delta T)^2\,\text{si}^2(\pi f \Delta T)$$

The statistical independence of two adjacent phase differences $\Delta\phi(t, \Delta T)$ and $\Delta\phi(t+\Delta T, \Delta T)$ is of significant importance and most advantageous when coherent communication systems are analysed by means of computer simulation. Here, random laser phases of both transmitter and local lasers have to be modelled by an appropriate software program. Beginning with a starting phase e.g., $\phi(0) = 0$, laser phase noise can easily be simulated by a step-by-step generation of independent Gaussian distributed phase noise changes to be added in successive stages i.e., $\phi([n+1]\Delta T) = \phi(n\cdot\Delta T) + \Delta\phi(n\cdot\Delta T, \Delta T)$. Thereby, a sample of phase noise after every ΔT is obtained. The parameter ΔT can be chosen arbitrarily for the desired simulation accuracy and run time. In order to generate the Gaussian distributed phase noise changes $\Delta\phi(n\cdot\Delta T, \Delta T)$, a simple computer-based random generator can be used. Finally, it may be mentioned that computer simulation represents a powerful and least expensive tool to analyse advanced optical communication systems. In this book, several simulation results are given.

3.2.4 STATISTICS OF LASER FREQUENCY NOISE

In this Section, statistics of laser frequency noise $\omega(t) = d\phi(t)/dt$ will be discussed. Any temporal derivation is generally defined by its differential quotient which is $\Delta\phi(t, \Delta T)/\Delta T$ with $\Delta T \to 0$. As $\Delta\phi(t, \Delta T)$ is a stationary random process, frequency noise also becomes a *stationary random process*. Spontaneous emission events which are responsible for $\phi(t)$ are, consequently, responsible also for $\omega(t)$. Fluctuations of laser centre frequency f_0 due to slow changes in temperature or in injection current are not included in $\omega(t)$. These fluctuations have to be considered separately. In this book, we presume that f_0 has been sufficiently stabilized by means of a frequency control circuit; for example, by a standard automatic frequency control (AFC).

(i) EXPECTED VALUE

As temporal derivation is a linear mathematical operation, expected value of $\omega(t)$ equals the expected value of $\Delta\phi(t, \Delta T)$. It leads to

$$E\{\omega(t)\} = 0 \tag{3.26}$$

(ii) VARIANCE

Direct analytical evaluation of variance of $\omega(t)$ similar to Eqs. (3.12b) and (3.20) is not possible. Thus, an alternative approach given by calculating the area under psd of $\omega(t)$ must be taken. Since this area is infinite (see acf and psd below), variance is also infinite i.e.,

$$\sigma_\omega^2 \to \infty \tag{3.27}$$

(iii) PROBABILITY DENSITY FUNCTION (pdf)

As $\Delta\phi(t, \Delta T)$ and $\omega(t)$ are interrelated by a linear mathematical operation, $\omega(t)$ is also Gaussian distributed. To describe the pdf by an equation is, however, not useful since variance is infinity.

(iv) AUTOCORRELATION FUNCTION (acf)

In principle, acf $R_\omega(\tau)$ as well as psd $G_\omega(f)$ of $\omega(t)$ could be derived directly from the statistical characteristics of $\phi(t)$ as given in Section 3.2.2. However, it becomes very comprehensive and difficult since the non-stationarity of $\phi(t)$ must be considered. In comparison to this, acf can be calculated much more easily when calculation is based on the stationary random process $\Delta\phi(t, \Delta T)$. Moreover, it is advantageous to derive psd first (see Subsection below). Then, acf

$$R_\omega(\tau) = E\{\omega(t)\ \omega(t+\tau)\} = 2\pi\Delta f\ \delta(\tau) \circ\!\!-\!\!\bullet\ G_\omega(f) \tag{3.28}$$

can simply be obtained by taking its inverse Fourier transform. It is shown in Fig. 3.9a.

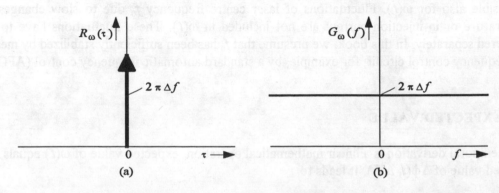

Fig. 3.9: (a) Autocorrelation function and (b) power spectral density of laser frequency noise $\omega(t)$

(v) POWER SPECTRAL DENSITY (psd)

The psd of $\omega(t)$ is given by

$$G_\omega(f) = G_\omega = 2\pi\Delta f \tag{3.29}$$

which is a constant (Fig. 3.9b). Thus, laser frequency noise is a *white noise process*.

Proof:
Starting with the equation

$$\Delta\omega(t, \Delta T)\,\Delta\omega(t+\tau, \Delta T) = \left[\omega(t+\Delta T)-\omega(t)\right]\left[\omega(t+\tau+\Delta T')-\omega(t+\tau)\right] \tag{3.30}$$

Its autocorrelation function can be written as

$$R_{\Delta\omega}(\tau) = 2R_\omega(\tau) - R_\omega(\tau+\Delta T) - R_\omega(\tau-\Delta T) \tag{3.31a}$$

The psd is obtained by taking Fourier transform on both sides. Thus,

$$|\,j2\pi f\,|^2 G_{\Delta\phi}(f) = 2\,G_\omega(f) - G_\omega(f)\,e^{j2\pi f\Delta T} - G_\omega(f)\,e^{-j2\pi f\Delta T} \tag{3.31b}$$

The factor $j2\pi f$ inside the absolute lines on the left side of above equation corresponds to the frequency response of a differentiator which converts the random process $\Delta\phi(t, \Delta T)$ at the input to a differential process $\Delta\omega(t)$ at the output. Substitution of psd $G_{\Delta\phi}(f)$ from Eq. (3.25a) and further simplification give Eq. (3.29).

Comparing the psd of $\omega(t)$ with the psd of $\phi(t)$, a typical feature of laser noise can be recognized: laser phase noise is predominant in the low frequency range though frequency noise is a white noise process. The reason for this feature can be explained by using the integral representation

$$\phi(t) = \int_0^t \omega(\tau)\,d\tau \tag{3.32}$$

which corresponds to low-pass filtering. Thus, laser phase noise is formally obtained at the output of a low-pass filter provided that white frequency noise is given at the input. Thereby, low frequencies will pass through the filter while high frequencies will be suppressed. As a result, the laser phase noise is a low-frequency process while the laser frequency noise is a white noise process. Finally, it will be mentioned that the integral representation given in Eq. (3.32) is in accordance with the fact that laser phase noise can simply be generated by a step-by-step addition of statistically independent and Gaussian distributed phase noise differences (Section 3.2.3).

Statistical properties of laser frequency noise $\omega(t)$ are summarized as follows:

(i)   stationary and white

(ii)   expected value: $E\{\omega\} = 0$

(iii)   variance: $\sigma_\omega^2 \rightarrow \infty$

(iv)   pdf: Gaussian

(v)   acf: $R_\omega(\tau) = 2\pi\Delta f\,\delta(\tau)$

(vi)   psd: $G_\omega(f) = 2\pi\Delta f$

3.2.5   STATISTICS OF HARMONIC OSCILLATIONS WITH PHASE NOISE

The purpose of this Section is to derive the statistical properties of the "harmonic" random process

$$w(t) = \cos\left(2\pi f_0 t + \phi_0 + \phi(t)\right) = \cos\left(\eta_\psi(t) + \phi(t)\right) = \cos\left(\psi(t)\right) \qquad (3.33)$$

As in the previous Sections, expected value, variance, pdf, acf and psd of random process $w(t)$ will be obtained and discussed.

Harmonic oscillations corrupted by phase noise exists at various points in coherent optical communication systems. Hence, the statistical properties of this random process are of prime importance to study such systems (Chapters nine and eleven). Depending on the position, centre frequency f_0, constant phase ϕ_0 and phase noise $\phi(t)$ of the random process $w(t)$ have to be replaced by the related quantities of local meaning (Table 3.1).

In the random process $\psi(t)$, phase $2\pi f_0 t + \phi_0 = \eta_\psi(t)$ is a deterministic signal instead of a random process. Hence, $\eta_\psi(t)$ can be regarded as a time-varying expected value of random phase $\psi(t) = \eta_\psi(t) + \phi(t)$. The phase noise $\phi(t)$ itself is a Gaussian process with zero mean as described in Section 3.2.2.

In order to obtain clear expressions and simplify calculations, time dependence of signals and random processes given in Table 3.1 will not be declared in our further considerations i.e., we substitute $\phi(t) \rightarrow \phi$, $\psi(t) \rightarrow \psi$, $\eta_\psi(t) \rightarrow \eta_\psi$, $\sigma_\psi(t) \rightarrow \sigma_\psi$, $\sigma_\phi(t) \rightarrow \sigma_\phi$ and so on.

Table 3.1: Phase noise in coherent optical communication systems

Signal	$\phi(t)$	$\eta_\psi(t)$	$\sigma_\psi^2(t) = \sigma_\phi^2(t)$	Remark
Carrier wave of transmitter laser	$\phi_t(t)$	$2\pi f_t\, t + \phi_{t0}$	$2\pi \Delta f_t\, t$	$\phi_t(t)$: phase noise of transmitter laser
Light wave of local laser	$\phi_l(t)$	$2\pi f_l\, t + \phi_{l0}$	$2\pi \Delta f_l\, t$	$\phi_l(t)$: phase noise of local laser
IF signal in heterodyne receiver	$\phi_l(t) - \phi_t(t)$	$2\pi[f_l - f_t]\, t$ $+ \phi_{l0} - \phi_{t0}$	$2\pi[\Delta f_l + \Delta f_t]\, t$	$\phi_t(t) = \phi_r(t)$ (see Section 9.4.2)
Baseband signal in homodyne receiver	$[\phi_l(t) - \phi_t(t)]$ $- \phi_{\text{PLL}}(t)$	0	see Section 11.1.3	$\phi_{\text{PLL}}(t)$: controlled local laser phase

(i) EXPECTED VALUE

As random variable ϕ and, hence, ψ are Gaussian distributed, expected value of new random variable $w = \cos(\psi)$ is given by

$$\eta_w = E\{w\} = E\{\cos(\psi)\} = \int_{-\infty}^{+\infty} \cos(\psi)\, f_\psi(\psi)\, d\psi$$

(3.34)

$$= \frac{1}{\sqrt{2\pi}\sigma_\phi} \int_{-\infty}^{+\infty} \cos(\psi)\, e^{-(\psi-\eta_\psi)^2/2\sigma_\phi^2}\, d\psi = \cos(\eta_\psi)\, e^{-\sigma_\phi^2/2}$$

Thus, the expected value η_w increases when σ_ϕ decreases. In absence of laser phase noise (i.e., $\sigma_\phi = 0$), expected value η_w reaches its maximum value $\eta_w = \cos(\eta_\psi)$.

It is remarkable that phase noise always decrease the signal (one-sided distortion), whereas AWGN may either increase or decrease a given signal (two-sided distortion). This important feature of phase noise becomes clear when an eye patten disturbed by phase noise is considered in Chapter twelve. Using Eq. (3.34), following interesting special cases can be derived:

$$\eta_\psi = 0: \quad w = \cos(\phi) \rightarrow \eta_w = e^{-\sigma_\phi^2/2} \tag{3.35}$$

$$\eta_\psi = -\frac{\pi}{2}: \quad w = \sin(\phi) \rightarrow \eta_w = 0 \tag{3.36}$$

$$w = e^{j\phi} \quad \rightarrow \eta_w = e^{-\sigma_\phi^2/2} \tag{3.37}$$

(ii) VARIANCE

Variance of random process $w(t)$ from Eq. (3.33) is given by

$$\sigma_w^2 = E\{w^2\} - \eta_w^2 = \frac{1}{\sqrt{2\pi}\sigma_\phi} \int_{-\infty}^{+\infty} \cos^2(\psi)\, e^{-(\psi-\eta_\psi)^2/2\sigma_\phi^2}\, d\psi - \eta_w^2$$

$$= \frac{1}{2}\left[1 - e^{-\sigma_\phi^2}\right]\left[1 - \cos(2\eta_\psi)\, e^{-\sigma_\phi^2}\right] \tag{3.38}$$

The following special cases are again of practical importance:

$$\eta_\psi = 0: \quad w = \cos(\phi) \rightarrow \sigma_w^2 = \frac{1}{2}\left[1 - e^{-\sigma_\phi^2}\right]^2 \tag{3.39}$$

$$\eta_\psi = -\frac{\pi}{2}: \quad w = \sin(\phi) \rightarrow \sigma_w^2 = \frac{1}{2}\left[1 - e^{-\sigma_\phi^2}\right]\left[1 + e^{-\sigma_\phi^2}\right] \tag{3.40}$$

$$w = e^{j\phi} \quad \rightarrow \sigma_w^2 = \left[1 - e^{-\sigma_\phi^2}\right] \tag{3.41}$$

The expected value of phase noise ϕ is zero and maximum gradient of sine function occurs at point $\phi = 0$. As a result, variance σ_w^2 of the random process $w = \sin(\phi)$ is always higher than variance of $w = \cos(\phi)$. If σ_ϕ approaches to infinity, then both variances are equal and approach to $\sigma_w^2 = 0.5$. This corresponds to the normalized average power of a sinusoidal or cosinusoidal harmonic oscillation. It may be mentioned that same result can also be obtained with uniformly distributed phase ϕ in the range $-\pi$ to $+\pi$ (or in the range 0 to 2π) instead of Gaussian distributed phase with $\sigma_\phi \rightarrow \infty$. To keep disturbance by laser phase noise low, σ_w^2 should be as small as possible. In the ideal case which practically can never occur, $\sigma_\phi = 0$.

(iii) PROBABILITY DENSITY FUNCTION (pdf)

In order to derive the pdf $f_w(w)$ of random variable $w = \cos(\psi) = \cos(\eta_\psi + \phi)$, it is advantageous to employ the statistical transformation

$$f_w(w) = f_\phi(\phi) \; \frac{1}{\left| \dfrac{d[\cos(\eta_\psi + \phi)]}{d\phi} \right|} \qquad \text{with} \quad \phi = \arccos(w) - \eta_\psi \tag{3.42}$$

It is valid for functions of one random variable only [22]. Due to the periodicity of cosine function $w = \cos(\psi)$ and the ambiguity of the inverse cosine function $\psi = \arccos(w)$, above transformation cannot be used without modifications. To avoid the ambiguity, we first have to split the cosine function and, consequently, the Gaussian pdf $f_\phi(\phi)$ as shown in Fig. 3.10. Second, we derive the pdf of random variable $w = \cos(\phi)$ presuming $\eta_\psi = 0$. This result will be frequently required in the analysis of coherent optical communication systems in Chapter eleven. Every section of cosine function is weighted by the Gaussian pdf of random phase ϕ. Thus, pdf $f_w(w)$ is determined by a sum of an infinite number of pdf components. The cosine function and Gaussian pdf are even functions. So, it is possible to restrain $\phi \geq 0$ and multiply the resulting pdf with a factor of two.

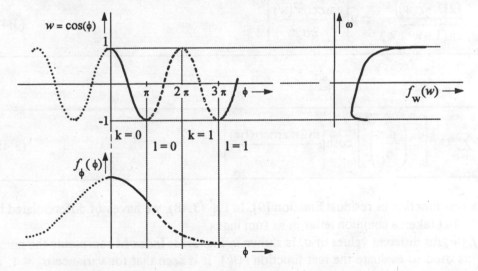

Fig. 3.10: Principle to evaluate the pdf $f_w(w)$ of random variable $w = \cos(\phi)$

The different phases in the sections outlined in Fig. 3.10 are defined by

$$\phi_k = \arccos(w) + k2\pi \tag{3.43}$$

and

$$\phi_1 = -\arccos(w) + (l+1)2\pi \tag{3.44}$$

where k and l are integers. With the above considerations and using the relationship

$$\left| \frac{d[\cos(\phi)]}{d\phi} \right|_{\phi=\phi_k} = \left| \frac{d[\cos(\phi)]}{d\phi} \right|_{\phi=\phi_l} = \sqrt{1-w^2} \tag{3.45}$$

pdf $f_w(w)$ can be written as

$$f_w(w) = \frac{1}{\sqrt{1-w^2}} \left[\sum_{k=-\infty}^{+\infty} f_\phi(k2\pi + \arccos(w)) + \sum_{l=-\infty}^{+\infty} f_\phi((l+1)2\pi - \arccos(w)) \right] \tag{3.46}$$

After going through some straightforward mathematical operations, the following result is finally obtained:

$$f_w(w) = \frac{2(1+r(w))}{\sqrt{2\pi(1-w^2)}\,\sigma_\phi} \exp\left[-\frac{\arccos^2(w)}{2\sigma_\phi^2} \right] \tag{3.47}$$

where

$$r(w) = 2\sum_{m=1}^{+\infty} \exp\left[-2\left(\frac{m\pi}{\sigma_\phi} \right)^2 \right] \cosh\left[\frac{m2\pi \arccos(w)}{\sigma_\phi^2} \right] \tag{3.48}$$

represents a rest function or residual function [9]. In Eq. (3.48), we have not differentiated between k and l and taken a common letter m as sum index.

The pdf $f_w(w)$ for different values of σ_ϕ is shown in Fig. 3.11. Index M represents the number of sum terms used to evaluate the rest function $r(w)$. It is seen that for variance $\sigma_\phi < 1$, rest function $r(w)$ is negligible i.e., $r(w) \ll 1$. In that case, $f_w(w)$ from Eq. (3.47) can be approximated by taking $r(w) = 0$ or M = 0. In coherent optical communication systems, for example in homodyne system, $\sigma_\phi < 0.24$ is required to achieve a BER of less than 10^{-10}. Thus, neglecting of rest function $r(w)$ is practically allowed. It becomes clear again from Fig. 3.11 that phase noise always attenuates the signal since $\cos(\phi) \le 1$ always appears as a random factor in the signal.

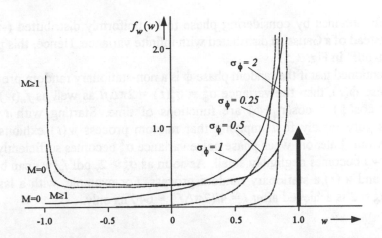

Fig. 3.11: Probability density function $f_w(w)$ of random variable $w = \cos(\phi)$

By replacing $\arccos(w) \to \arccos(w) - \eta_\psi$ in Eq. (3.46), η_ψ which has been neglected by now can be considered. After some mathematical manipulations, we obtain

$$
f_w(w) = \frac{2}{\sqrt{2\pi(1-w^2)}\,\sigma_\phi} \exp\left[-\frac{\arccos^2(w)}{2\,\sigma_\phi^2}\right] \cdot
$$
$$
\cdot \sum_{m=-\infty}^{+\infty} \exp\left[-\frac{(m2\pi-\eta_\psi)^2}{2\sigma_\phi^2}\right] \cosh\left[\frac{(m2\pi-\eta_\psi)\,\arccos(w)}{\sigma_\phi^2}\right]
$$

(3.49)

Following boundary cases can be derived from this generally valid result:

$$
w = \cos(\phi) \;\to\; f_w(w) = \begin{cases} \delta(w-1) & \text{if } \sigma_\phi = 0 \\[2mm] \dfrac{1}{\pi\sqrt{1-w^2}} & \text{if } \sigma_\phi \to \infty \end{cases}
$$

(3.50)

$$
w = \sin(\phi) \;\to\; f_w(w) = \begin{cases} \delta(w) & \text{if } \sigma_\phi = 0 \\[2mm] \dfrac{1}{\pi\sqrt{1-w^2}} & \text{if } \sigma_\phi \to \infty \end{cases}
$$

(3.51)

It can be deduced from the above equations that the pdf of random variables $\cos(\phi)$ and $\sin(\phi)$ are equal if phase noise variance becomes very large (i.e., $\sigma_\phi \to \infty$). In that case, pdf $f_w(w)$ becomes same as the pdf obtained by considering phase to be uniformly distributed ($-\pi \le \phi \le +\pi$ or $0 \le \phi \le 2\pi$) instead of a Gaussian distributed with infinite variance. Hence, this pdf becomes the so-called "start-pdf" in Fig. 3.5.

It may be mentioned that if the random phase ϕ is a non-stationary random process such as the laser phase noise $\phi(t)$, then the variance $\sigma_\phi^2 = \sigma_\phi^2(t) = 2\pi\Delta f t$ as well as $f_w(w)$ of the random variable $w = \cos(\psi) = \cos(\eta_\psi + \phi)$ are functions of time. Starting with $t = 0$, shape of $f_w(w)$ changes very rapidly first implying that random process $w(t)$ exhibits a strong non-stationary behaviour. Later on, when phase noise variance σ_ϕ^2 becomes sufficiently large, changes in shape of $f_w(w)$ becomes negligibly small. As soon as $\sigma_\phi^2 > 2$, pdf $f_w(w)$ can be considered as time-invariant and $w(t)$ a stationary random process. For example, with a laser of linewidth $\Delta f = 1$ MHz, $\sigma_\phi^2 = 2$ is achieved after $t = 4/(2\pi\Delta f) \approx 640$ ns only.

(iv) AUTOCORRELATION FUNCTION (acf)

Irrespective of whether the random process $w(t)$ is stationary or non-stationary, acf of $w(t)$ can be determined by

$$
\begin{aligned}
R_w(t_1, t_2) &= E\left\{\cos(\eta_\psi(t_1)+\phi(t_1))\,\cos(\eta_\psi(t_2)+\phi(t_2))\right\} \\
&= \frac{1}{2}E\left\{\cos\underbrace{\left(\eta_\psi(t_1)+\eta_\psi(t_2)+\phi(t_1)+\phi(t_2)\right)}_{u}\right\} + \frac{1}{2}E\left\{\cos\underbrace{\left(\eta_\psi(t_1))-\eta_\psi(t_2)+\phi(t_1)-\phi(t_2)\right)}_{v}\right\} \\
&= \frac{1}{2}\left[\cos\left(\eta_\psi(t_1)+\eta_\psi(t_2)\right)\exp\left(-\frac{1}{2}\sigma_u^2\right) + \cos\left(\eta_\psi(t_1)-\eta_\psi(t_2)\right)\exp\left(-\frac{1}{2}\sigma_v^2\right)\right]
\end{aligned}
\tag{3.52}
$$

The new random variables u and v which are Gaussian distributed with zero mean are characterized by their variances

$$
\begin{aligned}
\sigma_u^2 &= \sigma_\phi^2(t_1) + \sigma_\phi^2(t_2) + 2E\left\{\phi(t_1)\phi(t_2)\right\} \\
&= R_\phi(t_1, t_1) + R_\phi(t_2, t_2) + 2R_\phi(t_1, t_2)
\end{aligned}
\tag{3.53}
$$

and

$$
\begin{aligned}
\sigma_v^2 &= \sigma_\phi^2(t_1) + \sigma_\phi^2(t_2) - 2E\left\{\phi(t_1)\phi(t_2)\right\} \\
&= R_\phi(t_1, t_1) + R_\phi(t_2, t_2) - 2R_\phi(t_1, t_2)
\end{aligned}
\tag{3.54}
$$

If we substitute σ_u and σ_v in Eq. (3.52) from the Eqs. (3.53) and (3.54) respectively, we obtain the following result:

$$R_w(t_1,\ t_2) = \frac{1}{2}\exp\left[-\frac{1}{2}\left(R_\phi(t_1,\ t_1) + R_\phi(t_2,\ t_2)\right)\right]\cdot\left[\cos\left(\eta_\psi(t_1) + \eta_\psi(t_2)\right)\exp\left(-R_\phi(t_1,\ t_2)\right)\right.$$

$$\left. + \cos\left(\eta_\psi(t_1) - \eta_\psi(t_2)\right)\exp\left(+R_\phi(t_1,\ t_2)\right)\right]$$

(3.55)

This is a general result valid for both non-stationary as well as stationary phase noise $\phi(t)$. Let us now consider special random processes $\cos(\phi(t))$, i.e., $\eta_\psi = 0$, $\sin(\phi(t))$, i.e., $\eta_\psi = \pi/2$ and $\exp(j\phi(t))$. The phase noise $\phi(t)$ itself may be either non-stationary or stationary. For non-stationary phase noise, statistical properties of $\phi(t)$ have been derived in Section 3.2.2. Using Eqs. (3.13) and (3.15), we have

$$R_\phi(t_1,\ t_2) = K_\phi\ \min(t_1,\ t_2) = 2\pi\Delta f\ \min(t_1,\ t_2)$$

(3.56)

Depending on the physical meaning of random variable w, linewidth Δf has to be replaced by Δf_t, Δf_1 or $\Delta f_{IF} = \Delta f_t + \Delta f_1$ In case of stationary phase noise such as rest-phase noise in a homodyne receiver (Chapter eleven) with an optical phase-locked loop (OPLL), above equation simplifies to

$$R_w(t_1,\ t_2) = R_w(t_2 - t_1) = R_w(\tau) = \frac{1}{2}\exp\left(-R_\phi(0)\right)\left[\exp\left(+R_\phi(\tau)\right) + \exp\left(-R_\phi(\tau)\right)\right]$$

(3.57)

$$= \exp\left(-\sigma_\phi^2\right)\cosh\left(R_\phi(\tau)\right)$$

Taking Eqs. (3.55) and (3.56) for non-stationary processes and Eq. (3.57) for stationary processes, results summarized in Table 3.2 are obtained.

The variables of time t_1 and t_2 are related to time when the laser has been switched on and spontaneous emission and phase noise processes have started. Hence, they are always positive in Eqs. (3.58) to (3.60). As laser phase noise is a non-stationary process, "harmonic" processes $\cos(\phi(t))$ and $\sin(\phi(t))$ are also non-stationary. Therefore, their acfs are time-varying function. In the boundary case $t_1 \to \infty$ and $t_2 \to \infty$, that is a "long time" after the laser has been switched on, non-stationary processes $\cos(\phi(t))$ and $\sin(\phi(t))$ become stationary (see Subsection on pdf above). Thus, stationarity is sufficiently fulfilled when $t_1,\ t_2 \gg 1/(2\pi\Delta f)$. In that case, the acf of both the random processes $\cos(\phi(t))$ and $\sin(\phi(t))$ are function of the time difference $\tau = |t_2 - t_1|$ only.

Table 3.2: Autocorrelation function $R_w(t_1, t_2)$ of harmonic oscillations disturbed by phase noise

$\phi(t)$	$w(t)$	$R_w(t_1, t_2)$	$R_w(\tau)$ $\tau = t_2 - t_1$	
Non-stationary	$\cos(\phi(t))$	$\exp[\frac{1}{2}K_\phi(t_1+t_2)\cosh(K_\phi\min(t_1,t_2))]$	$\frac{1}{2}\exp(-\frac{1}{2}K_\phi\lvert\tau\rvert)$ $t_1\to\infty, t_2\to\infty$	(3.58)
$E\{\phi(t)\}= 0$	$\sin(\phi(t))$	$\exp[-\frac{1}{2}K_\phi(t_1+t_2)\sinh(K_\phi\min(t_1,t_2))]$	$\frac{1}{2}\exp(-\frac{1}{2}K_\phi\lvert\tau\rvert)$ $t_1\to\infty, t_2\to\infty$	(3.59)
$\sigma_\phi^2 = K_\phi t$	$\exp(j\phi(t))$	---	$\exp(-\frac{1}{2}K_\phi\lvert\tau\rvert)$	(3.60)
Stationary	$\cos(\phi(t))$	---	$\exp(-\sigma_\phi^2)\cdot$ $\cosh[R_\phi(\tau)]$	(3.61)
$E\{\phi(t)\}= 0$	$\sin(\phi(t))$	---	$\exp(-\sigma_\phi^2)\cdot$ $\sinh[R_\phi(\tau)]$	(3.62)
$\sigma_\phi^2 = \text{const.}$	$\exp(j\phi(t))$	---	$\exp(-\sigma_\phi^2)\cdot$ $\exp[R_\phi(\tau)]$	(3.63)

Irrespective of whether the phase noise is stationary such as the phase noise in PLL or non-stationary such as the laser phase noise itself, random process $\exp(j\phi(t))$ is always stationary (Eqs. 3.60 and 3.63).

Proof:

Consider the random process $\exp(j\phi(t))$. Its acf can simply be determined by the following equation

$$w(t) = e^{j\phi(t)} \rightarrow R_w(t_1, t_2) = E\{w(t_1)\, w^*(t_2)\} = E\{e^{j\phi(t_1)}e^{-j\phi(t_2)}\}$$

$$= E\{e^{j[\phi(t_1)-\phi(t_2)]}\} = E\{e^{j\Delta\phi}\} = e^{-\frac{1}{2}\sigma_{\Delta\phi}^2}$$

(3.64)

$$= e^{-\frac{1}{2}K_\phi\lvert t_2-t_1\rvert} = e^{-\frac{1}{2}K_\phi\lvert\tau\rvert}$$

Hence, the information of non-stationarity in the processes $\cos(\phi(t))$ and $\sin(\phi(t))$ is lost when the complex process $\exp(j\phi(t))$ is considered. The acf $R_w(\tau)$ of the random process $\exp(j\phi(t))$ is shown in Fig. 3.12a.

(v) POWER SPECTRAL DENSITY (psd)

The psd $G_w(f_1, f_2)$ of non-stationary random process can be obtained by two-dimensional Fourier transform of acf $R_w(t_1, t_2)$, whereas the psd $G_w(f)$ of stationary random process by a simple one-dimensional Fourier transform of acf $R_w(\tau)$.

In this Subsection, we focus our interest mainly on the random process $w = \exp(j\phi(t))$ which is an important harmonic random process in coherent optical communication systems. The physical representation of this process is the emission spectrum of laser light wave $\underline{E}(t)$ disturbed by phase noise $\phi(t)$ provided $f_0 = 0$ and $\phi_0 = 0$ (Eq. 3.8). This spectrum can simply be obtained by shifting the psd $G_w(f)$ to the laser centre frequency f_0 i.e., $G_w(f) \rightarrow G_w(f - f_0)$. Using Eq. (3.60), we get

$$G_w(f) = \int_{-\infty}^{+\infty} e^{-\frac{1}{2}K_\phi|\tau|} e^{-j2\pi f\tau} \, d\tau = \frac{4}{K_\phi} \frac{1}{1 + \left(\dfrac{4\pi f}{K_\phi}\right)^2} \tag{3.65}$$

The spectral bandwidth at 50% intensity points i.e., 3 dB bandwidth

$$\Delta f = \frac{K_\phi}{2\pi} = \frac{1}{4\pi} \frac{P_{sp}}{P_0} R_{sp} \quad \text{with } G_w\left(\frac{\Delta f}{2}\right) = \frac{1}{2}G_w(0) \tag{3.66}$$

is called the *laser linewidth*. The interrelation between Δf and K_ϕ given in the above expression has been used in the previous Sections. The relationship between linewidth Δf and power ratio P_{sp}/P_0 i.e., the ratio of average spontaneous emission power to the power level of the stimulated wave has already been given in Eq. (3.12). Combining Eqs. (3.65) and (3.66), we finally obtain the well-known result

$$G_w(f) = \frac{2}{\pi\Delta f} \frac{1}{1 + \left(\dfrac{f}{\Delta f/2}\right)^2} \tag{3.67}$$

This important formula, frequently called the *Lorentzian line shape*, describes the normalized emission spectrum of laser corrupted by phase noise. Here, the spectrum is shifted by laser centre frequency f_0 to the origin $f = 0$. Emission spectrum $G_w(f)$ is shown in Fig. 3.12b.

First measurement on the Lorentzian line shape of an AlGaAs injection semiconductor laser was carried out by Fleming and Mooradian in 1981 [8]. As shown by Eq. (3.66), a linewidth Δf that varies inversely with the laser output power P_0 was observed. It implies that laser linewidth is smaller when laser power is higher. This is generally valid irrespective of whether the laser is semiconductor, gas or solid-state type. Obviously, laser linewidth Δf should be as small as possible to achieve a high quality, high bit rate data transmission. The linewidth measured by Fleming and Mooradian was surprisingly about 50 times more than as predicted from Eq. (3.66). The reason for it will be explained briefly in the next Section.

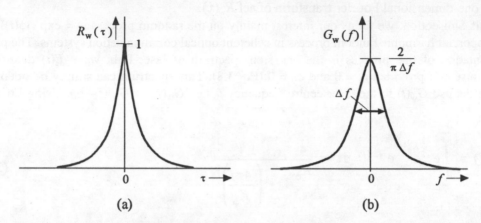

Fig. 3.12 (a) Autocorrelation function $R_w(\tau)$ and (b) power spectral density $G_w(f)$ of random process $w = \exp(j\phi(t))$

3.3   RELAXATION OSCILLATIONS

In the previous Section, formation of laser phase and laser amplitude noise has been explained and discussed by means of a simple model shown in Fig. 3.2. Based on this model, we have derived the statistical properties of phase noise, phase noise difference, laser frequency noise and harmonic oscillations disturbed by phase noise.

Except a minor modification which will be performed in this Section, justification of this model is given on the basis of good agreement of results obtained from this model and either by measurements [21, 25] or exact physical laser theory [16]. Here, the term exact laser theory means the mathematical solution of coupled differential equations with "laser phase" and "laser intensity" as the variables.

Like other physical models, for example, the atomic model by Niels Bohr, model of laser phase, intensity and amplitude noise is an approximate model, but very useful and advantageous to obtain a fast and good insight into the principle interrelations. Further, this model allows to obtain a first description by formulas. A detailed description of the actual microscopic processes in the system is not possible by using this simple model. However, this detailed knowledge is usually not required to analyse advanced optical communication systems for their behaviour in data transmission. For this, simple model given in Section 3.1 (Fig. 3.2) is normally sufficient and, in addition, often more powerful and practical based than a system analysis based on exact laser or quantum theory.

Concurrently to the development of advanced optical communication systems, intense practical and theoretical investigations had been performed in the area of semiconductor lasers, in particular to examine the spectral properties of laser diodes. A broader laser linewidth than predicted by the standard laser theory was measured. Further, a significant deviation in the Lorentzian line shape was observed. This deviation results in *satellite peaks* in the emission spectrum of laser.

In the following, reasons for line broadening and origin of satellite oscillations will be explained. For this, we shall take into consideration our simple noise model again. However, we have to perform now some appropriate modifications. A detailed mathematical analysis is not required and will not be covered in this book [4, 15, 16, 24, 26, 28, 30].

Up-to-now we have considered that the laser phase noise and amplitude or laser intensity noise are absolutely statistically independent. This is actually not correct since a close coupling of phase and amplitude noise can always be observed in semiconductor laser diodes (Fig. 3.13).

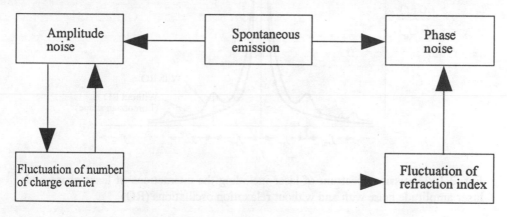

Fig. 3.13: Model of coupled phase and amplitude noise in laser

As shown in this figure, spontaneous emission is responsible for both laser phase and amplitude noise. Each spontaneously generated photon yields a *direct phase error* in the stimulated laser light wave since the phase of each photon is uniformly distributed in the range $-\pi$ to $+\pi$. Despite of this direct disturbance to the laser phase, an *additional delayed phase error* occurs as a response to the inherent direct amplitude disturbance. In order to restore the original steady state amplitude or intensity, laser performs relaxation oscillations for about 1 ns duration. During this characteristic time, real part $\text{Re}\{\underline{n}\}$ as well as imaginary part $\text{Im}\{\underline{n}\}$ of the refraction index $\underline{n}$ are changed due to fluctuation in the number of charge carriers, which again are caused by the amplitude fluctuations. Finally, the loop is closed by the spontaneous emission processes which are responsible for both phase as well as amplitude noise (Fig. 3.13).

The imaginary part $\text{Im}\{\underline{n}\}$ of the complex refraction index $\underline{n}$ is a measure of gain or loss in the active area of laser. Consequently, it is also a measure of the resulting amplification in this area. A change in the imaginary part $\text{Im}\{\underline{n}\}$ causes a change in the amplification, which finally yields the regeneration of the amplitude and, therefore, the intensity. After the relaxation oscillations are over, primary intensity is restored.

Due to relaxation oscillations of laser, emission spectrum exhibits satellite peaks after every f_{re} around the laser centre frequency f_0. The characteristic frequency f_{re} is called the *frequency of relaxation* and is in the order of 1 GHz to 2 GHz [15].

The real part $Re\{\underline{n}\}$ of complex refraction index $\underline{n}$ represents a measure of dispersion in the active area of laser. Hence, the real part $Re\{\underline{n}\}$ determines the phase and frequency of extendable laser waves in the laser cavity. It may be mentioned that real part $Re\{\underline{n}\}$ of refraction index $\underline{n}$ has already been used in Chapter two (represented by n) where the imaginary part was not required.

A change in the real part $Re\{\underline{n}\}$ causes delayed indirect phase changes which occur, in addition, to the direct phase changes due to spontaneous emission events. Hence, overall phase noise of a laser is increased two fold and linewidth is additionally broaden as shown in Fig. 3.14.

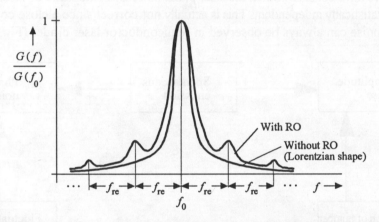

Fig. 3.14: Emission spectrum of laser including the interactions of laser phase and laser amplitude noise with and without relaxation oscillations (RO)

Relaxation oscillations influence the laser phase noise and the laser frequency noise. To make it more clear, Fig. 3.15a shows the variance $\sigma_\phi^2(t)$ of the laser phase noise and Fig. 3.15b the power spectral density $G_\omega(f)$ of the laser frequency noise.

In the absence of relaxation oscillations and line broadening, $\sigma_\phi^2(t)$ increases linearly with time (Section 3.2.2). In addition, delayed phase changes caused by changes in the real part $Re\{\underline{n}\}$ increases phase noise and $\sigma_\phi^2(t)$. Thus, $\sigma_\phi^2(t)$ now increases more rapidly with time. Finally, relaxation oscillations due to the interactions of laser phase and amplitude noise introduce variations in variance $\sigma_\phi^2(t)$ during the relaxation time of about 1 ns (Fig. 3.15a). It should be noted that line broadening and oscillation effects are always combined physically and cannot be split as shown in Fig. 3.15.

In case of frequency noise without relaxation oscillations and line broadening, $G_\omega(f)$ is constant. With additional delayed phase noise, $G_\omega(f)$ remains constant, but increases somewhat. Finally, relaxation oscillations cause a characteristic peak at the frequency of relaxation f_{re} as shown in Fig. 3.15b [24]. As the frequency of relaxation f_{re} is usually very high and an additional shift to still higher frequencies seems to be possible by means of appropriate technological steps, influence of relaxation oscillations can be neglected in most applications. However, the inherent line broadening must be considered, especially when highly sophisticated high bit rate systems, long-range links or coherent optical communication systems have to be used. Here, the actual laser linewidth Δf is of prime importance.

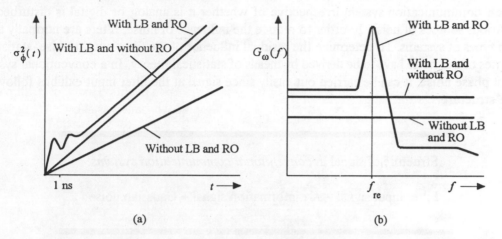

Fig. 3.15: (a) Variance $\sigma_\phi^2(t)$ of laser phase noise and (b) power spectral density $G_\omega(f)$ of laser frequency noise including laser phase-to-amplitude conversion (RO: Relaxation Oscillations, LB: Line Broadening)

In order to determine the laser linewidth mathematically, modification in Eq. (3.66) based on our simple model is required. It can be shown that this modification is given by the following simple substitution [15]

$$\Delta f \rightarrow \Delta f\left(1 + \alpha_e^2\right) \quad \text{with} \quad \alpha_e = \frac{\text{Re}\{\Delta n\}}{\text{Im}\{\Delta n\}} \tag{3.68}$$

Here, Δn represents the change in refraction index due to relaxation oscillations arising from the interaction of phase and amplitude fluctuations. The characteristic laser quantity α_e called the *enhancement factor* is a direct measure of linewidth broadening. Typical value of α_e is 6.

In conclusion, following results of primary importance can be given:

• spontaneous emission alter phase and amplitude of laser field (Section 3.2),

• changes in the laser amplitude induce relaxation oscillations which result in additional phase fluctuations while restoring the original steady state field amplitude,

• interactions of laser phase and amplitude noise produce serious linewidth broadening and side peaks (satellite peaks) in the laser emission spectrum.

3.4   INFLUENCE OF FILTERING

Each communication system irrespective of whether it is analog or digital is disturbed by various sources of noise. In order to reduce the influence of noise, filters are normally used in both types of systems. To determine the residual influence of the filter output noise, statistical properties of this noise have to be derived by means of statistical theory. In a conventional system without phase noise, it can be carried out easily since signal at the filter input exhibits following simple structure

> Structure of signal in *conventional communication systems*:
>
> Filter input signal = $K \cdot$ information signal + Gaussian noise

Here K is a constant. The systems with the above signal structure are frequently termed as Additive White Gaussian Noise (AWGN) systems. In an optical *direct detection system* with $K=R_0 \cdot M \cdot P_r$, filter input signal is the noisy signal at the input of the low-pass filter shown in Fig. 9.1a. Assuming a linear filter, filter response to the signal (i.e., information) and noise can be determined separately. As the filter input noise is usually Gaussian distributed, the noise at the filter output also has the same distribution. Hence, determining expected value (which is frequently zero) and standard deviation of the filter output noise is sufficient. The relevant formulas are given in Fig. 3.16a.

With a non-Gaussian noise at the filter input, calculation of statistical properties of the filter output noise becomes complicated (Fig. 3.16b). This problem, for example, in *coherent optical communication systems* with $K \sim \sqrt{P_r P_l}$ has to be analysed. Here, the information signal is additionally corrupted by laser phase noise. Thus, the characteristic signal structure is

> Structure of signal in *coherent optical communication systems*:
>
> Filter input signal = $K \cdot$ information $\cdot$ phase noise term + Gaussian noise

In contrast to the characteristic signal structure of conventional communication systems, an additional *phase noise term* has to be considered now. This term is *multiplicative* and *non-Gaussian* in nature.

In coherent optical communication systems with synchronous or coherent detection (for example, an optical homodyne system), noisy filter input signal is given by the baseband signal at the input of baseband filter (low-pass filter) shown in Fig. 9.1b. Here, the relevant phase noise term is mathematically represented by the cosine of phase noise i.e., $\cos(\phi(t))$. It will be explained in more detail in Section 11.1. Since the cosine function is a non-linear function, corresponding probability density function of the random process $\cos(\phi(t))$ is non-Gaussian (Section 3.2.5).

In an optical homodyne system, random process $\cos(\phi(t))$ is always a strongly correlated random process due to the required phase-locked loop circuit. Here, the so-called filter problem that is the determination of the statistical properties of the filter output noise with a non-Gaussian input noise can be avoided since strongly correlated phase noise changes very slowly. Hence, the phase noise term can be taken outside the convolution integral required to evaluate the filter output signal and noise (Section 11.1).

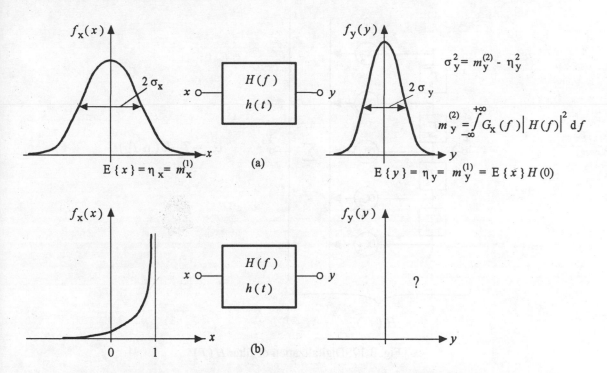

Fig. 3.16: Probability density function at the output of filter in case of (a) Gaussian and (b) non-Gaussian noise at the input

In coherent transmission systems with incoherent detection, for example, ASK heterodyne system with envelope detection, relevant filter input signal corresponds to noisy intermediate frequency signal at the input of IF filter. Here, the phase noise term can be expressed in terms of complex signal $\exp(j\phi(t)) = \cos(\phi(t)) + j\sin(\phi(t))$. As incoherent detection system does not require a phase control circuit, phase noise term $\exp(j\phi(t))$ is a relatively uncorrelated random process. It is not allowed to take out this term outside the convolution integral. Therefore, it is unavoidable to solve the filter problem (Section 11.3).

A filter only influences expected value and standard deviation of the pdf if the filter input noise is Gaussian. However, the shape of pdf may get fundamentally changed if the noise at the input is non-Gaussian. To obtain a complete description of this unknown non-Gaussian pdf, statistical moments of higher order are also required in addition to expected value and standard deviation.

In the following, various methods to solve the filter problem are presented and discussed in detail. As the numerical calculation is rather comprehensive, this book will focus mainly on the principle. A more detailed analysis is given e.g., in [2, 3, 13, 19]. Readers, who are not particularly interested in this theoretical Section may switch over to next Section or Chapter without loss of continuity.

The mathematical basis of most of the methods presented in this Section is the sampling theorem or digitalization of filter such as shown in Fig. 3.17 [20].

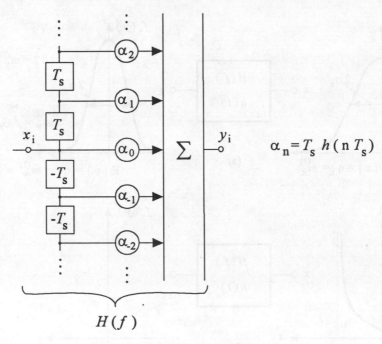

Fig. 3.17: Digitalization of filter $H(f)$

The sampled signal at the output of a digitalized filter after neglecting transit times can be described by

$$y(iT_s) = y_i = \sum_{n=-\infty}^{+\infty} x_{i-n}\, \alpha_n \qquad\qquad (3.69)$$

In this equation, T_s represents the sampling period which is required to fulfil the sampling theorem (Nyquist theorem), $\alpha_n = T_s h(nT_s)$ the sampled impulse response of the filter $H(f)$ weighted by T_s and $x_{i-n} = x([i-n]T_s)$ the sample values of the non-Gaussian distributed signal at the input of filter. To fulfil the sampling theorem, filter input signal $x(t)$ as well as filter frequency response (i.e., system function) $H(f)$ must be band-limited. Only in that case, sampled filter output signal $y(iT_s)$ equals the actual output signal $y(t)$ at the sampling points iT. Since this condition is normally not fulfilled in practice, an arbitrary cut-off frequency must be defined for the desired accuracy of calculation.

All the mathematical methods presented in the following Subsections are valid in principle for stationary as well as non-stationary random processes at the filter input. However, in order to avoid generality, a stationary random process will be considered. In that case, statistical properties of the random variables x_{i-n} and y_i are independent of sampling time $(i-n)T_s$ and iT_s respectively. For a non-stationary random process $x(t)$, calculation can be performed in the same way, but we have to incorporate some modifications which can be easily accomplished.

(i) METHOD OF CHARACTERISTIC FUNCTION

In the particular case of statistically independent sample values x_{i-n}, non-Gaussian pdf $f_y(y)$ at the filter output can be derived by means of characteristic functions $\Phi_x(\omega)$ of input signal $x(t)$. We obtain

$$f_y(y) = \frac{1}{2\pi} \int_{-\infty}^{+\infty} \prod_{n=-N}^{+N} \Phi_x(\alpha_n \omega) \, e^{-j\omega y} \, d\omega$$

$$= \frac{1}{|\alpha_{-N}|} f_x\left(\frac{y}{\alpha_{-N}}\right) \bullet \cdots \bullet \frac{1}{|\alpha_N|} f_x\left(\frac{y}{\alpha_N}\right) \quad \text{with} \quad N \to \infty$$

(3.70)

The characteristic function $\Phi_x(\omega)$ is given by

$$\Phi_x(\omega) = \int_{-\infty}^{+\infty} f_x(x) e^{j\omega x} dx$$

(3.71)

It corresponds to a Fourier transform of $f_x(x)$ with respect to the variable x. As seen from Eq. (3.70), $f_y(y)$ at the filter output can be determined in three steps: First, characteristic functions $\Phi_x(\omega)$ of the random variables x_{i-n} weighted by α_n have to be evaluated, second characteristic functions have to be multiplied with one another and third Fourier inversion must be performed to obtain the desired pdf $f_y(y)$. It should be noted that multiplication of characteristic functions is related to convolution of the corresponding probability density functions.

The method of characteristic function is a relatively simple solution of the filter problem, but requires the statistical independence of random variables x_{i-n} as mentioned above. The practical problem of this method is especially the large processing time required to evaluate Eq. (3.70) by means of computer. However, computer run time can at least partly be reduced when only a finite number N of sum terms instead of an infinite number are used in the calculation. It will impair the accuracy of calculation to some extent. Moreover, accuracy is also influenced by the bandwidth of filter $H(f)$. The higher the filter bandwidth, the faster the impulse response $h(t)$ decreases with time and higher is the accuracy of calculation.

(ii) METHOD OF NARROW-BAND APPROXIMATION

When the bandwidth of filter $H(f)$ is small in comparison to the bandwidth of filter input signal $x(t)$, impulse response $h(t)$ of the filter decreases very slowly while the acf $R_x(x)$ of the fast random input process $x(t)$ decreases very rapidly. In that case, sum in Eq. (3.69) includes an excessive number of terms during the impulse response of filter. Moreover, these terms are only weakly correlated. Hence, the necessary conditions for applying the central-limit theorem of statistics are fulfilled and the pdf of filter output process is approximately Gaussian [22]. It is sufficient to evaluate both the characteristic quantities which are required to describe a Gaussian pdf i.e., expected value η_y and standard deviation σ_y. Both quantities can easily be determined from the power spectral density $G_x(f)$ of the filter input process $x(t)$ and the filter transfer function $H(f)$ as given in Fig. 3.16. The method of narrow-band approximation requires filters of narrow bandwidth; for example, an intermediate frequency filter of bandwidth $B_{IF} \ll \Delta f$ if heterodyning is applied. Here, Δf represents the bandwidth of the IF signal at the input to filter due to laser phase noise alone. If phase noise is zero, then $\Delta f = 0$. However, the required relationship $B_{IF} \ll \Delta f$ is frequently not fulfilled in practical realizations.

(iii) METHOD OF STATISTICAL MOMENTS

In this method, all statistical moments $m_y^{(k)} = E\{y^k\}$, $k \in \mathbb{N}$ of the random filter output signal $y(t)$ are derived to calculate the pdf $f_y(y)$ by developing an appropriate series of sum terms. Here, three different cases have to be distinguished:

First, the filter input process $x(t)$ is a *Gaussian process* with *statistically independent* samples. In that case, calculation of pdf $f_y(y)$ is very easy since the moments of output process only depend on the moments of input process of same order (i.e., same k). Moreover, pdf $f_y(y)$ is also a function of the filter coefficients α_n. Since a Gaussian input process $x(t)$ always yields a Gaussian output process $y(t)$, only the moments of first and second order are actually required to describe the pdf $f_y(y)$. The filter output process $y(t)$ can also be completely described by the acf $R_y(\tau)$ or psd $G_y(f)$ as both functions already contains moments required to describe a Gaussian pdf i.e., expected value $\eta_y = m_y^{(1)}$ and variance $\sigma_y^2 = m_y^{(2)} - \eta_y^2$.

Second, the process $x(t)$ is a *non-Gaussian process* and samples are *statistically independent*. In that case, statistical moments of order k of the output process $y(t)$ are a function of filter coefficients α_n and all input moments of same *and* lower order. Method of statistical moments can be applied here also [2]. However, acf $R_y(\tau)$ and psd $G_y(f)$ are merely no longer sufficient to describe the pdf $f_y(y)$ completely.

The third case of importance is when *neither x(t) is Gaussian process nor the sample values are statistically independent*. Here, this method cannot be applied in general since further statistical information on the filter output process $y(t)$ are now required in addition to the moments $m_y^{(k)}$. For example, the expected value $E\{y(t_1)^k y(t_2)^l\}$ represents an autocorrelation function of higher order. Again, acf $R_y(\tau)$ and psd $G_y(f)$ are not sufficient to describe the random process $y(t)$ completely. For this, all linear and non-linear statistical interrelations in $y(t)$ must additionally be known, in particular all the moments and autocorrelation functions of higher order. Even if all these statistical quantities are known, calculation of the pdf $f_y(y)$ remains restricted to some special cases only (compare [2] and [3]). One such case of prime practical interest is, for example, given by the random process $x(t) = \cos(\phi(t))$. As already mentioned, this process is non-Gaussian due to non-linear cosine function, whereas the random phase $\phi(t)$ itself is Gaussian. By means of this process, which is of interest in coherent optical communication systems, method of statistical moments will be explained in more detail.

First, the moments $m_x^{(k)}$ of the random process $x(t) = \cos(\phi(t))$ at the input to the filter have to be calculated. These are given by

$$
\begin{aligned}
m_x^{(k)} &= E\{x^k\} = E\{\cos^k(\phi)\} = \frac{1}{\sqrt{2\pi}\sigma_\phi} \int_{-\infty}^{+\infty} \cos^k(\phi) \exp\left(-\frac{\phi^2}{2\sigma_\phi^2}\right) d\phi \\
&= \left(\frac{1}{2}\right)^k \sum_{j=0}^{j=k} \binom{k}{j} \exp\left(-\frac{(k-2j)^2\sigma_\phi^2}{2}\right)
\end{aligned}
\tag{3.72}
$$

Second, the acf

$$
R_x(\tau) = \frac{1}{2} e^{-\sigma_\phi^2} \left[e^{R_\phi(\tau)} + e^{-R_\phi(\tau)}\right] = e^{-\sigma_\phi^2} \cosh(R_\phi(\tau))
\tag{3.73}
$$

of the filter input process $x(t)$ is required. It may be remembered that $R_x(\tau)$ has already been derived in Section 3.2.5 (Eq. 3.61), where $x(t) = \cos(\phi(t))$ was represented by $w(t)$. To simplify calculations, following abbreviation will be used:

$$
R_\phi([i-n]T_s) := R_\phi(i-n)
\tag{3.74}
$$

The *first order moment* of random process $y(t)$ at the filter output is given by

$$m_y^{(1)} = E\left\{\sum_{n=-\infty}^{+\infty} x_{i-n}\alpha_n\right\} = \sum_{n=-\infty}^{+\infty} E\left\{x_{i-n}\right\}\alpha_n = m_x^{(1)}\sum_{n=-\infty}^{+\infty} \alpha_n = m_x^{(1)}H(0) \qquad (3.75)$$

As the random process $x(t)$ is assumed to be stationary, moments $m_x^{(1)} = \eta_x$ and $m_y^{(1)} = \eta_y$ are independent of sampling time iT_s. In the above equation, $H(0)$ represents the DC frequency response of the filter $H(f)$ at $f = 0$ (see Fig. 3.16). Using Eqs. (3.72) and (3.75), we obtain

$$m_y^{(1)} = e^{-\frac{1}{2}\sigma_\phi^2} H(0) \qquad (3.76)$$

For the *second order moment*, following expression can be derived:

$$m_y^{(2)} = \int_{-\infty}^{+\infty} G_x(f) |H(f)|^2 \, df \quad \text{with} \quad G_x(f) \bullet\!\!-\!\!\circ R_x(\tau) \qquad (3.77)$$

Here, the calculation is based only on the psd $G_x(f)$ or the acf $R_x(\tau)$ of random process $x(t) = \cos(\phi(t))$. As an alternative, calculation can also be based on acf $R_\phi(\tau)$ of Gaussian distributed random phase $\phi(t)$. In that case, we will get

$$m_y^{(2)} = \frac{1}{2}e^{-\sigma_\phi^2} \sum_{n=-\infty}^{+\infty}\sum_{m=-\infty}^{+\infty} \left[e^{R_\phi(n-m)} + e^{-R_\phi(n-m)}\right]\alpha_n\alpha_m \qquad (3.78)$$

Similarly, *third order moment* can be determined. With some simple trigonometrical relations, following comprehensive formula is obtained:

$$m_y^{(3)} = \frac{1}{4}e^{-\frac{3}{2}\sigma_\phi^2} \sum_{n=-\infty}^{+\infty}\sum_{m=-\infty}^{+\infty}\sum_{l=-\infty}^{+\infty} p(n,m,l)\,\alpha_n\alpha_m\alpha_l \qquad (3.79a)$$

where

$$
\begin{aligned}
p(n,m,l) = \ & e^{+R_\phi(n-m)\ +\ R_\phi(n-l)\ +\ R_\phi(m-l)} \ + \ e^{+R_\phi(n-m)\ -\ R_\phi(n-l)\ -\ R_\phi(m-l)} \\
+\ & e^{-R_\phi(n-m)\ -\ R_\phi(n-l)\ +\ R_\phi(m-l)} \ + \ e^{-R_\phi(n-m)\ +\ R_\phi(n-l)\ -\ R_\phi(m-l)}
\end{aligned}
\tag{3.79b}
$$

Moments of higher order can, in principle, be determined in the same way. The computer execution time increases rapidly if the order of moment is higher than three. It becomes clear from the above equation that terms $R_\phi(n-m)$, $R_\phi(n-l)$ and $R_\phi(m-l)$ frequently exhibit the same numerical value during the calculation of the sum with respect to n, m and l. Hence, the next step would be to reduce this redundancy existing manifold in the higher order moments.

When the required number of moments with desired accuracy are available, pdf $f_y(y)$ can be determined by developing an appropriate series of sum terms. For this, various series are available. As an example, characteristic functions and pdf calculated by means of Taylor series are given by the following equation [22]:

$$
\Phi_y(\omega) = 1 + \sum_{k=1}^{+\infty} \frac{m_y^{(k)}}{k!}(j\omega)^k \quad \text{and} \quad f_y(y) = \frac{1}{2\pi}\int_{-\infty}^{+\infty}\Phi_y(\omega)
\tag{3.80}
$$

Unfortunately, convergence of this series is very weak and a large number of moments of higher order are required to obtain sufficient accuracy. In contrast, the Gram-Charlier [1], the Edgeworth [5] and the Cornish-Fisher series are more convenient. A more detailed analysis of above series is not given in this book [7].

(iv) METHOD OF SHAPE FILTER

The idea of shape filter method is to trace back the correlation of input random process $x(t)$ characterized by dependent sample values x_i to a random process $u(t)$ first characterized by a psd $G_u(f)$ due to a band-limited white noise process and second by statistically independent samples u_i. The filter model including real filter and virtual shape filter is illustrated in Fig. 3.18. The task of shape filter $H_s(f)$ is to generate actual random process $x(t)$ at the input of real filter $H(f)$ by means of a virtual random process $u(t)$, defined by the statistical properties mentioned above. Hence, the *virtual shape filter $H_s(f)$* is only a mathematical tool to determine the pdf $f_y(y)$, but not a filter which really exists in optical communication systems.

Let us consider that $H_s(f)$ is mathematically existing. The pdf $f_y(y)$ at the output of real filter $H(f)$ can be calculated using the method of moments discussed above. It may be remembered that statistical independence of samples at the filter input have been the only presumption for applying this method. Here, statistical independence is fulfilled by the virtual samples u_i at the input of shape filter, whereas the samples x_i at the input of real filter are statistically dependent.

To clarify the calculation, characteristic features of random processes $u(t)$, $x(t)$ and $y(t)$ are summarized in Table 3.3 below.

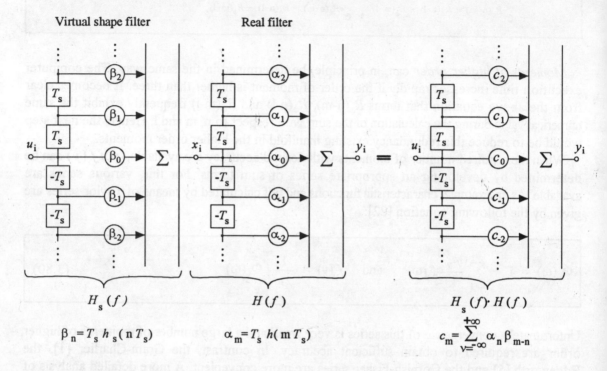

Fig. 3.18: Model of filter to calculate probability density function $f_y(y)$

Since newly defined random variables u_i are independent and psd of $u(t)$ is related to band-limited white noise, acf $R_u(\tau)$ and psd $G_u(f)$ are given by (see Fig. 3.19)

$$R_u(\tau) = \sigma_u^2 \, \text{si}\!\left(\frac{\pi\tau}{T_s}\right) \quad \circ\!\!-\!\!\bullet \quad G_u(f) = \begin{cases} \sigma_u^2 T_s & \text{if } |f| \leq \dfrac{1}{2T} \\[2ex] 0 & \text{otherwise} \end{cases} \qquad (3.81)$$

From this equation, it can be easily deduced that the random process $u(t)$ is band-limited by $|f| \leq 1/(2T_s)$. Thus, $u(t)$ is completely described by its samples $u_i = u(iT_s)$ using the sampling theorem. Moreover, samples u_i are statistically independent as acf $R_u(\tau)$ is zero at every T_s (refer to Fig. 3.19a).

Table 3.3: Statistical properties of random processes $u(t)$, $x(t)$ and $y(t)$ given in Fig. 3.18

Random process	Statistical properties
$u(t)$	• virtual random process • statistically independent samples • non-Gaussian distributed • correlated • acf related to a band-limited white noise (Fig. 3.19 and Eq. 3.81)
$x(t)$ and $y(t)$	• real random process • statistically dependent samples • non-Gaussian distributed • correlated • acf of $x(t)$: see Eq. (3.73) • acf of $y(t)$: $R_y(\tau) \; \circ\!\!-\!\!\bullet \; G_x(f)\,|H(f)|^2$

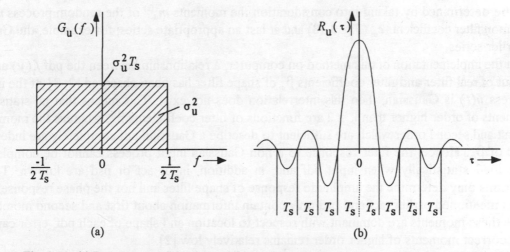

Fig. 3.19: (a) Power spectral density and (b) autocorrelation function of virtual random process $u(t)$ at the input of shape filter shown in Fig. 3.18

The random process $u(t)$ is now defined, whereas filter coefficients β_m of shape filter $H_s(f)$ and $f_u(u)$ or moments $m_u^{(k)}$ are still undefined. To determine the filter coefficients β_m, amplitude response

$$|H_s(f)| = \sqrt{\frac{G_x(f)}{G_u(f)}} \rightarrow h_s'(t) \; \circ\!\!-\!\!\bullet \; |H_s(f)| \rightarrow \beta_m' = T_s h_s'(mT_s) \tag{3.82}$$

of shape filter defined in terms of power spectral densities $G_x(f)$ and $G_u(f)$ has to be considered. The phase response of the filter still remains undetermined. As a result, impulse response $h_s(t)$ and coefficients β_m of shape filter cannot be unambiguously defined. Therefore, these quantities are marked with dash in Eq. (3.82). They only define the amplitude response of filter clearly, but not the phase response. Assuming a Gaussian input process $u(t)$, phase response of shape filter can, however, be chosen arbitrarily because a linear filter never changes the shape of Gaussian pdf.

Next, the moments $m_u^{(k)}$ of virtual input process $u(t)$ have to be determined. This can be done by taking into consideration the moments $m_x^{(k)}$ of real random process $x(t)$ and coefficients β_m of shape filter. Since the samples u_i are considered to be statistically independent, method of moments can be applied. The output moments $m_x^{(k)}$ will depend only on the input moments $m_u^{(k)}$ of same order (i.e., same k) and vice versa.

Finally, the moments $m_y^{(k)}$ and pdf $f_y(y)$ of the real filter output process

$$y_i = \sum_{m=-\infty}^{+\infty} u_m \, c_{i-m} \quad \text{with} \quad c_{i-m} = \sum_{n=-\infty}^{+\infty} \alpha_n \, \beta_{i-n-m} \tag{3.83}$$

can be determined by taking into consideration the moments $m_u^{(k)}$ of the random process $u(t)$, common filter coefficients c_m (Fig. 3.18) and at last an appropriate series; for example, the Gram-Charlier series.

In the implementation of this method on computer, a relationship between the pdf $f_y(y)$ at the output of real filter and filter coefficients β_m of shape filter has been observed [2, 3]. If the input process $u(t)$ is Gaussian, then this interrelation does not exist. This means that all statistical moments of order higher than k = 2 are functions of filter coefficients β_m, whereas the moments of first and second order which are sufficient to describe a Gaussian pdf completely are independent of β_m. Hence, the filter response to a non-Gaussian input process cannot be completely described statistically when input pdf and, in addition, input acf or psd are known. These functions only determine the amplitude response of shape filter and not the phase response.

As mentioned earlier, acf and psd only contain information about first and second moments. Since these moments are dominant with respect to location and shape of each pdf, error caused by incorrect moments of higher order remains relatively low [2].

(v) METHOD OF INTEGRAL

This method is based on the integration of multi-dimensional joint density function, briefly called the joint pdf. For determining the pdf $f_y(y)$, joint pdf has to be derived first. For this, a set of (2N+1) equations is required where (2N+1) equals the number of random variables on the right side of the following equation

$$y_i = \sum_{n=-N}^{+N} x_{i-n} \alpha_n \qquad (3.84)$$

It can be seen from Eq. (3.69) and (3.84) that N is infinite in Eq. (3.69). However, if the impulse response $h(t)$ of the filter $H(f)$ is practically restricted to a certain time interval, finite N can be employed. Of course, the accuracy of calculation will somewhat be reduced. For a straight forward calculation, unambiguous sets of equations are particularly suited since these are cha·acterized by only one single and perfectly specified solution with respect to all variables on the right hand side of set. However, special care has to be taken in case of random process $x(t) = \cos(\phi(t))$ which inherently exhibits an ambiguous inverse function (see Section 3.2.5). With new auxiliary random variables h_{-N+1} to h_N, following set of (2N+1) equations is obtained

$$
\begin{aligned}
y_i &= \sum_{n=-N}^{+N} x_{i-n} \, \alpha_n \\
h_{-N+1} &= x_{-N+1} \\
&\;\;\vdots \\
h_0 &= x_0 \\
&\;\;\vdots \\
h_N &= x_N
\end{aligned}
\qquad (3.85)
$$

It may be mentioned again that a large number of sets of equations are usually available, but only sets with a single perfectly specified solution yield a proper result. In the above equation, both sides exactly contain (2N+1) random variables. Using statistical transformation approach of multi-dimensional random variables [22], Eq. (3.85) yields the following multi-dimensional joint pdf

$$f_{y,\,h_{-N+1}\cdots h_N}(y,\,h_{-N+1}\cdots h_N) = \frac{f_{x_{-N}\cdots x_N}(x_{-N}^{(1)} \cdots x_N^{(1)})}{\left| \det J\left(x_{-N}^{(1)} \cdots x_N^{(1)}\right) \right|} + \cdots + \frac{f_{x_{-N}\cdots x_N}(x_{-N}^{(k)} \cdots x_N^{(k)})}{\left| \det J\left(x_{-N}^{(k)} \cdots x_N^{(k)}\right) \right|} \qquad (3.86)$$

Here, $x_n^{(1)}$ to $x_n^{(k)}$ with $-N \leq n \leq N$ represent the set of solutions for $(2N+1)$ random variables on the right hand side of Eq. (3.85). With an unambiguous set of equations, exactly one single perfectly specified solution instead of k solutions exists for each random variable (i.e., $k = 1$). In that case, joint pdf given in Eq. (3.86) is restricted to one single sum term only. The denominators in Eq. (3.86) are given by the determinant of Jacobian matrix [22]

$$J = \begin{vmatrix} \dfrac{\delta y}{\delta x_{-N}} & \cdots & \dfrac{\delta y}{\delta x_N} \\ \vdots & & \vdots \\ \dfrac{\delta h_N}{\delta x_{-N}} & \cdots & \dfrac{\delta h_N}{\delta x_N} \end{vmatrix} \qquad (3.87)$$

Due to proper selection of a set of equations (3.85), inverse set exhibits a perfectly specified solution for each of $(2N+1)$ variables. In this particular case, absolute of det J of Jacobian matrix exactly yields 1. Thus, the joint pdf simplifies to

$$f_{y,\, h_{-N+1} \cdots h_N}(y,\, h_{-N+1} \cdots h_N) = f_{x_{-N} \cdots x_N}\left(\frac{1}{\alpha_{-N}}\left[y - \sum_{n=N+1}^{N} x_n \alpha_n \right],\, h_{-N+1} \cdots h_N \right) \qquad (3.88)$$

Finally, we have to integrate (hence, the name method of integral) with respect to the random variables h_{-N+1} to h_N to obtain the desired pdf of random variable y. It is given by

$$f_y(y) = \int_{-\infty}^{+\infty} \cdots \int_{-\infty}^{+\infty} f_{x_{-N} \cdots x_N}\left(\frac{1}{\alpha_{-N}}\left[y - \sum_{n=N+1}^{N} x_n \alpha_n \right],\, h_{-N+1} \cdots h_N \right) dh_{-N+1} \cdots dh_N \qquad (3.89)$$

This equation gives an exact solution of pdf $f_y(y)$ i.e., no approximation is included. The problem in evaluating Eq. (3.89) primarily is due to multi-dimensional joint pdf in the integral which is to be determined. However, in case of random process $x(t) = \cos(\phi(t))$, this problem can be easily solved.

Beginning with an arbitrary "start-phase", for example, $\phi([i-N]T_a) = \phi_{i-N}$, all other phases can be determined by adding a phase difference step-by-step i.e., $\phi_{i-N+1} = \phi_{i-N} + \Delta\phi_{i-N}$. Hence, the joint pdf given by Eq. (3.89) can be replaced by a joint pdf of new random variables $\Delta\phi_{i-N}$. As the phase differences are statistically independent (see Section 3.2.3), joint pdf can be determined by a product of individual pdfs. By this approach, a compact analytical solution for the pdf $f_y(y)$ at the filter output is obtained. It will be taken up again when ASK heterodyne system is discussed in Section 11.2.

The interrelation between the method of integral and the method of statistical moments is given by the equation

$$
m_y^{(k)} = \int\limits_{-\infty}^{+\infty} \cdots \int\limits_{-\infty}^{+\infty} \left(\sum_{n=-N}^{N} x_n \alpha_n \right)^k f_{x_{-N} \cdots x_N}(x_{-N} \cdots x_N) \, dx_{-N} \cdots dx_N
\tag{3.90}
$$

where (2N+1) instead of 2N integrals have to be considered (compare with Eq. (3.89)).

(vi) METHOD OF QUASI-CONSTANT FREQUENCY

This approximation method, proposed first by Garrett and Jacobsen, presumes that the frequency noise $\phi(t) = \omega(t) = \omega_c$ is nearly constant during the time interval

$$
\Delta t_H = \frac{1}{h(0)} \int\limits_{-\infty}^{+\infty} h(t) \, dt = \frac{H(0)}{h(0)} = \frac{1}{\Delta f_H} \approx T
\tag{3.91}
$$

which is usually in the order of bit period T in practical systems [11, 17]. Physically, Δt_H represents the rectangular equivalent pulse width of filter time response $h(t)$ and Δf_H the double-sided rectangular equivalent bandwidth of filter frequency response $H(f)$. Thus, $\Delta f_H \cdot \Delta t_H = 1$. For a Gaussian filter $H(f) = \exp[-\pi(f/\Delta f_H)^2]$ with a bandwidth $\Delta f_H = 1/T$, rectangular equivalent pulse width exactly yields $\Delta t_H = T$.

As shown in Fig. 3.20, constant frequency ω_c during Δt_H is related to phase which increases or decreases (ω_c can also be negative) linearly with time i.e., $\phi(t) = \omega_c t + \phi_c$. It may be mentioned that ω_c is a zero mean, Gaussian distributed random variable that statistically changes its value from one time interval (or one bit) to the following interval (the next bit). The variance of Gaussian pdf is $\sigma_{\omega c}^2 = E\{[\Delta \phi(t, \Delta t_H)/\Delta t_H]^2\} = 2\pi \Delta f \Delta t_H$, where Δf represents the resulting laser linewidth of both transmitter and local lasers in a coherent system (Chapters nine and eleven).

In this method, filter output process $y(t)$ is first determined by convolution. For the special random process $x(t) = \exp(j\phi(t))$, we get

$$
y(t) = x(t) \star h(t) = \int\limits_{-\infty}^{+\infty} e^{j\phi(t-\tau)} h(\tau) \, d\tau = H(\omega_c) \, e^{j\phi_c} \, e^{j\omega_c t}
\tag{3.92}
$$

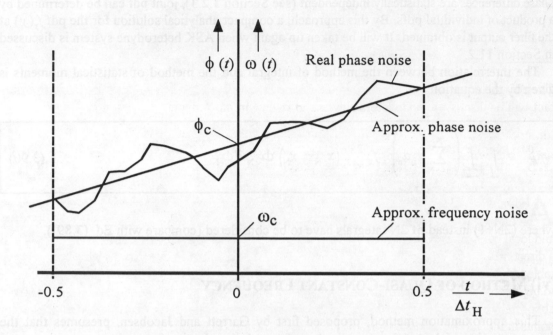

Fig. 3.20: Approximation of laser phase and frequency noise

The process $y(t)$ is a "harmonic" process changing frequency from one time interval to the other. In coherent optical communication systems with incoherent detection (for example envelope detection), absolute random process $|y(t)|$ is of prime importance. Since $|y(t)|$ is a function of random variable ω_c, its pdf can easily be determined by applying the statistical transformation of random variables. We obtain

$$f_{|y|}(|y|) = f_{\omega_c}(\omega_c) \; \frac{1}{\left| \dfrac{dH(\omega_c)}{d\omega_c} \right|_{\omega_c = H^{-1}(|y|)}}$$ (3.93)

The above equation represents a simple, but powerful method to evaluate the pdf of $|y|$ provided random frequency $\phi(t)$ is nearly constant within the filter time response $h(t)$. Hence, this equation is valid for $\Delta f_H \gg \Delta f$ (i.e., filter bandwidth $\gg$ laser linewidth).

Example
Consider a filter with Gaussian frequency response $H(f) = \exp[-\pi(f/\Delta f_H)^2]$ and a complex non-Gaussian random process $x(t) = \exp(j\phi(t))$ at the filter input with $\sigma_{\omega c}^2 = 2\pi \Delta f \Delta t_H$, pdf of the non-Gaussian random process $|y|$ at the filter output is

$$f_{|y|}(|y|) = \sqrt{\frac{\Delta f_H}{\Delta f}} \; |y|^{\frac{\Delta f_H}{\Delta f}-1} \; \frac{1}{\sqrt{-\pi \ln(|y|)}} \tag{3.94}$$

The pdf of $|y|$ primarily depends on the filter bandwidth-to-laser-linewidth ratio. This important fact will be discussed in Chapters eleven and twelve in more detail.

3.5   REDUCTION OF LASER NOISE

Advanced optical transmission systems are very sensitive to laser noise. Performance of coherent systems is, in particularly, very much affected by the phase noise of transmitter and local lasers as described in Chapters nine and eleven. As mentioned earlier, laser linewidth is a direct measure to assess the level of laser noise. The linewidth requirement ranges from comparable to bit rate to a small fraction of bit rate. In particular, when optical homodyne systems have to be used, laser sources with a very narrow linewidth and a high-quality emission spectrum are absolutely required (Section 11.1). Narrowing the laser linewidth is, therefore, one of the key points in the development of coherent optical communication systems. But in incoherent systems also, linewidth reduction is a most important issue since a broad laser linewidth increases dispersion-induced signal distortion and, hence, deteriorates overall system performance.

Despite of linewidth narrowing, high output power and uniform FM response are additional requirements of a transmitter laser. The latter one is important particularly when direct modulation instead of external modulation is used. If the laser is modulated in frequency with an analog signal, for example a multichannel cable TV signal, then an uniform FM response is an essential precondition. Lasers in WDM transmitters and local lasers in coherent receivers should additionally be tunable in wavelength, especially when high-density multichannel system have to be employed. Throughout the tuning range, fixed output power and constant linewidth should be maintained.

This Section is focussed on the techniques to reduce the influence of laser noise. Some typical physical approaches for narrowing the laser linewidth are briefly discussed. Most suitable are the DFB and DBR structure, in particularly to achieve single-mode operation, external cavity, optical reflection, coupled cavity technique and MQW structure. A more detailed discussion about these approaches is given, for example, in [6, 8, 12, 18]. In addition to these physical solutions concerning the laser source, various other system design techniques can also be employed to reduce the influence of laser intensity and phase noise; for example, error-correction coding and receiver-based phase noise cancellation circuit.

It may be mentioned that physical approaches described below are exclusively based on laser diodes since they are the most promising sources for practical systems due to their small size, mechanical stability and potential for integration. Furthermore, laser diodes can be ASK, FSK and DPSK modulated by direct modulation of the injection current.

Measurement of laser spectrum and linewidth can be performed, for example, by employing a Fabry-Perot interferometer or self heterodyning. The idea of the latter measurement technique is mixing the spectrum with itself, in order to produce a spectral image at low frequencies. This setup is basically same as a Mach-Zehnder interferometer. As described in detail in Chapter nine, mixing is simply obtained from a photodiode due to its quadratic field-to-current conversion. The low frequency spectrum is then observed and measured with a conventional electrical spectrum analyser.

(i)   DFB AND DBR LASER

As already described in Chapter two, the abbreviations DFB and DBR stand for *distributed feedback* and *distributed-Bragg reflection*. Both types of lasers use a periodically disturbed waveguide as reflector instead of a plane reflector (Fabry-Perot laser). A plane reflector (for example, a mirror) is broadband with respect to the optical spectrum, whereas a periodically disturbed waveguide acts as a wavelength selective grating. The grating period is defined by the ratio of emission wavelength to refraction index of active area. If wavelength is 1550 nm and refraction index is 3.3 then grating period is about 470 nm. In case of 1300 nm, grating period is only 394 nm.

Stable oscillation is only possible when all grating reflections interfere constructively. One specific mode is selected from the possible modes, while all others are suppressed. In a DFB diode laser, corrugated feedback grating is located within the active area, whereas in a DBR laser this grating is located outside i.e., directly in front of the active area. These lasers are in general single-mode and characterized by a highly stable wavelength. Hence, these are commonly used as high quality and powerful single-mode lasers. A recognizable reduction of laser linewidth in the order of some MHz can only be achieved by increasing the light power. When a source is operated at high power level, its life span will be reduced.

(ii)   EXTERNAL CAVITY LASER

A substantial reduction of linewidth can be achieved by means of an external passive cavity, in addition, to the real active resonator inside the laser. The higher the length ratio of passive to active cavity, the more is the improvement in linewidth narrowing. To obtain single-mode operation, a grating is mostly used as external reflector. For example, a linewidth of about 10 kHz was achieved by applying an external cavity with a length of 20 cm [18]. Acoustooptic disturbances inside the external resonator represent a significant drawback. Further, external cavity lasers are unsuitable for optical integration.

(iii) LASER WITH OPTICAL REFLECTION

It is well-known that a reflected light beam usually disturbs the emission spectrum of laser and decreases the overall system performance. This is not valid in general since reflected light may also decrease the linewidth of a laser and, hence, improve the system performance. The phase relationship between emitted and reflected light waves decides whether the first or second process is dominant. To reflect the light wave, a grating, mirror or fiber itself can be used. In the last case, reduction of laser linewidth is accidental and mostly not reproducible. Although linewidths in the order of 100 kHz can be achieved, this method is not suitable for commercial applications.

(iv) COUPLED-CAVITY LASER

Usually, a laser contains single optical cavity to generate the light wave. In principle, this resonator (cavity) can be divided in two subcavities which have to be coupled. It is called a *cleaved-coupled-cavity laser* or briefly a C^3-laser. Both cavities are separated by a small slit (gap) only. Length of both resonators approximately equals the length of a single cavity in a conventional laser. Therefore, size of a C^3-laser is essentially smaller than a laser with external cavity. The laser linewidth is, in principle, decreased by constructive interference between the fields of both cavities. Linewidths of less than 1 MHz are possible. In addition, wavelength stabilization is improved. The gain in linewidth reduction and stability primarily depends on the size of the slit between the resonators. However, realization and reproduction of slits of a well-defined size represent a technological problem.

(v) MQW LASER

As described in Chapter two, multi-quantum well (MQW) lasers show a much smaller linewidth than conventional lasers, such as Fabry-Perot and DFB bulk laser diodes. A quantum well is formed, when a very thin layer of low band-gap semiconductor is inserted between high band-gap barriers. Wells and barriers are often repeated in a laser structure to achieve cumulative effects. These structures are called MQW lasers. For comparison, linewidth of MQW lasers can be 200 to 300 times smaller than that of bulk DFB lasers. Even linewidths in the kHz-range are possible.

(vi) CHANNEL CODING

Besides the physical approaches mentioned above, channel coding or error correction coding is another well-suited possibility to reduce the influence of laser noise. Of course, coding can also be used, in addition, to the physical approaches. The physical approaches are related to laser sources, whereas coding is related to the system design.

It is well-known from coding theory that channel coding, such as block coding or convolution coding, can be used to correct bit errors, improve signal-to-noise ratio or both. Channel coding makes use of redundancy. If information bit rate is fixed, then redundancy increases the overall bit rate on the channel (e.g., the fiber), which is frequently called the channel bit rate. The ratio of information bit rate to channel bit rate is termed as code rate. As the coding increases channel bit rate, system bandwidth must be increased appropriately. However, a higher bandwidth results in a stronger influence of the additive noise (i.e., shot noise and thermal noise) and system performance degrades (first effect). On the other hand, coding allows to correct bit errors and system performance improves (second effect). If the second effect dominates the first effect, coding improves the overall system performance. For a fixed bit error rate, this improvement results in a lower required signal-to-noise ratio or a lower required optical power at the receiver input. This gain in receiver sensitivity or signal-to-noise ratio is termed as *coding gain*. Irrespective of whether the system is coherent or direct detection, a coding gain can always be achieved.

In a coherent optical communication system, third effect will also be observed in addition to earlier effects mentioned above. As will be discussed in Chapter eleven, influence of laser phase noise decreases rapidly with the decrease in the linewidth-bit duration product $\Delta f T$. Hence, the higher the bit rate $1/T$, the lower is the influence of laser phase noise. For a fixed information bit rate, channel coding always requires an increased channel bit rate and reduces the influence of phase noise. The coding gain obtained by error correcting codes offers an improvement in system performance with respect to signal-to-noise ratio as well as in laser linewidth requirements. By applying coding, maximum permissible linewidth may be approximately ten times broader than without coding [10].

(vii) PHASE NOISE CANCELLATION CIRCUITS

The phase noise in a coherent receiver can theoretically be suppressed completely when a reference signal having exactly the same phase noise as the carrier signal is available. Since phase noise is already existing in the transmitter, reference signal must be transmitted simultaneously to the information signal. This, for example, can be realized by means of an optical subcarrier generated by a simple frequency shift of the unmodulated laser wave using an acoustooptic modulator. In the receiver, phase noise is cancelled by generating the square of the sum of both reference and modulated carrier signal. As it is rather difficult to produce and transmit a reference signal which exactly contains and maintains a copy of the phase noise, phase noise cancellation circuits have a limited scope.

3.6 REFERENCES

[1]   Cramér, H.: Mathematical Methods of Statistics. 10. Aufl. Princeton University Press, Princeton 1963.

[2]   Cygan, D.: Berechnung der Wahrscheinlichkeitsdichtefunktion am Ausgang eines Filters bei beliebiger Eingangsverteilung und beliebiger Autokorrelationsfunktion. Diplomarbeit, TU München, Lehrstuhl für Nachrichtentechnik, 1986.

[3]   Cygan, D.; Franz, J.; Söder, G.: Einfluß eines Filters auf nicht-gaußverteilte Zufallsprozesse. AEÜ 40(1986)6, 377-384.

[4]   Daino, B.; Spano, P.; Tamburrini, M.; Piazolla, S.: Phase noise and spectral line shape in semiconductor laser. IEEE J. QE-19(1983)3, 266-270.

[5]   Draper, N. R.; Tierney, D. E.: Exact formulas for additional terms in some important series expansions. Communications in Statistics 1(1973), 495-524.

[6]   Favre, F.; Le Guen, D.: Emission frequency stability in single-mode-fibre optical feedback controlled semiconductor lasers. Electron. Lett. 19(1983)17, 663-665.

[7]   Fisher, R. A.; Cornish, E. A.: The percentile points of distributions having known comulants. Technometrics 2(1960), 209-225.

[8]   Fleming, M. W.; Mooradian, A.: Spectral characteristics of external-cavity controlled semiconductor lasers. IEEE J. QE 17(1981)1, 44-59.

[9]   Franz, J.: Evaluation of the probability density function and bit error rate in coherent optical transmission systems including laser phase noise and additive Gaussian noise. J. Opt. Commun. 6(1985)2, 51-57.

[10]   Franz, J.; Correll, C.; Dolainsky, F.; Schweikert, R.; Wandernoth, B.: Error correcting coding in optical transmission systems with direct detection and heterodyne detection. J. Opt. Commun. 14(1994)5, 194-199.

[11]   Garrett, I.; Jacobsen, G.: Theoretical analysis of heterodyne optical receivers for transmission systems using (semiconductor) lasers with nonneglible linewidth. IEEE J. LT-4(1986)3, 323-334.

[12]   Goldberg, L.; Taylor, H. F.; Dandrige, A.; Weller, J. F.; Miles, R. O.: Spectral characteristics of semiconductor lasers with optical feedback. IEEE J. QE-18(1982)4, 555-564.

[13]   Helnerus, U.: Der Einfluß des ZF-Filters auf das Laserphasenrauschen im optischen ASK-Heterodynempfänger mit Hüllkurvendemodulation. Diplomarbeit, TU München, Lehrstuhl für Nachrichtentechnik, 1986.

[14]   Henry, C.: Theory of the linewidth of semiconductor lasers. IEEE J. QE-18(1982)2, 259-264.

[15]   Henry, C.: Theory of the phase noise and power spectrum of a single mode injection laser. IEEE J. QE-19(1983)9, 1391-1397.

[16]   Henry, C.: Phase noise in semiconductor lasers. IEEE J. LT-4(1986)3, 298-311.

[17]   Jacobsen, G.; Garrett, I.: Error-rate floor in optical ASK heterodyne systems caused by nonzero (semiconductor) laser linewidth. Electron. Lett. 21(1985)7, 268-270.

[18]   Lee, T. P.: Linewidth of single-frequency semiconductor lasers for coherent lightwave communications. IOOC-ECOC (1985), 189-196.

[19]   Lutz, E.; Söder, G.; Tröndle, K.: Generation of discrete stochastic processes with given probability density and autocorrelation on a digital computer. 4. Seminar, Akademie der Wissenschaften der CSSR, Prag (1979), 308-329.

[20]   Marko, H.: Methoden der Systemtheorie. Springer-Verlag, 1977.

[21]   Okoshi, T.; Kikuchi, K.; Nakayama, A.: Novel method for high resolution measurement of laser output spectrum. Electron. Lett. 16(1980)16, 630-631.

[22]   Papoulis, A.: Probability, Random Variables and Stochastic Processes. McGraw-Hill, 1985.

[23]   Piazzolla, S.; Spano, P.; Tamburrini, M.: Characterization of phase noise in semiconductor lasers. Appl. Phys. Lett. 41(1982)8, 695-696.

[24]   Piazzolla, S.; Spano, P.: Analytical evaluation of the line shape of single-mode semiconductor lasers. Optics Commun. 51(1984)4, 278-280.

[25]   Saito S.; Yamamoto, Y.: Direct observation of Lorentzian lineshape of semiconductor laser and linewidth reduction with external grating feedback. Electron. Lett. 17 (1981)9, 325-327.

[26]   Spano, P.; Piazzolla, S.; Tamburrini, M.: Phase noise in semiconductor lasers: A theoretical approach. IEEE J. QE-19(1983)7, 1195-1199.

[27]   Vahala, K.; Yariv, A.: Semiclassical theory of noise in semiconductor lasers-Part I. IEEE J. QE-19(1983)6, 1096-1101.

[28]   Vahala, K.; Yariv, A.: Semiclassical theory of noise in semiconductor lasers-Part II. IEEE J. QE-19(1983)6, 1102-1109.

[29]   Wandernoth, B.: 1064 nm, 565 Mbit/s PSK transmission experiment with homodyne-receiver using synchronization bits. Electron. Lett., 27(1991)19, 1692-1693.

[30]   Yamamoto, Y.: AM and FM quantum noise in semiconductor lasers-Part I: Theoretical analysis. IEEE J. QE-19(1983)1, 34-46.

[31]   Yamamoto, Y.; Saito, S.; Mukai, T.: AM and FM quantum noise in semiconductor lasers-Part II: Comparison of theoretical and experimental results for AlGaAs lasers. IEEE J. QE-19(1983)1, 47-58.

4   OPTICAL FIBERS

In early sixties, preliminary experiments were carried out to study the propagation of information carrying light beam through the open atmosphere. It was soon realized that because of the vagaries of the atmosphere (like rain, fog, etc.), it was necessary to have a guiding medium through which light waves could be transmitted. The guiding medium after intensive investigations turn out to be the optical fiber which is hair-thin glass/plastic structure consisting of central core surrounded by a cladding of slightly lower refraction index (RI). The light propagation takes place because of the phenomenon of *total internal reflection* occurring at the core-cladding interface.

In addition to the capability of carrying a tremendous amount of information, fibers have *extremely low loss* of about 0.2 dB/km. It implies a loss of only 0.5% of power in a distance of 1 km. This loss is practically constant within a wide frequency range of some hundreds of GHz. As compared to this, attenuation of coaxial cable strongly depends on the frequency. A typical coaxial cable for TV distribution has a loss of the order of 20 dB/km at 50 MHz, 50 dB/km at 200 MHz and 100 dB/km at 500 MHz. Because of high information carrying capacity and low attenuation, now-a-days fibers are finding wide applications in telecommunications, local area networks, sensors, computer networks, cable TV etc.

There are many advantages of fiber optics. The *primary advantages* are following.

(i)   EXTREMELY LARGE BANDWIDTH

The bandwidth available with fiber optic is enormous (more than 100 GHz). With such a large bandwidth, it is possible to transmit thousands of voice conversations or dozens of video signals over the same fiber simultaneously. Irrespective of whether the information is voice, data or video or a combination of these, it can be transmitted easily over the optical fibers. More than clearing congestion on the communication channels, it will allow communication services which are not possible otherwise. As an example, memory-to-memory transfer between computers will now be possible even if the computer speed is doubled.

(ii)   SMALLER DIAMETER AND LIGHTER WEIGHT

The size reduction makes fiber optical cables the ideal transmission medium for ships, aircrafts and high rise buildings where bulky copper cables take up too much space. Together with the reduction in size goes an enormous reduction in weight. It is an important advantage in aircraft, missiles and satellites.

(iii) NO CROSSTALK BETWEEN PARALLEL FIBERS

In conventional communication circuits, signals often stray from one circuit to another, resulting in other calls being heard in the background. This crosstalk is negligible with fiber optics even if numerous fibers are cabled together.

(iv) IMMUNITY TO INDUCTIVE INTERFERENCE

As optical fibers are made of dielectric material rather than the metal, these do not act as antenna to pick up radio frequency interference (RFI), electromagnetic interference (EMI). The result is noise free transmission. Thus fibers can operate satisfactorily in noisy electrical environment.

(v) POTENTIAL OF DELIVERING SIGNALS AT A LOWER COST

In very short distance applications, it is difficult for fiber optics to compete economically with copper wires. However, when the communication system capacity requirement is high or where interference would require special shielding for metallic wire, fiber links can be competitive. As fiber optics systems become more and more common, their price is expected to drop significantly.

These primary advantages are sufficient to justify the use of fiber optics in a number of applications. However, the *secondary advantages* must not be overlooked.

(vi) GREATER SECURITY

As the light in an optical fiber does not radiate outside the cable, the only way to eavesdrop is to couple a tap directly to the fiber. But in that case, the loss would be so much high that at the receiving end an alarm would be sounded.

(vii) GREATER SAFETY

In fiber optics, only light not electricity is being conducted. If the fiber optic cable is damaged, there is no spark from a short circuit. Consequently, fiber optic cables can be routed through areas such as chemical plants and coal mines with highly volatile gases without fear of causing fire or explosion. In addition, there is no shock hazard with fiber optic cables. Fibers can be repaired in the field even when the transmission is on.

(viii)   LONGER LIFE SPAN

A longer life span of 20 to30 years is predicted for the fiber optics cables as compared to 12 to 15 years for the conventional cables. The main reason for this is that glass does not corrode as metal does.

(ix) HIGH TOLERANCE TO TEMPERATURE EXTREMES, LIQUIDS AND CORROSIVE GASES

As optical fiber cables are made of glass, these have a high tolerance to temperature extremes as well as to liquids and corrosive gases.

(x) GREATER RELIABILITY AND EASE OF MAINTENANCE

In fiber optic cables, transmission losses are lower than in coaxial cables. Therefore, it allows substantial increase in the distance between repeaters. Obviously, fewer the repeaters, lesser will be the likelihood for failure. The reliability increases accordingly.

(xi) NO EXTERNALLY RADIATED SIGNAL

As fiber optic cables do not radiate signals, the fiber optic transmission does not interfere with other services. In fiber optics, signal confinement is excellent.

(xii) EASE OF EXPANSION OF SYSTEM CAPABILITY

Many fiber optic systems can be easily upgraded to expand system capability by simply replacing light emitting diode (LED) sources by injection lasers and PIN photodiodes by avalanche photodiodes (APDs). One can upgrade the existing fiber optic systems without replacing the original cables. Improved modulation techniques can also accomplish the same goal.

(xiii) USE OF COMMON NATURAL RESOURCES

By using common natural resources like sand and plastic rather than scarce resources like copper and iron, fiber optics is helping to conserve the dwindling world resources.

In the following, various types of optical fibers, their important characteristics, fiber materials, fabrication and cabling, splices and connectors, non-linear optical effects and fiber gratings are discussed.

4.1   Types of Optical Fibers

An optical fiber is a long cylindrical dielectric structure usually of circular cross section. The structure and the RI distribution is generally uniform along the length of the fiber. The RI distribution in the transverse direction varies in such a way that the index is relatively higher in the central region, termed as the core, with respect to the surrounding region, termed as the cladding. The cladding is also surrounded by a protective jacket, usually of plastic material (refer to Fig. 4.1). The cladding is sufficiently thick and the propagating light does not interact with the jacket. Hence, the jacket does not play any role in the propagation characteristics.

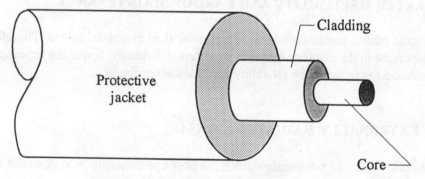

Fig. 4.1: Schematic of optical fiber

The material used in telecommunication fibers is either *silica* or *plastic*. In the silica fibers, cladding is usually of pure silica while the core is doped with other material such as germanium to increase the RI. The plastic fibers may consist of glass core with plastic sheath (cladding), plastic core with plastic cladding. Of these, the plastic core with plastic cladding fibers are commonly used because in this, both the core and cladding have similar softening points which simplify the production process. Plastic fibers have the advantage of more flexibility than glass fibers and are preferred in systems that require flexibility (say inter-building fiber connection). The attenuation characteristics of plastic fiber are worse as compared to that of glass fibers. But even then these are frequently used for short distance computer applications with information capabilities of about 6 Mbit/s over a distance of 50-200 metres. In addition to flexibility, the mechanical handling of plastic fiber such as gripping, crimping etc. is much easier. Plastic fibers can get damaged easily than the glass fibers and show a fatigue effect on repeated bending/jointing. Plastic fibers and cables have lesser operational temperature range, less than those of wire and coaxial cables. Hence for high temperatures and long distance applications, plastic fibers (or cables) can not be used.

Ordinary glass, as generally known, is brittle that is it can be easily broken or cracked. Optical fibers, in contrast, have high tensile strength and are quite tough. The fibers are as strong as stainless steel wires of the same diameter. In comparison with copper wire, an optical fiber has the tensile strength of a copper wire twice as thick. To produce fibers so tough, glass must be extremely pure. Further, core and cladding should be as free from microscopic cracks on the surface or flaws in the interior as possible. When a fiber is under stress, it can break at any of these flaws.

Depending on the refraction index of the core whether it is constant or varying, fibers can be classified into two types. These are shown in Fig. 4.2.

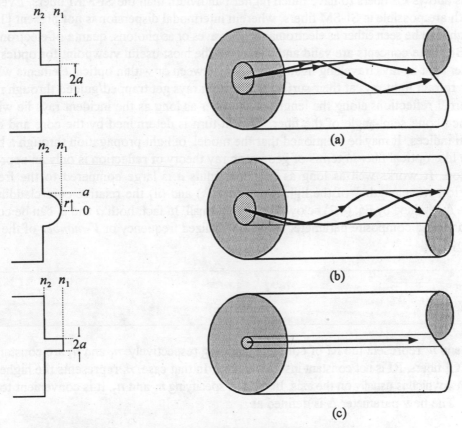

Fig. 4.2: Cross-section and ray paths in (a) multimode step index fiber,
(b) multimode graded-index fiber and (c) single-mode step index fiber

A fiber which has a uniform RI in the core and a step jump of RI at the core-cladding interface is known as a *step index* (SI) fiber. On the other hand, a fiber in which the core RI varies continuously is termed as *graded-index* (GI) fiber. Both the SI and GI fibers can be further divided into *single-mode* (SM) and *multimode* (MM) classes. As the name implies, a SM fiber (or monomode fiber) sustains only one mode of propagation, whereas a MM fiber many modes. In

SM and MM fibers, typical outer cladding diameter is 125 μm and core diameter in SM fiber is about 8 μm whereas in MM fiber it lies between 50-100 μm depending upon the application. Following advantages are offered by the MM fibers over the SM fibers: (i) It is easier to launch optical power into the fiber and connecting together of similar fibers because of large core radii, (ii) LEDs can be used to launch optical power into MM fibers, whereas SM fibers must generally be excited with laser diodes (LDs). Although LEDs have less output power than the LDs, these are easier to make, less expensive, require less complex circuitry and have longer life (Chapter two). A disadvantage of MM fibers is the intermodal dispersion (explained later in Section 4.4.1). This disadvantage can be overcome to a great extent by using graded-index profile in the fiber core. This allows GI fibers to have much higher bandwidth than the SI-MM fibers. Even higher bandwidth are possible in SI-SM fibers, wherein intermodal dispersion is not present [15].

The light can be seen either as electromagnetic waves or as photons, quanta of electromagnetic energy. Both the concepts are valid and valuable. The most useful viewpoint for optics is often to consider light as rays travelling in straight lines between or within optical elements which can reflect or refract light rays at their surfaces. The light rays get trapped/guided through repeated total internal reflections along the length of the fiber as long as the incident rays lie within the acceptance cone semi-angle of the fiber. This in turn is determined by the core and cladding refraction indices. It may be mentioned that the model of light propagation through a bounded structure like optical fiber in terms of geometric ray theory of reflection is only an approximate description. It works well as long as (i) core radius a is large compared to the free-space propagation wavelength λ of the light wave ($a > \lambda$) and (ii) the relative core-cladding index difference Δ (defined by Eq. (4.2) below) is not very small. In fact, both a and Δ can be combined with λ to yield a composite parameter called normalized frequency or V-$number$ of the fiber. It is given by

$$V = \frac{2\pi}{\lambda} a \sqrt{n_1^2 - n_2^2}$$
(4.1)

where n_1 and n_2 represent the RI of core and cladding respectively. n_1 and n_2 are constant for SI fibers. In GI fibers, RI is not constant inside the core. In that case, n_1 represents the highest value of the index which is usually on the axis. Instead of specifying n_1 and n_2, it is convenient to specify n_2 and Δ. The new parameter Δ is defined as

$$\Delta = \frac{n_1^2 - n_2^2}{2n_1^2} \approx \frac{n_1 - n_2}{n_1}$$
(4.2)

Since n_1 is usually very close to n_2, the above approximation is quite valid. The V-number of the fiber can then be written in terms of Δ as

$$V = \frac{2\pi}{\lambda} a n_1 \sqrt{2\Delta} \approx \frac{2\pi}{\lambda} a n_2 \sqrt{2\Delta} \tag{4.3}$$

When $V > 10$, geometrical optics results can be used to describe the propagation effects in optical fibers. For $V \leq 10$, geometrical optics cannot explain propagation effects. One has to make use of electromagnetic analysis to investigate propagation effects in the fibers. What is required in the latter approach is to express the wave equation in cylindrical coordinates and solve it subject to boundary conditions that represent core-cladding interface. In addition, if the core has a graded RI profile, the entire index profile $n(r)$ becomes part of the boundary condition.

4.2   Optical Propagation Theory

In the following Section, ray optics and electromagnetic or mode theories of propagation for the optical fibers are described in brief.

(i)   RAY OPTICS

In an optical fiber, light rays are guided by *total internal reflection* at the interface between fiber core and cladding as shown in Fig. 4.3. However, only those rays are guided which fulfil the condition of total internal reflection determined by Snell's law discussed below. In this case, rays follow a *zigzag path* along the fiber core with ongoing internal reflections. As these rays intersect the axis (center line) of the fiber after each reflection, they are called *meridional rays*. As shown in Fig. 4.3, input ray is refracted when entering the fiber core because RI n_0 of the surrounding medium is different from that of fiber core ($n_1 > n_0$). If input angle is too large, ray in fiber core enters the cladding and is no longer guided.

In contrast to meridional rays, *skew rays* propagate by forming a spiral (helical path) around the fiber axis as illustrated in Fig. 4.4. Skew rays are three dimensional rays, whereas meridional rays propagate in a two dimensional plane. It may be seen from Fig. 4.4 that the helical path traced through the fiber gives a change in direction of 2γ at each reflection where γ is the angle between the projection of the ray in two dimensions and the radius of fiber core at the point of reflection. Unlike meridional rays, the point of emergence of skew rays from the fiber in air will depend upon the number of reflections these rays undergo rather than the input conditions to the fiber.

It may be mentioned that skew rays greatly outnumbered the meridional rays. However, as explained later, it is sufficient to consider the meridional rays for all practical purposes.

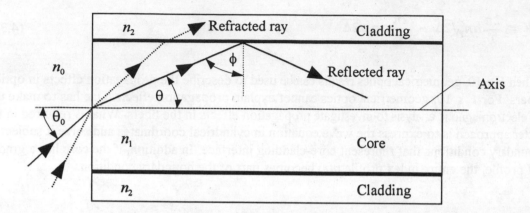

Fig.4.3: Meridional ray propagation in a step index fiber

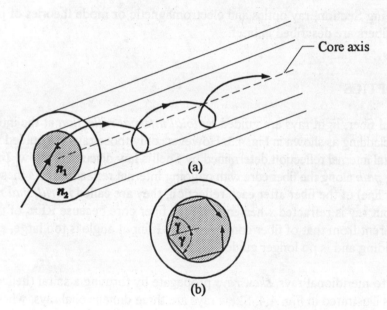

Fig.4.4: (a) Skew ray path along the fiber length and (b) cross-sectional view of the fiber

Let us now consider the propagation of meridional rays in a step index fiber as shown in Fig. 4.3. From Snell's law, the minimum angle ϕ_{min} that supports total internal reflection for the meridional ray is given by

$$\sin\left(\phi_{min}\right) = \frac{n_2}{n_1} \tag{4.4}$$

The rays for which the incident angle at the core-cladding interface is less than ϕ_{min} will be refracted out to the cladding and be lost there. The maximum incident angle $\theta_{0,max}$ for rays to propagate in the core through the mechanism of total reflection from Eq. (4.4) is given by

$$n_0 \sin\left(\theta_{0,max}\right) = n_1 \sin\left(\theta_c\right) = \sqrt{n_1^2 - n_2^2} \tag{4.5}$$

where θ_c is the critical angle and n_0 the RI of outside medium which is generally air (for air $n_0 = 1$). The above equation also defines the *numerical aperture (NA)* of a SI fiber for meridional rays

$$NA = n_0 \sin\left(\theta_{0,max}\right) = \sqrt{n_1^2 - n_2^2} = n_1 \sqrt{2\Delta} \tag{4.6}$$

The *NA* is an important parameter for an optical fiber as it describes the light acceptance or gathering capability of a fiber. It is a dimensionless quantity which is normally less than unity and ranges from 0.14 to 0.50. Typically, *NA* of telecommunication fibers is 0.1 to 0.2 which implies an acceptance angle of 5.7° to 11.5°. However, for non-telecommunication applications like in endoscope, *NA* could be much higher than 0.5 (i.e., $\theta_0 > 30°$). For skew rays, Eq. (4.6) gets modified as [26]

$$NA = n_0 \sin\left(\theta_{s,max}\right)\cos(\gamma) = \sqrt{n_1^2 - n_2^2} = n_1 \sqrt{2\Delta} \tag{4.7}$$

where $\theta_{s,max}$ is the maximum input angle within which the incident rays will be totally internally reflected. It is clear from the above equation that the skew rays are accepted at larger axial angles in a fiber of given *NA* than the meridional rays depending upon the value of $\cos(\gamma)$. In fact, for the meridional rays, $\cos(\gamma)$ is equal to unity and $\theta_{s,max}$ becomes equal to $\theta_{0,max}$. Though $\theta_{0,max}$ is the maximum critical half angle for the acceptance of meridional rays, it defines the minimum input angle for the skew rays [26]. For most communication design purposes, the expression for *NA* given in Eq. (4.6) for meridional rays is adequate.

(ii) MODE THEORY

In order to describe the propagation of light waves through an optical fiber in terms of electromagnetic waves, we have to solve the *Maxwell's equations* subjected to the cylindrical boundary conditions of the fiber. A complete treatment is beyond the scope of this book, only a general outline of the analysis will be given that too for the SI fiber.

First of all, let us examine the appearance of modal fields in the planar dielectric slab waveguide. Later, the results will be extended to the circular optical fibers. As shown in Fig. 4.5, the planar waveguide is composed of dielectric slab of RI n_1 sandwiched between dielectric material of RI $n_2 < n_1$. The cross-sectional view of the slab waveguide looks the same as the cross-sectional view of an optical fiber cut along its axis. It is clear from the field pattern of several lower-order modes shown in Fig.4.5 that the order of mode is equal to the number of field zeros across the guide. Further, the electric fields of guided modes are not totally confined to the central dielectric slab. Instead, the fields extend partially into the cladding. The fields vary harmonically in the guiding region of RI n_1 and decay exponentially outside of this region. For lower-order modes, the fields are highly concentrated near the centre of the slab (or the fiber axis) with little penetration into the cladding region. In contrast to this, fields for the higher-order modes are distributed more toward the edges of the guide and penetrate farther into the cladding region.

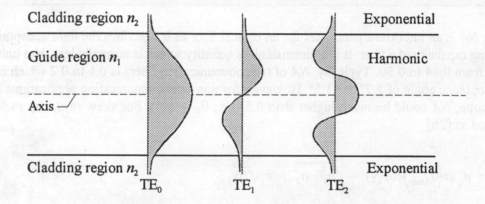

Fig. 4.5: Electrical field distribution in a symmetrical-slab waveguide for several
           lower-order guided modes

In addition to the guided modes, there are radiation modes also. These modes basically results from the radiation that is outside the fiber acceptance angle and being refracted out of the core. Some of these radiation get trapped in the cladding due to finite radius of the cladding. These are called cladding modes. As the core and cladding modes propagate along the fiber, mode coupling occurs between the cladding modes and the higher-order core modes. The reason for this coupling is the extension of electric fields of the core modes to the cladding and likewise for the cladding modes. Diffusion of power back and forth between core and cladding modes results in a loss of power from the core modes. In practice, cladding modes will be suppressed by a lossy coating on the fiber or these modes will scatter out of the fiber after travelling a certain distance because of roughness on the cladding surface.

In addition to core (bounded) and cladding (refracted) modes, a third category of modes called leaky modes is also present in optical fibers. The leaky modes are partially confined to the core region and attenuate by continuously radiating their power out of the core when propagate along the fiber. The analysis of leaky mode is fairly complex and beyond the scope of this book. It may be mentioned that a mode remains guided as long as $n_2 k < \beta < n_1 k$ where β is the propagation constant of the guided mode and $k = 2\pi/\lambda$ the magnitude of propagation vector k or phase propagation constant in the vacuum. The boundary between bounded modes and leaky modes is defined by the cut-off condition $n_2 k = \beta$. As soon as β becomes smaller than $n_2 k$, power leaks out of the core into the cladding region. Leaky modes can carry significant amount of optical power in short fibers. Most of these modes, disappear after travelling a distance of few centimetres, but a few have sufficiently low losses to persist up to a fiber length of one kilometre.

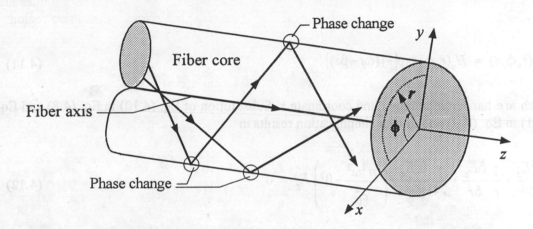

Fig. 4.6: Cylindrical coordinate system for an optical fiber to analyse electromagnetic wave propagation

To extend the plane dielectric slab waveguide theory to circular optical fibers, let us consider electromagnetic waves propagating along a cylindrical fiber as shown in Fig.4.6. For this fiber, a cylindrical coordinate system (r, ϕ, z) is defined with z-axis lying along the axis of the fiber. The wave equation from the Maxwell's equations in cylindrical coordinate [4] is given by

$$\frac{\delta^2 E_z}{\delta r^2} + \frac{1}{r}\frac{\delta E_z}{\delta r} + \frac{1}{r^2}\frac{\delta^2 E_z}{\delta\phi^2} + \frac{\delta^2 E_z}{\delta z^2} - \frac{n^2}{c_0^2}\frac{\delta^2 E_z}{\delta t^2} = 0 \qquad (4.8)$$

and

$$\frac{\delta^2 H_z}{\delta r^2} + \frac{1}{r}\frac{\delta H_z}{\delta r} + \frac{1}{r^2}\frac{\delta^2 H_z}{\delta\phi^2} + \frac{\delta^2 H_z}{\delta z^2} - \frac{n^2}{c_0^2}\frac{\delta^2 H_z}{\delta t^2} = 0 \qquad (4.9)$$

Here E_z and H_z are the z-components of electric and magnetic fields respectively, c_0 the velocity of light in vacuum and n the RI of medium (equals n_1 for core and n_2 for cladding). As the electromagnetic waves are to propagate along the z-axis, these will have a functional dependence of the form

$$E_z(r,\phi,z) = E_z(r,\phi)\exp[-j(\omega t - \beta z)] \qquad (4.10)$$

and

$$H_z(r,\phi,z) = H_z(r,\phi)\exp[-j(\omega t - \beta z)] \qquad (4.11)$$

which are harmonic in time t and coordinate z. Substitution of Eq. (4.10) in Eq. (4.8) and Eq. (4.11) in Eq. (4.9) and further simplification results in

$$\frac{\delta^2 E_z}{\delta r^2} + \frac{1}{r}\frac{\delta E_z}{\delta r} + \frac{1}{r^2}\frac{\delta^2 E_z}{\delta\phi^2} + \left(\frac{n^2\omega^2}{c_0^2} - \beta^2\right)E_z = 0 \qquad (4.12)$$

and

$$\frac{\delta^2 H_z}{\delta r^2} + \frac{1}{r}\frac{\delta H_z}{\delta r} + \frac{1}{r^2}\frac{\delta^2 H_z}{\delta\phi^2} + \left(\frac{n^2\omega^2}{c_0^2} - \beta^2\right)H_z = 0 \qquad (4.13)$$

In general, boundary conditions of the electromagnetic field components (tangential components of E_ϕ and E_z inside and outside of the dielectric interface at $r = a$ must be equal and similarly the tangential components of H_ϕ and H_z) require the coupling of E_z and H_z. If the boundary conditions do not lead to coupling between the field components, mode solutions can be obtained in which either $E_z = 0$ or $H_z = 0$. When $E_z = 0$, the modes are called transverse electric or TE modes and when $H_z = 0$, transverse magnetic or TM modes. Hybrid modes exist if both E_z and E_z are nonzero. These are designated as HE or EH modes depending on whether H_z or E_z, respectively makes a larger contribution to the transverse field [15].

The cylindrical waveguide is bounded in two dimensions rather than one. Thus, two integers l and m are necessary to specify the modes, in contrast to the single integer m required for the planar guide. Thus in a cylindrical waveguide, TE_{lm} and TM_{lm} modes will exist. These modes correspond to meridional rays travelling within a fiber. The modes which result from skew rays propagation are HE_{lm} and EH_{lm}. An exact description of modal field even in the SI fiber is somewhat complicated. Fortunately, the analysis becomes simplified for the fibers used in communications wherein $\Delta \ll 1$ (usually less than 0.03). Therefore, optical fibers can be considered as a weakly guided structure. In a weakly guiding structure such as in optical fibers, HE-EH mode pairs occur and they have almost identical propagation constant. Such modes are said to be degenerate. The superposition of these degenerating modes characterized by a common propagation constant corresponds to a particular linearly polarized (LP) mode regardless of their HE, EH, TE or TM field configurations. It may be observed from the field configurations of modes that the field strength in the transverse direction (E_x or E_y) is identical for the modes which belong to the same LP mode. It explains the origin of the term linearly polarized [26].

The relationships between the traditional HE, EH, TE and TM modes and the LP_{lm} modes are given in Table 4.1. The subscript l represents the number of multiples of 2π that the E field seems to rotate to a non-rotating observer in making one round trip in azimuthal angle ϕ around a circumference within the core. The subscript m is the number of semi-sinusoidal half-cycles going radially in r from core centre to the core-cladding interface [9].

Table 4.1: Composition of the lower-order LP modes

LP mode	Traditional mode
LP_{01}	HE_{11}
LP_{11}	HE_{21}, TE_{01}, TM_{01}
LP_{21}	HE_{31}, EH_{11}
LP_{02}	HE_{12}
LP_{31}	HE_{41}, EH_{21}
LP_{12}	HE_{22}, TE_{02}, TM_{02}
LP_{lm}	HE_{2m}, TE_{0m}, TM_{0m}
LP_{lm} ($l \neq 0$ or 1)	$HE_{l+1,m}$, $EH_{l-1,m}$

The Eqs. (4.10) and (4.11) constitute waves travelling in the z-direction with propagation constant β and z dependence of the E_z and H_z fields is of the form exp(-jβz). Also E_z and H_z must be periodic in ϕ so that their ϕ dependence will be exp(-jlϕ), where l is an integer as mentioned just above. The r dependence is still not known. Therefore, Eqs.(4.10) and (4.11) can be written as

$$E_z(r, \phi, z) = E_z(r) \exp(-j1\phi) \exp[-j(\omega t - \beta z)],$$

$$1 = \pm 1, \pm 2, \pm 3, \ldots\ldots$$

(4.14a)

and

$$H_z(r, \phi, z) = H_z(r) \exp(-j1\phi) \exp[-j(\omega t - \beta z)],$$

$$1 = \pm 1, \pm 2, \pm 3, \ldots\ldots$$

(4.14b)

Substitution of Eq. (4.14a) in Eq. (4.12) gives

$$\frac{\delta^2 E_z}{\delta r^2} + \frac{1}{r}\frac{\delta E_z}{\delta r} + \left(\frac{n^2 \omega^2}{c_0^2} - \beta^2 - \frac{1^2}{r^2} \right) E_z(r) = 0$$

(4.15)

which is a Bessel differential equation. In SI fibers, change in the RI is at the core-cladding boundary $r = a$ where n changes from n_1 to n_2. Let us define the propagation constant $\beta_1 = n_1 \omega / c_0$ for the core and $\beta_2 = n_2 \omega / c_0$ for the cladding. The condition for the wave to be guided is that $\beta_1 < \beta_{lm} < \beta_2$ where β_{lm} is the propagation constant of the wave. It is convenient to define

$$u_{lm}^2 = \beta_1^2 - \beta_{lm}^2$$

(4.16)

and

$$w_{lm}^2 = \beta_{lm}^2 - \beta_2^2$$

(4.17)

for the core and cladding respectively. Since the right sides of Eqs.(4.16) and (4.17) are positive, both u_{lm} and w_{lm} are real. The Eq. (4.15) can then be written as two Bessel differential equations one for the core

$$\frac{\delta^2 E_z}{\delta r^2} + \frac{1}{r}\frac{\delta E_z}{\delta r} + \left(u_{lm}^2 - \frac{1^2}{r^2} \right) E_z(r) = 0 \qquad r < a$$

(4.18a)

and other for the cladding

$$\frac{\delta^2 E_z}{\delta r^2} + \frac{1}{r}\frac{\delta E_z}{\delta r} - \left(w_{lm}^2 + \frac{l^2}{r^2}\right) E_z(r) = 0 \qquad r > a \qquad (4.18b)$$

Solutions to the above equations excluding functions that approach infinity at $r = 0$ are given by

$$E_z(r,\phi) \sim \begin{cases} J_l\,(u_{lm}r)\;\cos(l\phi) & r < a \quad \text{core} \\ K_l(w_{lm}r)\;\cos(l\phi) & r > a \quad \text{cladding} \end{cases} \qquad (4.19)$$

where $J_l(x)$ is the Bessel function of the first kind and order l. $K_l(x)$ is the modified Bessel function of the first kind and order l.

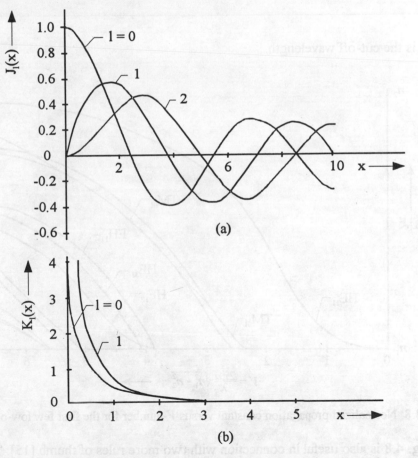

Fig. 4.7: (a) Bessel function $J_l(x)$ and (b) modified Bessel function $K_l(x)$

As shown in Fig. 4.7, J_l functions look like decaying sinusoid, while the K_l functions decaying exponential. This is what would be expected from the earlier discussion on planar dielectric waveguide. Following the same approach, $H_z(r,\phi)$ is evaluated. Once the E_z and H_z components are known, other components $E_r(r,\phi)$, $H_r(r,\phi)$, $E_\phi(r,\phi)$ and $H_\phi(r,\phi)$ can be determined from the Maxwell's equations.

A plot of propagation constant β of each mode versus the V-number $[= a(u_{lm}^2 + w_{lm}^2)^{1/2}]$ is shown in Fig. 4.8. This figure shows that each mode can exist only for values of V which exceed a certain limiting value. Further, if the V-number is less than the magic number 2.405, only SM propagation can exist. All the modes except HE_{11} have higher cut-off wavelengths when no energy can propagate. For $V < 2.405$, the optical wavelengths of all modes except HE_{11} are below cut-off. The HE_{11} mode has no cut-off and cease to exist only when the core diameter goes to zero. Thus for a SM operation

$$V = \frac{2\pi a}{\lambda}\sqrt{n_1^2 - n_2^2} \leq 2.405 \quad \text{or} \quad \lambda_c = \frac{2\pi a}{2.405}\sqrt{n_1^2 - n_2^2} \qquad (4.20)$$

where λ_c is the cut-off wavelength.

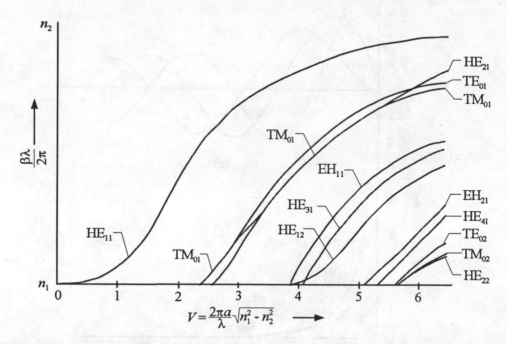

Fig. 4.8: Normalized propagation constant versus V-number for the first few low-order modes

The Fig. 4.8 is also useful in connection with two more rules of thumb [15]. The first one is that number of modes that can be excited in a MM fiber is approximately given by

$$M \approx \frac{V^2}{2} \qquad (4.21)$$

for large V, if the two degenerate states of the same mode is counted separately. The second says that the ratio of power carried by the cladding to carried by both core and cladding is given by

$$\frac{P_{clad}}{P_{total}} = \frac{2\sqrt{2}}{3V} \qquad (4.22)$$

In order to predict the performance characteristics of SM fibers, a parameter of fundamental importance is the *mode field diameter* (MFD). This parameter describes the geometrical distribution of light in the propagating mode. In SM fibers, not all the light which propagate through the fiber is carried in the core. Therefore, the MFD for these fibers becomes analogous to the core diameter in MM fibers. The MFD can be determined from the mode field distribution of the fundamental LP_{01} mode. There are several models available for characterizing and measuring MFD. In all these models, main consideration is the approximation used to represent the electric field distribution. In one of the commonly used approximation, distribution is considered to be Gaussian

$$E(r) = E_o \exp\left(-r^2/W_o^2\right) \qquad (4.23)$$

In the above equation, r represents the radius, E_o the field at zero radius and W_o the width of the electric field distribution. The MFD $2W_o$ for the LP_{01} mode can now be defined as

$$2W_o = 2\left[\frac{2\int_0^\infty r^3 E^2(r)\,dr}{\int_0^\infty r E^2(r)\,dr}\right]^{1/2} \qquad (4.24)$$

It may be mentioned that this definition is not unique and several others have been proposed [15]. In general, the mode field varies with the RI profile and its distribution can deviate from a Gaussian distribution.

4.3   FIBER ATTENUATION

Fiber attenuation arises due to various losses caused by impurities and irregularities in the fiber core and cladding. Losses are categorized as (i) absorption losses, (ii) scattering losses and (iii) bending losses. Absorption and scattering are the dominant mechanisms responsible for the transmission loss in optical fibers. The loss mechanisms are briefly discussed below.

4.3.1 ABSORPTION LOSSES

The propagating light interacts with the fiber material to cause electrons to undergo transitions. These electrons give up the absorbed energy by emitting light at other wavelengths or in the form of mechanical vibrations (heat). Thus, in absorption, energy is removed from the signal and given up later in some other form.

The oxygen ions in pure silica have very tightly bonded electrons and only ultraviolet (UV) light photons have enough energy to be absorbed. However, the dopants and transitional metal impurities such as iron (Fe^{++}), chromium (Cr^{++}), cobalt (Co^{++}) copper (Cu^{++}) etc. have electrons that can be excited in the visible and near infrared (IR) regions. The transition metal impurities which are present in the starting materials used for direct-melt fibers ranges between 1 and 10 parts per billion (ppb) causing losses between 1dB/km to 10 dB/km. The impurity levels in vapour phase deposition processes are usually one to two orders of magnitude lower. The absorption peaks of the various transition metal impurities tend to be broad and several peaks may overlap, which further broadens the absorption region [15].

Early fibers had high levels of OH ions which resulted in large absorption peaks occurring at 1.4 μm, 0.95 μm and 0.725 μm. These are the first, second and third overtones respectively of the fundamental peak of water at 2.7 μm [15]. Since most resonances are sharply peaked, narrow windows exist in the long wavelength region around 1.3 μm and 1.55 μm which are essentially unaffected by OH ion absorption once the impurity level has been reduced below 1 ppb [26]. It may be mentioned that in 1979 lowest attenuation for an ultra-low-loss SM fiber at 1550 nm already was 0.2 dB/km [20]. At present, fibers with a loss of 0.18 dB/km and also 0.17 dB/km are available which is quite close to the minimum possible attenuation of 0.154 dB/km at this wavelength.

4.3.2   SCATTERING LOSSES

Geometrical imperfections in the optical fibers (on a scale which can be small or large compared to wavelength) cause light to be redirected out of the fiber. Thus, in scattering the propagating energy leaves the fiber at the same wavelength at which it arrives at the geometrical imperfections.

The scattering in fiber arises from the microscopic variations in the material density, compositional fluctuations, structural inhomogeneities or defects occurring during manufacturing process. Rayleigh scattering arises when the size of the inhomogeneities is smaller than the wavelength of light. These inhomogeneities can not be totally eliminated and set the lower limit of the fiber loss. Rayleigh scattering loss, as shown in Fig. 4.9, is proportional to the fourth power of the wavelength.

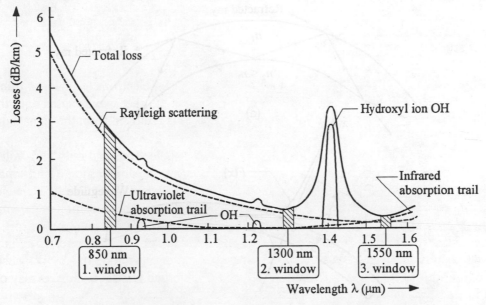

Fig. 4.9: Variations of loss with wavelength in a typical optical fiber

When the scattering inhomogeneity size is greater than $\lambda/10$, scattering caused by such inhomogeneities is mainly in the forward direction and is called Mie scattering. Depending upon the fiber material, design and manufacturing process, this scattering can cause significant loss. However, by careful fiber design, it is possible to reduce Mie scattering to insignificant level.

Other scattering mechanisms are stimulated Raman and Brillouin scattering. Both are non-linear effects and have optical power thresholds. The threshold power densities necessary for stimulated scattering are well above the normal power densities used in multimode fiber systems. However, it can be of concern for some single-mode systems. A brief review of non-linear effects in fibers is given in Section 4.8.

4.3.3 RADIATION LOSSES

Radiation losses occur whenever an optical fiber undergoes a bend of finite radius of curvature. The fiber bending can be labelled as microbending or macrobending. Microbending is due to imperfections in the geometry of the fiber such as core-cladding interface irregularities, bubbles, diameter fluctuations and axis misalignment. Other forms of microbending arise from external influences such as mechanical stress caused by pressure, tension or twist. Scattering losses due to these mechanisms are called microbending losses. These losses can be reduced by increasing the index difference between the core and cladding or by careful fiber drawing and cabling.

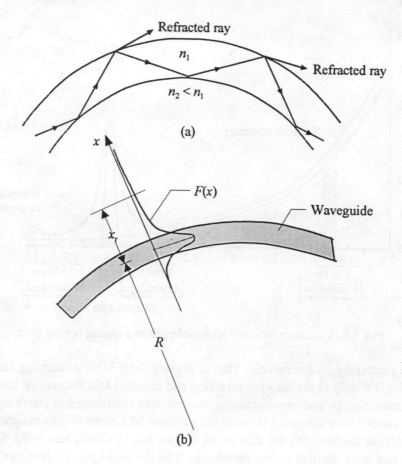

Fig. 4.10: Mechanisms of microbending loss in (a) multimode and (b) single-mode fibers

Losses due to macrobending are caused by large fiber curvatures. A MM fiber will radiate energy as rays near the critical angle will be refracted out of the core as shown in Fig. 4.10a [11]. It results in some loss. For SM fibers, loss due to microbending can be explained as follows (refer to Fig. 4.10b). The cladding field of the fundamental mode extends out to infinity. When the fiber is bent, the field at large core radii will be forced to propagate faster than that the field inside so that a wavefront perpendicular to the direction of propagation is maintained. Hence, part of the modes in cladding need to travel faster than the speed of light. As this is physically impossible, the energy associated with this part of the modes is lost through radiation. A fiber with a high index difference is less susceptible to bending losses.

4.4   FIBER DISPERSION

Dispersion is a measure of the temporal spreading that occurs when a light pulse propagates through an optical fiber. Dispersion is sometimes referred to as *delay distortion* in the sense that the propagation time delay causes the pulse to broaden. The broaden pulse overlaps with its neighbours, eventually becoming indistinguishable at the receiver input. This effect is known as *intersymbol interference* (ISI). As the ISI becomes more pronounced, increasing number of errors may be encountered at the receiver (Chapter nine). It may be mentioned that signal dispersion alone limits the maximum possible bandwidth or the data rate attainable with a particular optical fiber.

To make a conservative estimate of the practical transmission rate, let a digital signal in which bits are represented by a pulse of duration τ is transmitted over the fiber. At the receiver, it is possible to recognize the pulses if each pulse spread to $\tau/2$ on both sides i.e., pulse duration becomes 2τ. Therefore, the practical transmission bit rate R_T on such a fiber is limited to

$$R_T \leq \frac{1}{2\tau} \text{bits} \tag{4.25}$$

Another more accurate estimate of the practical bit rate for an optical fiber can be made by considering the light pulse at the output to have a Gaussian shape with an rms width of σ_T. The variations in the normalized optical output power with time is described as

$$p_o(t) = \exp\left[-\frac{t^2}{2\sigma_T^2}\right] \tag{4.26}$$

The Fourier transform of above equation is given by

$$P(\omega) = \sqrt{2\pi}\,\sigma_T \exp\left[-\frac{\omega^2 \sigma_T^2}{2}\right] \tag{4.27}$$

The *single-sided 3 dB optical bandwidth* B_{opt} is the modulation frequency at which the received optical power falls to one half of its peak value. This condition gives

$$2\pi B_{opt} = \frac{\sqrt{2}}{\sigma_T}\left[\ln(2)\right]^{\frac{1}{2}} \tag{4.28}$$

or

$$B_{opt} = \frac{1}{2\pi} \frac{\sqrt{2}}{\sigma_T} \left[\ln(2)\right]^{\frac{1}{2}} = \frac{0.187}{\sigma_T} \qquad (4.29)$$

The practical bit rate R_T when unipolar return to zero pulses are employed is, therefore, given by

$$R_T \approx \frac{0.2}{\sigma_T} \text{ bits} \qquad (4.30)$$

This equation gives a reasonably good approximation for the practical transmission bit rate. The maximum bit rate (i.e., the Nyquist bit rate) is twice this value as shown in Chapter twelve. Pulse broadening associated with the three common optical fibers viz., MM-SI, MM-GI and SM-SI fibers are shown in Fig. 4.11.

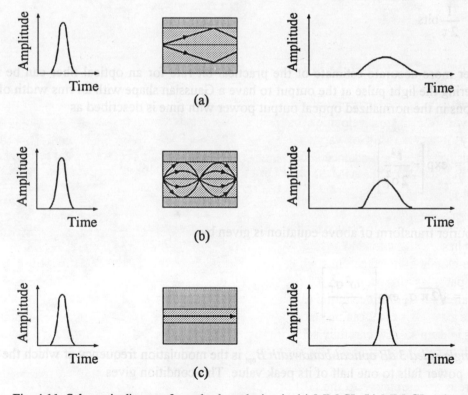

Fig. 4.11: Schematic diagram for pulse broadening in (a) MM-SI, (b) MM-GI and (c) SM-SI fibers due to intermodal dispersion

It may be observed that the MM-SI fibers (Fig. 4.11a) exhibit the greatest dispersion of a transmitted light pulse and the MM-GI fibers (Fig. 4.11b) give a considerably reduced dispersion.

The SM-SI fibers (Fig. 4.11c) give the minimum pulse broadening and thus are capable of giving the greatest transmission bandwidth. The amount of pulse broadening is dependent upon the distance a pulse travels within the fiber. In the absence of mode coupling or filtering, pulse broadening increases linearly with the fiber length and therefore the bandwidth is inversely proportional to distance. The dispersive properties of a particular fiber may be specified as the pulse broadening in time over a unit length of the fiber i.e., ns/km. It may also be specified in terms of another useful parameter viz., the *bandwidth-length product* $B_{opt} \cdot L$. This parameter is referred to as the information carrying capacity of an optical fiber. The typical values of this parameter for the three fibers MM-SI, MM-GI and SM-SI shown in Figs. 4.11 are 20 MHz·km, 1 GHz·km and 100 GHz·km respectively.

Three mechanisms are responsible for the pulse broadening in fibers: *modal* (or mode) *dispersion, material dispersion* and *waveguide dispersion*. Modal dispersion is often referred to as intermodal dispersion. The combination of material dispersion and waveguide dispersion is often called *chromatic dispersion* because both these depend on the wavelength. It is also referred to as intramodal dispersion. In the following, various dispersion mechanisms are discussed.

4.4.1 INTERMODAL DISPERSION

Intermodal or *modal dispersion* arises due to difference in the propagation times of modes with the slowest and fastest velocities. In SI fibers, where each mode has approximately the same velocity, it is the result of varying path lengths among the different modes. This dispersion mechanism creates the fundamental difference in the overall dispersion for the three types of fibers shown in Fig. 4.11. The MM-SI fibers give rise to largest modal dispersion which produces greatest pulse broadening.

There are two ways to reduce the modal dispersion. One is to use so called GI fiber. This type of fiber has a core whose RI is highest on the axis and tapers off roughly parabolically towards the core-cladding interface. In such fibers, rays do not travel the zig-zag paths (Fig. 4.3) rather helical paths as shown in Fig. 4.12. Clearly ray 1 travels a shorter path than ray 2 in order to propagate a given distance in the axial direction. However, the RI decreases as one move away from the axis. Thus ray 2 can make up in speed (light travels faster in a less dense medium) what it losses in distance. By carefully selecting the RI profile, one can arrange for all the guided rays to reach the fiber end at nearly the same time.

There is an alternative to the GI fiber for reducing pulse spreading. One can use a fiber with a very small core diameter i.e., SM fiber. In such fibers, only a single mode is allowed to propagate. Hence these fibers exhibit the least pulse broadening and have the greatest possible bandwidth. Such fibers have some disadvantages also. One of the main disadvantage is that due to small core diameter, it is difficult to couple the light into the fiber and splice (or connect) such fibers. Inspite of this, these fibers are becoming increasingly popular in many applications because of their ability to carry very narrowly spaced pulses.

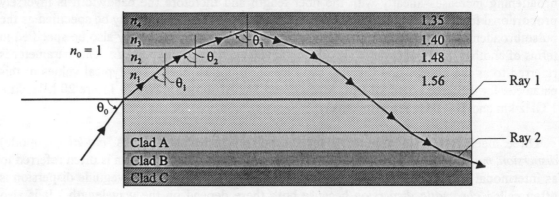

Fig. 4.12: Ray path in a graded-index fiber

In order to determine the intermodal pulse broadening in MM-SI and MM-GI fibers, it is useful to consider the geometric ray optics picture for these fibers.

(a) Multimode Step Index Fiber

Using the ray theory model, the fastest and slowest propagation in a SI fiber may be represented by the axial ray and the extreme meridional ray (which is incident on the core-cladding interface at the critical angle ϕ_c). As both rays are travelling at the same velocity in a constant RI core, the delay difference is directly related to their respective path lengths within the fiber. The time taken by an axial ray will be minimum and in a fiber of length L, it is given by

$$T_{min} = \frac{L}{(c_0/n_1)} = \frac{Ln_1}{c_0} \tag{4.31}$$

where c_0 as defined earlier is the velocity of light in vacuum. For the meridional ray, time taken will be maximum and is given by

$$T_{max} = \frac{L/\cos(\theta)}{c_0/n_1} = \frac{Ln_1}{c_0 \cos(\theta)} = \frac{Ln_1^2}{c_0 n_2} \tag{4.32}$$

since from Snell's law $\sin(\phi_c) = n_2/n_1 = \cos(\theta)$. The delay difference δT_s between the extreme meridional and the axial rays from Eqs. (4.31) and (4.32) will be

$$\delta T_s = T_{\max} - T_{\min} \approx \frac{L n_1 \Delta}{c_0} \tag{4.33}$$

where Δ is the relative RI difference and is defined in Eq. (4.2). Substituting Δ from Eq. (4.6) we obtain

$$\delta T_s \approx \frac{L(NA)^2}{2 n_1 c_0} \tag{4.34}$$

The Eqs. (4.33) and (4.34) are usually employed to estimate the pulse broadening in time due to intermodal dispersion in MM-SI fibers. The rms pulse broadening resulting from this dispersion mechanism can be determined by considering the input pulse $p_i(t)$ to MM-SI fibers is a pulse of unit area as shown in Fig. 4.13. It means

$$\int_{-\infty}^{\infty} p_i(t)\, \mathrm{d}t = 1 \tag{4.35}$$

Obviously $p_i(t)$ has a normalized constant amplitude $1/\delta T_s$ over the range

$$-\delta T_s/2 \leq t \leq \delta T_s/2 \tag{4.36}$$

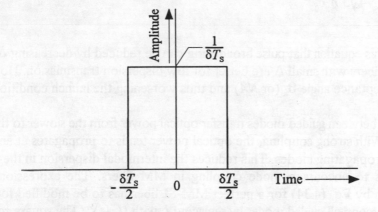

Fig. 4.13: A unit area optical input pulse as an input to MM-SI fiber

The rms pulse broadening σ_s at the fiber output due to intermodal dispersion can be obtained from

$$\sigma_s^2 = M_2 - M_1^2 \tag{4.37}$$

where

$$M_1 = \int_{-\infty}^{\infty} t\, p_i(t)\, \mathrm{d}t \tag{4.38}$$

and

$$M_2 = \int_{-\infty}^{\infty} t^2 p_i(t)\, \mathrm{d}t \tag{4.39}$$

As the mean value of pulse M_1 is zero and assuming that this is maintained for the output pulse, σ_s^2 from Eqs. (4.36), (4.37) and (4.39) is given by

$$\sigma_s^2 = \frac{1}{3}\left(\frac{\delta T_s}{2}\right)^2 \tag{4.40}$$

Substitution of δT_s from Eqs. (4.33) and (4.34) gives

$$\sigma_s \approx \frac{L n_1 \Delta}{2\sqrt{3}\, c_0} \approx \frac{L (NA)^2}{4\sqrt{3}\, n_1 c_0} \tag{4.41}$$

It is clear from this equation that pulse broadening can be reduced by decreasing Δ. It means that weakly guiding fibers with small Δ are better for low dispersion transmission. However, small Δ reduces the acceptance angle θ_a (or NA) and thus worsening the launch conditions.

The coupling between guided modes transfer optical power from the slower to the faster modes and vice versa. With strong coupling, the optical power tends to propagates at an average speed of the various propagating modes. This reduces the intermodal dispersion in the link and makes it advantageous to encourage mode coupling in MM fibers. The expression for the delay difference given by Eq. (4.34) for a perfect MM-SI fiber has to be modified for the fiber with mode coupling among all guided modes by replacing L with $(L \cdot L_c)^{1/2}$. The square root dependence on L arises from the mode mixing. Over short path lengths, mixing of power among the modes is incomplete. After a certain length, an equilibrium mode power distribution is reached. Mixing continues, but the power in any one mode remains the same. Under this condition, the $L^{1/2}$ dependence is observed. Here L_c represents the length at which equilibrium mode power distribution is reached. In typical fibers, it may be around one km [22].

(b) Multimode Graded-Index Fiber

As mentioned earlier, MM-GI fibers show substantial bandwidth improvement over MM-SI fibers. A typical RI profile with a maximum at the core axis for the GI fiber, as shown in Fig. 4.2b, is given by

$$n(r) = \begin{cases} n_1 \left[1 - 2\Delta \left(\dfrac{r}{a} \right)^g \right]^{1/2} & r < a \\ \\ n_1 \left[1 - 2\Delta \right]^{1/2} = n_2 & r \geq a \end{cases} \tag{4.42}$$

where g is the profile parameter which characterizes RI profile of the fiber core. Variations of g will give rise to different RI profiles e.g., step index profile when $g = \infty$, a parabolic profile when $g = 2$ and a triangular profile when $g = 1$. In a GI fiber, local group velocity is inversely proportional to the local RI. The longer ray paths are compensated for by higher speeds in the lower RI medium away from the axis. Hence there is an equalization of the transmission times of various trajectories towards the transmission time of the axial ray which travels exclusively in the high index region along the core axis at the lowest speed.

For a parabolic or near parabolic RI profile, delay difference between the fastest and slowest modes using rigorous analysis of electromagnetic theory is given by [7, 18]

$$\delta T_g = \frac{L n_1 \Delta^2}{8 c_0} \tag{4.43}$$

The best minimum theoretical intermodal rms pulse broadening for a GI fiber can be obtained by using an optimum RI profile [18, 21] with

$$g_{op} = 2 - \frac{12\Delta}{5} \tag{4.44}$$

With this profile, minimum rms pulse broadening is given by

$$\sigma_s = \frac{L n_1 \Delta^2}{20\sqrt{3}\, c_0} \tag{4.45}$$

A comparison of Eqs. (4.41) and (4.45) reveals that in a GI fiber rms pulse dispersion will decrease by a factor of $\Delta/10$ which is 1,000 for 1% difference in RI. This level of improvement is not usually achieved in practice due to difficulties in controlling the RI profile radially over long lengths of fiber. Any deviation in the RI profile from the optimum value results in increased intermodal dispersion. This may be observed from the curve shown in Fig. 4.14 which shows the variations of δT_g with g for typical GI fibers ($\Delta = 1\%$). The curve displays a sharp minimum at $g = 1.98$. This corresponds to the optimum value of g to minimize intermodal dispersion. The extreme sensitivity of the pulse broadening to slight variations in g from the optimum value is evident. Because of this reason, improvement factor for practical GI fibers over the corresponding SI fibers is around 100.

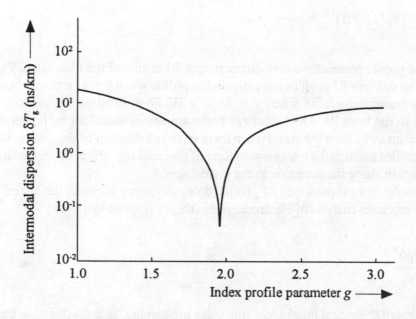

Fig. 4.14: Intermodal pulse broadening δT_g versus index profile parameter g for a GI fiber with $\Delta = 1\%$

4.4.2   INTRAMODAL DISPERSION

(a)   Material Dispersion

It arises due to wavelength dependence of the RI of fiber material. For example, fused silica has a RI of 1.453 at 820 nm, 1.447 at 1300 nm and 1.444 at 1550 nm. The light sources are not monochromatic and produce a band of wavelengths. Therefore, different wavelength components of the light pulse will travel at different speeds. It broaden the pulse as it propagates along the fiber. The rms pulse broadening due to material dispersion [26] is given by

$$\sigma_m \approx -\frac{\sigma_\lambda L}{c_0} \lambda \frac{d^2 n_1}{d\lambda^2} \tag{4.46}$$

where σ_λ represents the rms spectral width of the source. Sometimes, material dispersion is given in terms of the material *dispersion parameter* D which is defined as

$$D = -\frac{\lambda}{c_0} \frac{d^2 n_1}{d\lambda^2} \tag{4.47}$$

It is often expressed in ps nm^{-1} km^{-1} in fiber data sheets. Therefore,

$$\sigma_m \approx \sigma_\lambda L D \tag{4.48}$$

Variations of parameter D with λ for pure silica [5] is shown in Fig. 4.15.

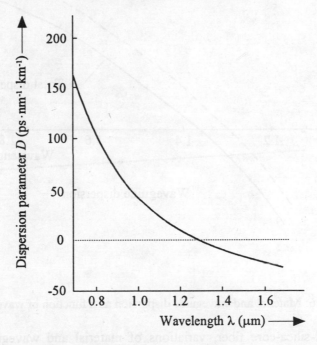

Fig. 4.15: Material dispersion parameter D as a function of wavelength λ for silica fiber

It may be observed that material dispersion tends to zero in the longer wavelength region around 1.3 μm. Fortunately, this is a wavelength where fiber attenuation can be quite low ($\approx$ 0.5 dB/km).

For this reason, 1.3 µm became very popular wavelength in fiber optic communication systems. Also, even in the shorter wavelength region, substantial reduction in the pulse broadening due to material dispersion can be achieved by using an injection laser with a narrow spectral width.

(b) Waveguide Dispersion

The energy guided by the fundamental mode is divided between core and cladding. For example, a SM fiber only confines about 80 percent of the optical power to the core [15]. The propagation velocity of a mode depends on the distribution of its energy between core and cladding. In the single-mode regime ($V < 2.405$) as the wavelength of light increases, the field extends farther into the cladding and the propagation constant increases or the propagation velocity decreases. Therefore, the use of a light source of varying wavelengths will give rise to dispersion [11]. The amount of waveguide dispersion depends on the fiber design since the modal propagation constant is a function of a/λ.

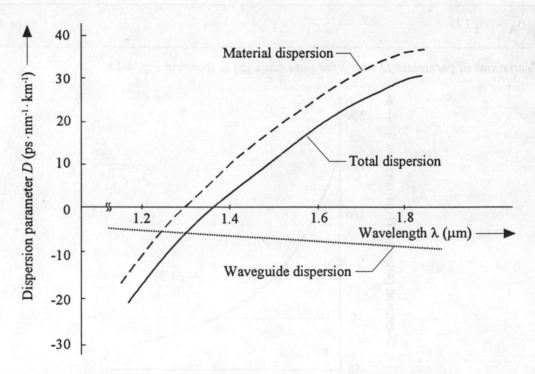

Fig. 4.16: Material and waveguide dispersion as a function of wavelength λ

For a SM fused-silica-core fiber, variations of material and waveguide dispersion with wavelength is shown in Fig. 4.16. In the same figure, total dispersion under the presumption that material and waveguide dispersion are additive is also shown. It is observed from this figure that the total dispersion can be reduced to zero at a particular wavelength through the mutual cancellation of material dispersion and waveguide dispersion. It may be mentioned that the particular wavelength at which the total dispersion is reduced to a minimum can be selected anywhere in the 1.3 μm - 1.7 μm spectral range through the proper fiber design.

4.4.3   POLARIZATION DISPERSION

Single-mode fibers support two orthogonally polarized degenerate modes (Chapter five). In a practical SM fiber, various perturbations in the fiber geometry and composition are present. The perturbations may occur at the time of fiber manufacturing and cabling and are difficult to eradicate. These perturbations plus environmental disturbances such as strain can remove the degeneracy of modes. In that case, orthogonally polarized modes will have different propagation constants which will lead to pulse broadening. This form of dispersion is called polarization mode dispersion δT_p. It is given by

$$\delta T_p = T_{px} - T_{py} \tag{4.49}$$

where T_{px} and T_{py} are the propagation delay for the two orthogonal polarization modes. Measured values of δT_p is significantly less than 1 ps km^{-1} in conventional SM fibers. Therefore, it is considered to be negligibly small as compared to other dispersion in the system design [23], particularly when the data rate is less than 10 Gbit/s.

4.4.4   OVERALL FIBER DISPERSION

In this Section, overall fiber dispersion for MM and SM fibers is discussed.

(a) Multimode Fibers

The overall dispersion in MM fibers comprises both intramodal and intermodal terms. The total rms pulse broadening σ_T is given by

$$\sigma_T = \sqrt{\sigma_c^2 + \sigma_n^2} \tag{4.50}$$

where σ_n is the rms intermodal dispersion and σ_c the rms intramodal or chromatic dispersion. The intramodal term σ_c consists of pulse broadening due to both material and waveguide dispersion. In MM fibers, waveguide dispersion is generally negligibly small as compared to material dispersion and therefore $\sigma_c \approx \sigma_m$.

(b) Single-Mode Fibers

In SM fibers, pulse broadening results almost entirely from the intramodal dispersion as only a single-mode is allowed to propagate and intermodal dispersion is almost zero. Of course, there will be polarization dispersion also. But it is negligibly small at a data rate of less than 10 Gbit/s. Unlike MM fibers, the mechanisms giving intramodal dispersion in SM fibers are interrelated in a complex manner [26].

4.5   FIBER DESIGN AND SELECTION ISSUES

Telecommunication companies use SM fibers as a principal optical transmission medium in their long-haul point-to-point links. Such fibers are quite important in many optical network applications also. In this Section, some important types of SM fibers are discussed.

4.5.1   DISPERSION-SHIFTED FIBERS

In the design of SM fibers, dispersion behaviour is a major distinguishing feature which limits long-distance and high-speed transmission. As mentioned earlier, dispersion of a SM silica fiber is lowest at 1.3 μm, but the attenuation is minimum at 1.55 μm. At 1.55 μm dispersion is higher. Ideally, for achieving a maximum transmission distance in a high-capacity link, dispersion null should be at the wavelength of minimum attenuation. This may be achieved by mechanisms like reduction in fiber core diameter with an accompanying increase in the relative or fractional RI difference or altering material composition of the fiber to create dispersion-shifted fibers (DSFs).

A wide variety of SM fiber RI profiles are capable of shifting the zero-dispersion wavelength point λ_0 to a specific wavelength. In the simplest case, the step index profile shown in Fig. 4.17 gives a shift to longer wavelength by reducing the core radius and increasing the fractional index difference. Typical values for the two parameters are 2.2 μm and 0.012 respectively [1]. The λ_0 could be shifted to longer wavelength by altering the material composition of the SM fiber also. For suitable power confinement of the fundamental mode, the V-number should be maintained in the range of 1.5 to 2.4. To keep V-number constant, fractional index difference must be increased as a square function whilst the core diameter is linearly reduced. The increase in fraction index difference can be achieved by substantially increasing the level of germanium doping in the fiber core. Such fibers may cause excess optical loss of the order of 2 dB/km.

Several graded RI profiles can be used for DSFs. Two such profiles are shown in Fig. 4.18. The triangular profile shown in Fig. 4.18a is the simplest one and was the first to exhibit the same loss (i.e., 0.24 dB/km) at a wavelength of 1.55 μm as conventional non-shifted SM fibers [6]. Furthermore, it provides an increased MFD which assists in fiber splicing. However, this design cause the LP_{11} mode to cut-off in the 0.85 μm - 0.9 μm wavelength region. Thus the fiber must be operated far off from the cut-off which produces sensitivity to bend induced losses (in particular microbending) at the 1.55 μm wavelength. The segmented-core triangular profile design shown in Fig. 4.18b reduces the sensitivity to microbending by shifting the LP$_{11}$ mode cut-off to longer wavelength whilst maintaining a MFD of around 9 µm at a wavelength of 1.55 µm. Such fibers are commercially available and have been utilized in the telecommunication networks with losses as low as 0.17 dB/km at 1.55 µm.

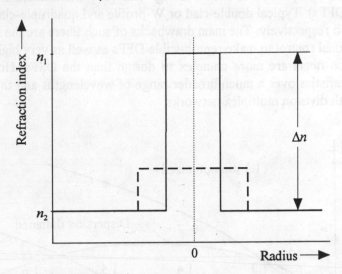

Fig. 4.17: Refraction index profile of a step index dispersion-shifted fiber (solid) and conventional non-shifted fiber (dashed)

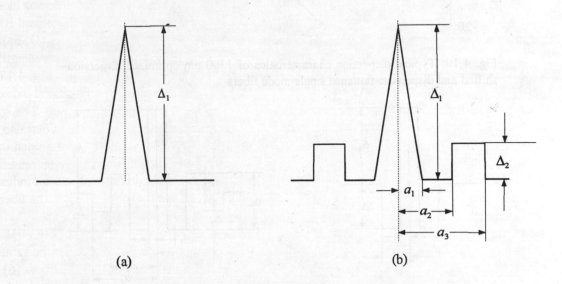

(a)   (b)

Fig. 4.18: Typical refraction index profiles for dispersion-shifted fibers (a) triangular and (b) segmented-core triangular

4.5.2   DISPERSION-FLATTENED FIBERS

By making a more complex index profile, it is possible to have SM fibers which give two wavelengths of zero dispersion as shown in Fig. 4.19 [15]. Such fibers are called dispersion-flattened fibers (DFFs). Typical double-clad or W-profile and quadruple-clad profile are shown in Fig. 4.20a and b respectively. The main drawbacks of such fibers are the requirement of high degree of dimensional control to make reproducible DFFs as well as very high sensitivity to fiber bend losses. Such fibers are more complex to design than the DSFs. However, DFFs offer desirable characteristics over a much broader range of wavelengths and thus could be used in optical wavelength division multiplex networks.

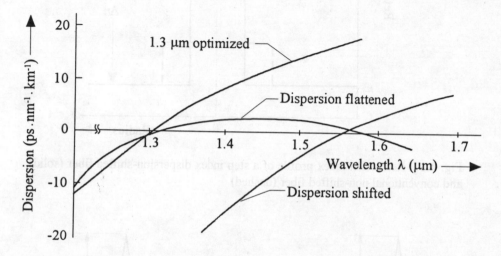

Fig. 4.19: Typical dispersion characteristics of 1300 nm optimized dispersion-shifted and dispersion-flattened single-mode fibers

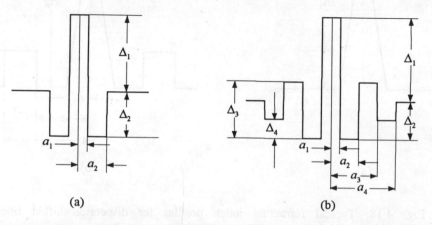

Fig. 4.20: Refraction index profile (a) double clad or W-profile and (b) quadruple-clad profile for dispersion-flattened fibers

4.5.3 DISPERSION-COMPENSATING FIBERS

In many countries, existing millions of kilometres of conventional SM fiber in the underground ducts operate at 1.3 μm. If these fibers are operated at 1.55 μm, there will be a significant residual dispersion. Further, replacing these fibers by dispersion-shifted fibers would involve huge costs. A practical approach to this problem is to use a fiber whose dispersion coefficient D is negative and large at 1.55 μm. These fibers are known as dispersion-compensating fibers (DCFs). A short length of DCF can be used in conjunction with the 1.3 μm optimized fiber length so as to have much lower total dispersion over the entire length. It may be mentioned that a few hundred metres to a kilometre of DCF can be used to compensate for dispersion over tens of kilometres of the fiber length.

To achieve a very high value of negative D, the core of the compensating fiber has to be doped relatively higher as compared to conventional fibers. Unfortunately, the fiber attenuation coefficient α increases due to this doping. Hence for DCFs, a figure of merit (FOM) is defined which is the ratio of the dispersion coefficient to the fiber attenuation coefficient in dB/km and has a unit of ps/(dB nm).

$$FOM = \frac{|D|}{\alpha} \qquad (4.51)$$

4.5.4 POLARIZATION-MAINTAINING FIBERS

An electromagnetic wave contains two fields-one electric and other magnetic oscillating perpendicular to each other and propagating in a direction perpendicular to both. Ordinary unpolarized light is made up of many waves with their electric and magnetic fields oriented randomly (although perpendicular to each other for each wave). If all the electric fields (and hence the magnetic fields as well) are aligned parallel to one another, the light will be linearly polarized. The light can also be circularly and elliptically polarized if the electric field vectors rotate in a circle or ellipse respectively. In optical fibers, each propagating mode will have its own polarization properties and hence light in a MM fiber will be of mixed polarization. To retain the light in a definable polarization state, a SM fiber must be used. In SM fiber, the single mode that propagates is doubly degenerated with two orthogonal polarizations, the LP_{11}^x mode and LP_{11}^y mode. For a truly symmetric SM fiber, these modes have almost identical propagation velocities and polarization state is maintained. However, as soon as any cross-sectional asymmetry is encountered, the degeneracy is removed. All practical fibers have some asymmetry due to non-circular core, linear strain or twist strain in the core. Fiber bending will introduce linear strain, while twisting will cause circular strain. The mechanism of polarization fluctuations in practical fibers is discussed in detail in Chapter five.

The single-mode single-polarization (SMSP) fibers are designed to maintain the polarization of the launched wave. By deliberately introducing birefringence, two waves have significantly different propagation characteristics. This keeps the waves from exchanging the energy during propagation through the fiber. The SMSP fibers are fabricated by designing asymmetries into the fiber. Examples include fibers with elliptical cores (which cause waves polarized along the major and minor axes of the ellipse to have different effective refraction indices) and fibers that contain non-symmetrical stress-producing mechanism. These are shown in Figs. 4.21a and 4.21b respectively. The shaded region in the bow-tie fiber is highly doped with a material such as boron. The thermal expansion of this doped region is different from that of the pure silica cladding. As a consequence, a non-symmetrical stress is exerted on the core. It produces stress-induced birefringence which in turn decouple the two orthogonal propagating modes in the SM fibers [22].

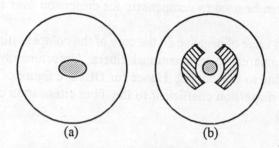

(a)   (b)

Fig. 4.21: (a) Elliptical core and (b) bow-tie polarization-preserving fibers

The SMSP fibers are required in applications like fiber optic gyroscope, coherent optical communications and these are used as fiber pigtails in polarization-sensitive components e.g., in external waveguide modulators. A more detailed description on SMSP fibers and polarization-control devices is given in Chapter five.

4.5.5   FIBER SELECTION

In the following, some issues relating to the fiber selection are discussed.

(i) MULTIMODE GRADED-INDEX AND STEP INDEX FIBERS

The MM-GI fibers can transmit at higher information rates than the MM-SI fibers. The SI fibers to source coupling is normally more efficient even if losses for the two fibers are the same. The GI fibers are designed for low dispersion, making them appropriate for medium distance and high data rate applications.

(ii) MULTIMODE AND SINGLE-MODE STEP INDEX FIBERS

Some systems perform adequately with MM-SI fibers. These are easier to handle than the systems with SM fibers. The advantage of SM-SI fibers is their large information carrying capacity resulting from the absence of modal dispersion. Long-haul and large information capacity systems require these fibers.

(iii) MATERIALS

The materials generally available for choice are glass, plastic cladded glass and plastic. Glass has the lowest attenuation, making it the choice for long-haul communication systems. Although plastic cladded glass fibers have higher losses, their larger NAs make source to fiber coupling more efficient. Such fibers are used for a moderate distance. Plastic fibers because of their higher losses and larger NAs are convenient to use for a short distance.

(iv) OPERATION WAVELENGTH

Operation in the shorter wavelength range (800 nm - 900 nm) has been quite practical. Though attenuation and pulse spreading in this range are high, still low enough for short distance and low data rate systems. Moreover, sources and detectors are readily available in this range. These are quite robust and inexpensive.

In the longer wavelength range (1300 nm - 1350 nm), dispersion is almost absent in SM fibers and very less (only intermodal dispersion) in MM fibers. The low dispersion makes this range useful in medium distance and moderate data rate systems.

In the still longer wavelength range (1540 nm - 1560 nm), SM dispersion-shifted and dispersion-flattened fibers have low attenuation and dispersion which make this range very attractive in long distance and high data rate systems.

4.6   FIBER MATERIALS, FABRICATION AND CABLING

In this Section, fiber materials, fabrication and cabling process are briefly discussed.

(i) FIBER MATERIALS

The material used for fiber manufacturing must satisfy the following requirements: (i) it must be possible to make long, thin and flexible fibers, (ii) the material must be transparent (implying low attenuation) at some wavelength and (iii) physically compatible materials with slightly different RI for the core and cladding must be available. Materials which satisfy the above requirements are glass and plastic. The majority of fibers are made of glass consisting either of silica (SiO_2) or a silicate. Plastic fibers are not so widely used because of their substantially higher attenuation coefficients as compared to glass fibers. The main application of plastic fibers is in short distance communications and in abusive environments where the greater mechanical strength of plastic fibers offers an advantage over the glass fibers.

(a) Glass Fibers

Most of the glass fibers are made of *silica*, which has a RI of 1.458 at 850 nm. The principal raw material for silica is sand. To produce two similar materials having slightly different RIs for core and cladding, dopants such as B_2O_3, GeO_2 or P_2O_5 are added to the silica.

Major constituents of a fluoride glass fiber referred to as ZBLAN (after its elements) are ZrF_4-54%, B_aF_2-20%, LaF_3-4.5%, AlF_3-3.5% and NaF-18%. This material forms the core of the fiber. To make a lower RI glass, ZrF_4 is partially replaced by HaF_4 to get a ZHBLAN cladding. Fluoride glass fibers offer intrinsic minimum losses of 0.01 to 0.001 dB/km. The main problems with these fibers are that the fabrication of long length is difficult and very precise control during fabrication is required. Otherwise, there may a drastic increase in the scattering loss and hence fiber attenuation coefficient [15].

Glass core and glass cladding fibers have wide applications in long distance links wherein very low loss achievable in these fibers is needed. Some parameters of typical glass fibers are given below.

Core diameter:	5 μm - 10 μm for SM
	30 μm - 100 μm for MM
	(50 μm, 62.5 μm, 85 μm, 100 μm)
Cladding diameter:	Generally 125 μm for SM
	125 μm - 500 μm for MM
Numerical aperture:	0.16 - 0.5 for MM
Attenuation coefficient:	0.17 dB/km - 2 dB/km

(b) Plastic Clad Glass Fibers

Plastic clad glass fibers are less expensive as compared to glass core and glass cladding fibers. These fibers are used in applications where higher losses are tolerable. Such fibers are composed of silica core with the lower RI cladding being a polymer (plastic). Plastic clad fibers are multimode and have either a step index or graded-index profile. These are often referred to as plastic clad silica (PCS) fibers.

In these fibers, large difference in the core and cladding RIs results in a high *NA*. This allows low cost and large area light sources to be used for coupling optical power into these fibers. It leads to comparatively inexpensive and lower data rate systems for many short distance applications. Typical parameters of GI PCS fibers are as follows [26].

Core diameter:	50 μm - 100 μm
Cladding diameter:	125 μm - 150 μm
Numerical aperture:	0.2 - 0.5
Attenuation coefficient:	1 dB/km - 100 dB/km

(c) Plastic Fibers

All plastic or polymeric fibers are exclusively MM-SI type with large core and cladding diameters. Although such fibers exhibit higher signal attenuation, these are usually cheaper to produce and easier to handle than the glass fibers. The higher RI difference that can be achieved between the core and cladding materials yields NAs as high as 0.6 and large acceptance angle up to 34°. It allows easier coupling of light into the fiber from a multimode source. Therefore, LEDs in conjunction with the less expensive plastic fibers make economically attractive systems. Typical parameters of plastic fibers are given below.

Core diameter:	200 μm - 600 μm
Cladding diameter:	450 μm - 1000 μm
Numerical aperture:	0.25 - 0.6
Attenuation coefficient:	50 dB/km - 300 dB/km

(ii) FIBER FABRICATION

There are basically two approaches which are followed to produce optical fibers

- Preform fabrication and then fiber drawing and
- Direct drawing from the melt.

The classification of fiber production processes is shown in Fig. 4.22. The first approach is most widely used for the production of (i) high quality silica fibers and (ii) plastic clad silica fibers.

For making preform, either vapour phase reaction/oxidation or sol-gel process can be used. In the vapour phase oxidation process as shown in Fig. 4.23, highly pure vapours of silicon tetrachloride ($SiCl_4$) and germanium tetrachloride ($GeCl_4$) are burnt in a flame producing a fine soot of silica (SiO_2) and germanium (GeO_2). The soot or particles are then collected on the surface of a bulk glass by one of the four different commonly used processes (refer to Fig. 4.22).

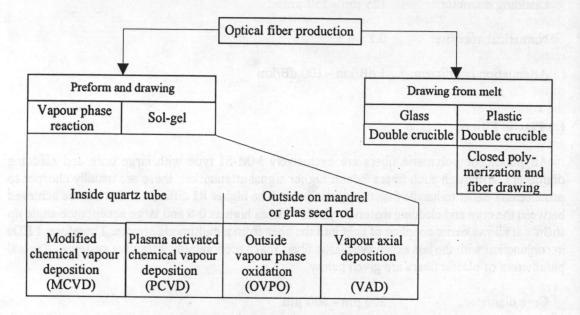

Fig. 4.22: Classification of fiber production processes

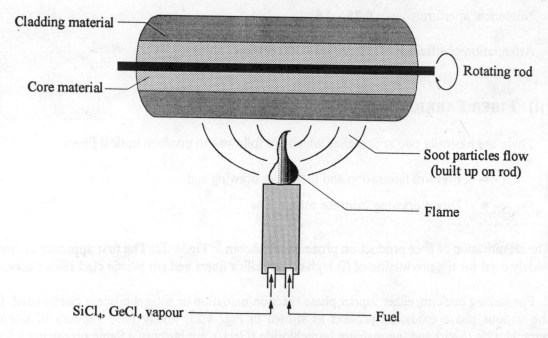

Fig. 4.23: Outside vapour phase oxidation process to make a preform

In the OVPO process (refer to Fig. 4.23), first of all glass is deposited on a rotating rod, then the composition is changed to deposit the cladding. Finally, the rod is removed and the remaining glass is heated and collapsed into a dense glass to form a clear glass rod or tube (depending upon the process). This rod or tube is called preform. It is typically 10 mm to 25 mm in diameter and 60 cm to 120 cm long [15]. In the sol-gel process, following steps are involved (i) hydrolysis of metal alkoxides to make a gel containing water and a solvent (methanol) e.g., tetra methoxysilane [$Si(OCH_3)_4$]:water:methanol in molar ratio 1:4:4.5 cast into cylindrical glass containers, (ii) dying of the gel to form a porous gel body (one week at 70° C), (iii) a chlorination process to reduce the initially high (~1000 ppm) OH content and (iv) sintering of the porous gel body to produce a transparent glass preform [2]. The main advantages of this method are low temperature and potential of mass production. The process has been used so far in the laboratory only. Fibers are made from the preform through fiber drawing set-up. In the set-up as shown in Fig. 4.24, the preform is mounted vertically in a furnace and heated until molten glass can be pulled from it in a fine fiber. The fiber diameter is monitored and the fiber is coated with a protective plastic layer as it is pulled from the preform. As fibers produced from preform based on the VAD process show minimum attenuation, this process is commonly used in practice.

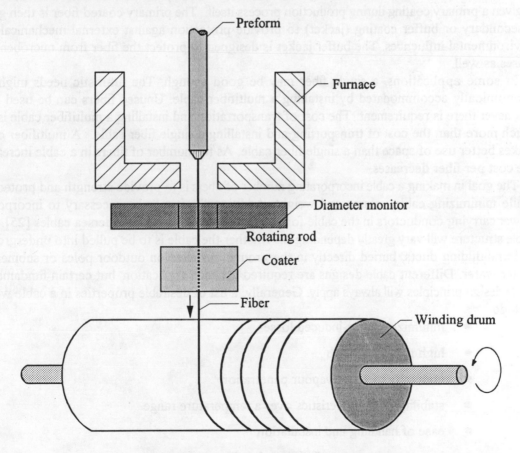

Fig. 4.24: Process of drawing glass fibers from preform

The second method relies on melting basic raw materials and pulling the optical fiber directly from the melt. This method can be used for both glass and plastic fibers. In this method, glass rods for the core and cladding materials are first made separately by melting mixtures of purified powders to make the appropriate glass composition with lower melting temperature than the pure silica. These rods are than used as feedstock for each of two concentric crucible. The inner crucible contains molten core glass and the outer one the cladding glass. The fibers are drawn continuously from the molten state through orifices in the bottom of the crucibles. Although this method often referred to as double-crucible method has the advantage of being a continuous process, due precautions must be taken to avoid contamination during the melting process. In case of plastic fibers, due to lower melting points of the starting materials than the glass, variant of this method called "closed polymerization and fiber drawing" is used [2].

(iii) FIBER CABLES

In order to prevent abrasion of the glass material and subsequent flaws in the material, a fiber is given a primary coating during production process itself. The primary coated fiber is then given a secondary or buffer coating (jacket) to provide protection against external mechanical and environmental influences. The buffer jacket is designed to protect the fiber from microbending losses as well.

In some applications, a single fiber may be good enough. The futuristic needs might be economically accommodated by installing a multifiber cable. Unused fibers can be used later whenever there is requirement. The cost of transportation and installing a multifiber cable is not much more than the cost of transporting and installing a single fiber cable. A multifiber cable makes better use of space than a single fiber cable. As the number of fibers in a cable increases, the cost per fiber decreases.

The goal in making a cable incorporating number of fibers is to provide strength and protection while minimizing cable volume and weight. Additionally, it may be necessary to incorporate power carrying conductors in the cable for some applications such as undersea cables [25]. The cable structure will vary greatly depending on whether the cable is to be pulled into underground or intrabuilding ducts, buried directly in the ground, installed on outdoor poles or submerged under water. Different cable designs are required for each application, but certain fundamental cable design principles will always apply. Generally, a list of desirable properties in a cable would include

- minimized stress-induced losses
- high tensile strength
- immunity to water vapour penetration
- stability of characteristics over a temperature range
- ease of handling and installation
- low acquisition, installation and maintenance costs

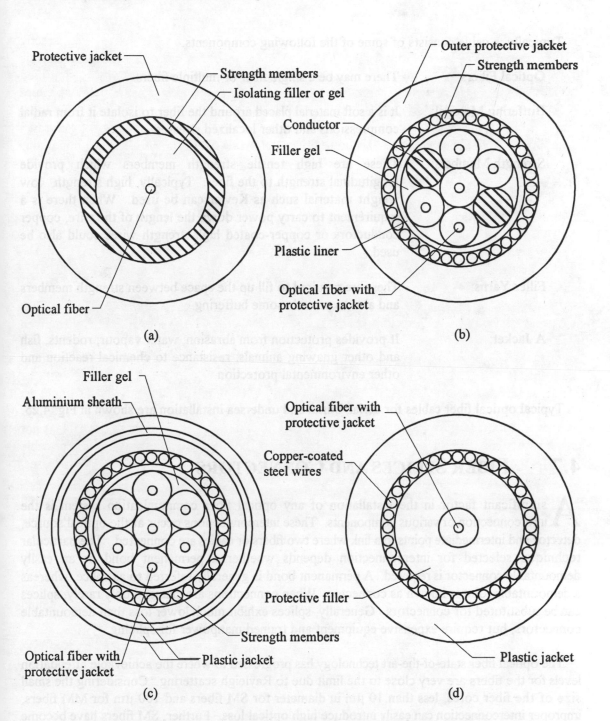

Fig. 4.25: Typical structures of (a) single fiber cable, (b) aerial multifiber cable,
(c) underground multifiber cable and (d) undersea single fiber cable

Typically, a cable consists of some of the following components.

Optical Fibers: There may be a single fiber or multiple fibers.

Buffering Material: It is a soft material placed around the fiber to isolate it from radial compressions and other localized stresses.

Strength Members: These are high tensile strength members which provide longitudinal strength to the fiber. Typically, high strength, low weight material such as Kevlar can be used. When there is a requirement to carry power down the length of the wire, copper conductors or copper-coated high-strength wires could also be used.

Filler Yarns: These are required to fill up the space between strength members and also to provide some buffering.

A Jacket: It provides protection from abrasion; water vapour; rodents; fish and other gnawing animals; resistance to chemical reaction and other environmental protection.

Typical optical fiber cables for aerial, burial and undersea installation are shown in Fig. 4.25.

4.7   FIBER SPLICES AND CONNECTORS

A significant factor in the installation of any optical fiber communication system is the interconnection of various components. These interconnections occur at the optical source, detector and intermediate points in a link where two fibers or cables are connected. The particular technique selected for interconnection depends whether a permanent bond or an easily demountable connector is required. A permanent bond is generally referred to as splice, whereas a demountable joint is known as connector. When connections are changed only rarely, splices can be substituted for connectors. Generally, splices exhibit much lower loss than demountable connectors, but require expensive equipment and trained manpower for splicing.

The optical fiber state-of-the-art technology has progressed to where the achievable attenuation levels for the fibers are very close to the limit due to Rayleigh scattering. Considering the small size of the fiber cores, less than 10 µm in diameter for SM fibers and 100 µm for MM fibers, improper interconnection can easily introduce high optical loss. Further, SM fibers have become the favourite choice for most of the light wave system being designed for telecommunication networks and in the future may be used in local area networks because of their practically unlimited bandwidth. To provide low loss connectors and splices for these SM fibers, alignment accuracies in the submicrometre range are required and the alignments must be reliable and cost effective. To achieve these goals is a challenging task for the R&D personnel.

In this Section, technologies used for both demountable connectors and splices are reviewed. As SM fibers require the greatest precision, more emphasis has been placed on the interconnection of these fibers. Factors causing losses in both connectors and splices can be divided into two types i.e., extrinsic and intrinsic losses. The extrinsic losses arise due to factors external to the fibers such as mechanical misalignments, end-face quality and reflections at the fiber-air interfaces.

Basically there are three types of misalignments viz., longitudinal, lateral and angular (Fig. 4.26). In longitudinal misalignment, fibers have same axis with a gap s between the end faces. Lateral misalignment (or axial misalignment) occurs when the lateral axes of the two fibers are separated by a distance d. In angular misalignment, fiber end faces are not parallel and fiber axes make an angle θ. Whenever there is a longitudinal misalignment (implying a gap between the fibers), there will be loss due to Fresnel reflections in addition to the misalignment loss. In any fiber joint, the fiber ends must be smooth and perpendicular to the fiber axis. The next step of aligning the fiber ends is very crucial because any kind of misalignment or the combination of misalignments would lead to a transmission loss. The intrinsic losses arise due to intrinsic factors like mismatches in fiber core diameters, index profiles, ellipticity of the cores etc. At a splice, major contributors to the loss are lateral offset, mismatches in core radii and relative core-cladding index difference (and hence NA).

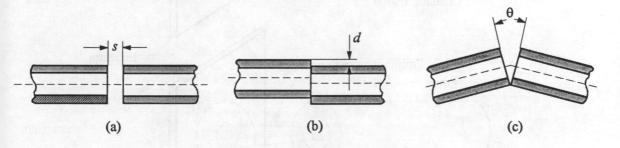

Fig. 4.26: Mechanical misalignments between two fibers (a) longitudinal (b) lateral and (a) angular

In addition to above extrinsic and intrinsic factors, there may be loss due to poor coupling efficiency. The coupling efficiency of a joint depends on the characteristics of source such as centre wavelength and coherence length. In case of MM fibers, it may also depend on the relative location of the optical source and the joint as well as the characteristics of fiber such as mode mixing and differential mode attenuation.

For example, let us consider a transmitting fiber which is to be coupled to an identical receiving fiber under two conditions: (i) when all modes are equally excited (Fig. 4.27a) and (ii) when only lower-order modes are excited (Fig. 4.27b). In the first case, emitting beam fills the entire *NA* of the fiber. For the receiving fiber (not shown in the figure) to except all the emitted energy, it must be in perfect alignment with the transmitting fiber and their characteristics viz., geometrical and waveguide should match. In the second case which normally occurs when there is a steady-state modal equilibrium, input *NA* of the receiving fiber is larger than the equilibrium *NA* of the transmitting fiber. Therefore, in this case requirements of mechanical alignment and matching of characteristics are not so critical. It may be mentioned that a steady-state modal equilibrium generally occurs in long fiber lengths. In the estimation of joint loss between two long fibers, calculations based on uniform modal power distribution may be too pessimistic. However, the estimate may be too optimistic if modal equilibrium is presumed.

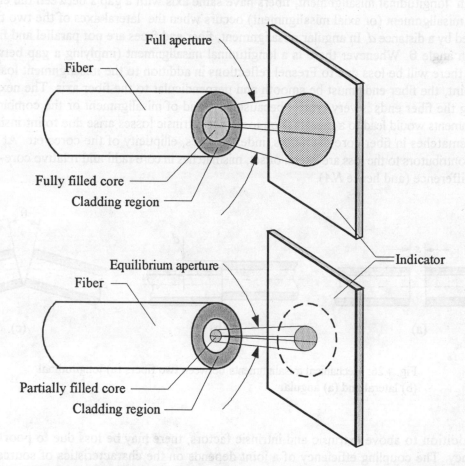

Fig. 4.27: Modal distribution of the optical beam (a) when all modes are equally excited and (b) when modes are in steady state

4.7.1   EXTRINSIC LOSSES

On the basis of various experiments and analytical models, variations of coupling loss as a function of longitudinal, lateral and angular misalignments for both GI-MM and SI-SM fibers are shown in Figs. 4.28 and 4.29 respectively.

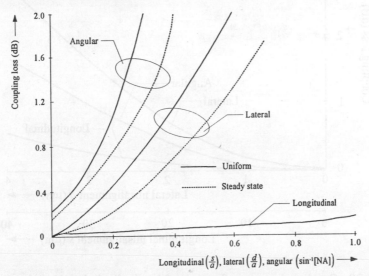

Fig. 4.28: Coupling loss as a function of normalized longitudinal, lateral and angular misalignments for GI-MM fibers

In MM fibers, effect of the misalignments for both uniform and steady-state launch conditions are included [17]. It can be seen from Fig. 4.28 that the coupling loss for MM fibers is significantly more sensitive to lateral misalignment than the longitudinal misalignment. Further, for SM fibers (refer to Fig. 4.29) lateral misalignment loss is more than the longitudinal and angular misalignments. Therefore, for low loss joint (≤ 0.5 dB), lateral offsets must be controlled to submicrometre accuracies for SM fibers and to micrometre accuracies for MM fibers.

With regard to fiber end-face quality, surface finish that is generally acceptable for low loss connectors is about 0.025 µm for non-index matching connectors and can be as much as 0.18 µm for connectors employing index matching fluid. The acceptable tolerance in end-face angle (i.e., the angle between the end-face and the plane perpendicular to the axis of the fiber), however depends on a particular application. An acceptable value for the end-face angle may be less than $\pm 1°$. Although cleaved ends are generally used in splicing, lately there has been a trend to employ cleaved fibers in connector installation also.

The last extrinsic factor to be discussed is the effect of reflections on the coupling efficiency. Fresnel reflections occur when there is an air gap between the adjacent fibers. For two air-glass interfaces, the loss amounts to about 0.3 dB for silica fibers. It may be mentioned that Fresnel reflections loss can be minimized by using an index-matching medium (fluid, gel etc.) between the fiber end faces. This approach is quite common in the splicing of fibers and usually not followed in demountable connectors due to practical considerations like cleanliness and contamination.

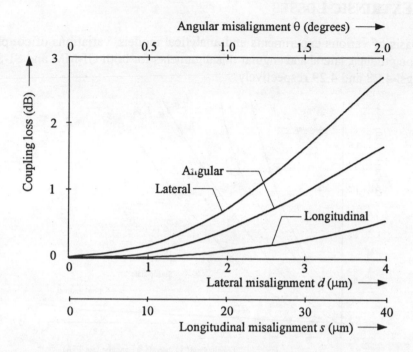

Fig. 4.29: Coupling loss as a function of longitudinal, lateral and angular misalignments for SM fibers with a MFD of 10 μm

4.7.2 INTRINSIC LOSSES

The intrinsic factors have a more significant effect on the loss of MM fibers joints than on SM fiber joints. However, the extrinsic features have the opposite effect i.e., these are less significant in MM fibers than in SM fibers. In the SM fibers, difference in the characteristics of the fibers have practically insignificant effect on the coupling efficiency as compared to the effect caused by lateral misalignment. Also, with practical SM fiber joints, the law of reciprocity applies. Therefore, the coupling efficiency is independent of the direction of propagation through the joint. On the other hand, in case of MM fiber joints one must consider the characteristics of the fibers on either side of the joint and the direction of propagation of the optical power through the joint. Dependence of the coupling loss on mismatches in NAs and core radii for GI-MM fibers is shown in Fig. 4.30. It may be mentioned that this figure shows the dependence of loss as the optical power propagates from a fiber of larger NA to smaller NA and from larger core radii to smaller core radii. There is no associated loss when the direction of propagation is from smaller to larger NAs and core radii.

Unfortunately for MM fiber joints, there exists no unique insertion loss value. The optical source, number of joints and their location along the fiber, mode-mixing properties and differential mode attenuation of fibers all play an important role in the performance of joints. Compared to MM fibers, it is straight forward to evaluate and predict the insertion loss of SM fiber joints [17].

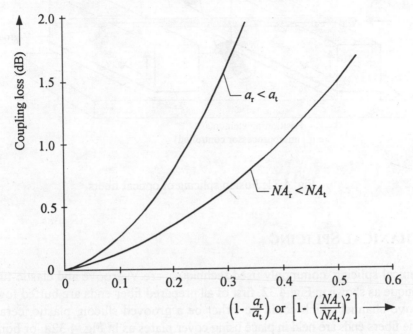

Fig. 4.30: Coupling loss as a function of mismatch in NAs and core radii for GI-MM fibers

4.7.3   FIBER SPLICING

In joining single fibers or arrays of fibers, one has to make use of a splicing mechanism. The splicing mechanisms presently being used can be characterized as (i) fusion type and (ii) mechanical type. These are discussed below.

(i)   FUSION SPLICING

In the fusion splicing, first step is the preparation of fiber ends which should be smooth and perpendicular to the fiber axis. These ends are then prealigned in a V-groove and butted together as shown in Fig. 4.31. Subsequently, precision alignment is made by positioning elements which can be controlled mechanically, by piezo-electric devices or automatically by means of a microprocessor. Then butt joint is heated with an electric light arc or a laser pulse so that fiber ends are momentarily melted and bonded together. This technique can be used to splice both SM and MM fibers with splice loss less than 0.1 dB. Further, fused fibers can be recoated to nearly the original coated dimension of fiber as is done for factory splicing of fibers prior to cabling.

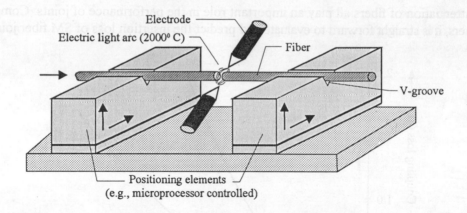

Fig. 4.31: Fusion splicing of optical fibers

(ii) MECHANICAL SPLICING

In mechanical splicing, commonly used techniques are V-groove and elastic-tube. In the V-groove technique as shown in Fig. 4.32, first of all prepared fiber ends are butted together in a V-shaped groove channel which could be either be a grooved silicon, plastic, ceramic or metal substrate. The fibers ends are held in place using cover plates as in Fig. 4.32a or bonded together with an adhesive. The splice loss in this technique depends largely on the parameters like outside dimension of the fiber, core diameter and core eccentricity to the centre of fiber.

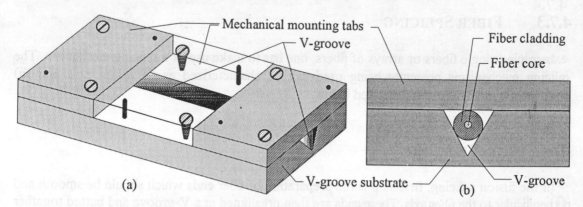

Fig.4.32: V-groove optical fiber splice (a) top view (without fiber) and (b) side view (with fiber)

The elastic-tube technique as shown in Fig. 4.33 makes use of a capillary tube made of an elastic material. The central hole diameter is made smaller than the diameters of the fibers. The reason for this is that when the fibers for splicing are inserted in this hole, elastic material exerts a lateral symmetrical pressure on the fibers which yields a self-alignment (Fig. 4.33a). Further, the hole is tapered on both sides for easy fiber insertion. In this technique, fibers to be spliced need not to be equal in diameter as each fiber moves into position independently relative to the tube axis. A wide range of fiber diameters can be inserted into the elastic tube.

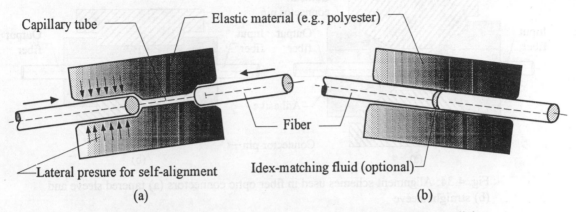

Fig. 4.33: Schematic of an elastomeric splice (a) during and (b) after splicing

The main disadvantage of mechanical splicing is the build up in the size of the fiber or cable being spliced. In some cases, increase in size can be tolerated, but in other cases it can not be. For example, when cables have a large fiber count, the enclosure required to contain mechanical splices can become excessively large.

4.7.4 FIBER CONNECTORS

Low-loss connectors for both MM and SM fibers require precise alignments to control the external factors. These alignments must also be maintained under various operating conditions such as shock, vibration, repetitive engagement and separation as well as when changes occur in the environmental conditions such as temperature and humidity. Some of the principal requirements of a good connector design are low coupling loss, interchangeability, low cost, reliability, ease of assembly and connection. To achieve these, two types of connector designs are available: the butt-joint and expanded-beam. These are described below.

(i) BUTT-JOINT CONNECTORS

These optical connectors employ a precision sleeve into which the connector pins fits. The fiber is epoxied with some adhesive in a precision hole which has been drilled into the pin. Two types of connectors shown in Fig. 4.34 are the most popular type of butt-joint connector used in both SM and MM fibers. In these connectors, sleeve length and its edges maintain the fiber ends separation. In the tapered sleeve connector, a tapered sleeve accept and guide the conical connector pin (Fig. 4.34a), while in the straight sleeve connector it is the cylindrical connector pin (Fig. 4.34b).

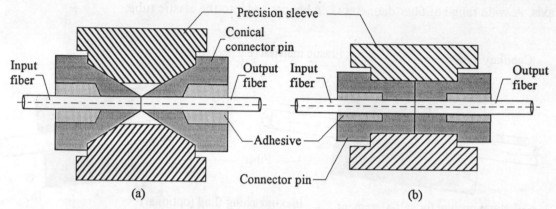

Fig. 4.34: Alignment schemes used in fiber optic connectors (a) tapered sleeve and (b) straight sleeve

(ii) EXPANDED-BEAM CONNECTORS

A typical expanded-beam connector shown in Fig. 4.35 consists of optical elements which collimate the beam radiated from the transmitting fiber and focus the expanded beam onto the core of the receiving fiber.

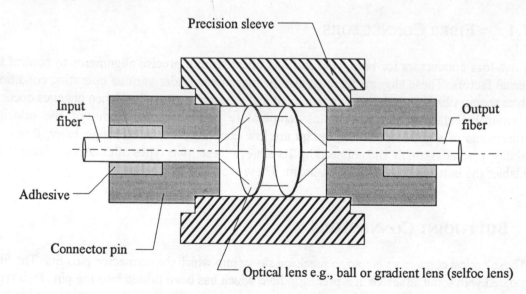

Fig. 4.35: Schematic of a typical expanded-beam connector

The first advantage of this connector is that it is more tolerant to lateral misalignment, dirt or any other contamination on the end faces. The second advantage arises due to incorporation of optical processing element for operation like beam splitting in the connector itself. Although expanded-beam design do not require alignments to submicrometre or micrometre accuracy, the fibers must be positioned with respect to the axes of the collimating and focussing element to the same accuracy. Also angular alignment of the expanded-beam is critical as misalignment cause the focussing point to be displaced laterally from the fiber core. For these reasons, major trade-off in butt-joint and expanded-beam connectors design is in the control of either lateral or angular alignment. Inspite of many advantages of expanded-beam connectors, majority of connectors used are of butt-joint type.

Some of the most common types of connectors used in practice are shown in Fig. 4.36. The SMA connector is generally used with MM fibers. A typical insertion loss is about 1.0 dB. But mechanical wear and the likelihood of fiber end surface becoming contaminated and abrased in working environments during a life (which may involve hundreds or thousands of disconnections and reconnections) cause this loss to increase considerably.

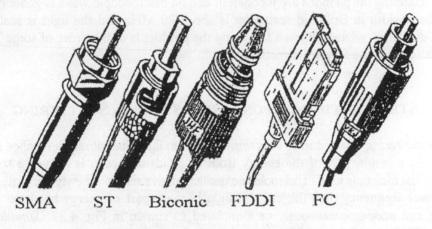

SMA   ST   Biconic   FDDI   FC

Fig. 4.36: Commonly used fiber optic connectors

With physical contact (PC) type connectors, great care to maintain cleanliness is essential for the proper connection. The ST™ connector is also intended for use with MM fibers. It is easy to terminate and employ a bayonet coupling collar somewhat similar to the BNC co-axial connector. It has a typical insertion loss of 0.5 dB. Some relatively inexpensive, biconical plastic snap-in connector has been developed and Type SC is one such connector. The ISO, the International Electrotechnical Commission and the Telecommunication Industry Association endorsed SC connector [10]. The SC and ST connectors offer quite similar optical attenuation and back-reflections. Both work with SM and MM fibers. The major differences are in mechanical characteristics that affect installation, packaging density and durability [10]. FDDI two fiber connector is meant for high capacity local area networks. The FC connector can be used with SM or MM fibers in high capacity systems. It is available in PC (physical contact) and APC (angled physical contact) forms for those applications which may be very sensitive to low levels of back-reflections (< -30 dB). It may give a coupling loss of as low as 0.2 dB [8].

4.8   NON-LINEAR EFFECTS IN FIBERS

Non-linear effects in optical fibers are due to either change in the refraction index (RI) of the medium with optical power or scattering phenomena. The power dependence of the RI is responsible for the *Kerr effect*. Depending upon the type of input signal, the Kerr non-linearity manifests itself in different effects such as *self-phase modulation* (SPM), *cross-phase modulation* (XPM) and *four-wave mixing* (FWM) [12]. The scattering phenomena are responsible for *Brillouin* and *Raman* effects At high power levels, these can induce stimulated effects and the intensity of scattered light grows exponentially once the incident power exceeds a threshold level. A fundamental difference between Brillouin and Raman scattering is that the Brillouin-generated phonons are coherent and give rise to a macroscopic acoustic wave in the fiber. In contrast to this, in Raman scattering the phonons are incoherent and no macroscopic wave is generated. Further, the gain bandwidth in Brillouin scattering is about 20 MHz and the light is scattered in the backward direction, while in Raman scattering the product is of the order of some THz and the light is scattered in the forward direction.

4.8.1   STIMULATED AND SPONTANEOUS RAMAN SCATTERING

Raman scattering arises due to interaction between light and vibration of silica molecules in the fiber. As a result, part of the energy from an incident photon is delivered to mechanical vibration of the molecule itself. This molecule oscillation give rise to a newly generated light wave with a lower frequency than the incident light because part of energy has been lost. Raman scattering can arise spontaneously or stimulated as shown in Fig. 4.37. *Stimulated Raman scattering* (SRS) is used for amplification purpose in Raman amplifiers discussed in Section 7.1.3. SRS is more or less isotropic i.e., a propagating light wave signal can be affected by other light wave signals e.g., in a WDM system, propagating in the same or in the opposite direction [9].

In quantum mechanical terms, Raman scattering can be described as a molecule absorbing the incident photon at the original frequency f_s whilst emitting a photon at a frequency f_R which is about 15 THz below f_s. If molecules were already exited, then photons at a frequency shifted above that of the incident photons are emitted. The former scattered wave is known as the *Stokes* whereas the latter as the *anti-Stokes* wave [26].

4.8.2   STIMULATED AND SPONTANEOUS BRILLOUIN SCATTERING

Brillouin scattering arises due to reflections at periodical variations in the density of fiber material. As a result, an incident photon of certain frequency f_s travelling at light velocity can produce a phonon travelling at sound velocity in the other direction and a Stokes photon with frequency f_B which is about 10 GHz below f_R. Thus, in a fiber essentially all of the scattered energy is counter propagating with respect to the signal. Brillouin scattering can arise spontaneously or stimulated as shown in Fig. 4.37. Stimulated Brillouin scattering (SBS) is extremely useful for narrowband amplification and is used in Brillouin amplifier discussed in Section 7.1.4. SBS may pose a problem in WDM systems where many transmitter lasers with narrow linewidths are used.

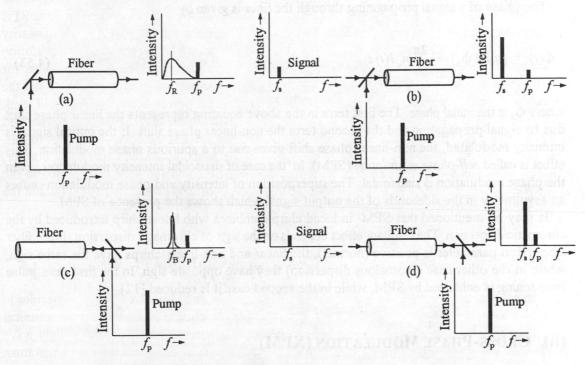

Fig. 4.37: (a) Spontaneous and (b) stimulated Raman, (c) spontaneous and (d) stimulated Brillouin Scattering

The *Rayleigh scattering* discussed earlier is elastic because the scattered wave has the same frequency as the incident wave. In contrast to this, non-linear scattering processes (SRS and SBS) are inelastic. Inelastic scattering does not only result in a frequency shift, but for sufficiently high incident power level it provides gain at the shifted frequency as shown in Fig. 4.37. The threshold power level required for SRS may be 100 mW to 200 mW, while for SBS it may be 10 mW.

4.8.3   KERR EFFECT

Due to the non-linear behaviour of fiber, the RI can be written as

$$n = n_1 + n_n I \tag{4.52}$$

where n_l and n_n are the linear and non-linear RIs respectively and I the field intensity. The different non-linear effects depends on the shape of the field injected into the fiber. In the following, the main effects due to the Kerr non-linearity are reviewed.

(i) SELF-PHASE MODULATION (SPM)

The phase of a signal propagating through the fiber is given by

$$\phi(t) = \left(n_l z + \phi_0\right) + \frac{2\pi}{\lambda} n_n I(t) z \qquad (4.53)$$

where ϕ_0 is the initial phase. The first term in the above equation represents the linear phase shift due to signal propagation and the second term the non-linear phase shift. If the optical signal is intensity modulated, the non-linear phase shift gives rise to a spurious phase modulation. This effect is called *self-phase modulation* (SPM). In the case of sinusoidal intensity modulation , even the phase modulation is sinusoidal. The superposition of intensity and phase modulation causes an asymmetry in the sidebands of the output signal which shows the presence of SPM.

It may be mentioned that SPM- induced chirp combines with linear chirp introduced by the chromatic dispersion. The resultant effect depends on the sign of chromatic dispersion. If the fiber dispersion parameter is positive (normal), the linear and non-linear chirps have the same sign, while in the other case (anomalous dispersion) they have opposite sign. In the first case, pulse broadening is enhanced by SPM, while in the second case it is reduced [12].

(ii) CROSS-PHASE MODULATION (XPM)

When a large number N of signals propagate though the fiber, the phase of the signal at frequency f_i depends not only on its power, but also on the power of the signals at other frequencies as well. It is given by

$$\Delta\phi_i(t) = \frac{2\pi n_n z}{\lambda} \left[I_i(t) + 2\sum_{j\neq i}^{N} I_j(t) \right] \qquad (4.54)$$

The first term in the above equation represents the contribution of SPM and the second term the XPM. The XPM causes a further non-linear chirp which interacts with fiber chirp as in the case of SPM. It can be deduced from the above equation that XPM is effective only when the interactive signals superimpose in time [12].

(iii) FOUR-WAVE MIXING (FWM)

When signals at different frequencies propagate through the fiber, besides XPM another effect called four-wave mixing (FWM) is also present. If three carrier frequencies f_1, f_2 and f_3 copropagate inside the fiber simultaneously, a fourth wave with frequency $f_4 = f_1 \pm f_2 \pm f_3$ is generated. Several frequencies corresponding to different plus and minus sign combinations are possible in principle. Normally, most of these frequencies do not build up due to phase matching requirement. The cross-product term like $f_1 + f_2 - f_3$ is the most trouble some for multichannel communication systems, especially when channel spacing is relatively small (~ 1 GHz) and the optical power per channel is high. The phase matching condition is nearly satisfied in this case. It may lead to power transfer among various channels. In a multichannel system, there are a number of such cross-product terms and severe contamination of any one channel. The term FWM comes from the fact that when there are three interfering waves, a fourth one is generated.

4.8.4   SOLITON TRANSMISSION

The existence of fiber soliton is due to a balance between chromatic dispersion (CMD) and SPM. The CMD broadens the pulse during propagation inside a fiber. As explained above, intensity dependent RI imposes a chirp on the optical pulse and because of that there may be pulse compression. Under certain conditions, these two effects will balance out and the optical pulse would then propagate undistorted in the form of soliton. In principle, solitons open the way to extremely high speed long distance light wave communication systems.

In soliton propagation, pulses must be of not only the right shape, but of the right amplitude too. It means that the pulses must be amplified periodically along the path. This, of course, can be achieved by using the in-line Erbium-doped fiber amplifiers (EDFAs). It may be mentioned that a perturbed soliton recovers its initial shape in its evalution along a fiber. What should exactly be that pulse shape? It is obvious that such a pulse is ought to be one that is richest in harmonics in the vicinity of the amplitude peak and very smoothly varying elsewhere. After intensive investigations reciprocal hyperbolic cosine is considered to the right shape [19].

4.9   FIBER BRAGG-GRATINGS

Fiber Bragg-gratings can be formed by using an ultraviolet (UV) source and an electron beam phase mask. The UV source can be either an argon fluoride (193 nm) or krypton fluoride (248 nm) excimer laser or a frequency doubled argon-ion laser (244 nm). A phase mask is essentially a binary grating in which the groove profile and depth have been specially designed to diffract most of the incident UV energy into the plus and minus first order diffraction orders while minimizing the energy in the zero and higher orders [12]. During fiber grating fabrication, phase mask diffracts incident UV light into the various diffraction orders (refer to Fig. 4.38).

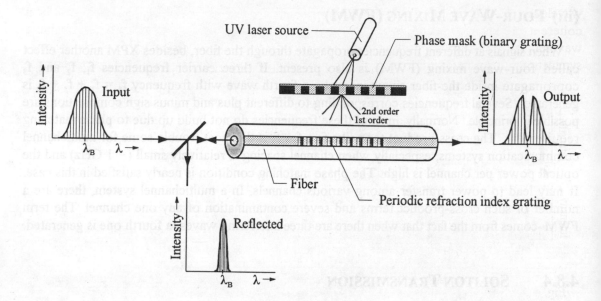

Fig. 4.38: Operation principle of fiber Bragg-grating

The interference of the plus and minus first order beams cause a fringe pattern to form in the near field. When a photosensitive fiber is placed in close proximity to the phase mask, permanent increase in the RI of the fiber, creating a periodic modulation of the RI in the fiber core can be achieved. RI increase of the order of 10^{-3} is possible. A fiber can be made photosensitive by doping the core with germanium, phosphorous or boron doping during fabrication of the preform. The fibers can also be sensitized by loading them with hydrogen under temperature and/or pressure i.e., by putting them in a chamber pressurized with hydrogen. This method yields variable results and is not suitable for mass production [12].

Fiber gratings selectively transmit some wavelengths and reflect others, depending on the spacing of pattern and RI of glass. A grating will reflect light with a wavelength corresponding to twice the grating period Λ multiplied by the effective index of refraction n_{eff}. This is called the Bragg condition. Light at other wavelengths will be transmitted without significant attenuation. In fact, the grating operates as a narrow band filter. By using a single wideband source and distributing it to many different gratings, each with a different fringe spacing and hence reflection wavelength, it is possible to generate signals of different wavelengths [14]. Changes in the RI of the glass due to environmental effects such as changes in temperature or pressure and spacing by strain along the length of the fiber can change the wavelengths of peak reflection and/or transmission. Measurement of these wavelength changes can be the basis of fiber grating sensors.

In communication, in addition to filtering, fiber Bragg-gratings can be used as add-drop multiplexer and dispersion compensator. Add-drop multiplexer can be fabricated by placing two identical gratings, one in each arm of the Mach-Zehnder interferometer as shown in Fig. 4.39. Presuming the reflection wavelength of the Bragg-grating to be λ_2, light of different wavelengths including wavelength λ_2 will be split equally between the two arms of the coupler. However, λ_2

wavelength signal will be reflected by the two identical gratings. In the coupler, there will be a coherent recombination and λ_2 wavelength signal will exit through port B. The remaining wavelengths will continue to propagate, recombine and exit through port D. As the device is symmetric, wavelength λ_2 can be added at port C to come out with other wavelengths at port D.

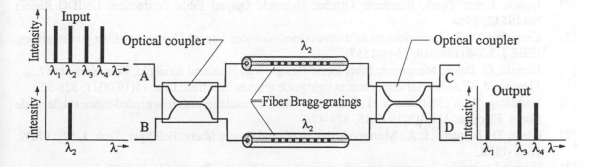

Fig. 4.39: Add-drop multiplexer with identical Bragg-gratings in a Mach-Zehnder Interferometer

The basic principle of operation of a chirped fiber grating as a dispersion-compensating element is shown in Fig. 4.40. Different wavelength components of a broadened pulse are reflecting at different locations along the grating resulting in a differential group delay. The maximum delay is proportional to the length of the grating.

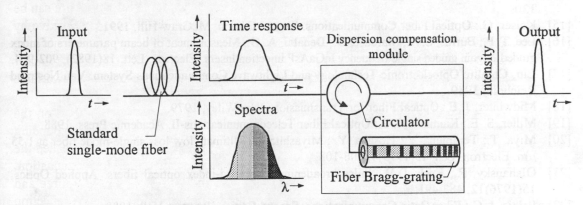

Fig. 4.40: Chirped fiber grating as a dispersion-compensating element

The DCFs are becoming impractical for dispersion compensation because of high non-linearity, loss and bulky size. In contrast to this, recent advances in Bragg-gratings technology have made the gratings an attractive alternative because these are compact, passive and commercially available at a reasonable cost [14]. Fiber Bragg-gratings were looked upon as a mere laboratory curiosity, but now numerous applications are found in the field of light wave communications. As the understanding of this technology spreads, fiber Bragg-gratings will play a significant role in the telecommunication networks.

4.10 REFERENCES

[1]   Ainslie, B. J.; Day, C. R.: A review of single-mode fibers with modified dispersion characteristics. IEEE J. LT-4(1986)8, 967-979.

[2]   Bonek, Ernst; Furch, Bernhard; Otruba, Heinrich: Optical Fiber Production. UNIDO Report No.IS542, 1985.

[3]   Chraplycy, A. R.: Limitations on lightwave communications imposed by optical-fiber nonlinearities. IEEE J. LT-8(1990)10, 1548-1557.

[4]   Donald, G. Baker: Monomode Fiber Optic Design. Van Nostrand Reinhold, New York, 1987.

[5]   Fleming, J. W.: Material dispersion in lightguide glasses. Electron. Lett. 14(1978)11, 326-328.

[6]   Gambling, W. A.; Matsumura, H.; Ragdale, C.M.: Zero total dispersion in graded-index single mode fibers. Electron. Lett. 15(1979)15, 474-476.

[7]   Gloge, D.; Marcatili, E.A.: Multimode theory of graded-core fibers: Bell Syst. Tech. J. 52(1973)9, 1563-1578.

[8]   Gower J.: Optical Communication Systems. Second Edition, Prentice-Hall, 1993.

[9]   Green, Paul E. Jr.: Fiber Optic Networks. Prentice-Hall, 1993.

[10] Hecht, Jeff: Understanding Fiber Optics. Second Edition, Howard W. Sams&Co., 1997.

[11] Hunsperger, Robert G.: Photonic Devices and Systems. Marcel Dekker, Inc., 1994.

[12] Iannone, E.; Matera, F.; Mecozzi, A.; Settembre, M.: Nonlinear optical communication networks. John Wiley, 1998.

[13] Juma Salim: Bragg gratings boost data transmission rates. Laser Focus Word (1996)11, S5-S9.

[14] Juma Salim: Chirped Bragg gratings compensate for dispersion. Laser Focus World (1997)10, 125-130.

[15] Keiser, G.: Optical Fiber Communications. Second Edition, McGraw-Hill, 1991.

[16] Lee, T. P.; Burrus, C. A.; Marcuse, D.; Denatai, A. G.: Measurement of beam parameters of index guided & gain guided single frequency InGaAsP injection lasers. Electron. Lett. 18(1982), 902-904.

[17] Lin, Chinlin: Optoelectronic Technology and Lightwave Communications System. Van Nostrand Reinhold, 1989.

[18] Midwinter, J. E.: Optical Fiber for Transmission. John Wiley, 1979.

[19] Miller, S. E.; Kaminow, I. P.: Optical Fiber Telecommunications-II. Academic Press, 1988.

[20] Miya, T.; Teramuna, Y.; Hosaka, Y. ; Miyashita, T.: Ultimate low-loss single-mode fiber at 1.55 μm. Electron. Lett. 15(1979)4, 106-108.

[21] Olshansky, R.; Keck, D.B.: Pulse broadening in graded-index optical fibers. Applied Optics, 15(1976)12, 483-491.

[22] Palais, J. C.: Fiber Optic Communications. Second Edition, Prentice-Hall, 1988.

[23] Payne, D. N.; Barlow, A. J.; Hansen Ramskov, J. J.: Development of low-and-high-birefringence optical fibers. IEEE J. QE-18(1982)4, 477-487.

[24] Personick, S. D.: Fiber Optics. Plenum Publishing Corp, 1985.

[25] Powers, J. P.: An Introduction to Fiber Optic System. Aksen Associates Inc., 1993.

[26] Senior, J. M.: Optical Fiber Communication-Principles and Practice. Second Edition, Prentice-Hall, 1992.

[27] White, K. I.; Nelson, B.P.: Zero total dispersion in step index monomode fibers at 1.30 and 1.55 μm. Electron. Lett. 15(1979)13, 396-397.

5   POLARIZATION FLUCTUATIONS

O rdinary single-mode fibers do not normally preserve the polarization state of the propagating light. As a consequence, polarization fluctuations can be observed at the output of fiber. Conventional optical communication systems with intensity modulation of light and direct detection (IM/DD) are insensitive to polarization fluctuations, which is one of their significant advantage. However, polarization dispersion becomes more and more important in long-range, high-speed IM/DD communication links. In contrast, polarization fluctuations as well as polarization dispersion are of great concern in coherent optical communication systems. As explained in Chapter nine, polarization fluctuations in the received light wave will produce fluctuations in the photodiode current and, therefore, in the detected signal at the input of sample and hold circuit. As a result, bit error rate increases. In the worst-case, polarization fluctuations may even extinguish the detected signal completely. Thus, either a polarization stabilization or an active control of *state of polarization* (SOP) is indispensable. It will not be true in polarization-diversity receiver.

In order to find appropriate techniques to solve the polarization problem, a profound knowledge about the reasons of fluctuations and their effects on the polarization of optical wave is required. This topic has been discussed in detail in Sections 5.1 and 5.2. Readers who are not especially interested in the theoretical analysis of polarization fluctuations may skip these Sections. Finally, Section 5.3 presents various technical approaches to stabilize or to control the polarization of a light wave. In this Section, polarization-diversity receiver which represents an attractive alternative to control and stabilization circuits is discussed.

It may be mentioned that the optical wave is frequently referred as light wave or simply "light". However, most light waves used in optical communication systems have their wavelengths in the range 800 nm to 1.5 μm. These light waves are invisible and, hence, not actually a light in true sense.

5.1   POLARIZATION PROPAGATION IN SINGLE-MODE FIBER

I n an ideal circular single-mode fiber - other types of fibers such as the graded-index fiber are unsuitable for advanced IM/DD and coherent optical communication systems and will not be considered here - two orthogonal independent *degenerate modes* (i.e., the orthogonal polarizations) are always existing. These modes are degenerated since both are defined by the same propagation constant or same velocity of propagation. In general, the electric field in an optical fiber is always a linear superposition of these *two eigenpolarizations* or *eigenmodes*. As eigenmodes are independent, they propagate without any mutual disturbance.

5.1.1   EIGENMODES

In a real single-mode fiber, various asymmetries can be observed; for example, non-circular fiber core or asymmetrical lateral pressures. Thereby, the degeneration of both orthogonal polarizations is removed and they propagate with different velocities instead of same velocity. Thus, these modes actually become two different modes called the *eigenmodes*.

Besides non-circular fiber core and asymmetrical lateral stress, various other internal and external imperfections can be observed along the fiber. Typical additional imperfections are bendings, torsions and asymmetrical distributions of the refraction index. Moreover, variations in temperature may also influence the quality of light propagation through the optical fiber since different thermal coefficients of expansion with respect to fiber core and jacket yield again asymmetrical internal stresses. Independent of type, all these perturbations break the circular waveguide geometry and influence the velocity of propagation of both the eigenmodes.

In addition to change in the velocity of propagation, fiber distortion such as mentioned above also yield an undesired *mode coupling*. Because of this, a reciprocal exchange of energy takes place. As a result, both modes influence each other and they are no longer independent. They are not eigenmodes now since only independent modes are called eigenmodes. However, it can be shown that coupled modes can always be expressed in terms of two independent eigenmodes again, which now also take into account the perturbations of the fiber (Section 5.1.2). The orthogonal polarizations of these new defined eigenmodes differ from the orthogonal polarizations of the eigenmodes which were existing in the absence of fiber perturbations. The orthogonal directions of polarization are called the *principal axes*. If, for example, fiber exhibits an elliptical core, principal axes are the small and the large half-axes of the elliptical cross-section of the fiber core (Section 5.3).

This Section first presumed that mode coupling does not exist. This means that one eigenmode propagates more slowly as compared to other in the single-mode fiber. To focus our discussion on the polarization effects only, we presume that the principal axes of the eigenmodes are same as the x- and y-directions of a rectangular cartesian coordinate system. In this coordinate system, propagation constants are represented by β_x and β_y respectively. The axis of fiber that is the direction of propagation is the z-direction. We also presume that the fiber is perturbed only by an axial asymmetry which does not change with the position variable z.

An important quantity to assess how polarization is changed or maintained in a single-mode fiber is given by the difference

$$\Delta\beta = \beta_x - \beta_y \tag{5.1}$$

of both propagation constants β_x and β_y [38]. Normalizing this difference by $2\pi/\lambda$, where λ denotes the wavelength of the light, we obtain the dimensionless quantity

$$D = \frac{\Delta\beta\,\lambda}{2\pi} \tag{5.2}$$

called the *double refraction* or *birefringence*. Thus, single-mode fibers with different velocities of propagation of both their eigenmodes are normally called *birefringent fibers*.

In the following, we examine the polarization propagation as a function of double refraction $\Delta\beta$ and location z. To avoid generality, we consider a plane optical wave with a linear polarization at the input to the fiber i.e., $z = 0$. This wave can be described by the electric field

$$\vec{E}(t) = \begin{pmatrix} E_x(t) \\ E_y(t) \end{pmatrix} = \hat{E}\, e^{j2\pi ft} \begin{pmatrix} \cos(\theta) \\ \sin(\theta) \end{pmatrix} = \hat{E}\, e^{j2\pi ft}\, \vec{e}$$

$$= \underline{\hat{E}\, \cos(\theta)\, e^{j2\pi ft}\, \vec{e}_x} + \underline{\hat{E}\, \cos(\theta)\, e^{j2\pi ft}\, \vec{e}_y} \tag{5.3}$$

$$\text{eigenmode x} \qquad\qquad \text{eigenmode y}$$

Here, $f = c/\lambda$ represents the frequency of light and $\hat{E}$ the field amplitude which is assumed to be constant. Thus, laser phase and amplitude noise are neglected in this Chapter. The linear polarization of the fiber input field is defined by the *unit polarization vector* $\vec{e}$ with $|\vec{e}| = 1$. The direction (i.e., the orientation) of this vector is determined by the angle θ called the *polarization* angle. Finally, $\vec{e}_x$ and $\vec{e}_y$ represent the unit vector in x- and y-directions respectively.

The locus curve (i.e., the polarization) of the real part of electric field vector $\vec{E}(t)$ in the xy-plane is shown in Fig. 5.1. At the input to the fiber ($z = 0$), this vector describes a straight line in accordance with the linear input polarization discussed above.

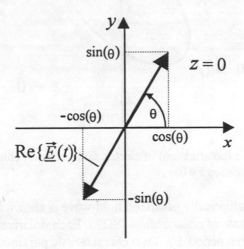

Fig. 5.1: Locus curve (i.e., polarization) of electric field vector of linearly polarized light wave at the fiber input ($z = 0$)

When we consider the shape of the locus curve at an arbitrary other fiber location $z \neq 0$, a polarization that is usually not linear can be observed. Instead, it is generally elliptical with an electric field given by

$$\vec{\underline{E}}(z,t) = \begin{pmatrix} \underline{E}_x(z,t) \\ \underline{E}_y(z,t) \end{pmatrix} = \hat{E}\ e^{j2\pi ft} \begin{pmatrix} \cos(\theta)\ e^{-j\beta_x z} \\ \sin(\theta)\ e^{-j\beta_y z} \end{pmatrix} = \hat{E}\ e^{j2\pi ft} \underline{\vec{e}}$$

$$= \underbrace{\hat{E}\ \cos(\theta)\ e^{j2\pi ft}\ e^{-j\beta_x z}\ \vec{e}_x}_{\text{eigenmode } x} + \underbrace{\hat{E}\ \sin(\theta)\ e^{j2\pi ft}\ e^{-j\beta_y z}}_{\text{eigenmode } y}$$

$$(5.4)$$

In contrast to Eq. (5.3), *unit polarization vector* $\underline{\vec{e}}(z)$ is complex and a function of location in addition. Optical waves characterized by Eq. (5.4) are termed as *elliptically polarized*.

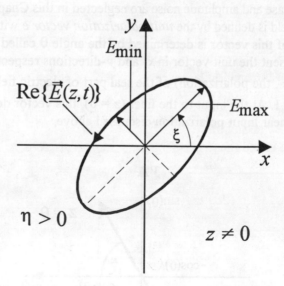

Fig. 5.2: Locus curve (polarization) of electric field vector of elliptically polarized light wave at fiber location $z \neq 0$

Typical locus curve of an elliptically polarized light wave is shown in Fig. 5.2. It is called the *polarization ellipse* or the *state of polarization* (SOP). Each polarization ellipse is periodically orbited by the field vector with a period $1/f$. Two characteristic parameters of polarization ellipse can be defined: *elevation angle* ξ between x-direction and large half-axis of the ellipse and *ellipticity* η. The ellipticity also represents an angle since η is defined by the arctangent of the half-axes ratio E_{min}/E_{max}. To calculate elevation angle Θ and ellipticity η, phase difference

$$\Delta\phi(z) = \Delta\beta z \qquad (5.5)$$

between both orthogonal field components $\underline{E}_x(z, t)$ and $\underline{E}_y(z, t)$ of the electric field given in Eq. (5.4) is required. This phase difference changes with location z. Using some simple trigonometrical relations, elevation angle ξ and ellipticity η can finally be expressed by [6]

$$\eta = \pm\arctan\left[\frac{E_{min}}{E_{max}}\right] = \pm\arctan\left[\frac{\sin(2\theta)\ \sin(\Delta\phi(z))}{1 + \sqrt{1 - \sin^2(2\theta)\ \sin^2(\Delta\phi(z))}}\right] \qquad (5.6)$$

and

$$\xi = \frac{1}{2}\arctan\left[\frac{\sin(2\theta)\ \cos(\Delta\phi(z))}{\cos(2\theta)}\right] \qquad (5.7)$$

The range of η and ξ are given by $-\pi/4 \le \eta \le +\pi/4$ and $-\pi/2 \le \xi \le +\pi/2$ respectively. The ellipticity η is positive if the electric field vector orbits the polarization ellipse anticlockwise and negative if the rotation is clockwise. The view of the observer is directed opposite to the direction of propagating light wave. This means that the observer is looking into the fiber core from the receiver side. By taking into consideration ellipticity η, another important parameter of the polarization ellipse called the *degree of polarization* can be defined. It is given by

$$P = \frac{1 - \tan^2(\eta)}{1 + \tan^2(\eta)} = \sqrt{1 - \sin^2(2\theta)\ \sin^2(\Delta\phi(z))} \qquad (5.8)$$

The degree of polarization is extended over the range 0 to 1. Here, $P = 0$ represents circular polarization, $0 < P < 1$ elliptical polarization and $P = 1$ linear polarization. When the characteristic quantities η, ξ and P are analysed in more detail, a periodicity in the variable z of location can be observed. Its period

$$L_b = \frac{2\pi}{\Delta\beta} \qquad (5.9)$$

represents a measure of length and is called the *beat length* of the fiber. It becomes clear that after a distance of every L_b, same polarization as given at the fiber input i.e., at $z = 0$ can always be observed provided, of course, that no additional fiber imperfections occur along the whole fiber length.

In order to illustrate the variations of polarization during propagation through a single-mode fiber of length L_b, Fig. 5.3 shows the locus curve of the real part of the electric field vector given by Eq. (5.4). At the fiber input ($z = 0$), a linear polarization is presumed. Further, an equal energy division between the orthogonal linear eigenmodes is also presumed. Thus, the angle of linear polarization at the fiber input is $\theta = \pi/4$ as shown in Fig. 5.3. It becomes evident from this figure that the polarization of light exhibits large variations along the fiber. These variations are deterministic if the fiber deformations are fixed and random for deformations fluctuating with time and temperature, as in practical systems.

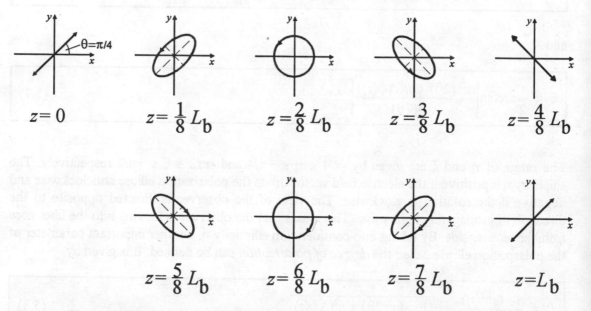

Fig. 5.3: Polarization in fiber at different locations z provided linear polarization with $\theta = \pi/4$ exists at fiber input ($z = 0$)

Let us now consider that one of the eigenmodes is excited more than the other at the fiber input. In such case, polarization during propagation is changed. As an example, Fig. 5.4 presumes that eigenmode of x-direction is more excited which means $\theta < \pi/4$. It becomes clear from this figure that the variations in state of polarization are less than that of given in Fig. 5.3. A circular polarization does not occur at all.

When only one of linear orthogonal eigenmodes is excited at the fiber input, a special case of significant importance is obtained. Here, the angle of linear fiber input polarization is determined by $\theta = k\pi/2$ where $k \in \{\cdots -1, 0, 1 \cdots\}$. In that case, all characteristic quantities of polarization ellipse, namely η, ξ and P are now independent of variable z of location. Thus, a linearly polarized light at the fiber input always remains linearly polarized along the entire fiber length. Eigenmodes always define two characteristic orthogonal directions (i.e., the principal axes) which are suitable to maintain polarization along the fiber in principle. However, it first requires that only one eigenmode is excited at the fiber input and second no additional unknown perturbations occur during propagation.

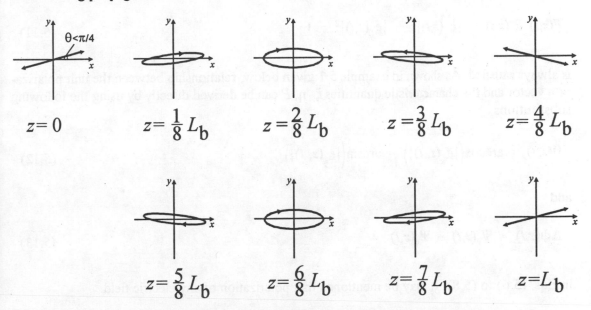

Fig. 5.4: Polarization in fiber at different locations z provided linear polarization with $\theta < \pi/4$ exists at the fiber input ($z = 0$)

In practice, various perturbations at different locations along the fiber can be observed, which all yield a certain fiber deformation. Each deformation changes the direction of the principal axes. Hence, even if one single eigenmode is excited at the fiber input, polarization usually remains constant in a short piece of fiber only. As soon as an additional fiber deformation occurs, new principal axes which are normally different from the principal axes at the fiber input determine the further polarization transmission. Therefore, stable transmission of polarization is usually not possible when standard single-mode fibers are used in the system. For this reason, fiber has to be modified by special technological processes (Section 5.3).

In order to summarize above discussions, Table 5.1 illustrates several types of polarizations and their relationship to the characteristic quantities viz., elevation angle ξ, ellipticity η and degree of polarization P. Shape of polarization ellipse and, consequently, the numerical values of their characteristic parameters are primarily defined by the unit polarization vector. In general, this vector is described by

$$\vec{\underline{e}}(z,t) = \begin{pmatrix} \underline{e}_x(z,t) \\ \underline{e}_y(z,t) \end{pmatrix} = \begin{pmatrix} |\underline{e}_x(z,t)| \ e^{j\Psi_x(z,t)} \\ |\underline{e}_y(z,t)| \ e^{j\Psi_y(z,t)} \end{pmatrix} \tag{5.10}$$

This means that it is now a function of time t as well as location z. As this vector is a unit vector, relationship

$$\vec{\underline{e}}(z,t)\,\vec{\underline{e}}^*(z,t) = |\underline{e}_x(z,t)|^2 + |\underline{e}_y(z,t)|^2 = 1 \qquad (5.11)$$

is always satisfied. As shown in example 5.1 given below, relationship between the unit polarization vector and the characteristic quantities ξ, η, P can be derived directly by using the following substitutions

$$\theta(z,t) = \arccos\left(|\underline{e}_x(z,\,t)|\right) = \arcsin\left(|\underline{e}_y(z,\,t)|\right) \qquad (5.12)$$

and

$$\Delta\phi(z,t) = \Psi_x(z,t) - \Psi_y(z,t) \qquad (5.13)$$

in Eqs. (5.6) to (5.8). It may be mentioned that polarization of any electric field

$$\vec{E}(z,t) = \begin{pmatrix} E_x(z,t) \\ E_y(z,t) \end{pmatrix} = \hat{E}\; e^{j2\pi f t}\; \vec{\underline{e}}(z,t) \qquad (5.14)$$

can be described completely by the unit polarization vector, irrespective of type and number of fiber imperfections.

Example 5.1

Consider a linearly polarized light wave at the input to a single-mode fiber and different propagation constants as the only imperfection of the fiber. In that case, unit polarization vector is given by

$$\vec{\underline{e}}(z,t) = \begin{pmatrix} \underline{e}_x(z,t) \\ \underline{e}_y(z,t) \end{pmatrix} = \begin{pmatrix} \cos(\theta)\; e^{-j\beta_x z} \\ \sin(\theta)\; e^{-j\beta_y z} \end{pmatrix} = \vec{\underline{e}}(z)$$

Hence, the following simple relations can be obtained: $|\underline{e}_x(z,t)| = \cos(\theta)$, $|\underline{e}_y(z,t)| = \sin(\theta)$, $\Psi_x(z,t) = \beta_x z$, and $\Psi_y(z,t) = \beta_y z$.

Table 5.1: Typical polarizations and their relationship to fundamental quantities ξ, η, P (Ψ_0: arbitrary, but constant angle)

	Polarization ellipse	Unit vector of polarization $\hat{e}$	Ellipticity η	Angle of elevation ξ	Polarization degree P	Coordinates of Poincaré (S_1, S_2, S_3)
(1)		$\begin{pmatrix} \frac{1}{\sqrt{2}} e^{j\Psi_0} \\ \frac{1}{\sqrt{2}} e^{(j\Psi_0 - 90°)} \end{pmatrix}$	45° anti-clockwise	undefined	0	$(0,\ 0,\ 1)$
(2)		$\begin{pmatrix} \frac{1}{\sqrt{2}} e^{j\Psi_0} \\ \frac{1}{\sqrt{2}} e^{(j\Psi_0 + 90°)} \end{pmatrix}$	-45° clockwise	undefined	0	$(0,\ 0,\ -1)$
(3)		$\begin{pmatrix} \frac{\sqrt{3}}{2} e^{j\Psi_0} \\ \frac{1}{2} e^{(j\Psi_0 - 90°)} \end{pmatrix}$	30°	0°	0.5	$\left(0.5,\ 0,\ \frac{\sqrt{3}}{2}\right)$
(4)		$\begin{pmatrix} \frac{1}{\sqrt{2}} e^{j\Psi_0} \\ \frac{1}{\sqrt{2}} e^{(j\Psi_0 - 45°)} \end{pmatrix}$	22.5°	45°	$1/\sqrt{2}$	$\left(0,\ \frac{1}{\sqrt{2}},\ \frac{1}{\sqrt{2}}\right)$
(5)		$\begin{pmatrix} \frac{1}{2} e^{j\Psi_0} \\ \frac{\sqrt{3}}{2} e^{(j\Psi_0 - 90°)} \end{pmatrix}$	30°	90°	0.5	$\left(-0.5,\ 0,\ \frac{\sqrt{3}}{2}\right)$
(6)		$\begin{pmatrix} 1\, e^{j\Psi_0} \\ 0 \end{pmatrix}$	0°	0°	1	$(1,\ 0,\ 0)$
(7)		$\begin{pmatrix} -\frac{1}{\sqrt{2}} e^{j\Psi_0} \\ \frac{1}{\sqrt{2}} e^{j\Psi_0} \end{pmatrix}$	0°	-45°	1	$(0,\ 1,\ 0)$

As mentioned earlier and explained in Chapter nine in detail, in coherent optical communication systems polarization of the received light wave at the output end of fiber is of significant importance. In that case, variable z of location is fixed and equals the length L of the fiber ($z = L$).

Therefore, it is no longer required to represent the fiber location explicitly. The unit polarization vector $\underline{e}(z,\,t) = \underline{e}(L,\,t) = \underline{e}(t)$ is only a function of time now due to the time dependence of different fiber perturbations; for example, temporal change in internal stress or temperature. In addition, various time-invariant deformations can be observed; for example, elliptical fiber core. Thus, polarization state is unstable at the fiber output and, therefore, unpredictable.

When all geometrical fiber deformations are assumed to be time-invariant and polarization ellipse to be unchanged, electric field vector at the input to the receiver ($z = L$) still changes direction. Since this vector orbits the polarization ellipse periodically, orientation and length of this vector are consequently changed. Therefore, polarization matching of both received and local laser light waves in coherent optical receiver is unavoidable even in this ideal case. For this reason, appropriate polarization-matching techniques are required (Section 5.3).

Another very useful tool to describe polarization fluctuations also exits. Every type of polarization is completely defined by three parameters: angle of elevation ξ, ellipticity η and rotating direction (clockwise or anticlockwise). These parameters can be expressed in terms of three orthogonal coordinates S_1, S_2 and S_3 which are called the *Stokes parameters*. It may be remembered that the degree of polarization P is an additional characteristic quantity, but not really a required parameter since P is a function of ξ or η as given in Eq. (5.8). Taking into consideration the normalized Stokes parameters

$$S_1 = |\underline{e}_x(z,t)|^2 - |\underline{e}_y(z,t)|^2 = \cos(2\eta)\cos(2\xi) \tag{5.15}$$

$$S_2 = \underline{e}_x(z,t)\underline{e}_y^*(z,t) + \underline{e}_x^*(z,t)\underline{e}_y(z,t) = \cos(2\eta)\sin(2\xi) \tag{5.16}$$

$$S_3 = -j\left(\underline{e}_x(z,t)\underline{e}_y^*(z,t) - \underline{e}_x^*(z,t)\underline{e}_y(z,t)\right) = \sin(2\eta) \tag{5.17}$$

characteristic quantities of each state of polarization can now be projected on the surface of a sphere called the *Poincaré sphere* [45]. Thereby, each possible state of polarization corresponds to exactly one characteristic point $\mathcal{P} := \mathcal{P}(S_1, S_2, S_3) = \mathcal{P}(2\xi, 2\eta)$ on the surface of this sphere as shown in Fig. 5.5.

For linearly polarized ($\eta = 0$) light wave, characteristic point $\mathcal{P}$ is always located on the "Equator" of the Poincaré sphere. The points H and V represent the horizontal and vertical polarizations respectively, whereas the points P and M represent linear polarization at an angle of elevation $\xi = \pm 45°$. If an optical wave is circularly polarized, then point $\mathcal{P}$ is either located at the "North Pole" (anticlockwise rotation) or "South Pole" (clockwise rotation) of the sphere. Elliptically polarized optical waves of same angle of elevation ξ are located on meridian, while same ellipticity η are always located on parallel. If both the states of polarization $\underline{e}_1(z,t)$ and $\underline{e}_2(z,t)$ are orthogonal, i.e., $\underline{e}_1(z,t) \cdot \underline{e}_2^*(z,t) = 0$, then they are diametrically opposite.

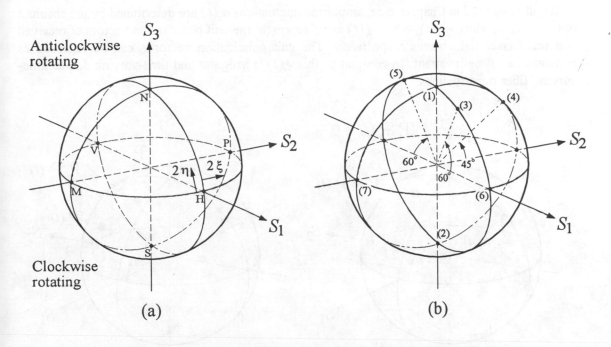

Fig. 5.5: Representation of state of polarization on Poincaré sphere (a) typical states
of polarization and (b) states of polarization corresponding to Table 5.1

As mentioned above, performance of a coherent optical communication system is considerably influenced by the states of polarization of both received light wave $\vec{E}_r(t)$ and local laser light wave $\vec{E}_l(t)$. In the following, performance degradation due to polarization fluctuations is briefly discussed. If a reader is not familiar with the basic principle of optical heterodyning which is simply the optical version of the well known electrical heterodyne technique used in every television and radio receiver, he can leave out following discussion and go to the next Section..

In the ideal case, both states of polarization are same and time-invariant. As discussed in Chapter nine, each deviation from this ideal case results in undesired amplitude and phase fluctuations $a_p(t)$ and $\phi_p(t)$ respectively in the photodiode current $i_{PD}(t)$ given by Eq. (9.67). Only in the ideal case, both fluctuations are always zero i.e., $\phi_p(t) = 0$ and $a_p(t) = a_{p,max} = 1$. It must be noted that $a_p(t)$ is a multiplicative factor in the photodiode current $i_{PD}(t)$. For this reason, $a_p(t)$ should be as large as possible in order to increase the signal power. Hence, any deviation from the maximum value $a_{p,max} = 1$ can be regarded as an attenuation. In practice, however, this deviation is time-varying since $a_p(t)$ is changing with time. In the worst-case i.e., $a_p(t) = 0$, both states of polarization are orthogonal. Here, the effective part of photodiode current $i_{PD}(t)$ becomes zero.

Phase fluctuations $\phi_p(t)$ caused by polarization fluctuations always occur in addition (i.e., additive) to the phase noise $\phi(t)$ of both transmitter and local lasers. However, phase fluctuations $\phi_p(t)$ are a rather slow random process as compared to laser phase noise $\phi(t)$. In practice, polarization matching is always required. Here, phase fluctuations $\phi_p(t)$ can usually be ignored in comparison with laser phase noise.

It will be derived in Chapter nine, amplitude fluctuations $a_p(t)$ are determined by the absolute of the scalar product $\vec{e}_l \cdot \vec{e}_r(t)$, where $\vec{e}_r(t)$ and $\vec{e}_l$ represent the unit polarization vectors of received and local laser light waves respectively. The unit polarization vector $\vec{e}_l$ can be regarded as constant i.e., time-invariant. In contrast to this, $\vec{e}_r(t)$ is irregular and time-varying due to time-varying fiber perturbations.

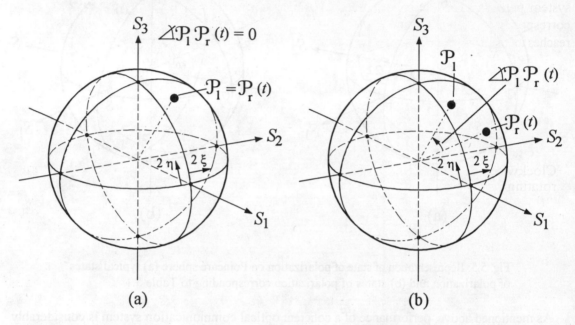

Fig. 5.6: States of polarization of received and local laser light waves in (a) ideal and (b) real case

The amplitude fluctuations $a_p(t)$ can be quickly analysed by considering the Poincaré sphere again. For this, unit polarization vectors $\vec{e}_l$ and $\vec{e}_r(t)$ must be described by two points on the Poincaré sphere (Fig. 5.6) using Eqs. (5.15) to (5.17). These points are

$$\mathcal{P}_l = \mathcal{P}_l(S_{1l}, S_{2l}, S_{3l}) \quad \text{and} \quad \mathcal{P}_r(t) = \mathcal{P}_r(S_{1r}(t), S_{2r}(t), S_{3r}(t)) \tag{5.18}$$

The point $\mathcal{P}_r(t)$ of unit polarization vector of $\vec{e}_r(t)$ is irregularly moving on the surface of the sphere according to its time dependence, whereas $\mathcal{P}_l$ remains fixed.

The spatial distance between the points $\mathcal{P}_r(t)$ and $\mathcal{P}_l$ expressed using the angle $\measuredangle \mathcal{P}_r(t)\mathcal{P}_l$ represents a direct measure to assess the strength of the amplitude fluctuations $a_p(t)$. After some mathematical simplifications, we obtain

$$a_p(t) = |\vec{e}_r(t)\,\vec{e}_l^*| = \cos\left(\frac{1}{2}\measuredangle \mathcal{P}_r(t)\,\mathcal{P}_l\right) \tag{5.19}$$

where

$$\Delta \mathcal{P}_r(t)\ \mathcal{P}_1 = S_{11}\,S_{1r}(t)\ +\ S_{21}\,S_{2r}(t)\ +\ S_{31}\,S_{3r}(t) \tag{5.20}$$

Performance of coherent optical communication systems rapidly deteriorates with the increase of angle $\Delta \mathcal{P}_r(t)\mathcal{P}_1$. Hence, as the spatial separation between both states of polarization increases, system performance deteriorates. When both states of polarization are same i.e., $\breve{e}_r(t) = \breve{e}_l$, corresponding points $\mathcal{P}_r(t)$ and $\mathcal{P}_1$ lie on top of each other as shown in Fig. 5.6a. Thereby, $a_p(t)$ reaches its maximum value $a_{p,\max} = 1$ irrespective of the location of common points on the sphere.

Table 5.2: Influence of deviation in states of polarization of received and local laser waves on $a_p(t)$

$\Delta \mathcal{P}_r \mathcal{P}_1$	a_p	Remarks
0°	1	Ideal case, same state of polarization
10°	0.996	Very little influence
15°	0.991	Very little influence
20°	0.985	Little influence
50°	0.906	Little influence
90°	0.707	Strong influence
180°	0	Extinguishing of signal, orthogonal polarizations

To make it more clear, Table 5.2 shows the influence of the spatial angle $\Delta \mathcal{P}_r(t)\mathcal{P}_1$ on $a_p(t)$. Here, $\Delta \mathcal{P}_r(t)\mathcal{P}_1 := \Delta \mathcal{P}_r \mathcal{P}_1$ is assumed to be constant. According to this table, influence of deviation in state of polarization is practically negligible when the spatial angle is less than about 15°. In that case, deviation in $a_p(t)$ from its maximum value $a_{p,\max} = 1$ is less than 1%. Even if this angle is relatively large, for example 50°, the deviation remains less than 10%. On the other hand, a considerable influence of deviation is observed if the angle $\Delta \mathcal{P}_R(t)\mathcal{P}_L$ exceeds 50°. Due to various disturbances in layed fiber cables, angle deviations of more than 50° are quite realistic. Therefore, either an appropriate polarization-matching technique or a polarization-diversity receiver is absolutely required when a coherent optical communication system has to be employed.

As explained above, the Poincaré sphere is a powerful tool offering a clear description of polarization fluctuations. Moreover, the Poincaré sphere simplifies the analysis of influence of deviation in the state of polarization of received and local laser light waves on the performance of coherent optical transmission systems. In combination with an appropriate software, the Poincaré sphere also represents a well-suited tool in polarization-measurement equipment to determine the statistics of polarization fluctuations or study the polarization behaviour of a single-mode fiber cable.

5.1.2   THEORY OF MODE COUPLING

In the previous Section, we have discussed the effects of deviation in velocity of propagation of both orthogonal polarizations. The main reasons for a velocity deviation are deformations of the fiber geometry, for example, an elliptical fiber core. Both polarizations have propagated independently without any mode coupling. However, this is not actually true since mode coupling is normally existent in standard fiber.

The main aim of this Section is to describe the reasons and effects of mode coupling which are caused, for example, by additional fiber perturbations. Due to mode coupling, a reciprocal energy exchange can be observed. As a result, both the modes i.e., orthogonal polarizations are no longer independent. Hence, coupled modes are not eigenmodes as eigenmodes always propagate independently as discussed in the previous Section.

In this Section, efforts are made to give an answer to the question: Is it possible to find out again two independent eigenmodes propagating without any energy exchange irrespective of type of fiber perturbation? Let us presume that these new eigenmodes are existent. Then we have to focus our interest on two further questions: Which are the new orthogonal orientations of the new eigenmodes and what are their velocities of propagation? It is clear that the new orthogonal eigenmodes as well as their velocities of propagation will primarily depend on the fiber characteristics especially on its geometrical deformations [24].

In order to avoid generality and simplify calculation, we first take a certain well-specified deformation (for example, an elliptical fiber core) that corresponds to two independent eigenmodes propagating with different velocities. Next, we assume a second (or even more) geometrical fiber perturbation which deforms the unperturbed fiber core in addition. As a result, both the orthogonal modes (mode x and mode y) are coupled now and they are no longer independent eigenmodes.

We consider that the first fiber deformation (for example, the elliptical fiber core as assumed above) as well as all further perturbations are independent of z-direction i.e., the fiber axis. The following results are valid only for geometrical deformations which are independent of location. Since this is normally valid over a very short distance, the results are applicable for short piece of fiber only.

Linear superposition of mode x and mode y to generate the optical fiber wave is again clarified by the following equations:

$$\vec{E}(z,\ t) = \begin{pmatrix} \underline{E}_x(z,\ t) \\ \underline{E}_y(z,\ t) \end{pmatrix} = \underline{E}_x(z,\ t)\ \vec{e}_x \quad + \quad \underline{E}_y(z,\ t)\ \vec{e}_y$$

(5.21)

$$= \hat{E}_x\ \underline{a}(z)\ e^{j2\pi ft}\ \vec{e}_x \quad + \quad \hat{E}_y\ \underline{b}(z)\ e^{j2\pi ft}\ \vec{e}_y$$

$\uparrow$ $\uparrow$

mode x ← dependent →   mode y

$\downarrow$ $\downarrow$

$$\vec{H}(z,\ t) = \begin{pmatrix} \underline{H}_x(z,\ t) \\ \underline{H}_y(z,\ t) \end{pmatrix} = \hat{H}_y\ \underline{a}(z)\ e^{j2\pi ft}\ \vec{e}_y \quad - \quad \hat{H}_x\ \underline{b}(z)\ e^{j2\pi ft}\ \vec{e}_x$$

(5.22)

$$= \underline{H}_y(z,\ t)\ \vec{e}_y \quad - \quad \underline{H}_x(z,\ t)\ \vec{e}_x$$

Eq. (5.21) describes the electric field and Eq. (5.22) the magnetic field of the light wave in the fiber. The modes are termed as mode x and mode y since the directions of the electric field components $\underline{E}_x(z,\ t)$ and $\underline{E}_y(z,\ t)$ are $\vec{e}_x$ and $\vec{e}_y$ respectively.

In previous Section 5.1.1, a plane wave is considered for both the modes which propagate in z-direction i.e., in the direction of fiber axis. The electric and magnetic field vectors of mode x and mode y are orthogonal and are located in the xy-plane (Fig. 5.7b). In addition, they are functions of time t and location z, but independent of location x and y. The light wave in a fiber is actually not a plane wave due to restricted size of the fiber core and the fact that the core is surrounded by a jacket with a refraction index lower than the index of the core (this is the fundamental condition for guided waves). However, in the discussion on principle of mode coupling, error introduced by taking a plane wave is less important. Instead, we take the advantages of plane waves to reach a clear description of mode coupling.

In order to describe the mutual energy exchange due to mode coupling, Eqs. (5.21) and (5.22) include two terms $\underline{a}(z)$ and $\underline{b}(z)$ which still have to be determined. These terms are complex since they additionally describe the periodical dependence of the fiber wave on the location z. It should be noted that $\underline{a}(z)$ and $\underline{b}(z)$ are dependent and cannot be chosen arbitrarily. In a fiber without loss, optical power flow S always remains constant i.e., independent of location z. We get

$$\vec{S} = \frac{1}{2}\mathrm{Re}\{\underline{\vec{E}}(z,\ t) \times \underline{\vec{H}}^*(z,\ t)\} = \vec{S}_x(z) + \vec{S}_y(z) = \left(S_x(z) + S_y(z)\right)\vec{e}_z$$

$$= \left(|\underline{a}(z)|^2 \underbrace{\frac{1}{2}\hat{E}_x\hat{H}_y}_{S_{x,\,max}} + |\underline{b}(z)|^2 \underbrace{\frac{1}{2}\hat{E}_y\hat{H}_x}_{S_{y,\,max}} \right)\vec{e}_z = \underbrace{\left(|\underline{a}(z)|^2 + |\underline{b}(z)|^2\right)}_{= 1} S\ \vec{e}_z = S\ \vec{e}_z \neq \vec{S}(z)$$

(5.23)

It is seen from the above equation that the terms $\underline{a}(z)$ and $\underline{b}(z)$ which describe the energy exchange as a function of location z are related by $|\underline{a}(z)|^2 + |\underline{b}(z)|^2 = 1$. Hence, no power is lost in the fiber as considered above. It should be noted that the sign "x" in the first line of Eq. (5.23) represents a vector product [3]. The optical power flow S given in Eq. (5.23) is measured in AV/m². Hence, this quantity physically describes the effective optical power floating through unit area perpendicular to the direction of propagation (z-direction). As shown in Eq. (5.23), power flow can be divided into two parts: one is the power flow $\vec{S}_x(z)$ of mode x and the other power flow $\vec{S}_y(z)$ of mode y. The maximum power flows $S_{x,max}$ and $S_{y,max}$ of both modes are equal. As $S_{x,max}$ and $S_{y,max}$ can never be more than the total power flow S of the light wave $\vec{E}(0, t)$ at the fiber input (i.e., $z = 0$), $S_{x,max}$ and $S_{y,max}$ are identical i.e., $S_{x,max} = S_{y,max} = S$.

In the following, $\underline{a}(z)$ and $\underline{b}(z)$ will be determined as a function of geometrical deformations which disturb the fiber. Afterwards, we shall analyse the field Eqs. (5.21) and (5.22). Finally, we shall study whether the field $\vec{E}(z, t)$ given in Eq. (5.21) can be divided into two independent modes again or not. For this, we follow the earlier procedure and take

$$\vec{E}(z, t) = \quad \underline{E}_u(z, t) \, \vec{e}_u \quad + \quad \underline{E}_v(z, t) \, \vec{e}_v$$

$$= \hat{E}_u \, e^{-j\beta_u z} \, e^{j2\pi ft} \, \vec{e}_u \quad + \quad \hat{E}_v \, e^{-j\beta_v z} \, e^{j2\pi ft} \, \vec{e}_v$$

(5.24)

$$\underset{\text{eigenmode } x}{\uparrow} \quad \leftarrow \text{independent} \rightarrow \quad \underset{\text{eigenmode } y}{\uparrow}$$

and

$$\vec{H}(z, t) = \quad \underline{H}_v(z, t) \, \vec{e}_v \quad + \quad \underline{H}_u(z, t) \, \vec{e}_u$$

$$= \hat{H}_v \, e^{-j\beta_u z} \, e^{j2\pi ft} \, \vec{e}_v \quad + \quad \hat{H}_u \, e^{-j\beta_v z} \, e^{j2\pi ft} \, \vec{e}_u$$

(5.25)

$$\underset{\text{eigenmode } x}{\uparrow} \quad \leftarrow \text{independent} \rightarrow \quad \underset{\text{eigenmode } y}{\uparrow}$$

which replace the field Eqs. (5.21) and (5.22). The unknown parameters in this new approach are the orthogonal directions $\vec{e}_u$ and $\vec{e}_v$ which define the principal axes of the new eigenmodes and their propagation constants β_u and β_v respectively. These quantities can be determined by a simple comparison of the electric field given in Eqs. (5.21) and (5.24), provided $\underline{a}(z)$ and $\underline{b}(z)$ have already been determined and substituted.

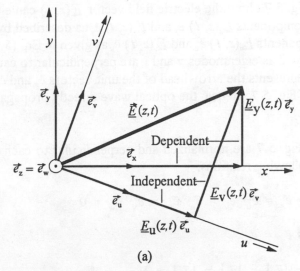

(a)

Mode x: $\underline{E}_X(z,t)\,\vec{e}_x$, $\underline{H}_y(z,t)\,\vec{e}_y$ Eigenmode u: $\underline{E}_u(z,t)\,\vec{e}_u$, $\underline{H}_v(z,t)\,\vec{e}_v$

Mode y: $\underline{E}_y(z,t)\,\vec{e}_y$, $-\underline{H}_X(z,t)\,\vec{e}_x$ Eigenmode v: $\underline{E}_v(z,t)\,\vec{e}_v$, $-\underline{H}_u(z,t)\,\vec{e}_u$

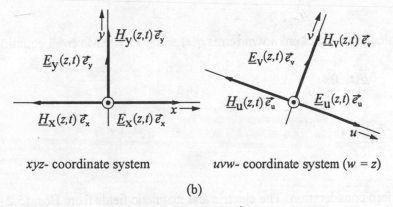

xyz- coordinate system uvw- coordinate system ($w = z$)

(b)

Fig. 5.7: (a) Splitting of electric field vector $\vec{E}(z, t)$ in modes x and y and eigenmodes u and v (b) orthogonality of modes

The amplitudes $\hat{E}_u$ and $\hat{E}_v$ of electric field depend on the input field of fiber being considered. These are completely determined by the boundary conditions (Section 5.2). The amplitudes $\hat{H}_u$ and $\hat{H}_v$ of magnetic field can be calculated directly by means of electric field since electric and magnetic fields are strongly correlated. They often differ only by a constant depending on fiber characteristics.

Similar to the coupled modes x and y, eigenmodes u and v are also plane waves propagating in the z-direction only. Hence, their electric and magnetic field vectors are again located in the xy-plane depending on time t and location z.

It is illustrated in Fig. 5.7a how the electric field vector $\vec{E}(z, t)$ can either be separated into two dependent field components $\underline{E}_x(z, t)\, \check{e}_x$ and $\underline{E}_y(z, t)\, \check{e}_y$ as described by Eq. (5.21) or in two independent field components $\underline{E}_u(z, t)\, \check{e}_u$ and $\underline{E}_v(z, t)\, \check{e}_v$ as given in Eq. (5.24). As shown in Fig. 5.7b, modes x and y as well as eigenmodes u and v are perpendicular to each other. The dot $\odot$ in the centre of Fig. 5.7 represents the arrowhead of the unit vectors $\check{e}_z$ and $\check{e}_w$ which are perpendicular to the surface of Fig. 5.7. Hence, the optical wave directly propagates in the direction of reader's eye (keep care).

All unit vectors in Fig. 5.7 are normalized and perpendicular to each other. Thus, they are characterized by the following relations:

$$\vec{e}_x \cdot \vec{e}_y = \vec{e}_x \cdot \vec{e}_z = \vec{e}_y \cdot \vec{e}_z = \vec{e}_u \cdot \vec{e}_v = \vec{e}_u \cdot \vec{e}_w = \vec{e}_v \cdot \vec{e}_w = 0 \tag{5.26}$$

$$\vec{e}_x \times \vec{e}_y = \vec{e}_z = \vec{e}_u \times \vec{e}_v = \vec{e}_w \tag{5.27}$$

$$|\vec{e}_x| = |\vec{e}_y| = |\vec{e}_z| = |\vec{e}_u| = |\vec{e}_v| = |\vec{e}_w| = 1 \tag{5.28}$$

In the above equations, symbols "$\times$" and "$\cdot$" represent the vector product and the scalar product respectively [3].

In order to calculate the still unknown terms $\underline{a}(z)$ and $\underline{b}(z)$, the Maxwell equations

$$\text{rot}\big(\vec{E}(z, t)\big) = -\frac{\delta \vec{\underline{B}}(z, t)}{\delta t} \tag{5.29}$$

$$\text{rot}\big(\vec{\underline{H}}(z, t)\big) = \frac{\delta \vec{\underline{D}}(z, t)}{\delta t} \tag{5.30}$$

have to be taken into consideration. The electric and magnetic fields from Eqs. (5.21) and (5.22) must be substituted in the following material equations [44]

$$\underline{\vec{D}}(z, t) = \epsilon\, \underline{\vec{E}}(z, t) = \epsilon_r \epsilon_0\, \underline{\vec{E}}(z, t) \tag{5.31}$$

$$\underline{\vec{B}}(z, t) = \mu\, \underline{\vec{H}}(z, t) = \mu_0\, \underline{\vec{H}}(z, t) \tag{5.32}$$

where $\underline{\vec{D}}(z, t)$ and $\underline{\vec{B}}(z, t)$ represent the electric and magnetic power flux respectively. Both the Maxwell and material equations are already adapted to dielectric waveguide i.e., optical fiber. Thus, Eq. (5.32) is valid only for a non-magnetic material i.e., $\mu_r = 1$ and Eq. (5.29) does not contain an electric current component [44].

In a completely isotropic material, the relative dielectric constant ϵ_r is either a real (i.e., material without any loss) or a complex quantity (i.e., material with loss). In that case, electric field $\vec{E}(z,t)$ and electric power flux $\vec{D}(z,t)$ vectors point in the same direction and are parallel. However, fiber is usually non-isotropic due to various inherent asymmetries as already mentioned above. In such cases, relative dielectric constant ϵ_r is not a simple scalar quantity. Instead it is a tensor [3, 44], describing how the field is changing in different directions in the fiber. Mathematically, a tensor describes the linear dependence of the components of two vectors. In the particular case of fiber, this tensor describes the linear correlation of electric field $\vec{E}(z,t)$ and electric power flux $\vec{D}(z,t)$ vectors. Due to non-isotropic fiber characteristics, both vectors are normally not parallel.

For studies based on system aspects, we are more interested on the effects of mode coupling than in the atomic reasons of mode coupling. Thus, it is not required for the reader to study tensor calculation in detail. Instead, it is sufficient to know that a tensor of dielectric constants contains nine (complex) scalar quantities similar to a 3×3 matrix, two tensors are added in the same way as two matrixes and a tensor is in principle multiplied by a vector in the same way as a matrix is multiplied by a vector [44]. The result of such a product is again a vector.

All nine components of a tensor are only defined unambiguously in a specified coordinate system. Here, it is the rectangular *xyz*-coordinate system. If this system is changed, for example rotated, then the numerical values of the tensor components are also changed. Hence, a tensor of dielectric constants varies with the change in the coordinate system. Optical fiber wave physically remains unchanged irrespective of kind of mathematical coordinate system chosen. Therefore, each change in coordinate system requires an appropriate tensor transformation which, however, will not be considered in this book. A more detailed description of tensor mathematics can be found in the literature; for example, in [3, 44].

In order to simplify the further calculations, it is useful to divide the tensor of dielectric constants

$$[\epsilon]_r = [\epsilon]_0 + [\epsilon]_m \qquad (5.33)$$

into two parts. Thereby

$$[\epsilon]_0 = \begin{bmatrix} \epsilon_{xx} & 0 & 0 \\ 0 & \epsilon_{yy} & 0 \\ 0 & 0 & \epsilon_{zz} \end{bmatrix} \qquad (5.34)$$

describes a fiber without any mode coupling and

$$[\epsilon]_m = \begin{bmatrix} \epsilon'_{xx} & \epsilon_{xy} & \epsilon_{xz} \\ \epsilon_{yx} & \epsilon'_{yy} & \epsilon_{yz} \\ \epsilon_{zx} & \epsilon_{zy} & \epsilon'_{zz} \end{bmatrix} \tag{5.35}$$

takes into account all the perturbations which are responsible for mode coupling. As mentioned in the beginning of this Section, $[\epsilon]_0$ takes into account the first fiber deformation (for example, the elliptical fiber core) and $[\epsilon]_m$ all additional perturbations of the fiber geometry. To distinguish the three diagonal tensor components of the first deformation and additional perturbations, these components are identified by dash in Eq. (5.35). It may be mentioned that a tensor of dielectric constants is normally symmetrical i.e., $\epsilon_{xy} = \epsilon_{yx}$, $\epsilon_{xz} = \epsilon_{zx}$ and $\epsilon_{yz} = \epsilon_{zy}$.

After this brief description of tensor of dielectric constants, we are now able to replace the field components in Maxwell Eqs. (5.29) and (5.30) by the field Eqs. (5.21) and (5.22). With some mathematical operations, following coupled system of differential equations is obtained:

$$\frac{d}{dz}\begin{bmatrix} \underline{a}(z) \\ \underline{b}(z) \end{bmatrix} = -j \begin{bmatrix} N_{11} & N_{12} \\ N_{21} & N_{22} \end{bmatrix} \cdot \begin{bmatrix} \underline{a}(z) \\ \underline{b}(z) \end{bmatrix} \tag{5.36}$$

This system of differential equations (DEQ) is called a normal, linear and homogeneous DEQ system in the variable of location z [3]. The four coefficients N_{11}, N_{12}, N_{21} and N_{22} which can be either real or complex (dependent on the type of fiber distortion) are a direct result of the substitution mentioned above. We obtain

$$N_{11} = \pi f \left(\epsilon_0 \frac{\hat{E}_x}{\hat{H}_y} \vec{e}_x \, [\epsilon]_r \, \vec{e}_x + \mu_0 \frac{\hat{H}_y}{\hat{E}_x} \right) = \beta_x \sqrt{1 + \frac{\vec{e}_x \, [\epsilon]_m \, \vec{e}_x}{\vec{e}_x \, [\epsilon]_0 \, \vec{e}_x}}$$

$$\tag{5.37}$$

$$= \beta_x \sqrt{1 + \frac{\epsilon'_{xx}}{\epsilon_{xx}}} = \beta'_x$$

$$N_{12} = \pi f \epsilon_0 \frac{\hat{E}_y}{\hat{H}_y} \vec{e}_x \, [\epsilon]_r \, \vec{e}_y = 0.5 \sqrt{\beta'_x \beta'_y} \; \frac{\vec{e}_x \, [\epsilon]_m \, \vec{e}_y}{\sqrt{\left(\vec{e}_x \, [\epsilon]_r \, \vec{e}_x \right) \left(\vec{e}_y \, [\epsilon]_r \, \vec{e}_y \right)}}$$

$$\tag{5.38}$$

$$= 0.5 \frac{\sqrt{\beta'_x \beta'_y} \, \epsilon_{xy}}{\sqrt{\left(\epsilon_{xx} + \epsilon'_{xx} \right) \left(\epsilon_{yy} + \epsilon'_{yy} \right)}}$$

$$N_{21} = \pi f \epsilon_0 \frac{\hat{E}_x}{\hat{H}_x} \, \vec{e}_y \, [\epsilon]_r \, \vec{e}_x = 0.5 \sqrt{\beta'_x \beta'_y} \, \frac{\vec{e}_y \, [\epsilon]_m \, \vec{e}_x}{\sqrt{\left(\vec{e}_x \, [\epsilon]_r \, \vec{e}_x\right)\left(\vec{e}_y \, [\epsilon]_r \, \vec{e}_y\right)}}$$

(5.39)

$$= 0.5 \sqrt{\beta'_x \beta'_y} \, \frac{\epsilon_{yx}}{\sqrt{\left(\epsilon_{xx} + \epsilon'_{xx}\right)\left(\epsilon_{yy} + \epsilon'_{yy}\right)}} = N_{12}^*$$

$$N_{22} = \pi f \left(\epsilon_0 \frac{\hat{E}_y}{\hat{H}_x} \, \vec{e}_y \, [\epsilon]_r \, \vec{e}_y + \mu_0 \frac{\hat{H}_x}{\hat{E}_y} \right) = \beta_y \sqrt{1 + \frac{\vec{e}_y \, [\epsilon]_m \, \vec{e}_y}{\vec{e}_y \, [\epsilon]_0 \, \vec{e}_y}}$$

(5.40)

$$= \beta_y \sqrt{1 + \frac{\epsilon'_{yy}}{\epsilon_{yy}}} = \beta'_y$$

In these equations, we have already taken into account that both the vector operations $\vec{e}_x [\epsilon]_0 \vec{e}_y$ and $\vec{e}_y [\epsilon]_0 \vec{e}_x$ yield zero.

Let us now consider the coefficients N_{11}, N_{12}, N_{21} and N_{22} in more detail. It becomes clear from Eqs. (5.37) to (5.40) that these coefficients are primarily determined by the tensors $[\epsilon]_0$ and $[\epsilon]_m$. Thus, they depend upon the fiber material and the type of geometrical perturbations. In addition, they are influenced by the frequency f of light. As will be shown below, both propagation constants β_x and β_y are dependent upon the fiber characteristics only. Hence, it is useful to evaluate and tabulate N_{11}, N_{12}, N_{21}, and N_{22} for some typical fiber deformations which are usually present in practice; for example, torsion, bending, transverse and axial pressure [35, 51]. The coefficients N_{12} and N_{21} are responsible for mode coupling. If the tensor of dielectric constants $[\epsilon]_r$ is real, then these coefficients are equal and real also. However, if $[\epsilon]_r$ is complex, then both coefficients are also complex and related by $N_{21} = N_{12}^*$.

When both the tensor components ϵ_{xy} and ϵ_{yx} and coefficients N_{12} and N_{21} are zero, mode coupling does not exist between mode x and mode y. Hence, these modes remain the characteristic eigenmodes of the single-mode fiber. Certainly, their propagation constants β_x and β_y and consequently velocities of propagation are changed when the diagonal components ϵ'_{xx}, ϵ'_{yy} and ϵ'_{zz} are changed. Propagation constants β_x and β_y are only related to the first primary fiber deformation, whereas β'_x and β'_y also consider the additional perturbations. From Eqs. (5.37) and (5.40), β'_x and β'_y are equal to the coefficients N_{11} and N_{22} respectively. If N_{12} and N_{21} are zero as considered above, then the coupled DEQ systems can be expressed in terms of two independent equations. In that case, first and second equations are only functions of $\underline{a}(z)$ and $\underline{b}(z)$ respectively. Moreover, the coefficients of coupled DEQ system in Eq. (5.36) and, therefore, the propagation of optical wave is not influenced by the six tensor elements ϵ_{zz}, ϵ'_{zz}, ϵ_{xz}, ϵ_{zx}, ϵ_{yz} and ϵ_{zy}. Therefore, mode x and mode y still propagate in the z-direction and their electric field vectors still contain no z-component. For the propagation constants β_x, β_y, β'_x, and β'_y as well as the ratios of field amplitudes $\hat{E}_x/\hat{H}_y$ and $\hat{E}_y/\hat{H}_x$, following relations can be derived:

$$\beta'_x = 2\pi f \mu_0 \frac{\hat{H}_y}{\hat{E}_x} = 2\pi f \epsilon_0 \vec{e}_x \,[\epsilon]_r\, \vec{e}_x \frac{\hat{E}_x}{\hat{H}_y} = 2\pi f \epsilon_0 \left(\epsilon_{xx} + \epsilon'_{xx}\right) \frac{\hat{E}_x}{\hat{H}_y}$$

$$= 2\pi f \sqrt{\mu_0 \,\epsilon_0 \left(\epsilon_{xx} + \epsilon'_{xx}\right)} = \underline{2\pi f \sqrt{\mu_0 \,\epsilon_0 \,\epsilon_{xx}}} \;\sqrt{1 + \epsilon'_{xx}/\epsilon_{xx}}$$

$$\uparrow$$
$$\beta_x$$

(5.41)

$$\beta'_y = 2\pi f \mu_0 \frac{\hat{H}_x}{\hat{E}_y} = 2\pi f \epsilon_0 \,\vec{e}_y \,[\epsilon]_r\, \vec{e}_y \frac{\hat{E}_y}{\hat{H}_x} = 2\pi f \epsilon_0 \left(\epsilon_{yy} + \epsilon'_{yy}\right) \frac{\hat{E}_y}{\hat{H}_x}$$

$$= 2\pi f \sqrt{\mu_0 \,\epsilon_0 \left(\epsilon_{yy} + \epsilon'_{yy}\right)} = \underline{2\pi f \sqrt{\mu_0 \,\epsilon_0 \,\epsilon_{yy}}} \;\sqrt{1 + \epsilon'_{yy}/\epsilon_{yy}}$$

$$\uparrow$$
$$\beta_y$$

(5.42)

It becomes clear again from the above equations that the propagation constants β_x and β_y are related to the main fiber deformation only (which is described by $[\epsilon]_0$), whereas the new propagation constants β'_x and β'_y now also take into account all additional fiber perturbations described by the tensor $[\epsilon]_m$. If $[\epsilon]_m = 0$, then $\beta'_x = \beta_x$ and $\beta'_y = \beta_y$.

In order to solve the coupled DEQ system given in Eq. (5.36), exponential formulation

$$\begin{bmatrix} \underline{a}(z) \\ \underline{b}(z) \end{bmatrix} = C_i e^{-j\beta_i z} \begin{bmatrix} e_{ix} \\ e_{iy} \end{bmatrix} = C_i e^{-j\beta_i z}\, \vec{e}_i$$

(5.43)

can be applied, where C_i is first taken as an arbitrary constant. Later on, this constant will be determined by means of appropriate boundary conditions. Taking now the coupled DEQ system and substituting $\underline{a}(z)$ and $\underline{b}(z)$ from the above expression, two independent solutions of equal significance are obtained. Finally, the general solution is simply obtained by a linear superposition of both the solutions. We get

$$\begin{bmatrix} \underline{a}(z) \\ \underline{b}(z) \end{bmatrix} = C_1 e^{-j\beta_1 z} \begin{bmatrix} e_{1x} \\ e_{1y} \end{bmatrix} + C_2 e^{-j\beta_2 z} \begin{bmatrix} e_{2x} \\ e_{2y} \end{bmatrix}$$

$$= C_1 e^{-j\beta_1 z}\, \vec{e}_1 + C_2 e^{-j\beta_2 z}\, \vec{e}_2$$

(5.44)

We call constants β_1 and β_2 the *eigenvalues* and unit vectors $\vec{e}_1$ and $\vec{e}_2$ the *eigenvectors* of coupled DEQ system. For the polarization propagation in single-mode fiber, these quantities are quite important as they define the four characteristic parameters β_u, β_v, $\vec{e}_u$, and $\vec{e}_v$ of new eigenmodes $\underline{E}_u(z, t)\, \vec{e}_u$ and $\underline{E}_v(z, t)\, \vec{e}_v$ given in Eq. (5.24). By using exponential formulation given in Eq. (5.43), these parameters are given by

$$\beta_1 = \beta_u = \frac{1}{2}\left[\left(N_{11} + N_{22}\right) + \sqrt{\left(N_{11} - N_{22}\right)^2 + 4\left|N_{12}\right|^2}\right] \tag{5.45}$$

$$\beta_2 = \beta_v = \frac{1}{2}\left[\left(N_{11} + N_{22}\right) - \sqrt{\left(N_{11} - N_{22}\right)^2 + 4\left|N_{12}\right|^2}\right] \tag{5.46}$$

$$\vec{e}_1 = \vec{e}_u = \frac{1}{\sqrt{1 + \left|\dfrac{\beta_u - N_{11}}{N_{12}}\right|^2}}\begin{bmatrix} 1 \\ \dfrac{\beta_u - N_{11}}{N_{12}} \end{bmatrix} \tag{5.47}$$

and

$$\vec{e}_2 = \vec{e}_v = \frac{1}{\sqrt{1 + \left|\dfrac{\beta_v - N_{11}}{N_{12}}\right|^2}}\begin{bmatrix} 1 \\ \dfrac{\beta_v - N_{11}}{N_{12}} \end{bmatrix} \tag{5.48}$$

Both eigenvectors $\vec{e}_1 = \vec{e}_u$ and $\vec{e}_2 = \vec{e}_v$ are perpendicular to each other and, in addition, normalized i.e., their length equals 1. Mathematically, this relationship can be expressed as

$$\vec{e}_i \cdot \vec{e}_j = \delta_{ij} = \begin{cases} 1 \text{ if } i = j \\ 0 \text{ if } i \neq j \end{cases} \tag{5.49}$$

where $i \in \{1, 2\}$ and $j \in \{1, 2\}$. The symbol δ_{ij} is the Kronecker symbol [3]. Considering both propagation constant β_u and β_v and, in addition, the coefficients N_{11}, N_{12}, N_{21}, and N_{22} of the coupled DEQ system given in (5.36), following relationships can be deduced:

$$\frac{\beta_u - N_{11}}{N_{12}} = \frac{N_{21}}{\beta_u - N_{22}} \quad \text{and} \quad \frac{\beta_v - N_{11}}{N_{12}} = \frac{N_{21}}{\beta_v - N_{22}} \tag{5.50}$$

The most important parameters to assess the stability of polarization propagation in a single-mode fiber are given by the difference

$$\Delta\beta_{uv} = \beta_u - \beta_v = \sqrt{(N_{11} - N_{22})^2 + 4\,|N_{12}|^2} = \sqrt{\Delta\beta'^2 + 4\,|N_{12}|^2} \qquad (5.51)$$

between the propagation constants β_u and β_v and beat length

$$L_b = \frac{2\pi}{\Delta\beta_{uv}} \qquad (5.52)$$

Difference $\Delta\beta_{uv}$ given in Eq. (5.51) now also includes all fiber perturbations which are responsible for mode coupling. This is in contrast to the difference $\Delta\beta = \Delta\beta_{xy} = \beta_x - \beta_y$ given in Eq. (5.1).

The efficiency of undesired mode coupling in a fiber is primarily determined by the coefficient N_{12}. Again, it should be noted that mode coupling does not occur when N_{12} approaches to zero. It becomes clear from Eq. (5.51) that the influence of mode coupling is less when difference $\Delta\beta'$ in propagation constants is large. Since $\Delta\beta'$ strongly depends on $\Delta\beta$ (Eqs. 5.37 and 5.40), $\Delta\beta'$ can be increased by increasing $\Delta\beta$. If $\Delta\beta'$ is large in comparison with $2N_{12}$, then the following simple approximation can be used:

$$\Delta\beta_{uv} \approx \Delta\beta' \qquad (5.53)$$

In order to prevent or even to reduce undesired mode coupling, the ratio $|\Delta\beta'/N_{12}|$ should be as large as possible. If polarization-maintaining single-mode fibers are to be realized, then this fact has to be taken into account. In practice, a high ratio $|\Delta\beta'/N_{12}|$ can be achieved by realizing a very strong desired principal fiber deformation; for example, a highly elliptical fiber core of very strong, but equal fiber torsion. Thereby, this perfectly specified principal deformation will always be dominant in comparison with all other undesired geometrical perturbations along the fiber (Section 5.3).

The principal fiber deformation is a well-defined and desired fiber distortion, whereas all other perturbations are of random in nature and undesired. In this particular case of prime importance, polarization propagation is completely defined and characterized by the principal fiber deformation only. Thus, state of polarization is maintained along the entire fiber length provided only one of the two eigenmodes has been stimulated at the fiber input. When the ratio $|\Delta\beta'/N_{12}|$ is low, second mode is also stimulated within a short distance from the fiber input since mode coupling is not negligible now. As already shown in Figs. 5.3 and 5.4, this results in a continuous change of polarization along the fiber.

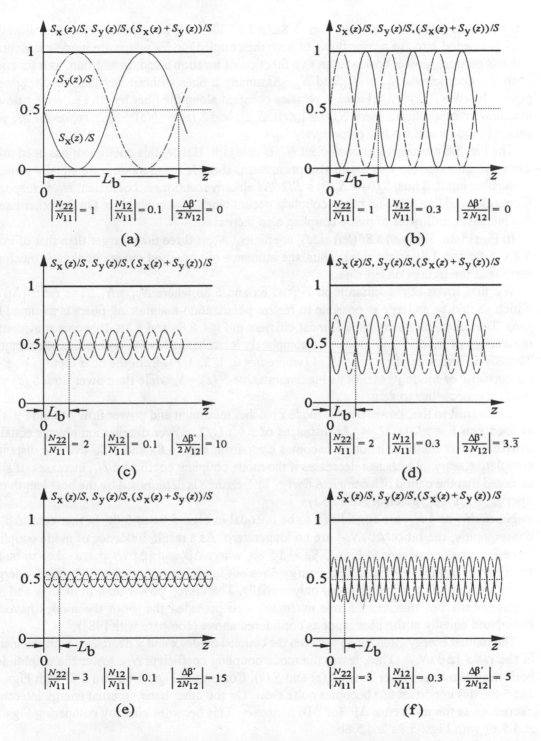

Fig. 5.8: Division of constant power flow $S = S_x(z) + S_y(z)$ in both coupled and orthogonal modes (polarizations) of single-mode fiber

In order to make it more clear, Figs. 5.8a to 5.8f illustrate how the optical power flow S in a fiber is divided into the power flows of both the coupled modes which are mode x and mode y. For this purpose, power flow is taken as a function of location z and, in addition, as a function of both the coefficients $N_{12} = N_{21}^*$ and N_{22}. Assuming a fiber without any loss, resulting optical power flow $S = S_x(z) + S_y(z)$ always remains constant along the fiber length i.e., power flow S is independent of location z. Here, $S_x(z) = |\underline{a}(z)|^2 \cdot S_{x,max}$ and $S_y(z) = |\underline{b}(z)|^2 \cdot S_{y,max}$ represent the power flow of mode x and mode y respectively.

The Fig. 5.8 presumes that coefficient N_{22} is constant. Hence, this coefficient has been used as a normalization factor. This figure also presumes that mode x and mode y are equally stimulated at the fiber input. Thus, $S_x(0) = S_y(0) = S/2$. As already mentioned, coefficient N_{12} is responsible for mode coupling since no mode coupling occurs when N_{12} equals zero. On the other hand, as N_{12} increases, influence of mode coupling also increases.

In Figs. 5.8b, 5.8d and 5.8f (left side), coefficient N_{12} is three times larger than that of in Figs. 5.8a, 5.8c and 5.8e (right side). Thus, the influence of undesired mode coupling is much more serious in the figures on left side.

We first focus our discussion on Figs. 5.8a and 5.8b where $N_{22} = N_{11}$. The ratio $|\Delta\beta'/N_{12}|$ which should be as large as possible to realize polarization-maintaining fibers is assumed to be zero. Therefore, mode coupling is most efficient in Figs. 5.8a and 5.8b. It becomes evident that the energy of mode x and mode y is completely interchanged twice within the beat length L_b. Thereby, in distances of $z = nL_b + L_b/4$ (where n = 0, 1, 2, 3, $\cdots$) from the fiber input (i.e., $z = 0$), power flow of mode y reaches its maximum value $S_y(z) = S$, while the power flow $S_x(z) = 0$ of mode x approaches to zero.

In contrast to this, power flow of mode x reaches maximum and power flow of mode y is zero at locations $z = nL_b + 3L_b/4$. At distances of $z = nL_b/2$, power distribution always equals the distribution at the fiber input. It becomes clear from Figs. 5.8a and 5.8b that the distance of complete energy interchange decreases if the mode coupling coefficient N_{12} increases. It should be noted that the period of a complete energy interchange is determined by the beat length of the fiber $L_b = 2\pi/\Delta\beta_{uv}$ given in Eq. (5.52).

Next, N_{11} and N_{22} are considered to be unequal in Figs. 5.8c to 5.8f. In that case, $\Delta\beta'$ and, consequently, the ratio $|\Delta\beta'/N_{12}|$ are no longer zero. As a result, influence of mode coupling is less efficient as compared to Figs. 5.8a and 5.8b, where $\Delta\beta'$ and $|\Delta\beta'/N_{12}|$ are taken to be zero. In contrast to Figs. 5.8a and 5.8b, energy does not interchange completely. Instead, energy of mode x and mode y is interchanged only partially. Therefore, power flow of mode x and mode y neither reaches maximum S nor minimum zero provided that both the modes have been stimulated equally at the fiber input as considered above (compare with [48]).

The mutual energy interchange between the coupled modes x and y decreases with the increase in the ratio $|\Delta\beta'/N_{12}|$. Thus, lesser the mode coupling coefficient N_{12}, lower the amplitudes of both the periodical power flows $S_x(z)$ and $S_y(z)$. Comparing Figs. 5.8d and 5.8f with Figs. 5.8c and 5.8e, this important fact becomes quite clear. On the other hand, mutual energy interchange decreases as the difference $\Delta\beta'$ (or $\Delta\beta$) increases. This becomes clear by comparing Figs. 5.8e and 5.8f with Figs. 5.8a and 5.8b.

The lowest energy interchange can be observed in Fig. 5.8e where the ratio $|\Delta\beta'/N_{12}|$ reaches its maximum value. Here, both the modes are only coupled very weakly and their power flows which are close to the average $S/2$ are only changed insignificantly. This means that mode coupling is practically negligible in Fig. 5.8e.

In the ideal case $|\Delta\beta'/N_{12}| \to \infty$ or $|\Delta\beta/N_{12}| \to \infty$ which can never be achieved in practice, no mode coupling exists and both the power flows remain constant. If both the modes are equally stimulated at the fiber input, constant power flow is given by the average $S/2$. If only one single mode is exactly stimulated e.g., mode x, then this mode retains the total, maximum power flow $S_x(z) = S_{x,max} = S$ along the entire fiber length. Thereby, mode y remains unstimulated i.e., $S_y(z) = 0$.

Now, both the terms $\underline{a}(z)$ and $\underline{b}(z)$ are completely determined and we are able to describe any energy interchange of coupled modes in a single-mode fiber. In the following, we have to come back again to the calculation of characteristic eigenmodes of single-mode fiber. For this purpose, we first consider Eq. (5.21) again where the substitution of the terms $\underline{a}(z)$ and $\underline{b}(z)$ must be made. In the second step, we shall compare this field equation with the eigenmode formulation given in Eq. (5.24). This will give us a generally valid solution for determining the optical field as well as characteristic eigenmodes of single-mode fiber. After discussing the generally valid solutions (see below), some special cases of practical importance will be used to explain the general results of this Section in more detail.

(i) GENERAL SOLUTION

In order to obtain a general valid solution, we have to combine solution given in Eq. (5.43) for coupled DEQ system (Eq. 5.36) and electric field Eq. (5.21) which describes the electric field vector of an optical fiber wave. By applying some straightforward mathematical operations, we obtain

$$
\begin{aligned}
\vec{E}(z,\,t) &= C_1\hat{E}_x\,e^{-j\beta_1 z}\,e^{j2\pi ft}\,\vec{e}_1 \;+\; C_2\hat{E}_y\,e^{-j\beta_2 z}\,e^{j2\pi ft}\,\vec{e}_2 \\[4pt]
&= \hat{E}_u\,e^{-j\beta_u z}\,e^{j2\pi ft}\,\vec{e}_u \quad\;\; + \hat{E}_v\,e^{-j\beta_v z}\,e^{j2\pi ft}\,\vec{e}_v \\[4pt]
&= \underline{E}_u(z,\,t)\,\vec{e}_u \qquad\qquad\;\; + \underline{E}_v(z,\,t)\,\vec{e}_v
\end{aligned}
\qquad (5.54)
$$

$$\uparrow \qquad\qquad\qquad\qquad \uparrow$$

eigenmode u ← **independent** → **eigenmode v**

which is exactly related to the eigenmode formulation given in Eq. (5.24). From the above equation, it becomes immediately clear that both the eigenvalues β_1 and β_2 and eigenvectors $\check{e}_1$ and $\check{e}_2$ of the coupled DEQ system are absolutely identical to the related characteristic quantities β_u, β_v, $\check{e}_u$ and $\check{e}_v$ of both the new eigenmodes (compare Eq. 5.24). As seen from Eq. (5.54), constants C_1 and C_2 which are still undefined have been combined with the electric field amplitudes $\hat{E}_x$ and $\hat{E}_y$. As a result, amplitudes $\hat{E}_u$ and $\hat{E}_v$ have to be determined now by appropriate boundary conditions instead of constants C_1 and C_2. A typical and frequently used boundary condition is given by a fixed and well-determined light wave at the input to the fiber at $z = 0$.

The amplitudes of the electric (Eq. 5.54) and magnetic fields (Eq. 5.25) are related by the following equation:

$$\frac{\hat{E}_u}{\hat{H}_v} = \frac{2\pi f \mu_0}{\beta_u} \qquad \frac{\hat{E}_v}{\hat{H}_u} = \frac{2\pi f \mu_0}{\beta_v} \qquad\qquad (5.55)$$

As seen from Eq. (5.54), it is always possible, in principle, to find two eigenmodes which propagate independently without any mode coupling either along the entire fiber or even along a small piece of fiber. The characteristic quantities β_u, β_v, $\vec{e}_u$ and $\vec{e}_v$ of both the eigenmodes are completely determined by the fiber material, type of fiber deformation and frequency of light as given in Eqs. (5.45) to (5.48). Remembering again our question in the beginning of this Section whether characteristic eigenmodes are existing in a geometrically perturbed single-mode fiber or not, this question can now be answered by yes.

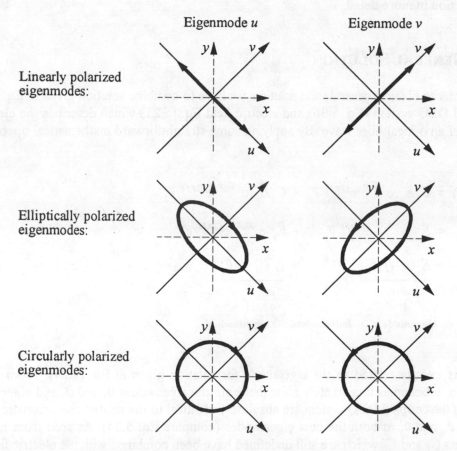

Fig. 5.9: Possible polarizations of both the orthogonal eigenmodes of single-mode fiber

It has been assumed so far that all four coefficients N_{11}, N_{12}, N_{21} and N_{22} of the coupled DEQ system (Eq. 5.36) are real. In such a case, eigenvectors $\vec{e}_u$ and $\vec{e}_v$ of the new eigenmodes (i.e., eigenmode u and eigenmode v) are real also. It can be simply proved from Eqs. (5.47) and (5.49). Thus, no phase difference exists between the vector components of each eigenvector. For this reason, eigenmodes of the optical fiber wave are called *linearly polarized eigenmodes* since the electric field vector $\underline{E}_u(z, t)\vec{e}_u$ of eigenmode u is always moving on a *linear locus curve* which is same as the u-axis of rectangular uv-coordinate system at a certain location z. Similarly, the electric field vector $\underline{E}_v(z, t)\vec{e}_v$ of eigenmode v is also moving on a linear locus curve which is, however, perpendicular to the locus curve of $\underline{E}_u(z, t)\vec{e}_u$. Thus, the locus curve of $\underline{E}_v(z, t)\vec{e}_v$ is same as the v-axis of the coordinate system as shown in Fig. 5.9 above.

It is, of course, not really necessary that a superposition of two linearly polarized eigenmodes yields again a linearly polarized wave. Because a difference in phase $\Phi(z) = \Delta\beta_{uv}z$ normally exists between the complex field vectors $\underline{E}_u(z, t)\vec{e}_u$ and $\underline{E}_v(z, t)\vec{e}_v$, polarization of superimposed optical wave usually varies with location z. This typical behaviour of single-mode fibers has already been discussed in Section 5.1.1 (Figs. 5.3 and 5.4). There, the linearly polarized eigenmodes are same as the eigenmode x and eigenmode y and the related locus curves of the perpendicular field vectors are identical to the x- and y-axes of the xy-coordinate system.

It should be noted that the orientation of the rectangular xy-coordinate system which is always located in the cross-section of the fiber core (xy-section) can be chosen arbitrarily. In contrast to this, z-direction is always same as the fiber axis (as considered in the beginning of this Section) and perpendicular to the xy-section. For this reason, we are able to turn the xy-section until x- and y-axes coincide with the characteristic principal axes of the fiber (i.e., u- and v-axes). In that case, eigenmodes x and y are identical with eigenmodes u and v. The main advantage of turning the coordinate system is that all formulas of the Section 5.1.1 can also be applied to the new eigenmodes of the perturbed single-mode fiber. For example, ellipticity η, angle of elevation ξ and degree of polarization P of both the new eigenmodes can easily be determined by substituting $\Delta\beta_{uv} \to \Delta\beta$ and $\Phi(z) = \Delta\beta_{uv}z \to \Phi(z) = \Delta\beta z$ and using Eqs. (5.6), (5.7) and (5.8). However, we have to consider that the angle of elevation ξ now describes the elevation of polarization ellipse to the turned x-axis (which is same as the u-axis) instead of the unturned x-axis. Similarly, all other characteristic parameters given in Section 5.1.1 can be determined for the new eigenmodes. Therefore, it is not required to give the modified equations.

As already mentioned, coefficients N_{11}, N_{12}, N_{21} and N_{22} of the coupled DEQ system (Eq. 5.36) can also be complex. In such a case, complex eigenvectors $\vec{e}_u$ and $\vec{e}_v$ are obtained and they exhibit a phase difference between the vector components of each eigenvector. Hence, both the eigenmodes are no longer linearly polarized, instead they are elliptically polarized in general. For this reason, these eigenmodes are called *elliptically polarized eigenmodes*. However, both the eigenmodes remain orthogonal to each other in this case also (Fig. 5.9).

When the phase difference between the vector components of eigenvectors $\vec{e}_u$ and $\vec{e}_v$ exactly equals $\pi/2$, both the eigenvectors describe a circle in the uv-plane. For this reason, eigenmodes are called *circularly polarized eigenmodes*. (Fig. 5.9).

The coefficients N_{11}, N_{12}, N_{21} and N_{22} are so far considered to be always constant. However, in practice these coefficients are usually constant in a short piece of fiber only. Moreover, they are functions of location z. This fact is normally taken into account when fiber perturbations are arbitrarily changing along the fiber. In that case, solving the DEQ system becomes complicated and comprehensive. Mostly, a solution can only be obtained by means of approximations and computer. However, an important exception is given when fiber perturbation changes periodically with location z; for example, when equal torsion is applied to the fiber (Section 5.3).

(ii) FIBER WITHOUT MODE COUPLING ($N_{12} = 0$)

Consider an ideal fiber without any mode coupling. In that case, all the coefficients of tensor $[\epsilon]_m$ are zero except the diagonal components $\epsilon'_{xx}\epsilon'_{yy}$ and ϵ'_{zz} As a result, mode coupling coefficients N_{12} and N_{21} of the coupled DEQ system (Eq. 3.36) are also zero. Thus, the DEQ system can be split into two independent uncoupled differential equations which can easily be solved by applying a simple exponential formulation. The solution obtained is the following:

$$\underline{a}(z) = C_1 \, e^{-j\beta_1 z} \, \vec{e}_1 \tag{5.56}$$

$$\underline{b}(z) = C_2 \, e^{-j\beta_2 z} \, \vec{e}_2 \tag{5.57}$$

This solution is a special case of the general solution given in Eq. (5.44). Hence, eigenvalues and eigenvectors of the DEQ system can again be determined from Eqs. (5.45) to (5.48) above. With $N_{12} = 0$, these equations get simplified to

$$\beta_1 = \beta_u = N_{11} = \beta'_x \tag{5.58}$$

$$\beta_2 = \beta_v = N_{22} = \beta'_y \tag{5.59}$$

$$\vec{e}_1 = \vec{e}_u = \vec{e}_x \tag{5.60}$$

$$\vec{e}_2 = \vec{e}_v = \vec{e}_y \tag{5.61}$$

Next we have to take into account the field as given in Eq. (5.21). Making substitutions for $\underline{a}(z)$, $\underline{b}(z)$, both the eigenvalues and eigenvectors as given from Eqs. (5.56) to (5.61), we get

$$\vec{E}(z, \, t) = \underline{\hat{E}_x \, e^{-j\beta'_x z} \, e^{j2\pi f t} \, \vec{e}_x} \quad + \quad \underline{\hat{E}_y \, e^{-j\beta'_y z} \, e^{j2\pi f t} \, \vec{e}_y} \tag{5.62}$$

$$\uparrow \qquad\qquad\qquad\qquad\qquad\qquad \uparrow$$

eigenmode u $\leftarrow$ independent $\rightarrow$ eigenmode v
= eigenmode x = eigenmode y

With the exception of new propagation constants β'_x and β'_y, above solution exactly yields the field Eq. (5.21). Thus, both the characteristic eigenmodes remain same. As the mode coupling does not disturb the optical fiber wave, change of the principal axes does not occur (Eqs. 5.60 and 5.61). However, the propagation constants and, so, the velocities of propagation are changed ($\beta_x \rightarrow \beta'_x$ and $\beta_y \rightarrow \beta'$), due to various geometrical fiber perturbations which are analytically defined by the diagonal tensor components ϵ'_{xx}, ϵ'_{yy}, and ϵ'_{zz}. It should be remembered that the above solution is a special case of general solution. This particular case is, however, not observed in practice since geometrical perturbations normally do not influence the diagonal components of $[\epsilon]_m$. Instead, the other tensor components are influenced which result in mode coupling.

(iii) FIBER WITHOUT PERTURBATIONS

In this Section, fiber without perturbations is assumed to be a fiber which is deformed by one well-defined and desired deformation only; for example, a highly elliptical fiber core. No additional undesired geometrical perturbations exist in the fiber. Thus, the term "without perturbations" mathematically means that all the nine components of tensor $[\epsilon]_m$ are zero, while $[\epsilon]_0$ describes the elliptical fiber core (Eq. 5.33). As a result, no mode coupling and also no change in velocity of propagation of eigenmodes are there. Hence, both the modes given in Eq. (5.21) always remain the eigenmodes along the entire fiber length. The propagation constants β_x and β_y are determined by the diagonal components of tensor $[\epsilon]_0$ and frequency of light from Eqs. (5.41) and (5.42). Therefore, the electric field from Eq. (5.21) is given by

$$\vec{E}(z,\ t) = \hat{E}_{\hat{x}}\ e^{-j\beta_x z}\ e^{j2\pi ft}\ \vec{e}_x \quad + \quad \hat{E}_y\ e^{-j\beta_y z}\ e^{j2\pi ft}\ \vec{e}_y \tag{5.63}$$

$$\uparrow \qquad\qquad\qquad\qquad\qquad\qquad \uparrow$$

eigenmode x ← independent → eigenmode y

(iv) STIMULATION OF SINGLE EIGENMODE

Let us now assume that additional geometrical imperfections impair the polarization propagation of single-mode fiber. In this case of practical importance, it is again possible to determine two new eigenmodes propagating independently without any undesired mode coupling. As explained earlier, these new eigenmodes (i.e., eigenmodes u and v) are completely and unambiguously determined by the eigenvalues and eigenvectors of the coupled DEQ system as given in Eq. (5.36).

Assuming further that only one eigenmode is stimulated at the fiber input (for example, eigenmode u), then this single mode propagates along the entire fiber. In contrast, the second eigenmode always remains unstimulated. Thus, the electric field is completely determined by the field of eigenmode u. If eigenmode v is stimulated, then the electric field of the optical fiber wave becomes same as the field of this mode. In conclusion, we can write

$$\vec{E}(z, t) = \begin{cases} \hat{E}_u \ e^{-j\beta_u z} \ e^{j2\pi ft} \ \vec{e}_u & \text{if eigenmode } u \text{ is stimulated} \\ \hat{E}_v \ e^{-j\beta_v z} \ e^{j2\pi ft} \ \vec{e}_v & \text{if eigenmode } v \text{ is stimulated} \end{cases} \qquad (5.64)$$

As seen from the above equation, polarization of an optical fiber wave is always same as the polarization of the eigenmode which has been stimulated at the fiber input. Since this polarization does not alter with location z, state of polarization remains unchanged along the entire fiber length. However, this special type of polarization-maintaining fiber first requires that all fiber deformations are known, second they are completely described by both the tensors $[\epsilon]_0$ and $[\epsilon]_m$ and third additional fiber perturbations do not occur.

When the fiber perturbations change randomly with time, principal axes $\vec{e}_u$ and $\vec{e}_v$ of both the eigenmodes and, consequently, the fiber polarization also changes randomly with time. Since this is generally true in practice, commercially available single-mode fibers are normally not suitable to maintain polarization.

In order to realize *polarization-maintaining fibers*, this particular problem must be solved. If a fiber is extremely disturbed by a fixed and perfectly specified deformation, for example, a highly-elliptical fiber core or a very strong, but equal torsion (Section 5.3), then all the other random perturbations are practically negligible in comparison with this dominant fiber deformation. In such a case, polarization propagation is completely characterized and determined by the dominant fiber deformation only. Hence, state of polarization always remains maintained provided only one eigenmode is stimulated at the fiber input.

5.2   MATRIX OF POLARIZATION PROPAGATION

In this Section, boundary conditions which are usually determined by the optical wave at the input to the fiber (i.e., $z = 0$) will be considered in more detail. For this purpose, stimulating optical wave

$$\vec{E}(0, t) = \begin{pmatrix} E_x(0, t) \\ E_y(0, t) \end{pmatrix} = \begin{pmatrix} E_{x0} \\ E_{y0} \end{pmatrix} e^{j2\pi ft} \qquad (5.65)$$

will be assumed at the fiber input. Substituting this equation in the general solution given in Eq. (5.54) for the electric field of an optical fiber wave, following result is obtained after some mathematical operations [6]:

$$\vec{E}(z,\ t) = \begin{pmatrix} \underline{E}_x(z,\ t) \\ \underline{E}_y(z,\ t) \end{pmatrix} = \begin{pmatrix} m_{11} & m_{12} \\ m_{21} & m_{22} \end{pmatrix} \begin{pmatrix} \underline{E}_{x0} \\ \underline{E}_{y0} \end{pmatrix} e^{-j0.5(N_{11}+N_{22})z}\ e^{j2\pi ft}$$

$$\uparrow$$
$$(m_{ij})$$

$$(5.66)$$

The matrix (m_{ij}) is usually termed as the *polarization-propagation matrix*. This matrix is defined by four coefficients which can be calculated by substituting Eq. (5.65) in Eq. (5.54). We obtain

$$m_{11} = \cos(0.5\Delta\beta_{uv}z) - j\frac{\Delta\beta_{uv}}{\Delta\beta'}\sin(0.5\Delta\beta_{uv}z) = m_{22}^* \qquad (5.67)$$

$$m_{12} = j\frac{2N_{12}}{\Delta\beta_{uv}}\sin(0.5\Delta\beta_{uv}z) = -m_{21}^* \qquad (5.68)$$

By taking into account matrix (m_{ij}), we can determine the electric field of an optical fiber wave at any location z as a function of any stimulating optical wave at the fiber input ($z = 0$). Subsequently, all its characteristic parameters can be determined by applying the appropriate formulas given in Section 5.1.1; for example, polarization ellipse, ellipticity, angle of elevation and degree of polarization. With the help of matrix (m_{ij}), each single-mode fiber can be represented as a simple two-port network characterized by the four matrix coefficients m_{11}, m_{12}, m_{21} and m_{22} and complex factor $\exp[-j0.5(N_{11} + N_{22})z]$. All kinds of geometrical imperfections are included in this network representation (Fig. 5.10).

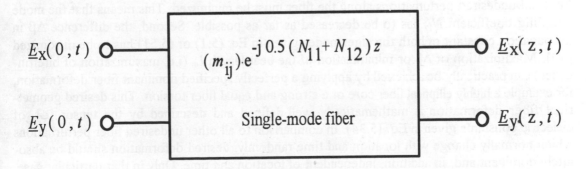

Fig. 5.10: Two-port network representation of single-mode fiber

5.3   REDUCTION OF POLARIZATION FLUCTUATIONS

This Section is focussed on technologies which are able to avoid or at least to reduce the influence of polarization fluctuations. As explained in the previous Sections, the two independent fundamental reasons for polarization fluctuations in a single-mode fiber are

- different velocities of propagation of both the orthogonal eigenmodes discussed in Section 5.1.1 and

- mode coupling explained in Section 5.1.2.

When only one of the two eigenmodes is excited at the fiber input, a stable transmission of polarization is possible in principle. In presence of mode coupling during transmission, second eigenmode is also excited. As a result, state of polarization of the optical fiber wave changes as a function of location z. Typical variations of polarization have already been shown in Figs. 5.3 and 5.4.

Further, we have discussed in Section 5.1.2 that always two independent and orthogonal eigenmodes are existing even if transmission is disturbed by mode coupling. If only one of these "new" eigenmodes is stimulated at the fiber input, then a polarization-maintaining transmission is possible again. However, all geometrical fiber perturbations and deformations must be absolutely time-invariant. Otherwise, the principal axes of the eigenmodes are temporally changed and polarization fluctuations occur again since both the eigenmodes are now stimulated in the fiber. Because this is normally true in practice, ordinary single-mode fibers are not suitable to preserve state of polarization.

5.3.1   POLARIZATION-MAINTAINING SINGLE-MODE FIBERS

For the reason mentioned above, the goal of realizing polarization-maintaining fibers, frequently called single-mode single-polarization fibers or briefly SMSP fibers, must be twofold: First, all undesired perturbations along the fiber must be minimized. This means that the mode coupling coefficient N_{12} has to be decreased as far as possible. Second, the difference $\Delta\beta$ in propagation constant of both the eigenmodes given in Eq. (5.1) or (5.51) has to be maximized [40]. Maximization of $\Delta\beta$ or minimization of the beat length L_b (i.e, maximization of birefringence) can practically be achieved by applying a perfectly specified dominant fiber deformation, for example a highly elliptical fiber core or a strong and equal fiber torsion. This desired geometrical fiber deformation is mathematically well-defined and described by the tensor $[\epsilon]_0$ of dielectric constants given in Eq. (5.34). In comparison to all other undesired fiber perturbations which normally change with location and time randomly, desired deformation should be absolutely dominant and, in addition, independent of location and time. Only in that particular case, all undesired and random fiber perturbations are negligible. As a result, eigenmodes and polarization propagation are completely characterized by the dominant fiber deformation only. Hence, state of polarization is absolutely maintained provided only one eigenmode is excited at the fiber input. Since maximization of birefringence is the basis of most polarization-maintaining fibers, these fibers are frequently called *birefringent fibers*.

Depending on the type of realized desired deformation, different eigenvalues and eigenvectors can be obtained (Eqs. 5.45 to 5.48). As a consequence, eigenmodes may exhibit different kinds of polarization. For this reason, polarization-maintaining fibers can be classified as follows:

- single-mode fibers with linearly polarized eigenmodes due to

  - an axial asymmetrical fiber core (e.g., elliptical-core fibers) or

  - an axial asymmetrical pressure on the fiber (e.g., stress-induced fibers),

- single-mode fibers with circularly polarized eigenmodes due to equal torsion of fiber and

- absolutely polarization-maintaining single-mode fibers.

In the following subsections, above fibers will be discussed in more detail.

(i) SINGLE-MODE FIBERS WITH LINEARLY POLARIZED EIGENMODES

These fibers are also called linear highly birefringent fibers since the difference $\Delta\beta$ in the propagation constants and, therefore, the birefringence is extremely large. Linear birefringent fibers are characterized by a well-defined deformation of the fiber core; for example, a highly elliptical fiber core as already mentioned. Fibers with a non-circular core have been first studied in 1978 [46]. Further examinations were primarily focussed on elliptical fiber cores [1, 5, 18, 59]. An important and common theoretical result of all these studies was that the beat length L_b which has to be minimized is inversely proportional to Δ^2 where

$$\Delta = \frac{n_1^2 - n_2^2}{2n_1^2} \tag{5.69}$$

represents the relative refraction index change between fiber core (n_1) and cladding (n_2). Thus, a short beat length L_b requires a large index change Δ. On the other hand, a large Δ is always combined with two significant drawbacks: first, fiber attenuation increases and second core diameter to provide single-mode operation decreases. Practically, beat lengths in the order of 1 mm can be achieved by applying elliptical fiber cores. In comparison, commercially available single-mode fibers without polarization-maintaining mechanism exhibit a beat length in the range of some 10 cm to 2 m. This range is also valid for all kinds of undesired random fiber perturbations which are always existing in addition to the desired elliptical fiber core. Taking into account a first order approximation, a characteristic ratio $|\Delta\beta/(2\,N_{12})|$ in the range of 100 to 1000 can be achieved in the polarization-maintaining fibers with elliptical fiber core (Fig. 5.8).

An improvement, especially in the fiber attenuation can be achieved when the fiber is deformed by an axial asymmetrical pressure [8, 11, 13, 47, 56, 57, 58]. Practically, this transverse stress can easily be realized by different temperature coefficients of expansion of fiber core and cladding. Thereby, beat lengths similar to elliptical core fibers can be obtained. However, fiber attenuation is much less as compared to the earlier case.

All single-mode linearly birefringent fibers show a common disadvantage of prime importance. If two fibers are to be connected, then the principal axes of both eigenmodes in the fibers must be exactly on top of each other. Otherwise, both the eigenmodes will be excited in the connected fiber and polarization fluctuations will occur again.

(ii) SINGLE-MODE FIBERS WITH CIRCULARLY POLARIZED EIGENMODES

A significant improvement in the polarization matching of coupled fibers can be achieved by using a circularly birefringent fiber which will be examined now in more detail [2, 10, 19, 26, 61]. Circularly birefringent fibers are realized by an equal torsion of the fiber about its axis as shown in Fig. 5.11.

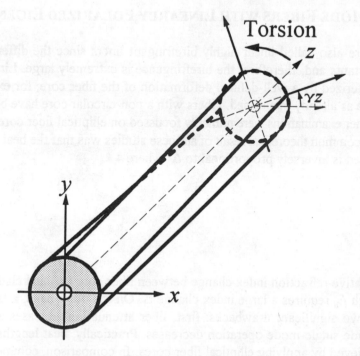

Fig. 5.11: Torsion of ideal circular single-mode fiber

To ensure that the fiber torsion is the dominant deformation along the fiber length, degree of torsion should be sufficiently large. Only in that case, all other undesired and random perturbations are negligibly small and polarization transmission is completely characterized by the fiber torsion only. For an ideal circular single-mode fiber with uniform torsion, coefficients N_{11}, N_{12}, N_{21} and N_{22} of the coupled DEQ system given in Eq. (5.36) are determined by [51]

$$N_{11} = N_{22} = \beta_0 = 2\pi f \sqrt{\mu_0 \epsilon_0} \qquad (5.70a)$$

$$N_{12} = -jp\gamma = N_{21}^* \qquad (5.70b)$$

Here, β_0 represents the propagation constant for an ideal circular single-mode fiber without any torsion. Twist rate γ given in unit of rad/m and constant p (e.g., $p \approx 0.07$ [50, 61]) describe the degree of torsion for a given fiber length. The coefficients N_{12} and N_{21} which are responsible for mode coupling are not real now, rather they are imaginary. This important fact results in a phase difference of exactly $\pi/2$ between the vector components (x- and y-components) of both eigenvectors $\vec{e}_u$ and $\vec{e}_v$ (see below). Therefore, both eigenmodes become circularly polarized as required above. In order to determine eigenvalues and eigenvectors using Eqs. (5.45) to (5.48), four coefficients given in Eq. (5.70) are substituted. We get

$$\beta_1 = \beta_u = \beta_0 + p\gamma \qquad (5.71)$$

$$\beta_2 = \beta_v = \beta_0 - p\gamma \qquad (5.72)$$

$$\vec{e}_1 = \vec{e}_u = \frac{1}{\sqrt{2}} \begin{bmatrix} 1 \\ +j \end{bmatrix} \qquad (5.73)$$

and

$$\vec{e}_2 = \vec{e}_v = \frac{1}{\sqrt{2}} \begin{bmatrix} 1 \\ -j \end{bmatrix} \qquad (5.74)$$

It becomes clear from Eqs. (5.71) and (5.72) above that the circular birefringence which is given by $D = \lambda/(2\pi)\Delta\beta = \lambda/(2\pi)[\beta_1 - \beta_2]$ is directly proportional to $\Delta\beta = \beta_1 - \beta_2$ and, hence, twist rate γ. Finally, we have to determine the matrix (m_{ij}) by taking into account either the eigenvectors and eigenvalues as just derived above or the coefficients N_{11}, N_{12}, N_{21} and N_{22} of the coupled DEQ-system given in Eq. (5.36). We obtain

$$(m_{ij}) = \begin{bmatrix} m_{11} & m_{12} \\ m_{21} & m_{22} \end{bmatrix} = \begin{bmatrix} \cos(p\gamma z) & -\sin(p\gamma z) \\ \sin(p\gamma z) & \cos(p\gamma z) \end{bmatrix} \qquad (5.75)$$

It is evident from the above equation that (m_{ij}) simply represents a matrix of transformation which turns the present system of coordinates (the xy-system) by an angle $p\gamma z$. Therefore, polarization at the fiber input is uniformly turned during propagation through the optical fiber. At the output of the fiber, polarization is finally turned by the angle mentioned above. Thereby, type of polarization remains exactly the same. This means that all the characteristic parameters of the polarization ellipse remain unchanged as shown in Fig. 5.12 i.e., η and P are not a function of z.

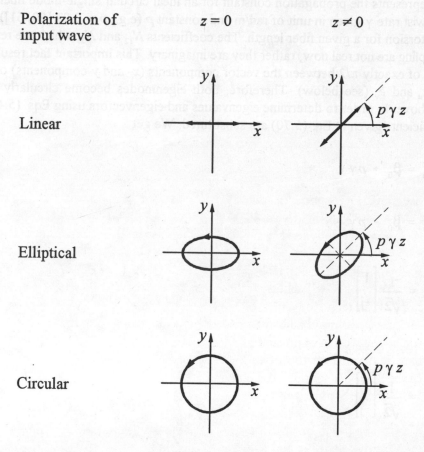

Fig. 5.12: Transmission of polarization along single-mode fiber with circularly polarized eigenmodes

The main advantage of a polarization-maintaining fiber with circularly polarized eigenmodes is its simple technique of connection. In contrast to a fiber with linearly polarized eigenmodes, no adjustment of the principal axes is required [23]. High stability of polarization transmission is second advantage since the influence of undesired geometrical fiber perturbations is much lower than that of linearly polarized eigenmodes. Moreover, realization of fibers with circularly polarized eigenmodes can be achieved more easily.

It is obvious that fiber torsion is practically limited to a certain degree. Hence, all additional fiber perturbations cannot be neglected usually. If, for example, a fiber has a highly elliptical core in addition, then eigenmodes are to be determined afresh. However, each additional geometrical fiber distortion makes calculation more and more comprehensive [51 - 55].

(iii) ABSOLUTELY POLARIZATION-MAINTAINING SINGLE-MODE FIBERS

In contrast to the polarization-maintaining single-mode fibers discussed above, which actually have two modes, this type of fiber really exhibits one single mode only. Here, the second mode is completely suppressed by means of a frequency shift. Thereby, the undesired mode is shifted beyond the cut-off frequency where a propagation is not possible. However, cut-off frequencies of both the eigenmodes are very close to each other [7, 36]. Hence, even a small undesired fiber deformation may either activate the second mode again or suppress both the modes. For this reason, realization of absolutely SMSP fibers is, of course, a very difficult task.

In order to summarize the results obtained in this Section, following important features of polarization-maintaining single-mode fibers are given:

- First, the attenuation of SMSP fiber is always more than the attenuation of ordinary single-mode fiber i.e., a fiber without any polarization maintaining. Hence, the gain in transmission range obtained by applying coherent optical receivers (heterodyne or homodyne) is partly reduced.

- Second, splicing or connecting SMSP fibers is rather difficult and a fundamental drawback of SMSP fibers. In contrast to standard fibers where the fiber cores have to be adjusted only, each connection of two SMSP fibers additionally requires an accurate adjustment of the principal axes of both the fibers. Thereby, the connection is optimized when both core and, in addition, principal axes are exactly aligned on top of each other. Hence, realizing appropriate mechanical connectors is a serious problem. One exception is, however, when SMSP fibers with circularly polarized eigenmodes are used.

- Third, fabrication of SMSP fibers requires a specific technology which is much more comprehensive and sophisticated than in the case of conventional single-mode fibers.

- Fourth, in contrast to ordinary single-mode fibers, influence of polarization dispersion is increased since difference in velocities of both orthogonal polarizations is very large in SMSP fibers (Section 5.3.2). However, polarization dispersion does not occur when only one eigenmode is excited along the entire fiber length or bit rate is low.

As further discussion on SMSP fibers will considerably exceed the scope of this book, some important references will finally be given [12, 19, 34, 37, 39, 40, 49].

5.3.2   POLARIZATION DISPERSION OF SMSP FIBERS

With polarization-maintaining fibers, a common problem related to the speed of information transmission has to be discussed. Since both eigenmodes always propagate with different velocities, transmitted signal reaches the fiber output at different points of time. Thus, a rectangular input pulse no longer remains rectangular at the fiber output. Rather, the pulse is deformed and in particular broaden. As it is well-known from telecommunication engineering, this effect is called dispersion (Chapter four). Since different velocities of both the polarizations are the reason for dispersion in SMSP fibers, this type of dispersion is referred as *polarization dispersion* (Section 4.4.3).

Irrespective of type, dispersion always degrades the performance of digital communication systems since a broaden pulse normally yields undesired intersymbol interference (Section 9.5.1). It is obvious that the influence of intersymbol interference strongly increases with the increase in the speed of information transmission i.e., the bit rate. Normally, polarization dispersion is negligible in comparison with waveguide dispersion and chromatic dispersion. When bit rate increases, polarization dispersion becomes more and more evident. As confirmed by various measurements, polarization dispersion must be taken into account when modulation frequencies exceed about 4 GHz [4, 9, 22, 25, 27, 28, 41 - 43]. Since modern optical communication systems are going to operate in the multigigabit range (i.e., 10 Gbit/s to 100 Gbit/s), polarization dispersion will become more and more serious problem. Further, influence of dispersion also increases with the increase in the transmission distance. Hence, polarization dispersion is, in particular, a very serious problem in long-range multigigabit systems; for example, in transoceanic transmission links.

The polarization dispersion is primarily determined by the difference in velocities of both the eigenmodes. Therefore, higher this difference more is the polarization dispersion. As the difference in velocities is proportional to the difference $\Delta\beta$ in the propagation constants, it is high, in particular, when SMSP fibers are used. It should be remembered that a large $\Delta\beta$ or a high birefringence D is a fundamental feature of most SMSP fibers. When only one eigenmode is stimulated at the fiber input and no undesired fiber perturbations occur in addition, polarization dispersion does not exist. However, this is generally not true in practice.

5.3.3 POLARIZATION CONTROL

When polarization is not stabilized by means of polarization-maintaining fiber, alternate techniques have to be employed to match the states of polarization of received and local laser light waves. One important approach is to adjust the polarization by appropriate components which are able to influence the polarization of light wave. These optical components are called *retarders*. A retarder enables to change the polarization of light wave by a specified amount. It is, in principle, independent of whether the polarization of local laser or received light wave is controlled. Due to practical reasons, it is, however, more useful to place the retarder in the light path of the local laser since each retarder also exhibits a certain amount of attenuation. In system performance, an additional attenuation of the received light wave will be much more evident than a somewhat decreased effective local laser power.

Retarders are based either on electrooptical, magnetooptical or mechanical effects. In the latter effect, fiber polarization is changed by applying external pressure to the fiber surface. Thereby, fiber is specifically deformed and polarization is changed as described in Section 5.2. An equal pressure of specified amount and, consequently, an equal fiber deformation can be obtained either by employing electromagnetic- or piezo-based fiber squeezers [60].

In order to match two polarizations e.g., polarizations of local laser and received light waves in a coherent receiver (Fig. 5.13), it is required that local laser polarization which is normally linear can be converted to every desired state of polarization. Hence, changing polarization by a retarder must always be a well-defined process. In case of electro- and magnetooptical retarders, this task is performed by controlling an electric driving current which is required to generate the electric or magnetic field [14, 16]. Retarders based on mechanical pressure can also be controlled by an electric current. Depending on the magnitude of electric current, mechanical force is generated and a piece of fiber of certain length is deformed accordingly.

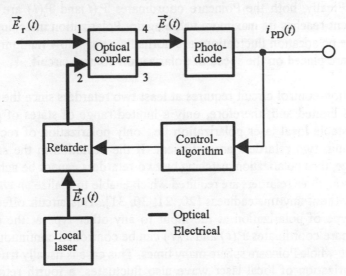

Fig. 5.13: Simplified block diagram of polarization-control circuit in single-diode coherent receiver

Each retarder acts as a control unit in polarization-control circuit. In addition, each polarization-control circuit requires a measure to determine the deviation of local laser polarization and polarization of received light wave. For this purpose, either the photodiode current $i_{PD}(t)$ or the intermediate frequency signal $i_{IF}(t)$ can be used. As explained in Section 5.1.1, current $i_{PD}(t)$ reaches its maximum value when the polarizations of local laser and received light waves are exactly same. This ideal case is characterized by two identical points $\mathcal{P}_r(t) = (S_{1r}(t), S_{2r}(t), S_{3r}(t))$ and $\mathcal{P}_l(t) = (S_{1l}(t), S_{2l}(t), S_{3l}(t))$ which are on top of each other on the Poincaré sphere. Any deviation from this ideal case decreases the effective part of the photodiode current $i_{PD}(t)$ given in Eq. (2.60). In the worst-case, where both the polarizations are perpendicular to each other, effective part of the photodiode current will become zero.

Deformation of the fiber is also the basis of another type of retarder called *Lefevre polarizer* [17]. Here, the fiber is carried around three cylindrical bodies (discs) which are sequentially located one after another. Number of coils (usually two, four and two on the first, second and third disc respectively) and diameter of the discs are chosen in such a way that the first and also the third disc represent a $\lambda/4$-plate and the second disc a $\lambda/2$-plate at a fixed wavelength (for example, 1.55 μm). State of polarization can be changed by turning the discs. Thereby, the fiber and, hence, the orientation of the principal axes are changed. Lefevre polarizers allow to transfer each type of polarization to every other desired type at the output (Chapter eight). However, such polarizers can only be controlled by hand and not automatically. Therefore, these polarizers are frequently used in laboratory experiments, but they are unsuitable for commercial use.

Maximization and stabilization of the amplitude of photodiode current must, therefore, be the goal of each polarization-control circuit. If, for example, the photodiode current decreases, then the retarder must change the polarization of local laser wave so long as both the polarizations are matched again. Finally, both the Poincaré coordinates $\mathcal{P}_r(t)$ and $\mathcal{P}_l(t)$ are congruent and the photodiode current reaches its maximum value again. Polarization matching always requires a certain time. Since polarization fluctuations are normally a very slow random process, no special strong demands are placed on the speed of polarization-control circuit.

Each polarization-control circuit requires at least two retarders since the dynamic range of a single retarder is limited and, therefore, only a limited range of states of polarization can be matched. With stable local laser polarization i.e., only polarization of received light wave is randomly changing, two retarders are sufficient. If the deviation in the state of polarization becomes very large, then polarization matching by two retarders cannot be achieved continuously [30]. For this reason, three retarders are required which enable to realize an *endless-polarization-control circuit* without any unsteadiness [20, 21, 30, 31]. This circuit offers the possibility to convert every type of polarization at the input to any other type at the output. Thus, each deviation in Poincaré coordinates $\mathcal{P}_r(t)$ and $\mathcal{P}_l(t)$ can be controlled continuously, even if $\mathcal{P}_r(t)$ is moving around the whole Poincaré sphere many times. This case is usually true in practice. When the state of polarization of local laser wave also fluctuates, a fourth retarder is required in addition to match both the random polarizations.

Control of all the retarders is accomplished by managing the related electric driving currents. This operation requires a complex and comprehensive control algorithm. The main task of this algorithm is to generate the appropriate driving currents for the retarders depending on the random changes in the photodiode current $i_{PD}(t)$. It can be performed, for example, by using a microprocessor. In that case, required algorithm is realized by a software program.

A polarization-control circuit reduces the influence of polarization fluctuations in the optical frequency range. This requires optical components such as retarders. However, realization of optical components is usually more difficult than a realization of electrical components. In addition, polarization control by means of permanent fiber deformation may finally yield a fiber break.

5.3.4 POLARIZATION-DIVERSITY RECEIVER

As mentioned above, a polarization-control circuit requires some critical optical components (for example, the retarder) which is a drawback. With a polarization-diversity receiver, this problem is shifted to the electrical frequency domain. Here, the influence of polarization fluctuations is reduced by a separate detection of both horizontal and vertical electric field components $E_x(t)$ and $\underline{E}_y(t)$ of the superimposed light wave $\vec{E}(t)$. For this purpose, two heterodyne receivers have to be realized as shown in Fig. 5.14; one for vertical field component and the other for horizontal component.

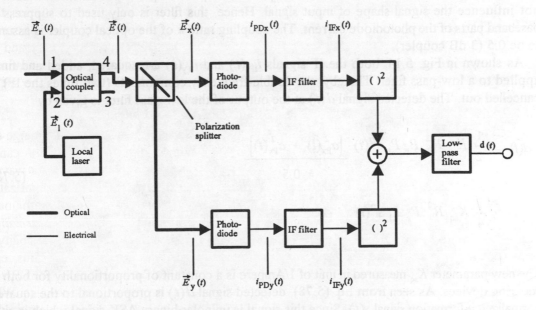

Fig. 5.14: Principle block diagram of polarization-diversity receiver

The output signals of both the receivers are finally combined. In the ideal case, this receiver configuration called the *polarization-diversity receiver* is able to extinguish the influence of polarization fluctuations completely. In comparison to polarization-control circuit, polarization-diversity receiver requires lesser number of optical components, but a higher number of electrical components. As an example, Fig. 5.14 shows the simplified block diagram of an ASK heterodyne polarization-diversity receiver. This receiver includes one local laser, one optical coupler, a polarization-beam splitter, two photodiodes, two identical electric signal paths (x-path and y-path), one sum up device and finally a low-pass filter. Similar to standard coherent receivers, detected signal $d(t)$ at the output of the receiver is applied to a sample and hold circuit and finally to a decision circuit.

The diversity receiver shown in Fig. 5.14 requires a stable and linear local laser polarization which shows an inclination of 45° to the horizontal x-axis. Thus, both the field components (x- and y-components) of the local laser light wave $\vec{E}(t)$ are equal. As shown in Chapter nine, intermediate frequency signals $i_{\mathrm{IFx}}(t)$ and $i_{\mathrm{IFy}}(t)$ are given by

$$i_{\mathrm{IFx}} = a_{\mathrm{Px}}(t) \, R_0 \, \sqrt{P_r P_1} \, s(t) \, \cos\left(2\pi f_{\mathrm{IF}} t + \phi_{\mathrm{Px}}(t)\right) \tag{5.76}$$

$$i_{\mathrm{IFy}} = a_{\mathrm{Py}}(t) \, R_0 \, \sqrt{P_r P_1} \, s(t) \, \cos\left(2\pi f_{\mathrm{IF}} t + \phi_{\mathrm{Py}}(t)\right) \tag{5.77}$$

For simplicity, laser phase noise and additive Gaussian noise (i.e., shot noise and thermal noise) has not been taken into consideration. Both the above equations presume that the IF filter does not influence the signal shape of input signal. Hence, this filter is only used to suppress the baseband parts of the photodiode current. The coupling ratio k of the optical coupler is assumed to be 0.5 (3 dB coupler).

As shown in Fig. 5.14, both the IF signals $i_{\mathrm{IFx}}(t)$ and $i_{\mathrm{IFy}}(t)$ are squared, added and finally applied to a low-pass filter. Thereby, the double-frequency components (two times the IF) are cancelled out. The detected signal $d(t)$ at the output of the low-pass filter is given by

$$d(t) = \frac{1}{2} \, K_{\mathrm{sq}} \, R_0^2 \, P_r \, P_1 \, s^2(t) \, \underbrace{\left[a_{\mathrm{Px}}^2(t) + a_{\mathrm{Py}}^2(t)\right]}_{= \, 0.5}$$

$$= \frac{1}{4} \, K_{\mathrm{sq}} \, R_0^2 \, P_r \, P_1 \, s^2(t) \tag{5.78}$$

The new parameter K_{sq} measured in unit of 1/Ampere is a constant of proportionality for both the squaring devices. As seen from Eq. (5.78), detected signal $d(t)$ is proportional to the square of normalized information signal $s(t)$. Since this signal is unipolar binary ASK signal which is either

0 or 1, the squaring process does not affect the result. Further, superposition of random factors $a_{Px}^2(t)$ and $a_{Py}^2(t)$ yields the constant 0.5. This will be proved in the following:

Proof:

As described in Section 9.4.3, polarization unit vectors $\vec{\varrho}_r(t)$ and $\vec{\varrho}_l(t)$ of the received and the local laser light waves are given by

$$\vec{\varrho}_r(t) = \begin{bmatrix} \varrho_{rx}(t) \\ \varrho_{ry}(t) \end{bmatrix} \quad \text{where } \vec{\varrho}_r(t)\, \vec{\varrho}_r^*(t) = |\varrho_{rx}(t)|^2 + |\varrho_{ry}(t)|^2 = 1 \tag{5.79}$$

$$\vec{\varrho}_l = \begin{bmatrix} \varrho_{lx} \\ \varrho_{ly} \end{bmatrix} = \begin{bmatrix} \dfrac{1}{\sqrt{2}} \\ \dfrac{1}{\sqrt{2}} \end{bmatrix} \quad \text{where } \vec{\varrho}_l(t)\, \vec{\varrho}_l^*(t) = 1 \tag{5.80}$$

As shown in Fig. 5.14, both the x-components are responsible for signal in the upper path, while the y-components in the lower path. The random amplitudes $a_{Px}(t)$ and $a_{Py}(t)$ as calculated in Chapter nine give rise to

$$a_{Px}^2(t) + a_{Py}^2(t) = \frac{1}{2}|\varrho_{rx}(t)|^2 + \frac{1}{2}|\varrho_{ry}(t)|^2 = \frac{1}{2} \tag{5.81}$$

Thus, polarization fluctuations in the received light wave $\vec{E}_r(t)$ can be completely suppressed by a polarization-diversity receiver. Similar results are obtained when FSK, DPSK or PSK modulation schemes instead of ASK are used [15, 29, 32]. Taking into account the additive Gaussian noise also, it can be shown that the receiver sensitivity of diversity receiver is only negligibly lower than that of conventional coherent receiver without polarization diversity.

5.3.5 POLARIZATION SWITCHING

Polarization switching represents a simple, but powerful solution to reduce the influence of polarization fluctuations. The most common technique is given by *data-induced polarization switching* which will be briefly discussed. Polarization switching can either be applied to the transmitter or receiver. Data-induced polarization switching is applicable only for FSK systems.

The principle block diagram is shown in Fig. 5.15 wherein the switching process is performed in the transmitter by a passive birefringent component. This component is placed right after the transmitter laser which is binary modulated in frequency by modulating the injection current or by an external frequency modulator. To avoid high insertion loss, it is advantageous to choose a piece of polarization-maintaining fiber as a birefringent component. Polarization switching requires a linearly polarized local laser light which has to be launched at 45° with respect to the principal axes of the polarization-maintaining fiber (Fig. 5.3). This means that the optical power is equally divided between the two eigenmodes.

As explained in Section 5.1, both eigenmodes of a polarization-maintaining fiber propagate with different velocities. Hence, state of polarization at the output of the birefringent component will generally differ from the linear polarization at the input. Depending on the phase difference between the eigenmodes (mode x and mode y), state of polarization at the output can be either linear, circular or elliptical. Since this difference in phase is also a function of frequency of light, it will change with the binary FSK modulation. If the product of time delay of both the eigenmodes and FSK frequency deviation is chosen to be 0.5, then the phase difference will changed by 180° as the frequency is switched. Because of this, polarization modulation is simultaneously added to the frequency modulation.

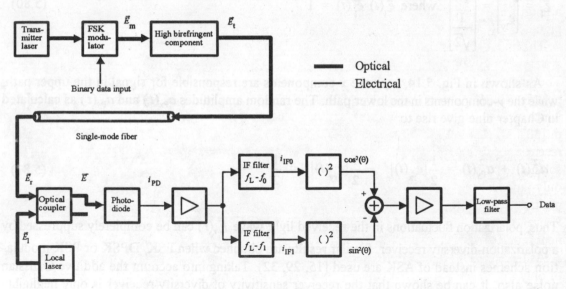

Fig. 5.15: Data-induced polarization switching in transmitter of coherent communication system

The optical signal which is now modulated in frequency *and* polarization is transmitted through a standard single-mode fiber to a conventional heterodyne receiver with a FSK demodulator; for example, a single- or dual-filter configuration with envelope or synchronous detection. However, due to polarization switching there is a permanent mismatch in state of polarizations between the received light wave and local laser light wave that results in an intrinsic loss of 3 dB.

The advantage of data-induced polarization switching is that the passive birefringent component is the only optical component required to overcome the polarization problem. No polarization control or diversity technique is necessary in addition. The possibility of applying polarization switching in the transmitter reduces the cost of receivers which, for example, is very advantageous if coherent optical multichannel distribution systems have to be realized.

Due to a strong polarization dispersion in the birefringent component in the transmitter, intersymbol interference can be observed which may limit the system performance and permissible bit rate.

5.3.6   COMPARISON

In the previous Sections, four different methods to tackle the polarization problem have been considered in detail. The simplest, but most expensive solution is given by the polarization-maintaining fiber (e.g., SMSP fiber). Due to higher cost, SMSP fibers are normally not used in long-haul optical fiber communication systems wherein the fiber length is usually in the range of few km to some hundreds of km. In addition, SMSP fibers exhibit higher attenuation than the standard single-mode fibers. Consequently, possible applications of SMSP fibers are restricted to a length of few cm only. The most common application is, in particular, given by SMSP fiber pigtails for polarization dependent optical components such as optical waveguide modulator.

The most competitive polarization-handling methods for coherent optical communication systems are polarization diversity, polarization control and data-induced polarization switching. The first two methods require a number of additional optical and electrical components, while polarization switching only requires a birefringent component in the transmitter or receiver, which can easily be realized by a rather short polarization-maintaining fiber of length about 100 m [33]. If the birefringent component is placed in the transmitter, then a standard heterodyne receiver can be employed. For this reason, polarization switching represents the least expensive solution to overcome the polarization problem. However, data-induced polarization switching is only applicable for FSK systems. Data-induced polarization switching is well-suited for multisubscriber systems such as coherent optical multichannel video distribution networks. However, in the multiple-access network, each optical transceiver requires a birefringent component in the transmitter.

As far as losses are concerned, polarization control is the most efficient since the insertion loss of only the retarder has to be considered. In the case of fiber retarder, this loss is typically less than 0.5 dB. Therefore, polarization control is well-suited for coherent optical long-range links where the sensitivity is a critical parameter. In comparison, polarization diversity and polarization switching exhibit an inherent power loss of 3 dB. In the former, this loss is additionally increased by the loss of optical components; for example, by the loss of polarization-beam splitters. In the latter, power loss of the birefringent component which is about 0.5 dB [33] has to be considered in addition. In the polarization-diversity receiver, inherent loss of 3 dB can, however, be avoided if two local lasers (one for each of the diversity receiver branches) are used.

Finally, the response time should be compared. The response time is rather slow if a polarization-diversity receiver or a polarization-switching component is used. In contrast, response time is in the order of some magnitudes higher if a polarization-control circuit is used in the system where state of polarization is tracked by means of a feedback loop and control algorithm. Hence, if a fast acquisition speed is required, polarization diversity and polarization switching are the most promising solutions. Summarizing the discussion above, it becomes clear that the most efficient technique to solve the polarization problem primarily depends on the applications of the coherent optical communication systems.

5.4 REFERENCES

[1]   Adams, M. J.; Payne, D. N.; Ragdale, C. M.: Birefringence in optical fibers with elliptical cross-section. Electron. Lett. 15(1979)10, 298-299.

[2]   Barlow, A. J.; Payne, D. N.: Polarisation maintenance in circularly birefringent fibres. Electron. Lett. 17(1981)11, 388-389.

[3]   Bronstein, I. N.; Semendjajew, K. A.: Taschenbuch der Mathematik, 19. Auflage.Harri Deutsch Verlag, Thun und Frankfurt/Main, 1980.

[4]   Burns, W. K.; Moeller, R. P.: Measurement of polarization mode dispersion in high-birefringence fibers. Optics Lett. 8(1983)3, 195-197.

[5]   Dyott, R. B.; Cozens, J. R.; Morris, D. G.: Preservation of polarisation in optical-fibre waveguides with elliptical cores. Electron. Lett. 15(1979)13, 380-382.

[6]   Franz, J.: Grundzüge des kohärent optischen Heterodynempfanges. TU München, Lehrstuhl für Nachrichtentechnik, Nachr.-techn. Ber. Band 14, 1984.

[7]   Hosaka, T.; Okamoto, K.; Sasaki, Y.; Edahiro, T.: Single mode fibres with asymmetrical refractive index pits on both sides of core. Electron. Lett. 17(1981)5, 191-193.

[8]   Hosaka, T.; Okamoto, K.; Miya, T.; Sasaki, Y.; Edahiro, T.: Low-loss single polarization fibres with asymmetrical strain birefringence. Electron. Lett. 17(1981)15, 530-531.

[9]   Imoto, N.; Ikeda, M.: Polarization dispersion measurement in long single-mode fibers with zero dispersion wavelength at 1,5 μm. IEEE QE-17(1981)4, 542-545.

[10]   Jeunhomme, L.; Monerie, M.: Polarisation-maintaining single-mode fibres cable design. Electron. Lett. 16(1980)24, 921-922.

[11]   Kaminov, I. P.; Ramaswamy, V.: Single-polarisation optical fibres: Slab model. Appl. Phys. Lett. 34(1979)4, 268-270.

[12]   Kaminov, I. P.: Polarisation in optical fibres. IEEE J. QE-17(1981)1, 15-22.

[13]   Katsuyama, T.; Matsumura, H.; Suganuma, T.: Low-loss single-polarisation fibers. Electron. Lett. 17(1981)13, 473-474.

[14]   Kidoh, Y.; Suematsu, Y.; Furuya, K.: Polarization control on output of single-mode optical fibres. IEEE J. QE-17(1981), 991-994.

[15]   Kreit, D.; Youngquist, R. C.: Polarisation-insensitive optical heterodyne receiver for coherent FSK communications. Electron. Lett. 23(1987)4, 168-169.

[16]   Kubota, M.; Oohara, T.; Furuya, K.; Suematsu, Y.: Electro-optical polarisation control on single-mode optical fibres. Electron. Lett. 16(1980)15, 573.

[17]   Lefevre, H. C.: Single-mode fibre fractional wave devices and polarisation controllers. Electron. Lett. 16(1980)20, 778-780.

[18]   Love, J. D.; Sammut, R. A.; Snyder, A. W.: Birefringence in elliptically deformed optical fibres. Electron. Lett. 15(1979)20, 615-616.

[19]   Machida, S.; Sakai, J.; Kimura, T.: Polarisation conservation in single-mode fibres. Electron. Lett. 17(1981)14, 494-495.

[20] Mahon, C. J.; Khoe, G. D.: Compensational deformation; new endless polarisation matching control schemes for optical homodyne or heterodyne receivers which require no mechanical drivers. IOOC-ECOC (1986), 267-270.

[21] Mahon, C. J.; Khoe, G. D.: Endless polarisation state matching control experiment using two controllers of finite control range. Electron. Lett. 23(1987)23, 1234-1235.

[22] Mochizuki, K.; Namihira, Y.; Wakabayashi, H.: Polarization mode dispersion measurements in long single mode fibres. Electron. Lett. 17(1981)4, 153-154.

[23] Monerie, M.: Polarisation-maintaining single-mode fibre cables: influence of joints. Appl. Optics 20(1980), 712-713.

[24] Monerie, M.; Jeunhomme, L.: Polarization mode coupling in long single-mode fibres. Optical and Quantum Electronics 12(1980), 449-461.

[25] Monerie, M.; Lamouler, P.; Jeunhomme, L.: Polarization mode dispersion measurements in long single mode fibres. Electron. Lett. 16(1980)24, 970-980.

[26] Monerie, M.; Lamouler, P.: Birefringence measurement in twisted single-mode fibres. Electron. Lett. 17(1981)7, 252-253.

[27] Namihira, Y.; Ryu, S.; Mochizuki, K.; Furusawa, K.; Iwamoto, Y.: Polarisation fluctuation in optical-fibre submarine cable under 8000 m deep sea environmental conditions. Electron. Lett. 23(1987)3, 100-101.

[28] Namihira, Y.; Horiuchi, Y.; Wakabayashi, H.: Dynamic polarisation fluctuation characteristics of optical-fibre submarine cable coupling under periodic variable tension. Electron. Lett. 23(1987)22, 1201-1202.

[29] Neidlinger, S.: DPSK polarisation diversity receiver with novel switching-demodulators and maximal-ratio combining. Electron. Lett. 26(1990)14, 1070-1071.

[30] Noé, R.: Endless polarization control in coherent optical communications. Electron. Lett. 22(1986)15, 772-773.

[31] Noé, R.: Endless polarizations control experiment with three elements of limited birefringence range. Electron. Lett. 22(1986)25, 1341-1343.

[32] Noé, R.: Sensitivity comparison of coherent optical heterodyne, phase diversity, and polarization diversity receivers. J. Opt. Commun. 10(1989)1, 11-18.

[33] Noé, R.; Rodler, H. J.; Ebberg, A.; Gaukel, G.; Noll, B.; Wittmann, J.; Auracher, F.: Comparison of polarisation handling methods in coherent optical systems. IEEE J. LT-9(1991)10, 1353-1365.

[34] Okamoto, K.; Sasaki, Y.; Miya, T.; Kawachi, M.; Edahiro, T.: Polarization characteristics in long length v.a.d. single-mode fibres. Electron. Lett. 16(1980)25, 768-769.

[35] Okamoto, K.; Hosaka, T.; Edahiro, T.: Stress analysis of single polarization fibers. Review of the Electrical Communication Laboratories, Vol.31(1983)3, 381-392.

[36] Okoshi, T.; Oyamada, K.: Single-polarisation single-mode optical fibre with refractive index-pits on both sides of core. Electron. Lett. 16(1980)18, 712-713.

[37] Okoshi, T.: Single-polarization single-mode optical fibers. IEEE J. QE-17(1981)6, 879-884.

[38] Okoshi, T.; Ryu, S.; Emura, K.: Measurement of polarization parameters of a single-mode optical fiber. J. Opt. Commun. 2(1981)4, 134-141.

[39] Okoshi, T.: Review of polarization-maintaining single-mode fiber. ECOC (1983), 57-59.

[40] Payne, D. N.; Barlow, A. J.; Ramskow Hansen, J. J.: Development of low-and high birefringence optical fibers. IEEE J. QE-18(1982)4, 477-488.

[41] Poole, C. D.; Bergano, N. S.; Schulte, H. J.; Wagner, R. E.; Nathu, V. P.; Amon, J. M.; Rosenberg, R. L.: Polarisation fluctuations in a 147 km undersea lightwave cable during installation. Electron. Lett. 23(1987)21, 1113-1115.

[42] Poole, C. D.; Bergano, N. S.; Wagner, R. E.; Schulte, H. J.: Polarisation dispersion in a 147 km undersea lightwave cable. ECOC (1987), 321-324.

[43] Poole, C. D.: Polarization dispersion and principal states in a 147 km undersea lightwave cable. IEEE J. LT-6(1988), 1185-1190.

[44] Purcell, E. M.: Elektrizität und Magnetismus. Berkeley Physik Kurs Band 2. Vieweg-Verlag, 1979.

[45] Ramachandran, G. N.; Ramaseshan, S.: Crystal optics. Handbuch der Physik Band 25/1 (S. Flügge), Springer-Verlag, 1962.

[46] Ramaswamy, V.; French, W. G.; Standley, R. D.: Polarization characteristics on noncircular core single-mode fibres. Appl. Optics 17(1978)18, 3014-3017.

[47] Ramaswamy, V.; Kaminov, I. P.; Kaiser, P.: Single polarization optical fibers: exposed cladding technique. Appl. Phys. Lett. 33(1978)9, 814-816.

[48] Ramaswamy, V.; Standley, R. D.; Sze, D.; French, W. G.: Polarization effects in short length, single mode fibers. Bell Systems Tech. J. 57(1978), 635-651.

[49] Rashleigh, S. C.; Stolen, R. H.: Preservation of polarization in single-mode fibers. Fiber-optic Tech. (1983)5, 155-161.

[50] Rocks, M.: Optischer Überlagerungsempfang: Die Technik der übernächsten Generation glasfasergebundener optischer Nachrichtensysteme. Der Fernmelde-Ingenieur 3(1985) 2.

[51] Sakai, J.-I.; Kimura, T.: Birefringence and polarization characteristics of single-mode optical fibers under elastic deformations. IEEE J. QE-17(1981)6, 1041-1051.

[52] Sakai, J.-I.; Machida, S.; Kimura, T.: Existence of eigen polarization modes in anisotropic single-mode optical fibers. Optics Lett. 6(1981)10, 496-498.

[53] Sakai, J.-I.; Kimura, T.: Polarization behavior in multiple perturbed single-mode fibers. IEEE J. QE-18(1982)1, 59-65.

[54] Sakai, J.-I.; Machida, S.; Kimura, T.: Degree of polarization in anisotropic single-mode optical fibers: theory. IEEE J. QE-18(1982)4, 488-495.

[55] Sakai, J.-I.; Machida, S.; Kimura, T.: Twisted single-mode optical fiber as polarization-maintaining fiber. Review of Electrical Communication Laboratories Vol.31(1983)3, 372-380.

[56] Sasaki, Y.; Shibata, N.; Hosaka, T.: Fabrication of polarization-maintaining and absorption-reducing optical fibers. Review of Electrical Communication Laboratories Vol. 31(1983)3, 400-409.

[57] Shibata, N.; Okamoto, K.; Sasaki, Y.: Structure design for polarization-maintaining and absorption-reducing optical fibers. Review of Electrical Communication Laboratories Vol. 31(1983)3, 393-399.

[58] Stolen, R. H.; Ramaswamy, V.; Kaiser, P.; Pleibel, W.: Linear polarization in birefringent single-mode fibers. Appl. Phys. Lett. 33(1978)8, 699-701.

[59] Tjaden, D. L. A.: Birefringence in single-mode optical fibres due to core ellipticity. Phillips J. Res. 33(1978)5/6, 254-263.

[60] Ulrich, R.: Polarization stabilization on single-mode fiber. Appl. Phys. Lett. 35(1979), 840-842.

[61] Ulrich, R.; Simon, A.: Polarization optics of twisted single-mode fibers. Appl. Optics 18(1979)13, 2241-2251.

6 OPTICAL DETECTORS AND RECEIVERS

In an optical communication system, the function of receiver is to convert a received optical signal into an electrical signal, which can serve as an input for other devices or communication systems. The basic functional elements of a receiver are as follows. (i) Detector - to convert the received signal from optical to electrical domain, (ii) Amplifier stages - to amplify the signal and (iii) Demodulator or decision circuit - to reproduce the original signal. When the incoming signal is very weak, the receiver may be preceded by an optical amplifier. In practice, functional distinctions given above can be hazy because detector may have in-built internal amplifiers. Some receivers do not have separate demodulation or decision circuit because the amplified signal is good enough for use by other electronic equipment.

Several stages of amplification may be needed to bring the signal to the desired level for further signal processing. The amplification function can be separated into two parts, preamplifier and main amplifier. The preamplifier brings the signal and noise amplitude to a level where the noise produced in the subsequent stages has a negligible effect on the overall *signal-to-noise (S/N) ratio*. Therefore, design of preamplifier which follows the photodiode is of utmost importance.

The basic elements of analog and digital optical receivers are shown in Fig. 6.1. In both the receivers, initial stages are same. The difference comes where the signal is converted into final form for inputting to other devices. Detection and amplification in digital receiver can distort the signal. For example, high and low frequencies may not be amplified by the same factor. The shaping/equalization circuit evens out these differences so that the amplified signal is closer to the original signal. Equalizer is required in analog receiver as well for high fidelity output signal.

For *analog systems*, the *S/N* ratio is a measure of the quality. Higher the ratio, better is the received signal. In many practical fiber optic systems, *S/N* ratio of 40 dB to 50 dB is considered to be good to excellent, but less than 30 dB is not acceptable for many applications. In *digital systems*, quality is measured in terms of probability of incorrect transmission or *bit error rate* (*BER*). The *BER* depends on several factors like received power level, detector sensitivity, noise level and data rate. In fact, *BER* and *S/N* ratio are interrelated. The exact relationship depends on the type of modulation and demodulation schemes used.

As mentioned above, detector is an essential component of optical receiver and is one of the crucial elements which dictates the overall system performance. Since the received optical signal is very weak, the photodetector must meet very high performance requirements. Among foremost are a high sensitivity in the wavelength range of interest, minimum addition of noise to the system and a fast response speed or sufficient bandwidth. In addition, photodetector should have insensitivity to temperature variations, compatible physical dimensions, reasonable cost and long operating life.

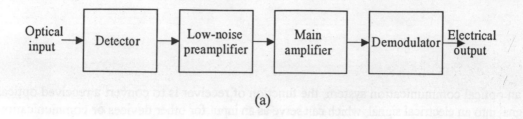

(a)

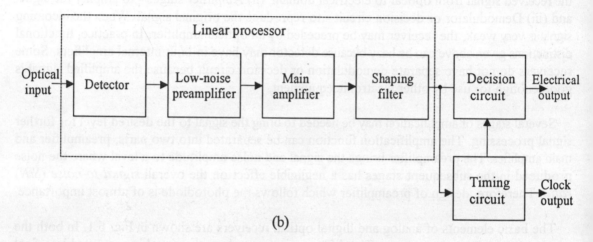

(b)

Fig.6.1: Block diagram of (a) analog and (b) digital receivers

Various types of detectors are in existence. Among these are photomultipliers, pyroelectric detectors and semiconductor-based photodetectors viz., photoconductors, photodiodes, phototransistors and photodarlington transistors. Some of these detectors do not meet one or more of the foregoing requirements. The photomultipliers, often referred to photomultiplier tubes (PMTs), consist of a photocathode and an electron multiplier packaged in a vacuum tube are capable of giving very high gain and very low noise. Their large size and high voltage requirements make them unsuitable for optical fiber systems. However, they may find applications in optical space communications (Chapter fifteen). Pyroelectric photodetectors involve the conversion of photons to heat. Photon absorption results in a temperature change of the detector material which is usually measured as a capacitance change. The response of these detectors is quite flat over a broad spectral range, but they are inefficient and relatively slow because of the time required to change their temperature. Consequently, these are not found suitable for optical fiber systems.

Of the semiconductor-based photodetectors, photodiodes are commonly used in the fiber optic systems because of their small size, suitable material, high sensitivity and fast response time. The two types of photodiodes used are the *PIN-photodiode* and *avalanche photodiode* (APD). In the following Sections, semiconductor-based photodetectors are dealt with.

6.1 PHOTODETECTORS

In this Section, some important issues related to photodetectors are discussed.

6.1.1 PHYSICAL PRINCIPLE

In PMTs, electrons excited by photons become free electrons and the effect is called *external photoeffect*. In contrast to this, in semiconductor-based photodetectors, electrons and holes created by the photons remain within the material and the effect is called *internal photoeffect*. Under *reverse bias condition*, electrons and holes drift in the electric field and induce a photocurrent in the external load. The number of electron-hole pairs per second so freed is dependent linearly on the power level of the incident optical field and hence the electric current is proportional to the optical power [4,11].

6.1.2 SPECTRAL RESPONSE

The amount of energy required to generate an electron-hole pair is the band-gap energy E_g of the material. Since the energy of a photon is hf where h is the Planck's constant and $f = c_0/\lambda$ the optical frequency the required condition is

$$hf \geq E_g \tag{6.1}$$

It means that the photodetector material will exhibit a long wavelength cut-off and is given by

$$\lambda_{max} = \frac{hc_0}{E_g} \tag{6.2}$$

The absorption coefficient of the photodetector material α_s will be essentially zero at $\lambda > \lambda_{max}$ i.e., no absorption of the incident photons takes place or the material is transparent. At $\lambda < \lambda_{max}$, absorption coefficient rises sharply. The high value of α_s at short wavelengths causes little power to penetrate into the depletion region. Therefore, there is a diminished photodetector response because of the increased surface absorption i.e., carriers recombine before they are collected [15].

6.1.3 DETECTOR MATERIALS

Silicon photodiodes have high sensitivity in the 0.8 μm to 0.9 μm wavelength range with adequate speed (hundreds of megahertz), negligible shunt conductance, low dark current and long term stability [7]. Therefore, these detectors are widely used in first generation systems and are commercially available. Since silicon has λ_{max} of 1.09 μm (band-gap energy E_g = 1.14 eV), it is suitable only up to this wavelength. Thus for higher generation systems in the long wavelength range 1.1 μm - 1.6 μm, research is devoted to the investigation of semiconductor materials which have narrower band gap. Interest has focussed on germanium (Ge) and III-V alloys. Ge may be used over the whole range of interest (0.5 μm - 1.8 μm), but such photodiodes have relatively large dark current due to their narrow band gaps in comparison to other semiconductor materials. This is a major disadvantage with the Ge photodiodes especially at shorter wavelengths (below 1.1 μm). III-V alloys are superior to Ge because their band gaps can be tailored to the desired wavelengths by changing the relative concentration of their constituents resulting in lower dark current. Further, such diodes can also be fabricated in a *heterojunction structure* (refer to Section 6.2.3) which enhances their high speed operations.

Ternary alloys such as InGaAs and GaAlSb deposited on InP and GaSb substrates respectively have been used to fabricate photodiodes in the long wavelength range. In particular, the alloy $In_{0.53}Ga_{0.47}As$ (E_g = 0.75 eV) lattice matched to InP responds to wavelengths up to around 1.7 μm and has been extensively used in the fabrication of photodiodes for operation at both 1.3 μm and 1.55 μm. *Quaternary alloys* such as InGaAsP grown on InP and GaAlAsSb grown on GaSb are also used for detection at these wavelengths [13].

In Table 6.1 below, useful wavelength ranges and other characteristics of commonly used photodetector materials [10,11] are given.

Table 6.1: Some characteristics of photodetector materials (Si, Ge, InGaAs and InGaAsP)

Material	Wavelength range (μm)	Wavelength of peak response (μm)	Peak responsivity (A/W)	Excess noise exponent x	Ratio of holes and electrons ionization rates k
Si	0.3-1.1	0.8	0.5	0.3-0.5	0.02-0.04
Ge	0.5-1.8	1.5	0.6	1.0	0.7-1.0
InGaAs	1.0-1.8	1.7	0.75	0.5-0.8	0.3-0.5
InGaAsP	1.0-1.6	1.4	0.7	0.4-0.9	0.2-0.6

6.2 TYPES OF PHOTODETECTORS

The choice of a photodetector for a specific optical receiver application depends on the performance and cost trade-offs. More mature technologies with simple fabrication processes in Si, Ge, InGaAs and InGaAsP give rise to lower cost devices with adequate performance. High performance devices have more involved fabrication process and are much more expensive. Speed of response, noise, quantum efficiency and gain are the parameters to be considered in the choice of a photodetector.

Semiconductor-based photodetectors are discussed below in details.

6.2.1 PHOTOCONDUCTOR

A photoconductor as shown in Fig. 6.2 is fabricated by placing ohmic contacts onto a semiconductor material [4].

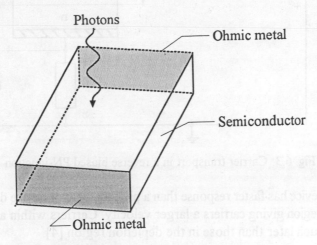

Fig. 6.2: Structure of a photoconductor

When the material is illuminated, electron-hole pairs are generated and the electrical conductivity of material increases. More the photon flux, greater the increase in conductivity [12,15]. When a voltage is applied to the contacts, electrons and holes move in opposite directions resulting in a photocurrent. Since the electron velocity is greater than that of the hole, the electron reaches the side of the photoconductor before the hole. Continuity of current requires that another electron enter the device through the external circuit. It means that one photon can cause several electrons to move through the device until recombination occurs. Photoconductive detectors have the advantages of conductive gain and ease of fabrication, but lack their speed capabilities.

6.2.2   PN-Photodiode

The PN-photodiode is basically a *reverse biased PN-junction* as shown in Fig. 6.3. When a photon is absorbed, it creates an electron-hole pair as in the photoconductive device. Under the influence of an electric field, electron moves toward the n-region while the hole drifts to the p-region. As the electric field is formed only in the depletion region of the device, carriers generated there or near there (within a diffusion length) have the ability to be moved in a specific direction and contribute to the photocurrent. Electron-hole pairs generated far away (greater than a diffusion length) from the depletion region diffuse around and eventually recombine not contributing to the photocurrent.

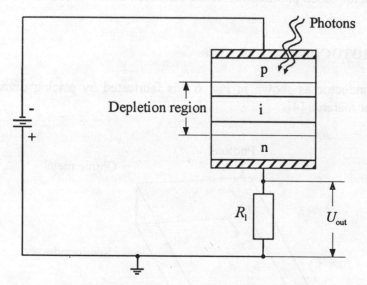

Fig. 6.3: Carrier transport in a reverse biased PN-junction

The PN-junction device has faster response than a photoconductor device due to strong electric field in the depletion region giving carriers a larger velocity. Carriers within a diffusion length are also collected, but much later than those in the depletion region [4].

6.2.3   PIN-Photodiode

At longer wavelengths ($\lambda \approx \lambda_{max}$), light penetrate more deeply in the semiconductor material. Therefore, there is need to have a wider depletion region. This can be achieved by doping n type material so lightly that it can be considered *intrinsic*. To make a low resistance contact, a highly doped n type (n^+) layer is added. This creates a p-i-n (or PIN) structure as shown in Fig. 6.4. The PIN-photodiode has advantages over the PN-junction device in terms of wider depletion region (more than the thickness of i-layer) and therefore a lower capacitance and lower RC time constant. The transit time, however, will increase [12]. High bandwidth in excess of 100 GHz can be achieved with a PIN-photodiode.

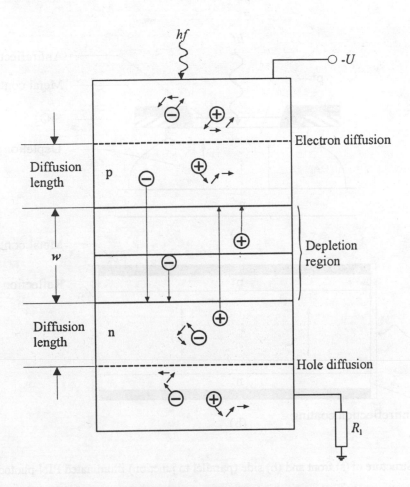

Fig. 6.4: Structure of a PIN-photodiode

When a PIN-photodiode is formed out of a single semiconductor material, as in Si PIN-photodiode, the device material is homogeneous and it is referred to as a *homojunction device*. A photodiode composed of two different semiconductor materials is referred to as a *hetero-junction device*.

The structures of two types of silicon PIN-photodiodes for operation in the shorter wavelength range (< 1.09 µm) is shown in Fig. 6.5. The front illuminated photodiode (Fig. 6.5a) when operated in the 0.8 µm - 0.9 µm range requires a depletion region of between 20 µm and 50 µm in order to attain high quantum efficiency (≈ 85%) together with fast response (< 1 ns) and low dark current (≈1 nA). The side illuminated structure (Fig. 6.5b), where light is injected parallel to the junction plane, exhibits a large absorption width (≈ 500 µm). It is particularly useful at wavelengths close to the band-gap limit (1.09 µm) where the absorption coefficient is relatively small [13].

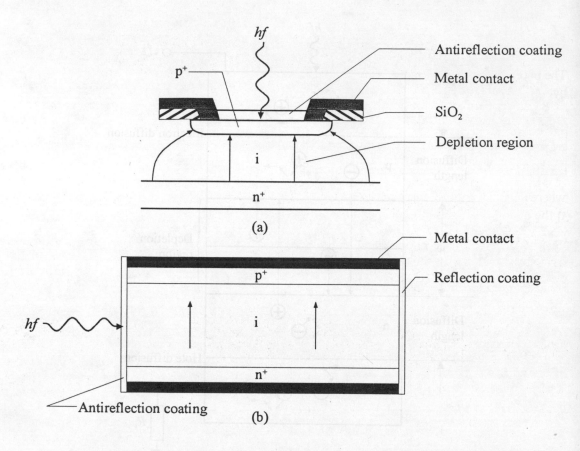

Fig. 6.5: Structure of (a) front and (b) side (parallel to junction) illuminated PIN-photodiodes

Germanium PIN-photodiodes which span the entire wavelength range of interest are also commercially available, but such photodiodes have relatively high dark current (typically 100 nA at 20° C increasing to 1 µA at 40° C). As outlined earlier, III-V semiconductor alloys are used in the fabrication of longer wavelength region photodiodes. The favoured material is lattice matched $In_{0.53}Ga_{0.47}As/InP$ [13] which can detect up to a wavelength of 1.67 µm. A typical planar device structure is shown in Fig. 6.6, which requires epitaxial growth of several layers on an n-type InP substrate. The higher band-gap material sits on the top, where light incident at $\lambda_g > hc_0/E_g$ is not absorbed, but passes through to the intrinsic layer. This top transparent layer is called the window layer. The incident light is, therefore, absorbed in the low doped n-type InGaAs layer. The PN-junction of the device is placed away from the surface and thus reduces surface recombination.

It is obvious from above that drift and diffusion currents make up the photocurrent of a photodiode [1]. As the photodiode is *reverse biased*, a small reverse saturation current, called the *dark current* I_{dark}, flows through the device even when no optical radiation is incident on it. Let I_p be the current generated by the incident photons. The total detector current I is the sum of I_{dark} and I_p i.e.,

$$I = I_p + I_{dark} \tag{6.3}$$

The photon generated current I_p is proportional to the incident optical power level P_r. It is given by

$$I_p = \frac{\eta e P_r}{hf} \tag{6.4}$$

where hf is the photon energy at frequency f, h the Planck's constant, e the electron charge and η the quantum efficiency explained below [1, 12].

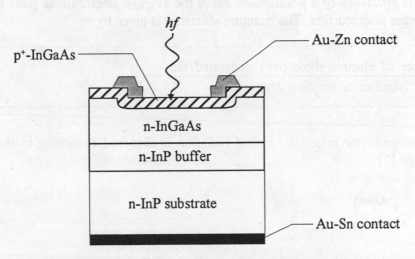

Fig. 6.6: Structure of a planar InGaAs PIN-photodiode

The equivalent circuit representation of a photodiode is shown in Fig. 6.7.

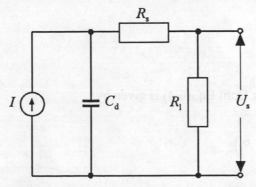

Fig. 6.7: Equivalent circuit representation of a reverse biased photodiode

The device behaves as a low-pass filter with a cut-off frequency (or bandwidth) $1/2\pi \cdot R_l \cdot C_d$ where R_l is the load resistance and C_d the internal capacitance of the device (≈ 1 pF). Here, R_s is the diode series resistance and usually $R_s \ll R_l$. To increase the bandwidth, R_l and C_d should be minimized. R_l can not be made arbitrary small because it will decrease the output signal voltage level and increase the thermal noise power. The junction/diode capacitance C_d can be reduced by decreasing the junction area and increasing the barrier thickness. The latter option increases the transit time of the charge carriers crossing the barrier which, in turn, reduces the response time (typically ≈ 1 ns). It may be mentioned that response speed of a photodiode is determined not only by the RC time constant, but also by the transit time and the diffusion time of charge carriers (discussed latter).

Quantum Efficiency: The number of electron-hole pairs generated per impinging photon determines the *efficiency of a photodiode*. More the average electron-hole pairs per photon, higher will be the photocurrent. The quantum efficiency is given by

$$\eta = \frac{\text{Number of electron-hole pairs generated/sec}}{\text{Number of incident photons/sec}} = \frac{I_p}{P_r} \qquad (6.5)$$

In the semiconductor materials, optical radiation is absorbed according to the following exponential law [5]

$$P(x) = P_r\left(1 - e^{-\alpha_s(\lambda)x}\right) \qquad (6.6)$$

where $\alpha_s(\lambda)$ is the absorption coefficient of the semiconductor material at wavelength λ and $P(x)$ the power absorbed in a distance x. If the width of the depletion region is w, the total power absorbed in distance w will be

$$P(w) = P_r\left(1 - e^{-\alpha_s w}\right) \qquad (6.7)$$

Therefore, the photocurrent from Eq. (6.4) is given by

$$I_P = \frac{e}{hf}P_r\left(1 - e^{-\alpha_s w}\right)\left(1 - R_f\right) \qquad (6.8)$$

where R_f is the reflectivity at the entrance face of the photodiode. From Eqs. (6.5) and (6.8), η is given by

$$\eta = \left(1 - e^{-\alpha_s w}\right)\left(1 - R_f\right) \qquad (6.9)$$

It means that to permit a large fraction of the incident light to be absorbed to achieve high η, the depletion layer must be thick enough. When the depletion layer is thicker, the photogenerated carriers will take more time to drift across the reverse biased junction. Longer is the drift time, slower is the speed of photodiode response. In practice, a compromise has to be made between speed of response and η [5].

In Si photodiodes, η is nearly 90% at a wavelength of about 0.8 μm, while Ge photodiodes have a maximum η of nearly 55% at 1.55 μm. The Ge photodiodes are usable over the wavelength range 1.1 μm - 1.6 μm. Photodiodes based on InGaAsP or InGaAs have an $\eta \approx 60\%$ - 80% in the wavelength range 1.0 μm - 1.7 μm [9].

Responsivity: One measure of photodiodes sensitivity is the responsivity R_0 which is defined as the ratio of the detector current to the input optical power level. It is usually expressed as A/W or μA/μW.

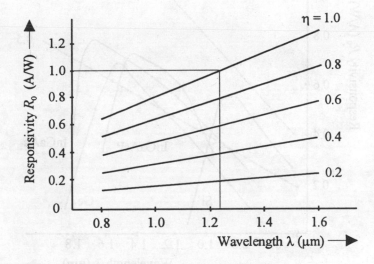

Fig. 6.8: Variations of responsivity R_0 with wavelength λ for different quantum efficiencies η

Using Eq. (6.4), responsivity R_0 is given by

$$R_0 = \frac{I_p}{P_r} = \frac{\eta e}{hf} \qquad (6.10a)$$

It is sometime useful to express R_0 as

$$R_0 = \frac{\eta e \lambda}{hc} = \frac{\eta \lambda/\mu m}{1.24} \frac{A}{W} \qquad\qquad (6.10b)$$

where λ is in μm. For some typical photodiode materials, peak responsivity is given in Table 6.1 above. It may be mentioned that the responsivity is a function of λ and photodiode material [5]. Linear increase of R_0 with λ for a fixed value of η is illustrated in Fig. 6.8. It may be mentioned that η is not constant in practice and depends on many factors like λ, detector material (see Eq. 6.9).

Variations of R_0 with λ for four typical devices are shown in Fig. 6.9. The dotted lines are for constant η. The long wavelength falloff is due to energy deficiency of the photons while the short wavelength falloff is due to increased absorption effect [11]. The quantum efficiency curve differs from that of a responsivity curve because photon energy changes with wavelength.

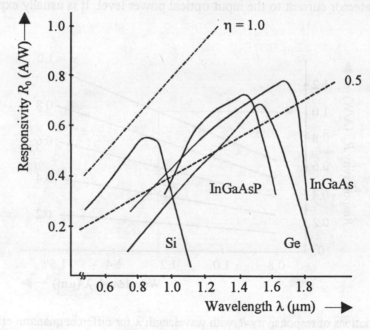

Fig. 6.9: Responsivity R_0 versus wavelength λ for some typical devices

Speed of Response: The speed of response of photodiodes is determined from their *rise time*. The rise time is normally defined as the time required by the output signal to rise from 10% to 90% of its final value for a step function optical input. Likewise, one can define the fall time of the photodiode response. The response time of a photodiode is generally taken as the larger of these two [9]. For Si photodiodes, response time is typically a few ns, while in specially prepared fast photodiodes it could be as low as few ps.

The speed of response of a photodiode together with its output circuit depends mainly on the following three factors.

Transit Time of Carriers through the Depletion Region: When field in the depletion region exceeds a saturation value, carriers may be assumed to travel at a constant (maximum) drift velocity v_d. The longest transit time t_{drift} is for carriers which must traverse the full depletion layer width w and is given by

$$t_{drift} = \frac{w}{v_d} \qquad (6.11)$$

A field strength above $2 \cdot 10^4$ Vcm^{-1} in Si gives maximum carrier velocities of approximately 10^7 cms^{-1}. Thus the transit time through a depletion layer of width 10 μm is around 0.1 ns [13].

Diffusion Time of Carriers Generated Outside the Depletion Region: Carrier diffusion is a comparatively slow process. The time taken t_{diff} for carriers to diffuse a distance d is given by

$$t_{diff} = \frac{d^2}{2D_c} \qquad (6.12)$$

where D_c is the minority carrier diffusion coefficient. For example, the hole diffusion time through 10 μm of Si is 40 ns, whereas the electron diffusion time over the same distance is around 8 ns [13].

RC Time Constant of Photodiode and its Associated Circuit: The junction capacitance $C_d = \epsilon A/w$ (where ϵ is the permittivity of the semiconductor material and A the diode junction area) must be minimized to reduce the RC time constant. Here $R = R_d \| R_l \| R_a$ and $C = C_d + C_a$ are the equivalent resistance and capacitance respectively. R_a and C_a are the resistance and capacitance of the following amplifier circuit respectively.

To maximize the speed of response, the transit time needs to be minimized. This can be achieved in two ways-by increasing the bias voltage and/or by decreasing the thickness (and width) of the intrinsic layer. Reducing intrinsic layer thickness will reduce the fraction of incident light absorbed and hence η value. Further, it will cause increase in C_d. The increased C_d will lead to rise in the RC time constant and therefore slow down the speed of response. To optimize the photodiode speed of response, trade-off must be made between transit time and RC time constant [4].

Typical responses of a photodiode to an input rectangular pulse are shown in Fig. 6.10. With low C_d and $w \gg 1/\alpha_s$, the response is as shown in Fig. 6.10b. The rise time and fall time of the photodiode are satisfactory and it follows the input pulse quite well. When C_d is large, the RC time constant will limit the response time. In that case, photodiode response will be as shown in Fig. 6.10a. If the width of the depletion layer is too narrow, then some of the charge carriers will be generated outside the depletion region. These carriers would have to diffuse back into the depletion region before recollection. It shows distinct slow (arises from the carriers in the undepleted material) and fast (result of absorption in the depletion region) response components as shown in Fig. 6.10c. For high η, the depletion layer width must be much larger than $1/\alpha_s$ so that most of the light is absorbed in the depletion region. A reasonable compromise between response speed and quantum efficiency is found for absorption region thickness between $1/\alpha_s$ and $2/\alpha_s$ [5].

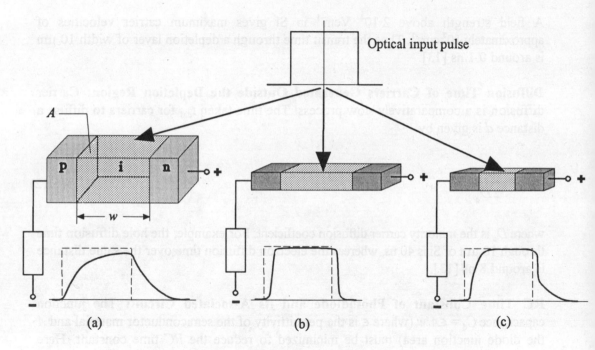

Fig. 6.10: Photodiode response to a rectangular optical input pulse for
(a) large width w and large area A, (b) large w and small A and (c) small
w and small A of depletion layer

Dynamic Range: The range of optical power within which a photodiode current shows linear variations with the input optical power levels determines the *dynamic range* of photodiode. The photodiode current normally varies linearly with input power over approximately six decades in well designed photodiodes. If the optical signal exceeds this dynamic range, the photodiode output current would no longer be linearly proportional to the input optical power level. It exhibits non-linear response. In analog communication systems, it will lead to severe distortion, while in digital communication systems increase in BER. Dynamic range of a particular photodiode is determined by the device parameters like its resistivity, contact resistance etc [9].

6.2.4   AVALANCHE PHOTODIODE

Unlike a PIN-photodiode which has a current gain of unity, an APD has been reported to have a gain as high as 1000, but a gain of several hundred is more typical. This is achieved through the process of avalanche multiplication. The APD is constructed such that carriers must traverse a region where there is a very high electric field. The intensity of the field is such that hole or electron can be imparted with sufficient energy. When these carriers collide with the crystal lattice, they lose some energy to the crystal. If the kinetic energy of a carrier is greater than the band-gap energy of the valence electrons, the collision can free a bound electron. The free electron and hole so created themselves acquire enough kinetic energy to cause further impact ionization. It results in an avalanche, with the number of free carriers growing exponentially as the process continues.

A schematic of reach-through avalanche photodiode (RAPD) is shown in Fig. 6.11 along with the associated field distribution. This structure is commonly used in practice as it leads to carrier multiplication with very little excess noise. It consists of p^+-π-p-n^+ layers. The π layer is basically intrinsic material and inadvertently has some p doping because of imperfect purification. When the diode is reverse biased, most of the applied voltage appears across the pn^+ junction because of negligibly small photocurrent. With the increase in the bias voltage, peak electric field at the junction and width of the depletion region increases. At a certain voltage level, this electric field is about 10% less than the avalanche breakdown limit and the depletion layer just reaches through to the nearly intrinsic π region. Because of this reason, it is referred to as the reach-through APD.

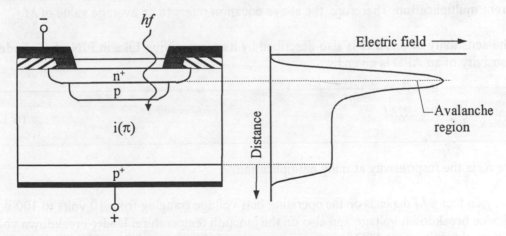

Fig. 6.11: APD structure and electric field distribution

The RAPD is normally operated in the fully depleted mode. Photons enter the device through the p^+ region and are absorbed by the high resistivity intrinsic p-type layer where electron-hole pairs are created. The relatively weak electric field in this region then separates the carriers causing the electrons and holes to drift into the high electric field region where avalanche multiplication occurs resulting in a current gain.

In practice, it is advantageous to have only one type of carrier for impact ionization. This is generally the electron [17]. Although presence of both types of carriers increases the gain of the device, it cause some undesirable device properties [12]. The continual cascading caused by both carriers could lead to instability and precipitate avalanche breakdown. The random process of collisions increases noise in the device and takes more time than a single multiplication process leading to reduction in the bandwidth.

The photodetector materials generally exhibit different ionization rates for electrons α_i and holes β_i. The ionization rate is defined as the average number of electron-hole pairs created by the carrier per unit distance travelled. The ratio $k = \beta_i/\alpha_i$ of the two ionization rates is a measure of the photodiode performance. The APDs constructed of materials in which one type of carrier largely dominates impact ionization exhibit low noise and large gain-bandwidth products [5].

The multiplication M for all carriers generated in the photodiode is defined by

$$M = \frac{I_M}{I_p} \qquad\qquad (6.13a)$$

Here, I_M is the average value of the multiplied output current and I_p the primary unmultiplied current. Because of the statistical nature of the multiplication process, each carrier undergoes different multiplication. Therefore, the above equation refers to an average value of M.

The sensitivity of an APD is also described by its responsivity. Like in PIN-photodiode, the responsivity of an APD is given by

$$R_{APD} = \frac{\eta e}{hf}M = R_0 M \qquad\qquad (6.13b)$$

where R_0 is the responsivity at unity multiplication.

The gain factor M depends on the operating bias voltage (ranging from 10 volts to 100 volts), the device breakdown voltage and also on the junction temperature. Under breakdown voltage condition, the gain of an APD is very temperature sensitive, especially at high bias voltage. An example of this is shown in Fig. 6.12 for a Si APD [8]. This makes it necessary to control the ambient temperature, use a thermoelectric cooler to control the photodiode temperature or to automatically compensate for change in gain by adjusting the bias. Ambient temperature control is costly and the latter two methods require additional electronic circuitry which decreases the reliability [1].

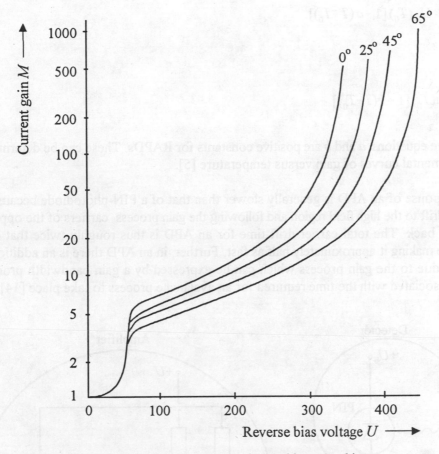

Fig. 6.12: Variations of APD current gain M with reverse bias
voltage U at different temperatures in centigrade

Variations of gain with temperature can be described by the following simple empirical
relationship [5]

$$M = \frac{1}{1-(U/U_B)^n} \qquad\qquad (6.14)$$

where U_B is the breakdown voltage at which M goes to infinity, n varies between 2.5-7 depending
on the photodiode material and U the reverse bias voltage available across the junction. The
breakdown voltage U_B and parameter n are known to vary with temperature as

$$U_B(T) = U_B(T_0)[1 + a(T-T_0)] \qquad (6.15)$$

and

$$n(T) = n(T_0)[1 + b(T-T_0)] \qquad (6.16)$$

In the above equations, a and b are positive constants for RAPDs. These can be determined from the experimental curves of gain versus temperature [5].

The response of an APD is generally slower than that of a PIN-photodiode because carriers must first drift to the high field region and following the gain process, carriers of the opposite type must drift back. The total carrier drift time for an APD is thus roughly twice that of a PIN-photodiode making it approximately half as fast. Further, in an APD there is an additional speed limitation due to the gain process which can be expressed by a gain-bandwidth product. This effect is associated with the time required for an avalanche process to take place [14].

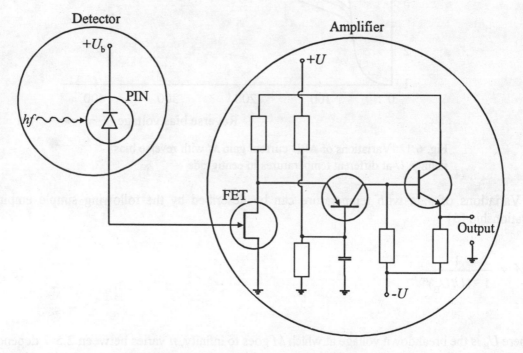

Fig. 6.13: Circuit diagram of a PIN-FET receiver

In contrast to an APD, there exists another variety of photodiodes in which an amplifier (or preamplifier) is integrated with a PIN-photodiode on the same semiconductor chip to reduce noise. These devices are called integrated detector-amplifiers, detector-amplifiers or PIN-FET, the last because the preamplifier uses field-effect transistor (FET) circuitry. The type of circuits used in a hybrid receiver with PIN-photodiode and low noise FET preamplifier is shown in Fig. 6.13. The amplifier circuit converts the current signal from the photodiode into a voltage signal as used by many electronic devices. Often, the circuit includes automatic gain control so that the voltage level is compatible with the later amplification stages. This circuit amplifies the electrical signal before it encounters the noise associated with the load resistor, increasing output signal power and S/N ratio.

These receivers have become popular for many moderate data rate applications because of their simplicity and reasonable cost. Their rise times tend to be slower than the discrete PIN-photodiodes or APDs, but adequate for transmitting hundreds of megabits per second. Further, unlike APDs such devices do not require voltage above 5V normally needed by semiconductor electronics [2, 9].

6.2.5 PHOTOTRANSISTOR

A phototransistor is an integrated combination of a photodiode and a transistor. A phototransistor circuit symbol along with an equivalent representation as a separate photodiode and transistor is shown in Fig. 6.14. It is clear from the equivalent circuit that the photocurrent I_p of the photodiode becomes the base current of the transistor. This is amplified by the current gain β of the transistor to become the collector current of βI_p. The total phototransistor current becomes $(\beta+1)I_p$. Thus, phototransistor provides an internal current gain of $(\beta+1)$. Since the β value of phototransistor normally ranges from 100 up to a value in excess of 1000, very large gains can indeed be obtained.

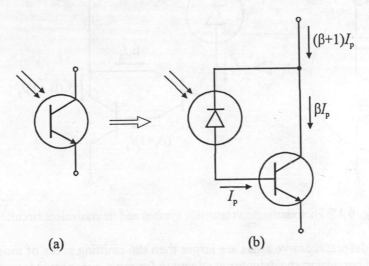

(a)   (b)

Fig. 6.14: Phototransistor symbol and its equivalent circuit

The main problem with the phototransistor is that the current gain β will not be a constant, but vary with current level as is the case with all transistors. As a result, there will no longer be linearity between the optical power and the resulting output current as was the case with a photodiode. In general, the phototransistor current will vary with the incident optical power level as $I \propto (P_r)^n$, where the exponent n will be about 2/3 at low-to-moderate power levels and may decrease to as low as 1/2 at very low power levels and only at relatively high power levels n will approach unity. It has much poorer frequency response as compared to that of a photodiode.

6.2.6 PHOTODARLINGTON TRANSISTOR

A photodarlington transistor is an integrated combination of a photodiode and a Darlington transistor configuration as shown in Fig. 6.15. As seen from the equivalent circuit, the photocurrent produced by the photodiode part of the structure is multiplied by the current gains of both transistors so that the resulting device current will be approximately $\beta_1\beta_2 I_p$. The photodarlington transistor thus has the advantage of providing an extremely large internal current gain. Since the individual transistor current gain will usually be in the range of about 100 to 1000, the overall current gain will be about 10,000 to as much as 10^6. The drawbacks of this device, however, are the same as for the phototransistor.

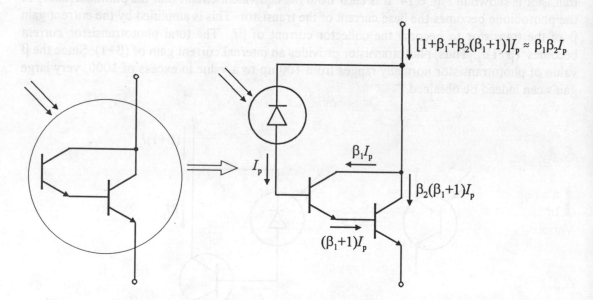

Fig. 6.15: Photodarlington transistor symbol and its equivalent circuit

Generally, the detectors active areas are larger than the emitting areas of most fibers. Thus, simply butting the fiber against the detector is adequate for most purposes. Many receivers come with integrated connectors that mate with fiber connectors. The connector case either delivers light directly to the detector or delivers the light to a fiber that carries it to the detector. The detector modules are packaged inside receivers just as light source modules are put inside transmitters. Generally, packaged receivers look very much like transmitters. One has to read the labels to identify them. Internal design constraints become increasingly severe at high frequencies because of the inherent problems in high frequency transmission.

The responsivity and rise time characteristics of PIN-photodiodes, APDs, PIN-FETs, phototransistor and photodarlington are given in Table 6.2 [2].

Table 6.2: Some typical detector characteristics

Device	Responsivity	Rise time
PIN-photodiode (Si)	0.5 A/W	0.1-5 ns
PIN-photodiode (InGaAs)	0.8 A/W	0.01-5 ns
Avalanche photodiode (Ge)	0.6 A/W	0.3-1 ns
Avalanche photodiode (InGaAs)	0.75 A/W	0.3 ns
Si PIN-FET (detector-amp)	15,000 V/W	10 ns
InGaAs PIN-FET (detector-amp)	5,000 V/W	1-10 ns
Phototransistor (Si)	18 A/W	2.5 μs
Photodarlington (Si)	500 A/W	40 μs

6.3 NOISE IN PHOTODIODES

In a photodiode, main sources of noise are *dark current noise*, *shot noise* and *thermal noise*. In an APD, there is one more source of noise due to random nature of the avalanche effect. Various types of noise are briefly described below.

6.3.1 DARK CURRENT NOISE

The dark current noise arises due to dark current which flows in the circuit when the photodiode is in unilluminated environment under bias condition. It is equal to the reverse saturation current of the photodiode. The magnitude of this current is strongly dependent on the operating temperature, the bias voltage and the type of detector. Variations of this current with temperature is essentially governed by the relationship $\exp[-E_g/k_B T_0]$ [14].

In an optical receiver, dark current sets a noise floor for the detectable signal power level. Therefore, it should be minimized by careful device design and fabrication. Dark current in optical telecommunication grade Si PIN-photodiodes is typically 100 pA, while in Si APDs it is typically 10 pA. In InGaAs based PIN-photodiodes and APDs, the dark current is typically 2 nA - 5 nA. In Ge APDs, the dark current is of the order of 100 nA and it could pose a serious problem unless the device is cooled to an appropriate temperature [6, 9].

6.3.2 SHOT NOISE

It arises from the statistical nature of the generation and collection of photoelectrons when an optical signal is incident on a photodiode. These statistics follow a Poisson process. Since the fluctuations in the number of photocarriers generated from the photoelectric effect are a fundamental property of the photodetection process, it will always exist. It thus set the lower limit on the receiver sensitivity when all other conditions are optimized. The shot noise current mean square value is given by

$$\langle i_{sh}^2 \rangle = 2 e I_p B \tag{6.17}$$

where I is the average photocurrent which is equal to the sum of average dark current and average signal current and B the single-sided (i.e., physical) receiver bandwidth (compare with Chapter nine where double-sided i.e., mathematical bandwidth is used). If dark noise current is large compared to signal current, signal current may be masked by the noise and therefore becomes unusable. On the other hand, if dark noise current is relatively small, it may have a negligible effect.

6.3.3 THERMAL NOISE

It originates within the photodiode load resistance R_l. Electrons within any resistor never remain stationary. They continuously move because of their thermal energy even with no applied voltage. The electron motion is random, so the net flow of charge could be towards one electrode or the other at any instant. Thus, a randomly varying current exists in the resistor. The mean square value of this current is given by

$$\langle i_{th}^2 \rangle = \frac{4 k_B T_0 B}{R_l} \tag{6.18}$$

where k_B is the Boltzmann's constant, T_0 the absolute temperature in degree Kelvin and R_l the load resistance.

6.3.4   EXCESS NOISE

The amplification process that provides the internal gain in photodetectors (such as APDs) is random. Each detected photon generates a random number of carriers with an average value M. The noise associated with APD signal current $I_{APD} = M \cdot I_p$ will consist of the sum of amplified shot noise of the primary photocurrent and excess noise produced by the multiplication process. The excess noise is usually incorporated by an excess noise factor $F(M)$. It is defined as the ratio of the total noise associated with I_{APD} to the noise that would exist in I_{APD} if the multiplication process produces no excess noise at all. The mean square shot noise current is $F(M)$ times the mean square multiplied shot noise current

$$\langle i_{sh}^2 \rangle = 2eI_p M^2 F(M) B \tag{6.19}$$

where $I_p = R_0 P_r$ is the primary photocurrent. The $F(M)$ from the following empirical relationship [13] can be approximated as

$$F(M) \approx M^x \tag{6.20a}$$

The parameter x, called excess noise exponent, is typically 0.4 for Si, 1.0 for Ge and 0.7 for InGaAs APDs. The resulting noise is assumed to be white with a Gaussian distribution.

In a second and more exact relationship, $F(M)$ is related to the multiplication factor M and the ratio of ionization rates k by the following relationship

$$F(M) \approx kM + \left(2 - \frac{1}{M} \right)(1-k) \tag{6.20b}$$

It is clear from Eq.(6.20) that $F(M)$ will be small if x or k is small. Typical range of values of x and k for Si, Ge, InGaAs is given in Table 6.1. For PIN-photodiodes, obviously M and $F(M)$ are unity.

In an APD, I_{dark} is a combination of bulk and surface dark currents. The bulk dark current I_{db} is due to the thermally generated (electrons and/or holes) in the photodiode. These carriers also undergo multiplication process in the same way as the signal carriers and generate a bulk dark current I_{db}. The mean square value of this current is given by

$$\langle i_{db}^2 \rangle = 2eI_{db} M^2 F(M) B \tag{6.21}$$

where I_{db} is the primary (unmultiplied) detector bulk dark current. The surface dark current I_{ds}, also referred to as a surface leakage current or simply the leakage current, is dependent on several factors like surface defects, cleanliness, bias voltage, surface area etc. The mean square value of this current is given by

$$\langle i_{ds}^2 \rangle = 2eI_{ds}B \tag{6.22}$$

It may be mentioned that avalanche multiplication is a bulk effect and the surface dark current is not affected by the avalanche gain. Similarly, the thermal noise current is also not amplified by the multiplication mechanism because it is not generated inside the photodetector.

6.3.5   SIGNAL-TO-NOISE RATIO

Let us presume that the received optical power level is P_r corresponding to bit "1" and 0 for bit "0". In an APD receiver, peak S/N ratio is given by

$$\frac{S}{N} = \frac{M^2 I_p^2}{2e(I_p + I_{db})M^2 F(M)B + 2eI_{ds}B + 4k_B T_0 B/R_1 + \langle i_a^2 \rangle} \tag{6.23a}$$

where $\langle i_a^2 \rangle$ is the mean square value of the following preamplifier noise current. The above equation can also be written as

$$\frac{S}{N} = \frac{M^2 I_p^2}{2e(I_p + I_{db})M^2 F(M)B + 2eI_{ds}B + 4k_B T_0 BF_t/R_1} \tag{6.23b}$$

where F_t is the noise figure of the following preamplifier. With $I_p = R_0 P_r$, above equation will be

$$\frac{S}{N} = \frac{M^2 R_0^2 P_r^2}{2e(R_0 P_r + I_{db})M^2 F(M)B + 2eI_{ds}B + 4k_B T_0 BF_t/R_1} \tag{6.24}$$

For high P_r, shot noise due to signal will dominates over all other noise sources. Therefore, the above equation can be approximated as

$$\frac{S}{N} \simeq \frac{M^2 R_0^2 P_r^2}{2eR_0 P_r M^2 F(M)B} = \frac{R_0 P_r}{2eF(M)B} \tag{6.25}$$

When P_r is low, thermal noise will become predominant. In that case Eq. (6.24) can be approximated as

$$\frac{S}{N} \simeq \frac{M^2 R_0^2 P_r^2}{4k_B T_0 BF_t/R_1} \tag{6.26}$$

For a PIN-photodiode, peak S/N ratio can be obtained by modifying Eq. (6.24). It is given by

$$\frac{S}{N} = \frac{R_0^2 P_r^2}{2e(R_0 P_r + I_{dark})B + 4k_B T_0 BF_t/R_1} \tag{6.27}$$

When P_r is high, following the same reasoning as for APD, it can be approximated as

$$\frac{S}{N} \simeq \frac{R_0^2 P_r^2}{2eR_0 P_r B} = \frac{R_0 P_r}{2eB} \tag{6.28}$$

When P_r is low, thermal noise will limit the S/N ratio. Under this condition, Eq. (6.27) can be approximated as

$$\frac{S}{N} \simeq \frac{R_0^2 P_r^2}{4k_B T_0 BF_t/R_1} \tag{6.29}$$

It is seen from Eqs. (6.25) - (6.29) that an APD receiver does not perform always better than a PIN receiver. When P_r is high, performance of APD receiver (Eq. (6.25)) becomes worse than the PIN-photodiode receiver (Eq. (6.28)) because of the term $F(M)$ in the denominator.

6.4   MINIMUM DETECTABLE POWER

For an optical receiver, minimum detectable power (MDP) is defined as the mean signal power that yields unity S/N ratio. For shot noise limited case i.e., other sources of noise are small as compared to shot noise, S/N ratio for a PIN-photodiode receiver from Eq. (6.27) is given by

$$\frac{S}{N} = \frac{I_p^2}{2e(I_p + I_{dark})B} \tag{6.30}$$

When $I_p \ll I_{dark}$ and using Eq. (6.10a), P_r for unity S/N ratio will be

$$P_r = \frac{\sqrt{2eI_{dark}}}{R_0}\sqrt{B} \tag{6.31}$$

or

$$MDP = NEP\sqrt{B} \tag{6.32a}$$

where

$$NEP = \frac{\sqrt{2eI_{dark}}}{R_0} \tag{6.32b}$$

Here, *NEP* stands for *noise-equivalent power*. The *NEP* depends on frequency of the modulating signal, bandwidth over which the noise is measured, area of the detector (when background radiation and thermal generation are the dominant cause of dark current rather than surface conduction) and operating temperature [2]. In Eq. (6.32b), dark current is considered in isolation, so it becomes a characteristic of photodetector. If other sources of noise are also considered, it will then be a characteristic of receiver.

6.5   OPTICAL RECEIVER CONFIGURATIONS

An optical receiver consists of a photodiode, an amplifier and a signal processing circuitry. Its task is first to convert the optical energy emerging from the end of a fiber into an electrical signal and then to amplify this signal to a large enough value so that it can be processed by the electronic circuits following the receiver amplifier. In these processes, various noise and distortions will unavoidably be introduced, which can lead to errors in the interpretation of the received signal. The design of amplifier which follows the photodiode, often referred to as preamplifier, is of critical importance because it is the amplifier where the major noise sources are expected to arise.

A simplified optical receiver circuit is shown in Fig. 6.16a. It consists of a reverse biased photodiode in series with a current limiting resistor R_l bypassed to ground by a large capacitor C_{bp} and feeding a high input impedance, low noise preamplifier. The photodiode capacitor $C_d \approx 5$ pF and its internal resistance $R_d \approx 500$ MΩ are also depicted in the figure.

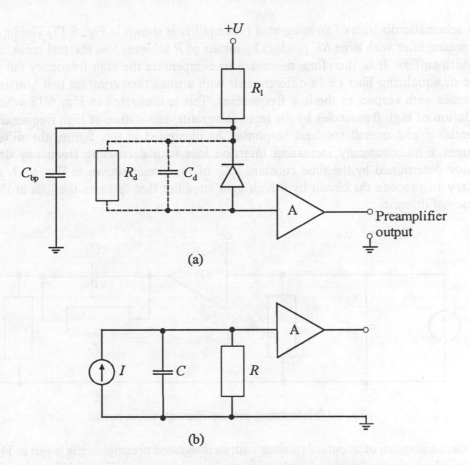

Fig. 6.16: (a) Optical receiver and (b) its equivalent circuit

The ac equivalent circuit is shown in Fig. 6.16b with the photodiode signal represented by an ideal current source I. The 3 dB electrical bandwidth of this circuit is $B = 1/2\pi RC$ where $R = R_l \| R_d \| R_a$ and $C = C_d + C_a$ are the equivalent resistance and capacitance respectively. Here, R_a and C_a represent the amplifier resistance and capacitance respectively. To keep the bandwidth large, R should be small. But making R small may introduce excessive thermal noise. Thus, we are faced with a dilemma:

- Low R to achieve large bandwidth or fast response time
- High R to achieve low noise

Four basic preamplifier designs provide a solution to the above dilemma.

6.5.1   INTEGRATED PREAMPLIFIER

The schematic diagram of an integrated preamplifier is shown in Fig. 6.17. The input circuit of this preamplifier with large RC product by means of R achieves low thermal noise. However, bandwidth suffers. It is, therefore, necessary to compensate the high frequency fall off by an inverse or equalizing filter i.e., a differentiator with a small time constant that boosts the high frequencies with respect to the low frequencies. This is illustrated in Fig. 6.18 which shows degradation of high frequencies by the input integrator, restoration of high frequencies by the differentiator and overall resultant response. As illustrated in this figure, the differentiator introduces a monotonically increasing insertion loss with decreasing frequency down to a frequency determined by the time constant r_2C_2 of the circuit shown in Fig. 6.17. It is thus necessary to precede the circuit by a high gain amplifier that restores the loss at the lowest frequency of interest.

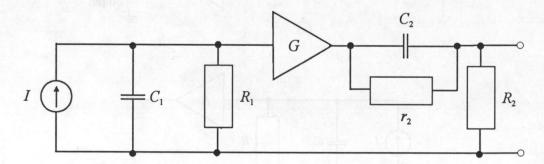

Fig. 6.17: Integrating preamplifier equivalent circuit

The circuit diagram of an optical receiver with an integrated preamplifier is given in Fig. 6.19a. The equivalent circuit for signal and noise analyses is given in Fig. 6.19b and a simplified equivalent circuit in Fig. 6.19c. For a PIN-photodiode receiver, shot noise power can be obtained from Eq. (6.17) and thermal noise power from Eq. (6.18) by replacing R_1 by $R = R_1\|R_d\|R_a$. The transistor in the amplifier may be either BJT or JFET.

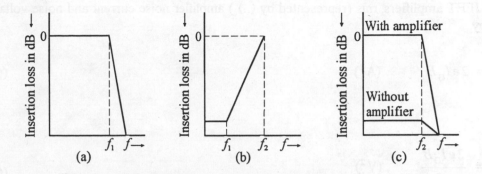

Fig. 6.18: Integrating preamplifier restoration of high frequencies ($f_1 = 1/2\pi r_2 C_2$ and $f_2 = 1/2\pi R_2 C_2$) (a) input integrator response, (b) differentiator response and (c) overall response without and with amplifier

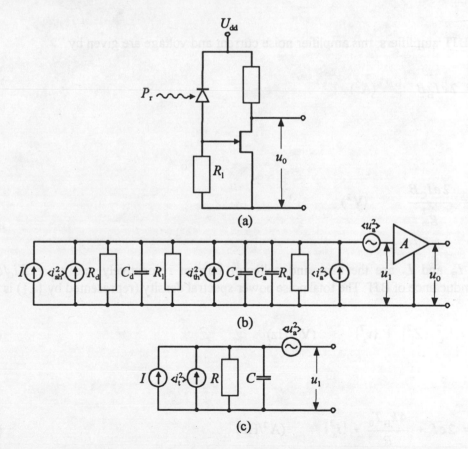

Fig. 6.19: Optical receiver with integrating preamplifier (a) circuit diagram, (b) equivalent circuit and (c) simplified equivalent circuit. R_d and C_d are the resistance and capacitance of the photodiode; $R = R_1 \| R_d \| R_a$ and $C = C_d + C_s + C_a$ are the equivalent resistance and capacitance

For JFET amplifiers, rms (represented by $\langle .. \rangle$) amplifier noise current and noise voltage are given by

$$\langle i_a^2 \rangle = 2eI_G B \qquad (\text{A}^2) \tag{6.33}$$

and

$$\langle v_a^2 \rangle = \frac{2eI_D B}{g_m^2} \qquad (\text{V}^2) \tag{6.34}$$

where I_G is the gate bias current, I_D the drain current and $g_m = \Delta i_D/\Delta v_{GS}$ the transconductance of JFET.

For BJT amplifiers, rms amplifier noise current and voltage are given by

$$\langle i_a^2 \rangle = 2eI_B B \qquad (\text{A}^2) \tag{6.35}$$

and

$$\langle v_a^2 \rangle = \frac{2eI_C B}{g_m^2} \qquad (\text{V}^2) \tag{6.36}$$

where I_B and I_C are the base and collector current respectively and $g_m = \Delta i_C/\Delta v_{BE}$ the transconductance of BJT. The total noise power spectral density (represented by $\{..\}$) is given by

$$\{v_n^2\} = \{i_t^2\}|Z_L^2| + \{v_a^2\} \qquad (\text{V}^2/\text{Hz}) \tag{6.37a}$$

where

$$\{i_t^2\} = 2eI + \frac{4k_B T_0}{R} + \{i_a^2\} \qquad (\text{A}^2/\text{Hz}) \tag{6.37b}$$

and

$$Z_L = \frac{R}{1 + j2\pi fRC} \qquad \text{(Ohms)} \qquad (6.37c)$$

Substitution of Eqs. (6.37b) and (6.37c) in Eq. (6.37a), gives

$$\{v_n^2\} = \left[2eI + \frac{4k_B T_0}{R} + \{i_a^2\}\right] \frac{R^2}{1 + 4\pi^2 f^2 R^2 C^2} + \{v_a^2\} \qquad (6.38)$$

The above equation gives the spectral density of noise voltage at the input of the amplifier. If the gain of the amplifier is A, the rms output noise voltage is given by

$$\langle v_n^2 \rangle = A^2 \int_0^\infty \{v_n^2\} df \qquad (V^2) \qquad (6.39)$$

It is clear from above that the increase in R will decrease the noise power. On the other hand, output signal voltage will increase with the increase in R. Both these will lead to improvement in the S/N ratio. At the same time, it will reduce the bandwidth of the receiver. Thus, bandwidth requirement for the receiver limits the extent to which S/N ratio can be improved.

The bandwidth limitation due to RC load impedance can be compensated to some extent by placing an equalizer in the receiver circuit. The transfer function of the equalizer is given by

$$H_{eq}(f) = 1 + j2\pi fRC \qquad (6.40)$$

With the above equalizer, the rms output noise voltage from Eq. (6.39) is given by

$$\langle v_n^2 \rangle = A^2 \int_0^\infty \{v_n^2\} |H_{eq}(f)|^2 df \qquad (V^2) \qquad (6.41)$$

Substitution of $\{v_n^2\}$ and $H_{eq}(f)$ from Eqs. (6.38) and (6.40) respectively in the above equation and further simplification gives

$$\langle v_n^2 \rangle = A^2 \left[\left(2eI + \frac{4k_B T_0}{R} + \{i_a^2\}\right) R^2 + \{v_a^2\}\left(1 + \frac{(2\pi RCB)^2}{3}\right)\right] B \qquad (6.42)$$

The output signal voltage is given by

$$U_s = RI_p A \quad \text{(V)} \tag{6.43}$$

Therefore, the output S/N ratio from Eqs. (6.42) and (6.43) will be

$$S/N = \cfrac{I_p^2}{\left[2eI + \cfrac{4k_B T_0}{R} + \{i_a^2\} + \{v_a^2\}\left(\cfrac{1}{R^2} + \cfrac{(2\pi CB)^2}{3}\right)\right]B} \tag{6.44}$$

It is clear from the above equation that as R is made larger, S/N ratio increases. A practical limit is that as R becomes larger, effective equalization becomes more and more difficult. The difficulty of equalization or the reduction of photodiode circuit bandwidth will limit the value of R. The above equation for S/N ratio is derived for digital intensity modulation signalling scheme in which bit "1" is represented by a power level P_r and bit "0" by zero power level. For other modulation schemes like finite extinction ratio on-off signalling or analog intensity modulation, it can be suitably modified. For example, when the modulation scheme is analog intensity modulation i.e.,

$$P(t) = P_r[1 + mf(t)] \tag{6.45}$$

where m is the modulation index and $f(t)$ the modulating signal. The Eq. (6.44) in this case will get modified as

$$S/N = \cfrac{I_p^2 m^2 <f(t)^2>}{\left[2eI + \cfrac{4k_B T_0}{R} + \{i_a^2\} + \{v_a^2\}\left(\cfrac{1}{R^2} + \cfrac{(2\pi CB)^2}{3}\right)\right]B} \tag{6.46}$$

For an APD receiver, S/N ratio can be derived in the same way. Presuming I_{ds} to be negligibly small, the current I becomes equal to sum of signal current I_p and bulk dark current I_{db}. In this case, the above equation will become

$$S/N = \cfrac{I_p^2 m^2 <f(t)^2>}{\left[2eIF(M) + \cfrac{4k_B T_0}{M^2 R} + \cfrac{\{i_a^2\}}{M^2} + \cfrac{\{v_a^2\}}{M^2}\left(\cfrac{1}{R^2} + \cfrac{(2\pi CB)^2}{3}\right)\right]B} \tag{6.47}$$

It may be mentioned that as M increases, $F(M)$ also increases. It means that with the increase in M, first term in the denominator increases, while the other three terms decreases. Thus, there is an optimum value of M that will produce a maximum S/N ratio. For a typical APD, variations of $F(M)$ and S/N with M are shown in Fig.6.20 [16].

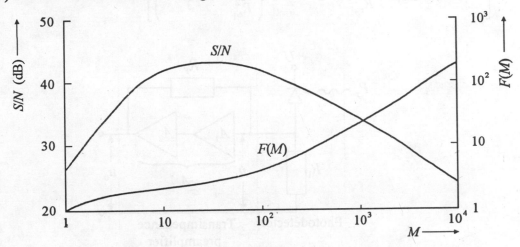

Fig. 6.20: Variations of $F(M)$ and S/N ratio with M

6.5.2   TRANSIMPEDANCE PREAMPLIFIER

A schematic circuit diagram and equivalent circuit for transimpedance preamplifier are shown in Fig. 6.21. It utilizes *resistive feedback* to increase bandwidth at the expense of gain for a constant gain-bandwidth product. Analysis for signal output from the equivalent circuit is same as for an operational amplifier. For determining the output rms noise voltage, current and voltage noise sources are considered separately and rms noise voltages are determined. The output rms voltage following the principle of superposition will be the sum of these rms voltages. Finally, the S/N ratio is given by

$$S/N = \frac{I_p^2}{\left[2eI + \frac{4k_B T_0}{R_{eq}} + \{i_a^2\} + \{v_a^2\} \left(\frac{1}{R_{eq}^2} + \frac{(2\pi CB)^2}{3} \right) \right] B} \tag{6.48a}$$

where

$$R_{eq} = R \| R_F \tag{6.48b}$$

The above equation after including dark noise current can be written as

$$S/N = \frac{\left(R_0 P_r\right)^2}{\left[2e\left(R_0 P_r + I_{\text{dark}}\right) + \frac{4k_b T_0}{R_{eq}} + \{i_a^2\} + \{v_a^2\}\left(\frac{1}{R_{eq}^2} + \frac{(2\pi CB)^2}{3}\right)\right]B} \qquad (6.49)$$

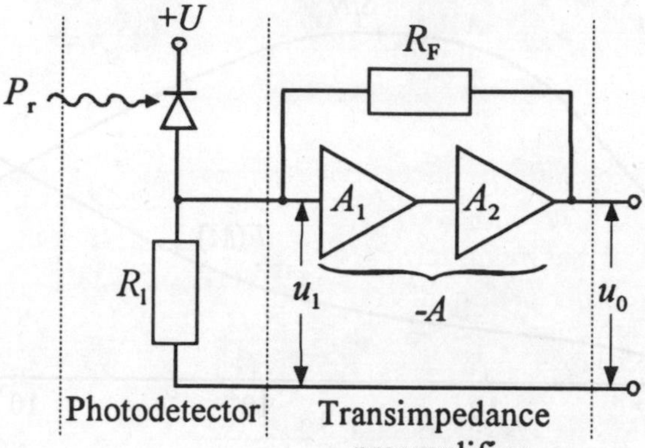

(a)

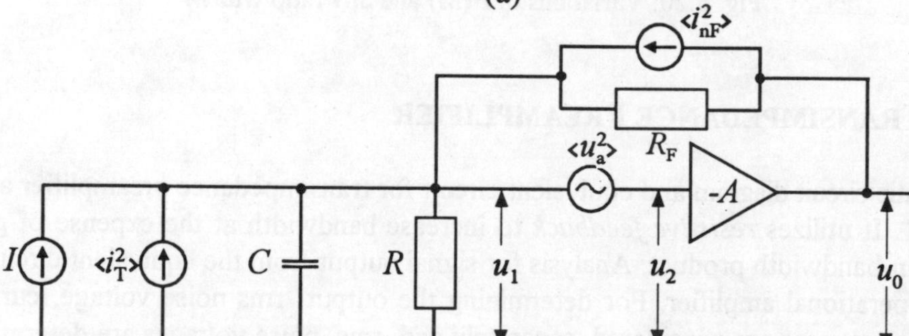

(b)

Fig. 6.21: Optical receiver with transimpedance preamplifier (a) circuit diagram and (b) equivalent circuit

The conditions which must be satisfied in using the above simplified equation are

$$A \gg 1 + \frac{R_F}{R} \qquad (6.50a)$$

and

$$B \leq \frac{A}{2\pi R_F C} \qquad (6.50b)$$

The first condition is easy to satisfy as long as $A \gg 1$. The second condition imply that the 3 dB bandwidth of the circuit ($=1/2\pi R_F C$) multiplied by A should be greater than the signal bandwidth B. The designer has to ensure that A is high and constant over the bandwidth B.

6.5.3   LOW-IMPEDANCE PREAMPLIFIER

In high-speed optical communication systems with bit rates above 1 Gbit/s (e.g., 2488.32 Mbit/s in STM-16 or 9953.28 Mbit/s in STM-64), the low-impedance preamplifier is the only alternative. In comparison to the other preamplifiers, it can be easily realized. But in practice a commercially available high-frequency or microwave 50 Ω amplifier is normally used as preamplifier. Besides a high signal bandwidth of 10 GHz and more, it shows a very good linearity. However, it has relatively more noise. In addition, output voltage level is less as compared to that of other preamplifiers.

6.5.4   HIGH-IMPEDANCE PREAMPLIFIER

The high-impedance preamplifiers are characterized by very low noise and smaller bandwidth (maximum 500 MHz). Therefore, such preamplifiers are well suited in optical space communication systems wherein bit rate is moderate (much less than the bit rate in fiber-based systems) and optical received input power level is low. The latter one is due to the extremely large distance normally involved in space (Chapter fifteen). For this application, a low-noise receiver is essentially required to obtain a desired S/N ratio. In addition to smaller bandwidth, another disadvantage of this preamplifier is the need of an equalizer since low RC product in this preamplifier acts as an integrator which distort the signal. In practice, high-impedance preamplifiers are normally realized by using a PIN-FET.

6.5.5   COMPARISON

The major disadvantage of the integrating preamplifier is that if there is a long sequence of "1", the receiver will saturate as the integrator builds up an appreciable DC component. Integrated preamplifier has two limitations: (i) requirement of equalization for broadband application and (ii) limited dynamic range. The transimpedance preamplifier also has a disadvantage. There is a limit to how large feedback resistance can be made in order to reduce noise. The output voltage equals input current times the feedback resistance. With a large input signal, the output may exceed the preamplifier dynamic range. The transimpedance preamplifier is less sensitive, but it has the advantage of a relatively large dynamic range and less need for equalization [3, 5, 16].

A comparison of transimpedance, low-impedance and high-impedance preamplifiers is given in the following table.

Table 6.3: Comparison of preamplifiers in terms of typical data

Preamplifier	Load Resistor	Signal bandwidth	Noise current
Low-impedance	50 Ω	10 GHz	1 mA
High-impedance	1 MΩ	500 MHz	1 nA
Transimpedance	1 kΩ	1 GHz	1 μA

It becomes clear from this table that choice of a preamplifier mainly depends on desired application. High-speed transmission links require the use of a 50 Ω amplifier, whereas in long-haul links with a moderate bit rate a high-impedance preamplifier may be the best option. As seen from this table, the relatively low noise (compared to the low-impedance) and relatively large bandwidth (compared to the high-impedance) of a transimpedance preamplifier is a compromise of both these preamplifiers.

6.6 REFERENCES

[1]   Christian, N.L.; Passauer, L.K.: Fiber Optics Component Design, Fabrication, Testing, Operation, Reliability and Maintainability. Noyes Data Corporation, 1989.
[2]   Hecht, Jeff: Understanding Fiber Optics. Second Edition, Howard W.Sams&Co., 1997.
[3]   Henri Hodara (Ed-in Chief): Fiber and Integrated Optics, Crane, Russak&Company Inc., 1983.
[4]   Hunsperger, Robert G.: Photonic Devices and System. Marcel Dekker Inc., 1994.
[5]   Keiser, Gerd: Optical Fiber Communications. Second Edition, McGraw-Hill, 1992.
[6]   Li, T.: Advances in optical fiber communications: A historical perspective. IEEE J. SAC-1(1983), 356-372.
[7]   Melchior, H.; Fisher, M.B.; Arams, F.R.: Photodetectors for optical communication systems. Proc. IEEE, 58(1970), 1446-1486.
[8]   Melchior, H.; Hartman, A.R. et. al.: Planar epitaxial silicon avalanche photodiode. Bell System Tech. J. 57(1978), 1791-1807.
[9]   Pal, B.P.(Editor): Fundamental of Fiber Optics in Telecommunication and Sensor. Wiley Eastern, 1994.
[10]   Palais, J.C.: Fiber Optic Communications. Second Edition, Prentice-Hall, 1988.
[11]   Powers, J.P.: An Introduction to Fiber Optic System. Aksen Associates Inc., 1993.
[12]   Saleh, B.E.A.; Teich, M.C.: Fundamentals of Photonics. Wiley, 1991.
[13]   Senior, John M.: Optical Fiber Communication-Principles and Practice. Prentice-Hall, 1992.
[14]   Smith R.G.: Photodetectors for fiber transmission system. Proc. IEEE, 68(1980)10,1247-1253.
[15]   Sze, S.M.: Physics of Semiconductor Devices. Wiley, 1981.
[16]   William B. Jones Jr.: Introduction to Optical Fiber Communication Systems. Holt, Rinehart&Winston Inc., 1987.
[17]   Wilson, J.; Hawkes, J.F.B.: Optoelectronics. Prentice-Hall, 1989.

7   OPTICAL AMPLIFIERS

One of the most exciting development in the field of optics has been the discovery of optical amplifier. It is probably the most significant development since the discovery of single-mode fiber. The availability of high performance optical amplifiers has revolutionized the world of light wave communications [7, 12, 20, 21, 26, 31, 39]. The first phase of this revolution brought into repeaterless transoceanic transmission (TOT) systems with bit rate-distance products in excess of about 100 Tbit/s·km. The second phase of the revolution is the upcoming wavelength multiplexed systems, which have become economical since several channels may be amplified by a single (sufficiently powerful) optical amplifier. The third phase of the revolution may bring the use of optical techniques in routing and networking. In this Chapter, different types of optical amplifiers have been described (Section 7.1), compared (Section 7.2) and various applications of optical amplifiers in optical communication systems have been discussed (Section 7.3). Noise of amplifiers is considered in Section 7.4.

The optical amplifiers play an important role in the optical communication systems as they enable the direct amplification of light with a minimum of electronics and thus eliminate the electronic bottleneck associated with many current systems. In a long-haul optical networks, system upgradation (in terms of increased data rate or users or improvement in performance or link length) can be achieved without the modification of hardware. In local area networks, optical amplifiers can be used to overcome the splitting loss associated with passive networks and increase the fan out capability.

At present, a large number of applications of optical amplifiers are existent. Most important are the following applications which are discussed in more detail in Section 7.3:

- power or booster amplifier,
- in-line amplifier,
- preamplifier,
- remotely pumped booster amplifier,
- remotely pumped preamplifier,
- loss cancellation in optical networks such as distribution networks, ring networks and WDM-based overlay networks,
- loss cancellation in optical components such as optical crossconnects and
- optical processing and switch.

The use of optical amplifiers give rise to substantial span increase in repeaterless systems. An optical postamplifier boosts the output power of a transmitter by about 20 dB and an optical preamplifier improves the receiver sensitivity by about 10 dB (Section 7.3). Another technique called remotely pumped amplification, which consists in sending the pump signal from the terminals (transmitter and receiver) towards an optical amplifier located several tens of kilometres away, allows further increase of power budgets. Currently, optical fiber amplifiers (Section 7.1.2) have been deployed as in-line amplifiers (Section 7.3) worldwide in terrestrial networks and undersea transmission systems, in point-to-point and network configurations.

Advanced all-optical networks will have optical amplifiers to compensate loss while capitalizing on their transparency to modulation format and signal wavelength. Considering WDM based networks (Chapter fourteen), fiberoptic amplifiers allow all wavelength channels to be amplified simultaneously. The primary disadvantage of the bus topology i.e., poor efficiency could be overcome with a suitable optical fiber amplifier to compensate the high signal attenuation in the network.

7.1 TYPES OF OPTICAL AMPLIFIERS

Optical amplifiers can be classified on the basis of device characteristics i.e., whether it is based on linear characteristic (semiconductor laser amplifiers and rare-earth doped fiber amplifiers) or non-linear characteristic (Raman amplifiers and Brillouin amplifiers). Optical amplifiers can also be classified on the basis of structure i.e., whether semiconductor based or optical fiber based. In this classification, semiconductor laser amplifiers (SLAs) will fall in the first category, while the other three i.e., rare-earth doped fiber amplifiers, Raman and Brillouin scattering amplifiers in the second category.

Different rare-earth ions, such as Erbium, Praseodymium, Holmium, Neodymium, Samarium, Thulium and Ytterbium can be used to realize fiber amplifiers operating at different wavelengths covering visible to infrared region (up to 2.8 µm). Erbium-doped fiber amplifiers (EDFAs) have attracted the most attention among them simply because they operate near 1550 nm, the wavelength region in which fiber attenuation is minimum. Raman and Brillouin amplifiers are based on the scattering of a photon to a lower energy photon such that the energy difference appears in the form of a phonon. The main difference between the two is that optical phonons participate in the Raman scattering, whereas acoustic phonons in the Brillouin scattering.

The SLAs are based on existing laser structure with anti-reflection (AR) coating applied to the facets. They exhibit high gain, low polarization sensitivity and high saturation power. Progress in fiber amplifier technology during the nineties has been very rapid and EDFAs have been demonstrated in many system experiments as well as in commercial applications such as TAT 12/13 (Appendix A1). The fiber amplifiers have a particular advantage that they can be spliced into the system fiber with very low loss and this is in contrast to large coupling loss associated with the SLAs and fiber coupling. Therefore, the fiber amplifiers exhibit higher gain than the SLAs [8].

In the following Sections, different types of optical amplifiers have been discussed in more detail.

7.1.1   SEMICONDUCTOR LASER AMPLIFIERS

The structure of SLA is similar to that of semiconductor laser source used in optical communi-cation systems. The basic difference is that in SLA, resonant cavity is essentially eliminated by substantially reducing the facet reflectivities down to levels of 10^{-3} to 10^{-4} or even less [33]. This, in turn, will reduce the optical feedback. An optical amplifier without feedback is referred as travelling wave amplifier (TWA). The optical gain for such an amplifier depends not only on the frequency (wavelength) of the incident signal, but also on the intensity of the signal at any point inside the amplifier. The relationship of gain with the frequency and intensity depends upon the amplifier medium. When the gain is modelled as homogeneously broaden two-level system, gain coefficient of such a medium is given by [4]

$$g(f) = \frac{g_0}{1 + 4\pi^2(f-f_0)^2T_2^2 + \dfrac{P_i}{P_s}} \tag{7.1}$$

Here, g_0 represents the peak value of gain determined by the pumping level of amplifier, f and P_i are the optical frequency and power of the incident signal respectively and f_0 the atomic transition frequency. P_s is the saturation power level (it depends on the gain medium) and T_2 is known as *dipole relaxation time* which typically varies from 0.1 ps to 1 ns. Depending on P_i/P_s ratio, there are two cases:

(i)   UNSATURATED REGION

In this region $P_i/P_s \ll 1$ and $g(f)$ from Eq. (7.1) is approximately given by

$$g(f) = \frac{g_0}{1 + 4\pi^2(f-f_0)^2T_2^2} \tag{7.2}$$

It is seen from this equation that the gain coefficient is maximum when $f = f_0$. Let us define the gain/amplification factor of the amplifier as

$$G = \frac{P_o}{P_i} \tag{7.3}$$

where P_o is the output power and P_i as defined earlier is the input power of the amplifier. This amplification factor for an amplifier of length L is given by [4]

$$G(f) = e^{g(f)L} \tag{7.4}$$

It is seen from the above equation that the $G(f)$ and $g(f)$ are maximum when $f = f_0$ and decreases when $f \neq f_0$. The maximum unsaturated gain G_0 of the amplifier will be

$$G_0 = e^{g_0 L} \tag{7.5}$$

The amplifier bandwidth defined as full width at half maximum (FWHM) is given by [4]

$$(\Delta f)_{\text{TWA}} = \left(\frac{1}{\pi T_2} \right) \cdot \left(\frac{\ln(2)}{g_0 L - \ln(2)} \right) \tag{7.6}$$

It is seen from Eqs. (7.5) and (7.6) that both G_0 and $(\Delta f)_{\text{TWA}}$ depend on g_0 and L.

(ii) SATURATED REGION

In this region $P_i/P_s \gg 1$ and $g(f)$ from Eq. (7.1) will be relatively lower and hence G also. For simplicity, let the frequency of incoming optical signal is exactly tuned to the atomic transition frequency i.e., $f = f_0$. Effect of detuning $(f \neq f_0)$ on the gain can be easily incorporated. The large signal amplifier gain is given by [15]

$$G = G_0 e^{(1-G)\frac{P_i}{P_s}} \tag{7.7}$$

This non-linear equation is to be solved for G, once G_0, P_i and P_s are known or specified. The above equation shows that G decreases from its unsaturated value G_0 when P_i becomes comparable with P_s. The gain saturation not only reduces the amplifier gain, but also leads to non-linear effects that can increase the crosstalk in a multichannel transmission system.

In semiconductor lasers, relatively large feedback is present due to reflections occurring at the cleaved facets (reflectivity $\approx 30\%$). Such lasers can also be used as amplifiers when biased below threshold and are referred as Fabry-Perot amplifiers (FPAs). The amplification factor for such amplifiers is given by [4, 42]

$$G_{FPA}(f) = \frac{(1-R_1)(1-R_2)G(f)}{\left(1 - G(f)\sqrt{R_1 R_2}\right)^2 + 4G(f)\sqrt{R_1 R_2}\,\sin^2(\phi)} \tag{7.8a}$$

where

$$\phi = \frac{\pi(f-f_r)}{\delta f} \tag{7.8b}$$

Here, R_1 and R_2 are the facet reflectivities, f_r represents the cavity resonance frequencies and $\delta f = c_0/2nL$ (n is the refraction index and c_0 the velocity of light) the longitudinal mode spacing. The parameter $G(f)$ is the single pass amplification factor and corresponds to that of TWA (refer to Eq. 7.4). The above equation shows that the amplifier transmission characteristics contain resonant peaks whose absolute wavelength and spacing depends on the cavity dimensions. Typical transmission characteristics of a FPA with $R_1 = R_2 = 0.3$ and a single pass gain of 25 dB is shown in Fig. 7.1.

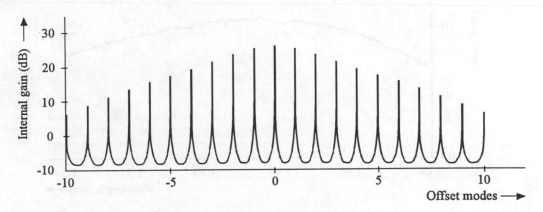

Fig. 7.1: FP amplifier passband, mode 0 corresponds to peak gain wavelength; mode spacing $\lambda^2/(2nL) = 1.5$ nm for $L = 200$ μm [27]

It is possible to determine the amplifier bandwidth by calculating the detuning $f - f_r$ for which G_{FPA} drops by 3 dB from its peak value. It is given by [29]

$$(\Delta f)_{FPA} = \frac{2\delta f}{\pi}\,\sin^{-1}\left[\frac{1 - G\sqrt{R_1 R_2}}{\left(4G\sqrt{R_1 R_2}\right)^{1/2}}\right] \tag{7.9}$$

For large amplification factor, $G\sqrt{R_1 R_2}$ should be close to unity. (refer to Eq. 7.8a). Under this condition, amplifier bandwidth from Eq. (7.9) will be quite small (< 10 GHz). Such a small bandwidth makes the FPAs unsuitable for light wave communication systems and they are primarily used for signal processing applications. The FPAs are characterized by narrow bandwidth (in the order of GHz) and a high sensitivity to temperature, bias current and signal polarization fluctuations. For wide bandwidth applications, amplifier bandwidth can be increased by suppressing the reflection feedback from the end facets. This is normally achieved by applying AR coating to the facets. When the reflectivity is sufficiently reduced (less than 1%), it will become a TWA. The bandwidth of TWA is three order of magnitudes larger than that of FPA. Nevertheless, passband in the TWA comprises peaks and troughs whose relative amplitudes are determined by the residual facet reflectivities, single pass gain (and hence the applied bias current) and the input signal power level. Variations of amplifier gain with signal wavelength for a SLA whose facets are anti-reflection coated to reduce reflectivity to about 0.04% (SLA is operating in a nearly travelling wave mode) are shown in Fig. 7.2 [28].

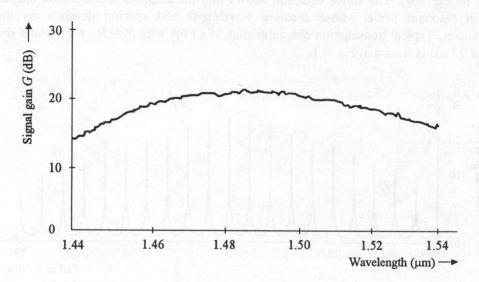

Fig. 7.2: Amplifier gain versus signal wavelength for semiconductor laser amplifier for transverse electric (TE) mode whose facets are anti-reflection coated to produce reflectivity of about 0.04% [27]

The gain undulation or peak-trough ratio of the ripple (ΔG) is given by [29]

$$\Delta G = \frac{1 + \sqrt{R_1 R_2}\ G}{1 - \sqrt{R_1 R_2}\ G} \tag{7.10}$$

For wide-band operation, ΔG must be small and is considered to be less than 3 dB (i.e., $\Delta G < 2$) for TWAs over their signal gain spectrum. A gain ripple of zero imply an ideal TWA. Hence, an amplifier whose gain ripple significantly exceeds 3 dB is usually categorized as a FPA.

The structures of FPA and TWA are shown in Fig. 7.3. The FPAs operating around a central wavelength of 1300 nm or 1500 nm with a facet-facet gain of about 30 dB are presently available [33]. Their gain-bandwidth product is about 40 nm (more than 4000 GHz). Devices based on special structure viz., multiple quantum well (MQW) have shown gain bandwidth of more than 200 nm. Noise figure (NF) of the SLAs is typically in the order of 7 dB [9]. In practical systems, it is necessary to couple the light to and from the amplifier and there are associated coupling losses. Typically, these losses are about 3.5 dB/facet. It means fiber-fiber gain will be 7 dB lower than the facet-facet gain. For this reason, maximum realistic gain of a wide-band amplifier is in the order of 20 dB.

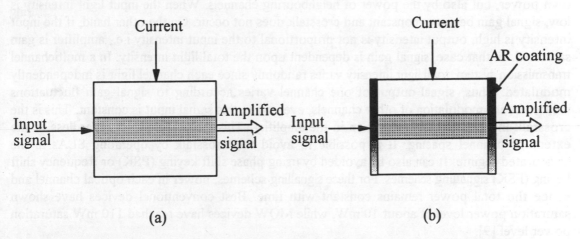

Fig. 7.3 Structure of semiconductor laser amplifiers (a) Fabry-Perot amplifier (FPA) and (b) travelling wave amplifier (TWA)

The optical amplifiers can be used to amplify several channels simultaneously if their carrier frequencies lie within the passband of amplifier. In practice, several non-linear phenomena in SLAs induce interchannel crosstalk. Two such non-linear phenomena of importance are cross-saturation and four-wave mixing (FWM). In a multichannel transmission system, total input optical power P_i is given by [4]

$$P_i = \sum_{j=1}^{N} P_j + \sum_{j=1}^{N} \sum_{k=1}^{N} \sqrt{P_j P_k} \cos\left(2\pi(f_j - f_k)t + \phi_j - \phi_k\right) \qquad (7.11)$$

Here, f_j is the frequency, ϕ_j the phase of 'j'th channel signal and N the total number of channels. The power as given by Eq. (7.11) cause the carrier population in the amplifier medium to oscillate at the beat frequency $f_j - f_k$. As a consequence, gain and refraction index are also modulated at the same frequency. Thus, multichannel signal creates the gain and index gratings. These gratings induce interchannel crosstalk by scattering a part of signal from one channel to another. This phenomena is referred as FWM [4]. If three optical fields with carrier frequencies f_1, f_2 and f_3 copropagate inside the amplifier simultaneously, new fields with frequencies $f_1 \pm f_2 \pm f_3$ are generated in principle due to FWM. In practice, most of these combinations are not build up due to phase-matching requirements for any FWM process. The frequency combination of the form $f_4 = f_1 + f_2 - f_3$ is most troublesome for multichannel communication systems especially when the channel spacing is relatively small (≈ 1GHz) and optical power per channel is high. The phase-matching condition is nearly satisfied in this case. It may lead to power transfer among various channels.

As evident from Eq. (7.11), amplifier gain for a particular channel is saturated not only by its own power, but also by the power of neighbouring channels. When the input light intensity is low, signal gain becomes constant and crosstalk does not occur. On the other hand, if the input intensity is high, output intensity is not proportional to the input intensity i.e., amplifier is gain saturated. In that case, signal gain is dependent upon the total light intensity. In a multichannel transmission system, total light intensity varies randomly since each channel light is independently modulated. Thus, signal output of one channel varies according to signal gain fluctuations induced by the modulation of other channels, even when this signal input is constant. This is the crosstalk induced by gain saturation in the amplifier. This crosstalk occurs regardless of the extent of channel spacing. It is possible to avoid this crosstalk by operating SLAs in the unsaturated regime. It can also be avoided by using phase shift keying (PSK) or frequency shift keying (FSK) signalling schemes. For these signalling schemes, power in each optical channel and hence the total power remains constant with time. Best conventional devices have shown saturation power level of about 10 mW, while MQW devices have reached 110 mW saturation power level [9].

The interchannel crosstalk induced by FWM can occur for all multichannel communication systems irrespective of the modulation scheme used. In the case of equispaced channels, new frequencies coincide with the existing frequencies and FWM leads to power transfer among various channels. When channels are not equispaced, new frequencies fall in between the channels and interfere with the detection process. In both the cases, system performance degrades because of a loss in the channel power and noise introduced by the interchannel interference. Impact of FWM on system performance can be reduced by decreasing the channel power or by increasing the channel spacing. It becomes negligibly small when the optical channel spacing exceeds 10 GHz [2, 4, 11, 16, 18].

While cascading a number of SLAs, one has to take into account the fact that the amplifier noise keeps on accumulating in the chain. It limits the number of cascadable devices because of saturation of latter amplifiers. Further, any undesirable reflections along the line may cause feedback into the amplifier cavity and generate oscillations and instabilities. In order to prevent the oscillations, optical isolators are required which are, however, quite expensive and bulky devices [9].

7.1.2   ERBIUM-DOPED FIBER AMPLIFIERS

The most important event in the nineties has been the emergence of the Erbium-doped fiber amplifier (EDFA). Advances have been so rapid that passage from first laboratory prototype to significant commercial applications has taken less than five years. Using EDFAs, up to 10,000 km transmission distance has already been achieved under the laboratory conditions [35]. The EDFA has brought new exceptional possibilities in the design of high-speed transmission and all-optical switching systems. It allows efforts to be focussed on the main remaining obstacles (fiber dispersion and non-linearities) for realizing the full potential of fiber communications.

Rare-earth doped fiber amplifier, particularly EDFA has emerged as a big competitor to SLAs [38]. It is based on glass fibers doped with an appropriate amount of Erbium. The energy for amplification is provided in the form of optical radiation at an appropriate wavelength around 1550 nm and power level tens to hundreds of milliwatt. A schematic view of EDFA is shown in Fig. 7.4.

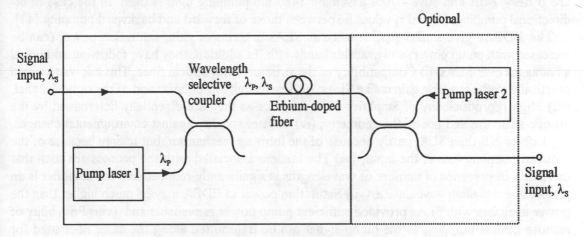

Fig. 7.4 Schematic of an Erbium-doped optical fiber amplifier

As shown in the figure above, EDFA consists of three basic components: length of suitable Erbium-doped fiber, pump laser and wavelength selective coupler to combine the signal and pump wavelengths. In some applications, it may be required to remove any unused pump light. For this purpose, an optical filter can be placed at the opposite end of amplifier. When the pump enters the amplifier at the same end as the signal, beams are copropagating. An alternate geometry is possible where the pump laser is introduced at the opposite end of fiber amplifier and pump light propagates in the opposite direction from that of the signal. The beams are then counter-propagating. The use of insufficient pump power or too long fiber can lead to part of the amplifier providing loss rather than gain. Alternatively, increasing pump power arbitrarily or using too short a fiber leads to insufficient use of pump power. There is, therefore, an optimum fiber length that has a strong dependence on pump power, input signal power, amount of rare-earth doping, pumping wavelength and typically spans from few metres to around 100 metres. In the extreme case of very low level doping, kilometre lengths may be used. In that case, amplifier will provide a kind of continuous amplification along the fiber. It may be quite useful for local loop distribution or sustaining non-linear propagation for soliton pulse transmission.

Various wavelengths are accessible for pumping, but the most useful are those that allow the use of semiconductor lasers as pumping source. These sources are the best choice because of size, ruggedness, power consumption and reliability. Use of lasers at 820 nm would have been the better choice, but it is not among the best from the technical point of view. The 980 nm wavelength has proved to be the best one in terms of efficiency (more than 10 dB gain per mW pump power) and better noise performance. The 1480 nm wavelength may be preferred because of higher source reliability as well as lower attenuation of communication fibers in this spectral range [9].

Generally, forward pumping leads to smaller NF and lower pumping light to signal light conversion efficiency η, than the backward pumping. Typically NF lies between 4 dB - 5 dB and η between 40% - 50% with forward pumping and the equivalent figures for backward pumping are 6 dB - 7 dB and 60% - 70% assuming 1480 nm pumping light is used. In the case of bi-directional pumping, NF and η values lie between those of forward and backward pumping [43].

The EDFAs give similar performance as SLAs in terms of gain, saturation power (can be increased with pump power) and available bandwidth. In addition, they have following additional advantages over SLAs: (i) Compatibility of the structure with optical fiber. This allows to retain practically all the available gain unlike SLAs, (ii) Insensitivity to polarization of the optical signal, (iii) High reproducibility of amplifier characteristics as they are essentially determined by the atomic structure and not by the geometry, (iv) Greater stability against environmental changes, (v) Lower NF than SLA, partly because of the intrinsic mechanism, but mainly because of the reduced coupling loss at the input, (vi) The lifetime associated with the process are such that crosstalk in presence of number of wavelengths is significantly reduced and the amplifier is an almost true travelling wave device, (vii) Saturation power in EDFA may be much higher than the power available with SLAs provided sufficient pump power is available and (viii) Possibility of remote optical pumping as the pump signal can be transmitted along the same fiber used for signal transmission. In conclusion, characteristic features of EDFA are high gain, high output power, low noise, polarization insensitiveness and ultra-low distortion. On the negative side, their weak points are: large size and difficulty of integration with the other optical components. Typical data of an EDFA taken from a commercial data sheet are given in following table [19].

Table 7.1: Typical data of an EDFA

Output power (> -3 dBm input)	10 dBm, 13 dBm, 15 dBm or 18 dBm
Gain (> -35 dBm input)	25 dB, 30 dB, 33 dB or 35 dB
Noise figure (> -35 dBm input)	< 7 dB
Bandwidth	about 30 nm

The bandwidth of an EDFA, i.e. 1.53 µm to 1.56 µm, is well matched to the minimum loss telecommunications window in silica fiber. However, most national light wave networks were designed to be used in the 1.3 µm window to take advantage of the zero dispersion at this wavelength. At present, it is estimated that more than 90% of the fiber in the world's telecommunications network has a dispersion zero at 1.3 µm.

Many nations throughout the world invested in light wave long-haul networks during the late '80s and the early '90s. By the end of the century, about 70 million kilometres of fiber have been deployed. At first, these networks contained conventional fibers almost exclusively, e.g.; fiber with step index waveguide, a zero in the dispersion near 1.31 µm and a fiber dispersion of 17 ps/(nm·km) in the 1.56 µm window. Now, some dispersion-shifted fibers (DSFs) are being deployed, but the national networks still predominantly contain conventional fibers. Since these networks are an enormous investment, there is strong incentive to design systems that use conventional fibers. Designers of terrestrial light wave transmission systems intend to use Erbium-doped optical amplifiers, which forces operation in the 1.56 µm transmission window and allows manifold increases in the regenerator span lengths. However, they are constrained to use embedded fiber base which often contains conventional fibers with a total dispersion of 17 ps/(nm·km) in the 1.56 µm transmission window and finally, as the cost of 10 Gbit/s electronics and electrooptics comes down, there will be a desire to operate at higher bit rates. In order to meet all these requirements, it is necessary to either eliminate or mitigate the impairments caused by chromatic dispersion.

Most operators prefer to upgrade the capacity of their existing fiber cable networks because of heavy investment rather than lay new fiber cables with dispersion zero tailored in the 1.56 µm window. The development of 1.3 µm fiber amplifiers based on Praseodymium-doped fluoride fiber now makes this option increasingly attractive [23, 24, 36, 40, 41].

7.1.3   RAMAN FIBER AMPLIFIER

In Raman fiber amplifier, amplification mechanism is the stimulated Raman scattering (SRS) in the fiber which causes an energy transfer from the pump to the signal. The vibrational spectra of core material defines the Raman shift. If the wavelength of signal laser beam is known, then the optimum wavelength for pump signal can be calculated i.e., Raman amplification can occur at any wavelength as long as appropriate pump laser is available. The structure of this amplifier is very much similar to that of EDFA. There are again three basic components: pump laser, wavelength selective coupler and fiber (Fig. 7.4). The major difference between these two amplifiers is that in the Raman amplifier standard single-mode optical fiber can be used, although for certain applications it may be desirable to use special fiber designs to enhance the amplifier performance. Gains of Raman amplifier for standard single-mode fiber of length 100 km in the 1550 nm transmission window and for 100 mW pump power are in the order of 2.5 dB for step index fiber and 5 dB for dispersion shifted fiber [13, 34]. The difference in gains between the two fibers arises mainly because of more germanium content in the fiber core of the dispersion shifted fiber. In both the cases, 95% of the gain is obtained in the first 50 km of fiber. Pump power of 100 mW at the appropriate wavelength from each end of above 100 km length of the dispersion shifted fiber reduce the loss of link from 20 dB to only 10 dB at 1550 nm. This imply an effective gain of 15 dB with dispersion shifted fiber. A Raman power amplifier can be made using a short (few km) length of specific fiber of optimum length for a given pump power and then it can be incorporated in the optical network based on standard optical fibers.

The main features of the Raman amplification are that it may be realized as a continuous amplification along the fiber (which could be an ordinary telecommunication fiber) thus never letting the signal to become too low; it is bidirectional in nature; it offers more stability and insensitivity to reflections; spectral range of gain can be chosen in a continuous way along all the optical spectrum of interest. The saturation optical power level is very high as it depends on the pump power. The only disadvantage which has greatly reduced interest in this kind of amplifier is the requirement of relatively high power pump laser (100 mW-200 mW) in comparison with SLAs and EDFAs.

7.1.4   BRILLOUIN FIBER AMPLIFIER

The operating principle of Brillouin fiber amplifier is essentially the same as for Raman fiber amplifier except that in this amplifier, optical gain is obtained by stimulated Brillouin scattering (SBS) instead of SRS. The Brillouin fiber amplifiers are also pumped optically and a part of the pump power is transmitted to the signal through SBS. Physically, each pump photon creates a signal photon and the remaining energy is used to excite an acoustic phonon. Classically, the pump beam get scattered from an acoustic wave moving through the medium at the speed of sound. Despite a formal similarity between SBS and SRS, SBS differs from SRS in three important aspects: (i) Amplification occurs only when the signal beam propagates in direction opposite to that of the pump beam (backward pumping), (ii) Stokes shift i.e., difference in the signal and pump frequency for SBS (≈ 10 GHz) is smaller by three order of magnitudes compared with that of SRS and (iii) Brillouin gain spectrum is extremely narrow with a bandwidth less than 100 MHz. Gains in the order of 25 dB have been obtained for pump power of only 5 mW [5]. The bandwidth available in Brillouin amplifier arises from the thermal distribution of the phonons in the fiber core material and is less than 100 MHz. The most efficient use of this amplification mechanism requires the use of narrow-band lasers for signal and pumping (<100 MHz), separated by about 10 GHz to an accuracy of only a few MHz. Only for a certain specialized application, this complexity is justified. Gain of Brillouin amplifier can be as high as 50 dB with a moderate pump power level. For a gain of about 20 dB, few mW pump source is good enough.

The narrow bandwidth of Brillouin fiber amplifiers makes them less suitable as power amplifier, preamplifier or in-line amplifier in light wave systems. The same feature can, however, be exploited for some other applications in coherent and multichannel communication systems.

7.2   COMPARISON BETWEEN SEMICONDUCTOR AND FIBER AMPLIFIERS

The typical characteristics and main features of optical amplifiers are given in Table 7.2. Further, a representative gain spectral curve is shown in Fig. 7.5.

Table 7.2: Comparison of optical amplifier characteristics (na: not applicable)

Property	Laser	Erbium fiber	Raman fiber	Brillouin fiber
Unsaturated device gain	> 20 dB	> 20 dB	5 dB - 15 dB	> 25 dB
Optical pump power	na	20 mW - 50 mW	100 mW - 200 mW	< 10 mW
Optical pump wavelength	na	820 nm, 980 nm, 1400 nm - 1500 nm	Stokes shift below signal	
Electrical bias current	50 mA	> 100 mA	> 500 mA	< 50 mA
Wavelength of operation	any	1525 nm - 1565 nm	any, but subject to pump	
Bandwidth	20 nm - 50 nm	10 nm - 40 nm	20 nm - 40 nm	0.001 nm
Coupling loss	5 dB - 6 dB	< 1 dB	< 1 dB	< 1 dB
Polarization sensitivity	less than few dB	0 dB	0 dB	0 dB
Saturated output	less than few mW	few mW	limited only by pump power	
Directions	bidirectional	bidirectional	bidirectional	unidirectional
Noise	low	low	very low	very low

Some of the additional points to be mentioned in connection with the optical amplifiers are following: (i) For SLAs and EDFAs, typical unsaturated gain are greater than 20 dB, while for Raman fiber amplifiers, gain is restricted to lower values by the stringent pump power requirement. If coupling losses of the amplifiers are also considered, gain of SLAs will get reduced by about 7 dB, (ii) The SLAs need an electrical bias supply at levels of around 50 mA, while the supply requirement is much more stringent in Erbium and Raman fiber amplifiers because of the high power pump laser requirement, (iii) In SLAs, gain spectrum is dependent upon the band gap of material used (which broadly defines the wavelength at which maximum gain is obtained), bias current supplied to the device (which defines the broad envelope of the gain) and residual reflectivity of the facets (which governs the fine structure on the overall envelope). The wavelength of optimum gain can be selected by the growth of appropriate semiconductor material.

In fiber amplifiers, pump laser and gain spectrum must be at or around specific wavelengths. The EDFAs require semiconductor pump laser at one of the 820 nm, 980 nm or 1400 nm - 1500 nm wavelength and the gain spectrum can extend across 1525 nm - 1565 nm by using alumina co-doped fibers. For Raman amplifiers, pump laser wavelength is usually 100 nm (Raman shift) below that of the signal. As the commercial high power lasers are available between 1470 nm and 1550 nm wavelength region, signal wavelength must be greater than 1570 nm to obtain optimized Raman gain from an amplifier. At this wavelength, fiber attenuation may not be minimum and therefore a part of the gain will be eaten away by the increased fiber attenuation.

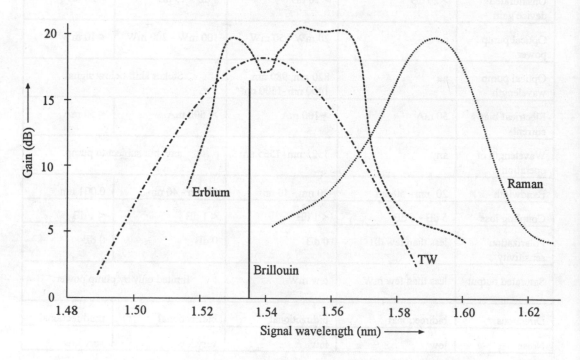

Fig. 7.5: Typical gain curves for various optical amplifiers

In Brillouin fiber amplifiers, problem is of matching the signal and pump source wavelengths within 0.1 nm, (iv) The polarization dependence of SLAs can be reduced to levels of 1 dB or less by using special device structures. Although fiber amplification mechanism is polarization insensitive, care must be taken to ensure that couplers used do not introduce polarization dependence into the amplifiers, (v) For both SLAs and EDFAs, saturated output power is typically a few milliwatt. In Raman and Brillouin amplifiers, saturated output power could be much higher as it is limited only by the pump power and (vi) Intermodulation distortion (arise because of departure from linearity and is similar in nature to FWM distortion) and saturation induced crosstalk in WDM system is negligibly small in fiber amplifiers as compared to SLAs. However, if the power in any one signal channel becomes large (more than tens of milliwatt), then that signal can become a pump for Raman amplification of all other signals at longer wavelengths. This is known as Raman-induced crosstalk and puts a fundamental limit on the high power multichannel transmission. In contrast to Raman-induced crosstalk, Brillouin-induced crosstalk is easily avoided because of the narrow frequency range (≈ 100 MHz) over which such crosstalk occur [3].

7.3   APPLICATIONS OF OPTICAL AMPLIFIERS

All optical amplifiers (except the Brillouin type) may find applications in optical communication systems. However, due to specific characteristics, a particular amplifier may be better suited for certain applications rather than others. For instance, EDFA would be a better choice as in-line amplifier because of its compatibility and other characteristics. Similarly, Raman amplifier could be an excellent power amplifier in view of its high saturation power. If the integrated-optic chip including optical amplifier has to be developed, SLA appears to be the best choice.

In this Section, possible applications of optical amplifiers in optical communication systems and suitability of various amplifiers for a particular application have been discussed.

(i)   POWER AMPLIFICATION

The maximum loss capability of an unrepeatered optical communication system is set by the difference between transmitter laser launch power and receiver sensitivity. To design a system with a large loss capability, one has to increase launch power and use highly sensitive receiver. For a given requirements, receiver sensitivity may be significantly improved by the use of avalanche photodiode (APD) instead of PIN-photodiode in the optical receiver or coherent detection techniques if the extra complexity is justified. However, shot noise sets a fundamental limit beyond which the sensitivity will not be further improved.

The power launched from the source into the fiber can be increased by using an optical power amplifier also referred as booster-amplifier (Fig. 7.6a). The upper limit to the transmitter power is set by non-linear optical effects in the fiber such as SRS and SBS. These effects can cause severe degradation in the system performance [10].

Though all types of amplifiers can be used as power amplifier, Raman fiber amplifier will be a better choice as the optical pump power required to drive this amplifier is the highest implying the highest saturated output power.

EDFA power amplifiers can either be located in the transmitter itself or some kilometres away from the transmitter. In the latter case, the amplifier must be remotely pumped by using a second fiber besides the fiber which carries the information. By employing remotely pumped power amplifiers, transmission distance is increased.

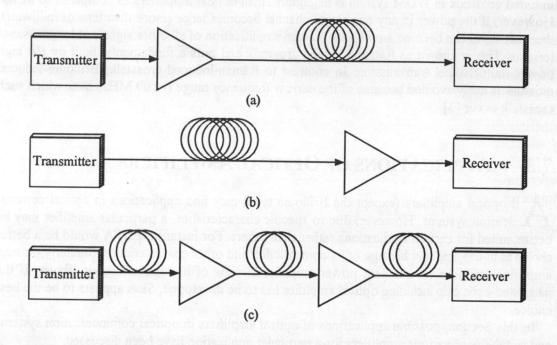

Fig. 7.6: Transmission systems with (a) optical power amplifier, (b) optical preamplifier and (c) cascaded optical in-line amplifiers

(ii) PREAMPLIFIER

When optical amplifier is used as preamplifier, it is placed directly in front of the receiver as shown in Fig. 7.6b. Both SLA and EDFA can be used for this purpose. The SLA has a slight disadvantage that residual Fabry-Perot resonances can give small changes in the sensitivity of receiver for different signal wavelengths. Therefore, it will not be suited for wavelength division multiplex systems. Similar to the power amplifier located at the transmitter side, a preamplifier can either be located in the receiver itself or some kilometres in front of the receiver. In the latter case which again increases transmission distance, the preamplifier is called a remotely pumped optical preamplifier. Remotely pumped optical amplifiers are frequently used in repeaterless submarine systems [32].

(iii) IN-LINE AMPLIFIER

Current medium- and long-haul optical communication systems use regenerative repeaters at regular intervals to compensate for both fiber loss and dispersion. When optical in-line amplifier is used as a direct replacement for optoelectronic repeater in the system, regeneration of optical signal is not required (Fig. 7.6c). For long-haul systems (e.g., trans-Atlantic optical fiber communication system), while a chain of amplifiers could compensate for the loss in the system, but the accumulation of spontaneous emission from amplifiers will set an upper limit on the number of amplifiers to be cascaded. The regenerative repeaters may still be required to prevent this build up of noise and also to restore the optical pulse distortion arising from the fiber dispersion. The Raman gain, which requires many kilometres of fiber, is most likely to be used to provide distributed gain in the transmission fiber itself [13]. However, SLA (< 1 mm long) and EDFA (tens of metres long) can both be used as discrete amplifiers.

In 1996, first trans-Atlantic fiber transmission system employing in-line EDFAs started operation [37]. This system was named trans-Atlantic transmission (TAT)-12/13. The undersea segment of this link is 5913 km long and contains 133 optical amplifiers spaced every 45 km. The system provides 10 Gbit/s (STM-16) capacity between United States, United Kingdom and France.

(iv) NON-LINEAR PROCESSING

In optical communications, there is a need to have an optical device which can be used for a variety of non-linear applications including pulse reshaping allowing the prospect of all-optical regeneration. The semiconductor laser amplifiers, particularly FPA is more suitable for such applications. In FPA, change in the refraction index with input signal is sufficient to observe non-linear behaviour [1]. This can be used, for example, to provide optical pulse shaping.

The FWM in wide-band SLA is also of current interest. This phenomenon can be used for frequency translation, an important function for multiwavelength systems. The SLA can be used as phase modulator in coherent systems [14]. This type of modulator provides gain in contrast to the loss associated with lithium niobate modulator and has lower modulation voltage requirement.

(v) OPTICAL SWITCH

All optical amplifiers require the supply of electrical current to drive the amplification mechanism. Removal of this bias removes the optical gain. In some cases, optical amplifier may become opaque to the signal wavelengths. Therefore, it behaves as an electrically controlled optical switch.

The SLA and EDFA which become strongly absorbing without any bias are suitable for such an application. The SLA responds on nanosecond time scale (carrier recombination time ≈ 1 ns) and can be used as a high speed switch. The speed of EDFA is governed by the life time of the Erbium ions (≈ 14 ms) and therefore it is only possible to make slow switches with this type of amplifier.

(vi) SOLITON TRANSMISSION

Soliton provides a means for balancing the pulse dispersion of silica with the non-linear process of self-phase modulation [17, 22]. As a consequence, very short pulses can propagate undistorted along the fiber and ultrahigh bit rates (greater than 10 Gbit/s) can be achieved. For this, peak power of the pulse must be kept constant at a level determined by its temporal width. As the soliton pulse propagates along the fiber, it slowly loses energy and can fall below the soliton level. At this point, an in-line EDFA can amplify the pulse and bring it back to the soliton level. High bit rate and long distance soliton transmission has been achieved with lumped EDFA separated by a distance of 25 km - 30 km [6]. The basic principle of soliton is explained in Chapter four.

7.4   NOISE IN OPTICAL AMPLIFIERS

An optical amplifier, in addition of amplifying the input signal, will add to it a signal due to spontaneous emission of light. A portion of the spontaneously emitted light is in the same direction as the signal and gets amplified along with the main signal. This added light, called the amplified spontaneous emission (ASE) noise is spread over a wide wavelength range (in the order of 50 nm) compared to signal as shown in Fig. 7.7.

The spectral density of ASE noise is nearly constant (white noise) in TWA and is given by [26]

$$\rho_{sp}(f) = (G-1)n_{sp}hf \tag{7.12}$$

where f is the optical frequency and n_{sp} the population inversion parameter. For amplifiers with complete population inversion (all atoms in the excite states), its value is minimum i.e., unity. For a two level system, it is given by

$$n_{sp} = \frac{N_2}{N_2 - N_1} \tag{7.13}$$

where N_1 and N_2 are the atomic populations in the ground and excited states respectively. Since EDFAs operate on the basis of a three-level pumping scheme, N_1 remains nonzero and $n_{sp} > 1$. Further, N_1 and N_2 vary along the fiber length because of their dependence on the pump and signal power and n_{sp} should be averaged along the fiber length. As a consequence, n_{sp} is expected to depend on both the amplifier length and pump power [4].

The amplifier output-signal plus ASE when detected by a photodiode consists of desired signal component along with the beat and shot noise components. Signal-spontaneous beat noise is generated by mixing between the amplified signal and ASE components. Similarly, spontaneous-spontaneous beat noise is generated by mixing within the ASE components themselves. In addition to these two noise components, there will be shot noise components produced by the signal and the spontaneous emission.

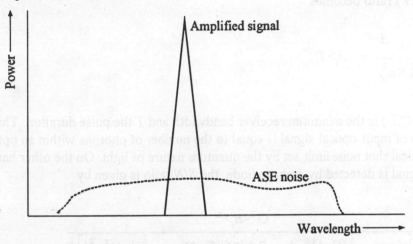

Fig. 7.7: Amplified optical signal and ASE noise at the amplifier output

The optical output (i.e., average number of photons) from an optical amplifier consists of an amplified optical signal and ASE noise. It is given by [25, 30]

$$\langle n_0 \rangle = G\langle n_i \rangle + (G-1)n_{sp}b \tag{7.14}$$

Here, $\langle n_i \rangle$ is the average number of incident photons, G the amplifier gain and b the bandwidth of optical band-pass filter inserted at the output end of amplifier to limit the ASE noise. The variance (σ_{n0}^2) of the number of photons after amplification is given by

$$\sigma_{n0}^2 = G\langle n_i \rangle + (G-1)n_{sp}b + 2\langle n_i \rangle G(G-1)n_{sp} + (G-1)^2$$

$$+ G^2\left(\langle n_i^2 \rangle - \langle n_i \rangle^2 - \langle n_i \rangle\right) \tag{7.15}$$

In the above equation, first and second terms correspond to shot noise due to signal and ASE respectively. The third and fourth terms are beat noise originating from signal-ASE and ASE-ASE interference respectively. The fifth term depends on the state of incident electric field. With a coherent incident light, average and variance of the photon number are equal and, therefore, the last term vanishes [25].

7.5   NOISE FIGURE OF AMPLIFIER

Without an amplifier i.e., when $G = 1$, variance σ_{n0}^2 from Eq. (7.15) equals $\langle n_i \rangle$. Thus, signal-to-noise (S/N) ratio becomes

$$\left(\frac{S}{N}\right)_i = \frac{\langle n_i \rangle^2}{2B\sigma_{n0}^2} = \langle n_i \rangle T \tag{7.16}$$

where $B = 1/(2T)$ is the minimum receiver bandwidth and T the pulse duration. This means that the S/N ratio of input optical signal is equal to the number of photons within an optical pulse. It is the theoretical shot noise limit set by the quantum nature of light. On the other hand, when the amplified signal is detected by a photodiode, the S/N ratio is given by

$$\left(\frac{S}{N}\right)_o = \frac{G^2\langle n_i \rangle^2}{2\left[G\langle n_i \rangle + (G-1)n_{sp}b + 2\langle n_i \rangle G(G-1)n_{sp} + (G-1)^2 n_{sp}^2 b\right]B} \tag{7.17}$$

For high G, shot noise terms (first and second) will be comparatively much smaller than the beat noise terms (third and fourth). Under this practical condition, above expression gets simplified to

$$\left(\frac{S}{N}\right)_o \simeq \frac{\langle n_i \rangle^2}{2\left[2\langle n_i \rangle n_{sp} + n_{sp}^2 b\right]B} \tag{7.18}$$

If b is so small that the ASE-ASE beat noise component is negligibly small as compared to signal-ASE beat noise component, then the second term in the denominator of above equation can be neglected. In that case, S/N ratio becomes

$$\left(\frac{S}{N}\right)_o = \frac{\langle n_i \rangle T}{2n_{sp}} \tag{7.19}$$

The above S/N ratio is called the "beat noise" limited. By comparing Eqs. (7.16) and (7.19), noise figure (NF) of optical amplifier is defined as follows:

$$NF = \frac{\left(\dfrac{S}{N}\right)_i}{\left(\dfrac{S}{N}\right)_o} = 2n_{sp} \qquad (7.20)$$

The best NF occurs when n_{sp} is unity. For an ideal TWA, n_{sp} is unity and NF will be 3 dB. For FPA, it ranges from 5 to 7 dB. EDFA with forward pumping has a NF of about 5 dB.

7.6 REFERENCES

[1] Adams, M. J.; Westlake, H. J.; O'Mahony, M. J.; Henning, I. D.: A comparison of active and passive bistability in semiconductors. IEEE J. QE-21(1985)9, 1498-1504.

[2] Agrawal, G. P.: Amplifier induced crosstalk in multichannel coherent lightwave systems. Electron. Lett. 23(1987)17, 1175-1177.

[3] Agrawal, G. P.: Nonlinear Fiber Optics. Academic Press Inc., 1989.

[4] Agrawal, G. P.: Fiber Optic Communication System. John Wiley & Sons Inc., 1992.

[5] Atkins, C. G.; Cotter D.; Smith, D. W.; Wyatt, R.: Application of Brillouin amplification in coherent optical transmission. Electron. Lett. 22(1986)10, 556-558.

[6] Becker, P. C.: Erbium-doped fiber makes promising amplifiers. Laser Focus World, 1970, 197-203.

[7] Bjarklev, A.: Optical Fiber Amplifiers: Design and System Applications. Artech House Publishers, 1993.

[8] Cochrane, P.: Future directions in long haul fiber optic systems. British Telecom.Tech. J. 8(1990)1, 5-17.

[9] Costa, B.: Towards a future all-optical telecommunication network. Lecture notes in third college on physics and technology of lasers and optical fibers held at I.C.T.P. Trieste, Italy, 27 January-21 February, 1992.

[10] Cotter, D.: Observation of stimulated Brillouin scattering in low-loss silica fiber at1.3 μm. Electron. Lett. 18(1982), 495-496.

[11] Darcie, T. E.; Jopson, R. M.; Tkach, R. W.: Intermodulation distortion in optical amplifier from carrier-density modulation. Electron. Lett. 23(1987)25, 1392-1394.

[12] Ehlers, H.; et. al.: Halbleiterlaser-Verstärker für optisches HDTV-Übertragungssystem. ntz 43(1990)1.

[13] France, P.W.: Optical Fiber Lasers and Amplifiers. CRC Press Inc., 1991.

[14] Franz, J.: Optische Übertragungssysteme mit Überlagerungsempfang. Springer-Verlag, 1988.

[15] Giles, C.R.; Desurvire, E.: Modelling erbium-doped fiber amplifiers. IEEE J. LT-9(1991)2, 271-283.

[16] Grosskopf, R; Ludwig, W.; Weber, H. G.: Cross-talk in optical amplifiers for two channel transmission. Electron. Lett. 22(1986)17, 900-902.

[17] Hasegawa, A.; Tappert, F.: Transmission of stationary nonlinear optical pulses in dispersive dielectric fibers: Anomalous dispersion. Appl. Phys. Letter, 23(1973)3, 142-146.

[18] Hodgkinson, T. G.: Receiver analysis for synchronous coherent optical fibre transmission systems. IEEE J. LT-5(1987)4, 573-586.

[19] JDS FITEL: ErFA 1100 Sereis. Ontario, Canada, 1994.

[20] Li, T.: Optical amplifiers transform lightwave communications. Photonics Spectra, (1995)1, 115-117.

[21] Ludwig, R.: Optische Halbleiterlaser-Verstärker in zukünftigen LWL-Übertragungssystemen. ntz 43 (1990)1.

[22] Mollenauer, L. F.; Smith, K.: Demonstration of soliton transmission over more than 400 km in fiber with loss periodically compensated by Raman gain. Opt. Lett. 13 (1988)8, 675-680.

[23] Nishida, Y.; et. al.: Efficient PDFA module uisng High-NA PbF_2/InF_3-based Fluoride Fiber. IEEE Photonics Techn. Letters 9(1997)3, 318-320.

[24] Ohishi, Y.; et. al.: Analysis of gain characteristics of bi-directionally pumped Praseodymium-doped fiber amplifiers. IEEE Photonics Techn. Letters 8(1996)4, 512-514.

[25] Okoshi, T.; Kikuchi, K.: Coherent Optical Fiber Communications. Kluwer Academic Publisher, 1988.

[26] Olsson, N. A.: Lightwave systems with optical amplifiers. IEEE J. LT-7(1989)7, 1071-1082.

[27] O'Mahony, M. J.: Semiconductor laser optical amplifiers for use in future fiber systems. IEEE J. LT-6(1988)4, 531-544.

[28] Saitoh, T.; Mukai, T.: 1.5 µm GaInAsP travelling-wave semiconductor laser amplifier. IEEE J. QE-23(1987)6, 1010-1020.

[29] Saitoh, T.; Mukai, T.: Recent progress in semiconductor laser amplifiers. IEEE J.LT-6(1988)11, 1656-1664.

[30] Shinda, S: Optical amplifiers for optical communications systems, IEIC Trans. E74(1991)1, 65-74.

[31] Sietmann, R.: Ein Verstärker für alle Kanäle. Teil 1 und 2. Funkschau 18/19, 1990.

[32] Silva, da V. L.; et. al.: Remotely pumped Erbium-doped fiber amplifiers for repeaterless submarine systems. IEEE Photonics Techn. Letters 7(1995)9, 1081-1083.

[33] Simon, J. C.: GaInAsP semiconductor laser amplifiers for single-mode fiber communications. IEEE J. LT-5(1987)9, 1286-1295.

[34] Spirit, D. M.; Blank, L. C.: Raman-assisted long-distance optical time domain reflectometry. Electron. Lett. 25(1989), 1687-1688.

[35] Taga, H.; et. al.: Recent progress in amplified undersea systems. IEEE J. LT-13(1995)5, 829-840.

[36] Tomita, N.; et. al.: Digital signal transmission experiment using a 1.3 µm-band Pr^{+3}-doped fluoride fiber amplifier. IEEE Photonics Techn. Letters 6 (1994)2, 258-259.

[37] Trischitta, P.; et. al.: The TAT-12/13 Cable Network. IEEE Commun. Magazine (1996)2, 24-28.

[38] Urquhart, P.: Review of rare-earth doped fiber lasers and amplifiers. IEE Proc. Pt. J, 135(1988)6, 385-407.

[39] Werth, D. L.: Increasing the options for fiber amplifiers. Photonics Spectra 1994 (2), 88-92.

[40] Whitley, T. J.: A review of recent system demonstrations incorporating 1.3-µm Praseodymium-doped fluoride fiber amplifers. IEEE J. LT-13(1995)5, 744-760.

[41] Yamada, M.; et. al.: Low-noise and high-power Pr^{3+}-doped fluoride fiber amplifier. IEEE Photonics Techn. Letters 7(1995)8, 869-871.

[42] Yamamoto, Y.: Noise and error rate performance of semiconductor laser amplifiers in PCM-IM optical transmission system. IEEE J. QE-16(1980)10, 1073-1081.

[43] Yoneda, E.; Suto, K.-I.; et. al.: Erbium-doped fiber amplifiers for all fiber video distribution (AFVD) systems. IEICE Trans. on Commun., E-75B(1992)9, 850-861.

8   OPTICAL NETWORK COMPONENTS

Optical communication systems are progressing from point-to-point links to networks. In optical networks, in addition to conventional components like sources, fibers and detectors which are sufficient to realize point-to-point links, various other highly sophisticated network components are required. In a fiber-based network, optical components more and more replace electronic components which finally result in an all-optical network as described in Chapter fourteen. In the latter network, routing and switching functions are also performed optically resulting in an end-to-end optical connection.

Two prerequisites for the realization of all-fiber components are easy access in the guided optical field and a suitable means of interaction with the exposed field in accordance with the desired component function. There are number of techniques available to access the optical field. Some of the mature ones are described in the following Sections.

8.1   FIELD ACCESS TECHNIQUES

A large number of optical fiber components are commercially available based on the different methods of accessing the fiber mode fields. These are semidestructive methods as the mode fields in the fiber core for standard fibers are well protected. The techniques commonly used for direct methods are *tapering*, *polishing*, *etching*, *cleaving* and *bending*, while the indirect methods are based on *acoustic*, *thermal*, *mechanical* and *piezo-electric effects* [1].

8.1.1   TAPERING

In single-mode (SM) or multimode (MM) fiber, tapering can be achieved with the aid of a constant source of heat which is used to soften the fiber while pulling. In tapering, the fiber and core radii are generally decreased. So the optical fields are forced to spread into cladding. The strength of the fiber is not significantly affected along the taper as the taper is adequately protected from humidity and other environmental perturbations.

8.1.2   POLISHING

Polishing is one of the commonly used technique for removing part of the fiber cladding. As shown in Fig. 8.1, a SM fiber is embedded in a circular groove of appropriate depth cut in a supporting silica substrate. The fiber is curved and fixed with an epoxy resign of hardness similar to silica. It ensures that the fiber, substrate and epoxy have a similar behaviour. The fiber is fixed so that part of the cladding extends above the surface of substrate block by roughly the extend to be polished away. Normally, after polishing a few microns of cladding are left above the core. Polishing is usually done in two stages. The first with coarse SiC (silicon carbide) powder and followed by a second with much finer powder to obtain an optically flat surface. The optical fields are then accessible from the polished surface without any reduction in the fiber strength. The polishing is controlled by monitoring the power of the transmitted light throughout the polishing process [1].

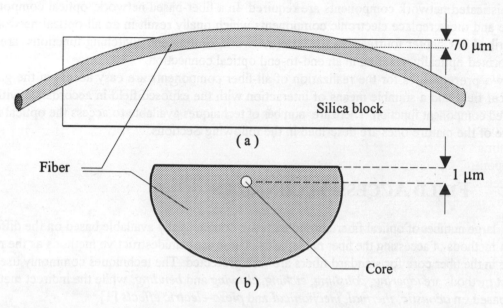

Fig. 8.1: (a) Cross-section of fiber embedded in a curved groove cut in a silica block for polishing and (b) fiber cross-section after polishing

Polishing is a lengthy and expensive technique for assessing the fiber fields. However, it has the advantage of not changing the core dimension as in tapering.

8.1.3 ETCHING

In etching, a part of the fiber cladding is removed by using chemical such as hydrofluoric acid. The acid erodes the surface of the fiber, reducing the diameter linearly with time and thus assessing the core fields. For a circular fiber, etching will remove cladding symmetrically and therefore result in a very thin fiber which is weaken. Etching is more appropriate for assessing the fields of eccentric core fields such as D shape [19]. Often, a combination of etching and tapering is used for fabricating devices such as couplers [1].

8.1.4   CLEAVING

This is a destructive method which is commonly used for fiber gap devices such as isolators, polarizers etc. It is possible to obtain clean 90° breaks with less than 0.5° off a true perpendicular cleave using commercially available cleaving tools. After cleaving, fiber ends act as a receiving antenna for high efficiency collection. A combination of taper and cleaving is also used to fabricate fiber gap devices [9].

8.1.5   BENDING

Large bend radius (macrobends) can be used in fiber components. A guided mode in a fiber has associated with it an evanescent field which propagates in the cladding. Bending loss occurs because the evanescent field would have to travel faster than the speed of light to keep up with the core field. As this is not physically realizable, the field beyond a critical distance radiates away. Although components based on bending are not used extensively, it is possible to demonstrate components based on this phenomenon. For example, an optical tap can be fabricated by placing a receiving fiber, the end of which has been cut to a specific angle, in contact with the bent region of the main fiber. Similarly, modal filters can be made by bending fibers because higher order modes which are bound less tightly to the core than lower order modes, will radiate away first [1].

A number of optical fiber components based on the above field access techniques exist. These are categorised according to the function performed (refer to Fig. 8.2). The power splitters or taps split the optical power from a single-input fiber into several (≥2) output fibers as shown in Fig. 8.2a. Power combiners perform the opposite function i.e., mixing the output of several transmitters and launching it into a single output fiber as shown in Fig. 8.2b. 2×2 couplers couple beam with a predetermined power distribution (refer to Fig. 8.2c). As an example, 3 dB coupler distributes half of light power at each input to each output port. Star couplers mix the optical signals injected at input fibers and divide each of these equally among the output fibers as shown in Fig. 8.2d. Wavelength multiplexers combine signals at different wavelengths and launch the composite signal into a single optical fiber as shown in Fig. 8.2e. Wavelength demultiplexers split the multichannel input into different output fibers according to their wavelengths as shown in Fig. 8.2f. The optical isolator protect the transmitter or any other sensitive device from undesired reflections of the optical radiations (Fig. 8.2g).

The implementation of multiaccess optical fiber networks, in addition to above passive components, also requires active components like tunable optical filters.

In the following Sections, both the passive and active components have been discussed.

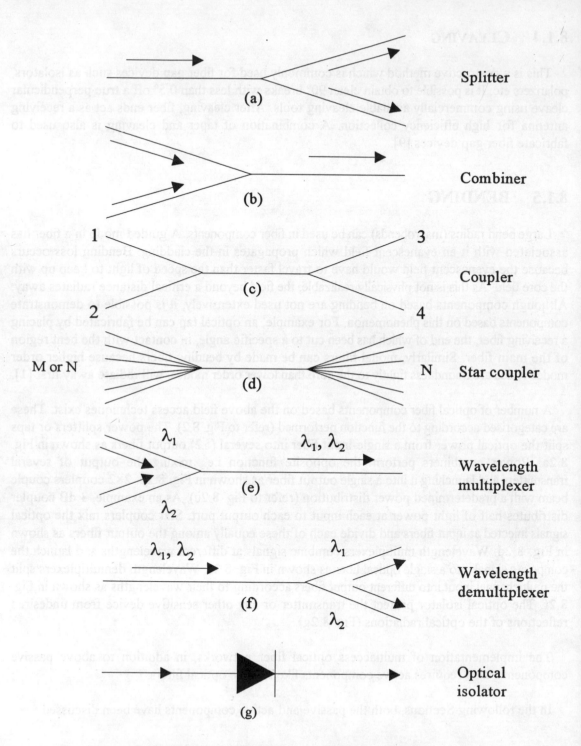

Fig. 8.2: Optical fiber (a) splitter, (b) combiner, (c) 2×2 coupler, (d) star coupler, (e) wavelength multiplexer, (f) wavelength demultiplexer and (g) optical isolator

8.2 DIRECTIONAL COUPLERS

The directional couplers can be classified into three major types: *bulk, fused-fiber* and *planar-waveguide*. The bulk-type coupler uses a beam splitter such as a half mirror (Fig. 8.3a). The other two types use a couple structure (Fig. 8.3b and c). For some devices, bulk technology is the only known practical solution e.g., Fabry-Perot interferometer (FPI) filter. Designers concentrate their efforts more on all-fiber optic and integrated-optic technologies. The main reasons are that fiber attachment is much easier with all-fiber optic devices, while integrated-optics technology has the potential of dramatic cost reduction when large scale production is envisaged. For both technologies, the physical mechanisms are basically the same except that the waveguide cross-section is circular in all-fiber devices, while usually rectangular in integrated-optic devices [40].

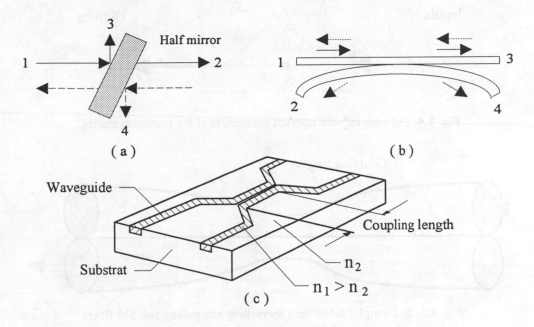

Fig. 8.3: Directional couplers (a) bulk, (b) fused-fiber and (c) integrated planar-waveguide type

8.2.1   2×2 COUPLER

Several useful properties of many optical devices can be obtained by considering the simple case of a 2×2 coupler (Fig. 8.4).

Let the 2×2 fused fiber coupler is obtained by melting and pulling two SM fibers together over a uniform section of length as shown in Fig. 8.5. Each input and output fiber has a large tapered section since the transversal dimensions are gradually reduced down to that of the coupling region. As the light propagates along the taper and into the coupling region, there is a decrease in V-number (the ratio a/λ decreases). It means much of the input field propagates outside the original fiber core. By carefully dimensioning of the coupling region, this decoupled field can be recoupled into the other fiber. The fraction of the launched power that appears at one of the two outputs can be varied between zero and unity. To make couplers that have accurate coupling ratios, optical power is launched into a fiber and the output power from a fiber is monitored during the heating and pulling.

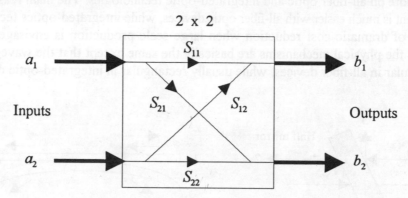

Fig. 8.4: 2×2 coupler with relevant parameters of the scattering matrix

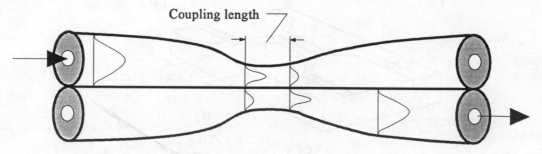

Fig. 8.5: 2×2 coupler fabrication by melting and pulling two SM fibers

Guided-wave 2×2 couplers, irrespective of whether all-fiber or integrated-optic type, can be analysed in terms of the scattering matrix S. It defines the relationship between the two input field strengths a_1 and a_2 and the two output field strengths b_1 and b_2

$$\begin{bmatrix} b_1 \\ b_2 \end{bmatrix} = \begin{bmatrix} S_{11} & S_{12} \\ S_{21} & S_{22} \end{bmatrix} \begin{bmatrix} a_1 \\ a_2 \end{bmatrix} \tag{8.1}$$

where S_{ij} represents the coupling coefficient from input port j to output port i. There are two fundamental physical restrictions on the scattering matrix S. The first one is reciprocity condition. For a SM operation and lossless device, it implies

$$S_{12} = S_{21} \qquad (8.2)$$

The second one is the energy conservation assuming a lossless device (no excess loss). It means that the sum of the output intensities

$$I_o = |b_1|^2 + |b_2|^2 = I_i = |a_1|^2 + |a_2|^2 \qquad (8.3)$$

is equal to the sum of the input intensities I_i. The above equation and Eq. (8.1) gives the following set of equations

$$|S_{11}|^2 + |S_{12}|^2 = 1 \qquad (8.4a)$$

$$|S_{22}|^2 + |S_{12}|^2 = 1 \qquad (8.4b)$$

and

$$S_{11}S_{12}^* + S_{12}S_{22}^* = 0 \qquad (8.4c)$$

Let us assume that

$$S_{11} = \sqrt{1-k} \qquad (8.5)$$

where k being a real number between 0 and 1 i.e., a fraction $(1-k)$ of the input power at port 1 appears at output port 1, while a fraction k of the same power appears at output port 2. Inserting S_{11} from Eq. (8.5) into Eq. (8.4) gives

$$S_{12} = S_{21} = \sqrt{k} \cdot \exp(j\phi_{12}) \qquad (8.6a)$$

and

$$S_{22} = \sqrt{1-k} \cdot \exp(j\phi_{22}) \qquad (8.6b)$$

Substitution of Eq. (8.6) in Eq. (8.4c) leads to

$$\exp\left[j(2\phi_{12} - \phi_{22})\right] = -1 \qquad (8.7)$$

It gives

$$\phi_{12} = \frac{\phi_{22}}{2} + (2n+1)\frac{\pi}{2} \qquad (n = 0,1,2,...) \qquad (8.8)$$

Without loss of generality, one can presume $\phi_{22} = 0$ and finally obtain

$$S = \begin{bmatrix} \sqrt{1-k} & j\sqrt{k} \\ j\sqrt{k} & \sqrt{1-k} \end{bmatrix} \qquad (8.9)$$

It means, irrespective of k value S_{ij} ($j \neq i$) has a $\pi/2$ phase shift relative to S_{ij} (i,j = 1,2). This property is used in balanced homodyne and heterodyne receivers. It is clear from the above equation that if a coupler is designed to deliver a large amount of power from input port 1 to output port 1, then the amount of power reaching output port 1 from input port 2 is small. It may be mentioned that when the role of input and output ports is reversed, the same matrix S still applies.

In the derivation of Eq. (8.9), the device is considered to be lossless. In practice, passive devices always have some associated excess loss β which is defined as the fraction of the total input power that appears as the total output power

$$\beta = -10\log_{10}\left[\frac{\Sigma P_{out}}{\Sigma P_{input}}\right] \qquad (8.10)$$

Normally, the value of β ranges 0.1 dB to 0.5 dB. It includes radiation, scattering, absorption losses and coupling loss to the isolated port. The directional isolation is the ratio of the backscattered power received at the second input port (also referred to as isolated port) to that of the input power. The isolation in decibels ranges over 30 dB - 50 dB.

Typical characteristics of fused-fiber type directional coupler for SM fiber are given in Table 8.1 [45]. Waveguide-type directional coupler typical characteristics are also included in the same table. The excess loss, coupling loss and isolation characteristics of fused-fiber couplers are better than the waveguide-type couplers. The size of both the couplers is nearly same. However, the size of a large port coupler obtained by cascading becomes larger in case of fused-fiber type couplers. It is one of the disadvantages of this coupler. The other disadvantage is the lack of integration with other optical devices.

Table 8.1: Characteristics of SM fiber directional couplers
* Depends on core size and Δ.

	Excess loss (dB)	Coupling loss to SM fiber (dB)	Isolation (dB)
Fused-fiber	0.1-0.2	Fusion splice < 0.1 Connector < 0.3	40-50
Waveguide	0.3-0.5	0.3-1*	30-40

In many applications, a wavelength-independent coupler is desirable. Wavelength-flattened or wide-bandwidth directional couplers are obtained by using two fibers with a slightly different propagation constant [43]. There are several methods which have been proposed for obtaining different propagation constants. These are (i) the use of fibers with different diameters, (ii) the use of fibers with different refraction index (RI) profiles and (iii) using different tapering ratios for two identical fibers. In the last method, a fiber is heated and elongated in advance (pretaper) to obtain nonequal dimensions. Subsequently, pretapered fiber and an ordinary fiber are heated and elongated. The fabricated coupler has 50%±9% coupling over the wavelength range from 1.23 μm to 1.57 μm [45].

Polarization-maintaining fiber couplers are required in some applications. Such couplers are made from polarization-maintaining fibers such as a PANDA fiber[1] and elliptical-core fiber. Angular alignment of such fibers are carefully made during the fabrication process. Two types of

---

[1]One method of introducing high birefringence in optical fiber is through asymmetric stress with two-fold symmetry in the core of the fiber. The required stress can be obtained by introducing two identical stress applying parts (SAPs) centred on a diameter, one on each side of the core. The SAPs have different thermal expansion coefficient than that of the cladding material due to which an asymmetrical stress is applied on the fiber core after it is drawn from the preform and cooled down. One of the shape used for SAPs is circular shape. The corresponding fiber is known as PANDA (polarization-maintaining and absorption reducing) fiber.

polarization-maintaining fiber couplers are shown in Fig. 8.6 [56]. In the coupler shown in Fig. 8.6a, y-polarized lights are output from two output ports when a y-polarized light is input, while x-polarized lights are output in case of x-polarized input light. In the second type (Fig. 8.6b), a y-polarized light is output from only one output port, while an x-polarized light is only output from the other output port in the case of an x-polarized input light. Two orthogonal polarization states are separated at the output ports for an input light with x- and y-polarized components. This type of coupler has been used as a polarization beam splitter. Their characteristics are generally wavelength dependent.

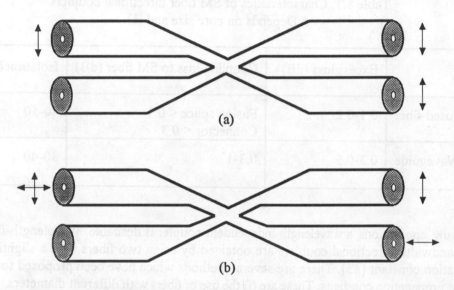

Fig. 8.6: Polarization-maintaining fiber couplers (a) same polarization and (b) different polarization at the output ports

8.2.2   Y-COUPLER

The structure of a simple Y-coupler is shown in Fig. 8.7. The coupler is used for several applications such as optical power divider, a combiner, an optical switch, an interferometer (with two Y-couplers) and a modulator. The main advantage of a Y-coupler is shorter realized device length as compared to a directional coupler. The drawback is the higher loss which is mainly the radiation loss at the branching point.

One method to decrease the loss is shown in Fig. 8.8 [24]. In a simple Y-coupler, the phase fronts of waves near the input of the two arms are almost perpendicular to the propagation direction of the incident waveguide. Therefore, the incident waves into the two arms are oblique to the two arm waveguides. It results in a significant radiation loss at the branching point. In an antenna-coupled Y-branch (Fig. 8.8), the fields spread into the n_3 region and the propagation ability of this n_3 region is weak since $n_3 < n_1$ ($n_3 < n_2$). Therefore, the field in the initial waveguide radiates and spreads into the n_3 region. As a result of field spreading, the phase fronts are perpendicular to the two arm waveguides and give rise to lower loss as compared to a simple Y-coupler. Experimentally, the branching loss of 3.4 dB and 1.0 dB respectively have been reported for simple and antenna-coupled Y-couplers [24].

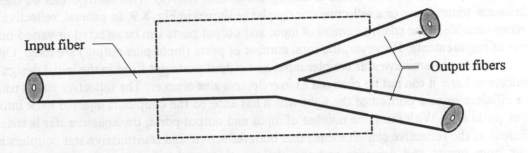

Fig. 8.7: Structure of a simple Y-coupler

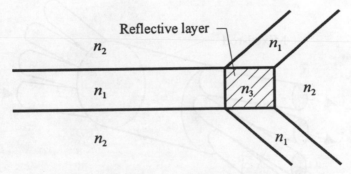

Fig. 8.8: Antenna-coupled Y-coupler

8.2.3   STAR COUPLER

A star coupler combines the optical signal(s) injected at the input port(s) and divide each signal equally among the output port(s). These couplers can be broadly classified into two types viz., N×M full star couplers and 1×M splitters/combiners. The latter types are a special case of former with N = 1.

The star coupler function is quite analogous to free-space in radio communications which mix input optical signals and mixed signals are transmitted to the output channels. The star couplers therefore introduce the merits of radio transmission system into fiber optic transmission system. However, such couplers also bring the disadvantages of radio transmission system. For example, interference of channels in the mixing results in crosstalk and a degradation in the receiver sensitivity. Further, in MM fiber star couplers modal noise may get enhanced [45].

The two principal manufacturing techniques for fabricating MM fiber star couplers are the mixed-rod and fusion methods. In the mixed-rod method, a thin platelet of glass mixes the light from the input fiber(s) and divide it among the output fiber(s). This method can be used to fabricate a transmissive or a reflective star coupler as shown in Fig. 8.9. In general, reflective star is more versatile as the relative number of input and output ports can be selected or varied on the basis of requirements. However, the total number of ports (input plus output) are fixed. On the other hand, in transmissive star coupler input and output ports get fixed in the initial design and fabrication. Later it can not be changed as in reflective star coupler. The reflective star is usually less efficient since a portion of the light which has entered the coupler is injected back into the input ports also. With the same number of input and output ports, transmissive star is twice as efficient as the reflective star. It means that both reflective and transmissive star couplers have their own merits and demerits and the choice of a coupler for a particular application is determined by the network topology [33].

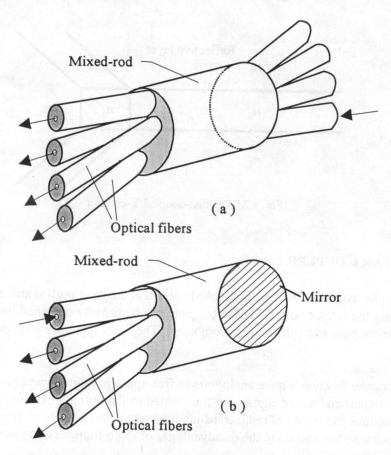

Fig. 8.9: Fiber optic star coupler using mixed-rod technique (a) transmissive and (b) reflective

The manufacturing process for the fused star coupler is similar to that of fused 2×2 coupler. Thus the fibers which constitute the star coupler are bundled, twisted, heated and pulled to form the device as shown in Fig. 8.10. With MM fibers, this method relies upon the coupling of higher order modes between the different fibers. It is highly mode dependent and give rise to significant variations in the output port power in comparison with the star couplers based on the mixed-rod technique [15].

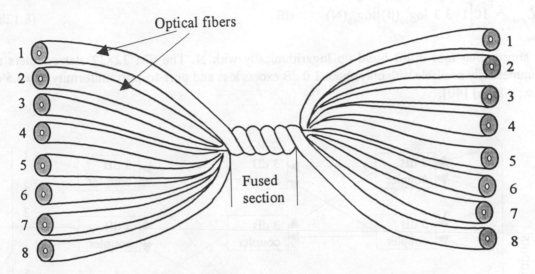

Fig. 8.10: Fiber optic star coupler using fusion technique

With fiber-fusion technique, N×N star couplers with N up to 7 and splitters/combiners (1×M/M×1) with M up to 19 have been constructed with excess loss as low as 0.4 dB and 0.85 dB respectively at 1.3 μm wavelength [4, 5]. Fabrication of large port couplers has been difficult as the coupling behaviour between the large number of fibers need to be controlled during the melting and pulling processes. Further, it is difficult to fabricate so-called achromatic or broadband star coupler since the coupling between different fibers is wavelength dependent.

A more attractive alternative consists of cascading 3 dB couplers as shown in Fig. 8.11 for an 8×8 star coupler [39]. The input signal passes through $\log_2(N)$ 3 dB couplers in a N×N star coupler. Further, required number of 3 dB couplers to form a N×N full-star coupler is

$$N_c = \frac{N}{2}\log_2(N) \tag{8.11}$$

and that a fraction 1/N of the launched power at each input port effectively appears at all output ports. Presuming β ($0 \le \beta \le 1$) be the excess loss of each 3 dB coupler, then the loss incurred by a signal passing through the N×N star coupler is

$$L_{star} = \frac{\beta^{\log_2(N)}}{N} \tag{8.12a}$$

This loss when expressed in dB is given by

$$L_{star} = 10\left[1 - 3.3 \log_{10}(\beta)\right]\log_{10}(N) \quad \text{dB} \tag{8.12b}$$

It shows that loss in dB build up logarithmically with N. The SM 32×32 star couplers are commercially available with only about 1.0 dB excess loss and port-to-port uniformity of ±0.5 dB (i.e., ±11%) [40].

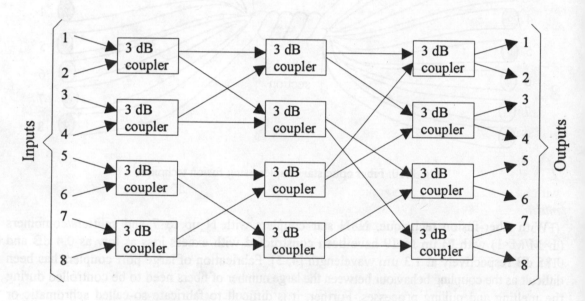

Fig. 8.11: 8×8 star coupler formed by cascaded 3 dB couplers

Another possibility with integrated-optic technology is the so-called planar-lens star coupler. It consists of a planar-input waveguide array that radiate into a free-space planar region, referred to as waveguide region (Fig. 8.12). This region works just like a free-space in radio systems and serves as a light wave mixer. The input power from any of the N input channels radiates into the slab waveguide region and the radiated power is received by the output array. Uniform branching or splitting among N output channels is realized when the uniform radiation pattern is produced over the output channel waveguide array. Some 19×19 star couplers fabricated on a silicon substrate by this technique have an excess loss of 3.5 dB and port-to-port uniformity of ±1.0 dB [16, 17]. The excess loss is reported to decrease to about 1.5 dB for 8×8 star couplers using silica planar waveguide technologies [46]. An integrated-optic 1×128 power splitter composed of a slab waveguide, funnel-shaped waveguide and output waveguide has been reported with an excess loss of 2.3 dB and a small standard deviation of 0.63 dB [53].

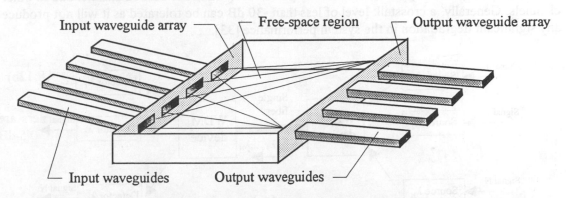

Fig. 8.12: Principle of a free-space star coupler

8.3   WAVELENGTH DIVISION MULTIPLEXER AND DEMULTIPLEXER

A dramatic increase in the information carrying capacity of a fiber can be achieved by the simultaneous transmission of many optical signals over the same fiber from different sources with properly spaced peak emission wavelengths. This is referred to as the *wavelength division multiplexing* (WDM). In principle, WDM scheme is same as frequency division multiplexing (FDM) used in microwave, radio and satellite systems. The difference between them is the spacing in the wavelength/frequency domain. The WDM is coarse, while optical FDM (OFDM) is fine. The wavelength spacing $\Delta\lambda$ of WDM systems is about 1 nm (Section 9.2.2) and that of OFDM systems is much less when coherent detection schemes are used.

The WDM devices can be unidirectional or bidirectional. A unidirectional device is used to combine different signal wavelengths onto a single fiber at one end and to separate them out into their corresponding detectors at the other end (Fig. 8.13a). In contrast to this, a bidirectional device is used to send information in one direction and simultaneously receive information in the other direction at different wavelengths (Fig. 8.13b). In principle, any *optical multiplexer* can also be used as a *demultiplexer*. The word "multiplexer" is often used as a general term to refer to both multiplexer and demultiplexer except when it is necessary to distinguish the two devices or functions.

The three basic performance criteria for wavelength division multiplexer (MUX)/demultiplexer (DEMUX) are insertion loss, channel width and crosstalk. The insertion loss of a WDM device includes connector/splice loss of the device to the fiber and the intrinsic losses within the device. Channel width i.e, the separation between adjacent carriers is the wavelength range that is assigned to a particular optical source. To ensure that no interchannel interference results from source instability (for example, drift of peak operating wavelength with temperature), channel widths of in the order of nanometre are required in case of laser diodes and hundreds of nanometres for LEDs. Crosstalk refers to the power coupled in the desired channel due to other channels. Generally, a crosstalk level of less than -30 dB can be tolerated as it will not produce any significant degradation in the system performance [33].

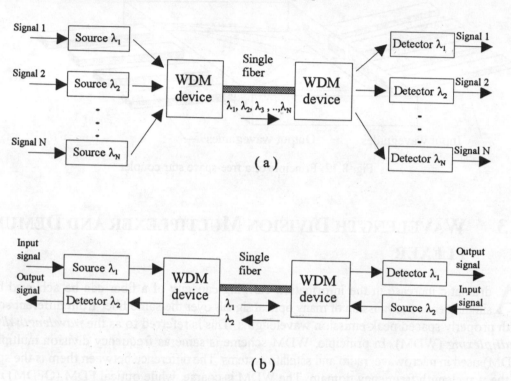

Fig. 8.13: (a) Unidirectional and (b) bidirectional WDM transmission systems

The optical signals that are combined generally do not emit a significant amount of optical power outside of their designated spectral width. Therefore, interchannel crosstalk factors are relatively unimportant at the transmitter end. At the receiver end, it is not so since photodetectors are usually sensitive over a broad range of wavelengths which could include all the WDM channels. Thus, to prevent significant amounts of the wrong signals from entering into each receiving channel, either demultiplexer must be carefully designed or very stable optical filters with sharp cutoffs must be used [33].

As in the case of directional couplers, WDM devices/filters can be bulk, fiber or waveguide/integrated-optic type. These are briefly discussed below.

8.3.1   BULK-TYPE

Bulk-type WDM devices frequently use dispersive devices such as a grating or a prism. The dispersive element transforms the wavelength difference to the light propagation direction difference (Fig. 8.14a and b). Some WDM devices are dielectric thin-film filters (Fig. 8.14c).

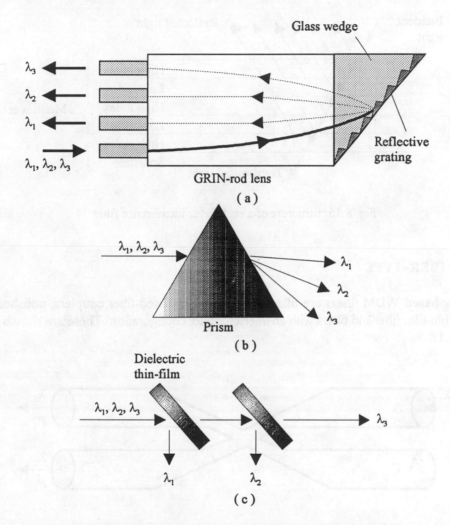

Fig. 8.14: Bulk-type WDM devices based on (a) grating, (b) prism and (c) dielectric thin-film filters

Optical filters using multilayer dielectric thin-films are well known in optics. This type of filters uses the interference phenomenon of optical wave [8]. Such filters, as shown in Fig. 8.15, can be constructed from alternate layers of high RI (e.g., zinc sulphide) and low RI (e.g., magnesium fluoride), each of which is one quarter wavelength thick [21]. In this structure, light wave which is reflected within the high index layer does not suffer any phase shift on reflection, while those rays reflected within low index layers undergo a phase shift of 180°. Thus the successive reflected beams recombine constructively at the filter front face and produce a high reflectance over a limited wavelength range. This range is dependent upon the ratio of high and low RIs. Outside this high reflectance range, the reflectance changes abruptly to a low value [48].

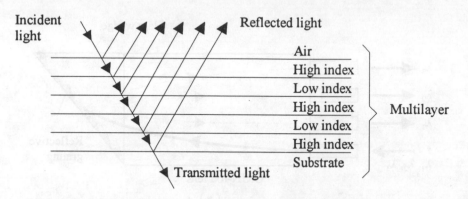

Fig. 8.15: Structure of a multilayer interference filter

8.3.2   FIBER-TYPE

The fiber based WDM filters are of several types viz., fused-fiber couplers, polished fibers, fibers with thin-film filter and fibers with an interferometer configuration. These are shown in Figs. 8.16 and 8.18.

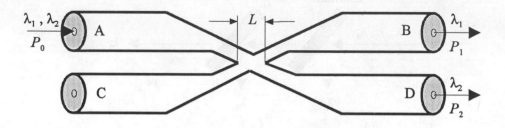

Fig. 8.16: Fiber-type WDM device using fused-fiber coupler

A fused-fiber coupler can be used as a fiber-type WDM filter when the coupling coefficient k is designed for WDM operation. This coefficient depends on the waveguide structure such as distance between two cores, shape of core and RI profile. The value of k for a WDM device is different from that of a 3 dB coupler. For a 3 dB coupler, the value of k is nearly constant for a wavelength range. On the other hand, k is variable for a wavelength range in a WDM filter. The power at the output ports of a WDM filter is given by

$$P_1(\lambda) = P_0 \cos^2(|k(\lambda)|L) \tag{8.13}$$

and

$$P_2(\lambda) = P_0 \sin^2(|k(\lambda)|L) \tag{8.14}$$

where L is the coupling length as shown in Fig. 8.16 and $k(\lambda)$ is designed such that $k(\lambda_1)L = 2\pi n + \pi/2$ for wavelength λ_1 and $k(\lambda_2)L = 2\pi m + \pi/2$ for wavelength λ_2, where n and m are integers. Then $P_1(\lambda) = 0$ and $P_2(\lambda) = P_0$ for $\lambda = \lambda_1$; $P_1(\lambda) = P_0$ and $P_2(\lambda) = 0$ for $\lambda = \lambda_2$ [45]. A typical transmission curve of WDM devices based on this directional coupler is shown in Fig. 8.17. Generally, wavelength selective and broadband fused-fiber couplers have very small insertion loss and stable properties against temperature change. For a fused-fiber coupler, isolation and insertion losses for a 1.32 μm/1.55 μm WDM coupler have been reported to be 16 dB to 18 dB and 0.05 dB respectively [37].

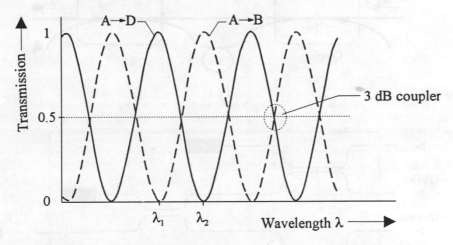

Fig. 8.17: Typical transmission curve for coupler-based WDM devices

A similar coupling mechanism takes place in the polished-fiber type of device as shown in Fig. 8.18a. By polishing the fiber side, cores in two fibers become closer and coupling takes place between them. For one such coupler, reported isolation ranges from 10 dB to 50 dB depending on the wavelength spacing of 35 nm to 200 nm [14].

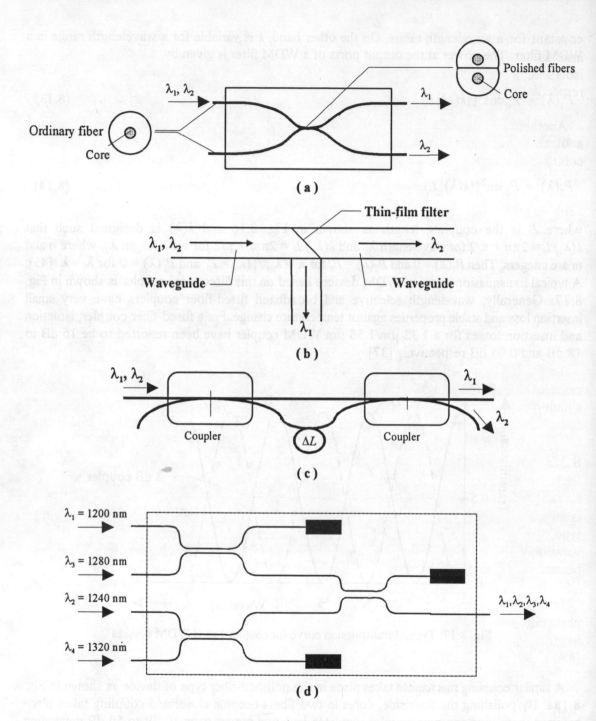

Fig. 8.18: Fiber-type WDM devices using (a) polished fibers, (b) thin-film filter, (c) interferometer and (d) cascade of two wavelength WDM devices to form a four wavelength WDM device

A fiber-type WDM device using a dielectric thin-film filter is possible provided the filter is inserted between the fibers. One example of this type is shown in Fig. 8.18b. The insertion and isolation losses of this device for a wavelength of 0.82 μm/1.2 μm are about 1 dB and 40 dB respectively [42].

Another fiber-type WDM device using a SM fiber is shown in Fig. 8.18c [49]. It consists of a Mach-Zehnder interferometer (MZI) using two couplers and fibers. The fiber length ΔL contributes a phase difference

$$\Delta \phi = 2\pi n_{eff} \frac{\Delta L}{\lambda} \tag{8.15}$$

where n_{eff} is the effective RI which depends on the wavelength. Two light waves with a phase difference interfere at the second coupler. This interference results in the wavelength dependent property used for wavelength division MUX/DEMUX. It is reported that the output shows instability with time when constructed by two couplers with a 0.5 m long fiber between them [49]. Miniaturization of the device will help in realizing stable properties [45].

To multiplex or demultiplex more than two wavelengths, WDM devices can be placed in cascade. An example is given in Fig. 8.16d in which three fused-fiber WDMs are cascaded to form a multiplexer at 1200 nm, 1240 nm, 1280 nm and 1320 nm [40].

8.3.3 WAVEGUIDE-TYPE

The principle of waveguide-type filters is similar to those of fiber-type. The basic configuration is either a simple directional coupler as shown in Fig. 8.19a or a MZI as in Fig. 8.19b. In the latter type, the wavelength response is determined mainly by the length difference ΔL, not by the wavelength dependence of two directional couplers used in MZI. This device is commonly used because a desirable k value is easily obtained with the interferometer design.

In the MZI device (Fig. 8.19b), incident light is split by the first directional coupler and the phase difference of the two light waves is $\pi/2$. There is an additional phase difference due to the length difference ΔL. The total phase difference $\Delta\theta$ at the input of the second coupler using Eq. (8.15) is obtained as

$$\Delta\theta = \frac{\pi}{2} + 2\pi n_{eff} \frac{\Delta L}{\lambda} \tag{8.16}$$

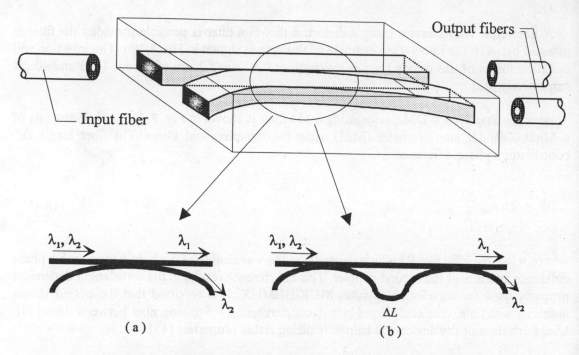

Fig. 8.19: Waveguide-type WDM devices using (a) directional coupler
and (b) Mach-Zehnder interferometer

With a phase shift of $\pi/2$ by the second directional coupler, the required $\Delta\theta$ for the two
wavelengths λ_1 and λ_2 are

$$\Delta\theta = 2n\pi + \pi/2 \quad \text{for} \quad \lambda_1 \tag{8.17a}$$

and

$$\Delta\theta = 2m\pi + \pi/2 \quad \text{for} \quad \lambda_2 \tag{8.17b}$$

where n and m are integers. By combining Eqs. (8.16) and (8.17), following equations for the
filter design are obtained

$$n_{\text{eff}}\Delta L = n\lambda_1 \tag{8.18a}$$

and

$$n_{\text{eff}}\Delta L = (m + 1/2)\lambda_2 \qquad (8.18b)$$

As an example, ΔL values are 1.27 µm for a 1.3 µm/1.55 µm MUX/DEMUX and 15.5 µm for a 1.5 µm/1.55 µm MUX/DEMUX respectively [35]. For a 1.48 µm/1.55 µm MUX/DEMUX, an insertion loss of 0.7 dB (coupling loss to a fiber is not included) and isolation and polarization dependent losses of 30 dB and 0.03 dB respectively have been reported [3].

8.4   Attenuators, Isolators, Circulators and Polarizers

These components are used for improving the system performance of point-to-point optical communication links (IM/DD and coherent) as well as optical networks. These are briefly discussed below.

8.4.1   ATTENUATORS

The function of optical attenuators is to decrease the light intensity. These are used in several applications like in receiver to prevent its saturation due to strong intensity input signal. Several types of attenuators are shown in Fig. 8.20.

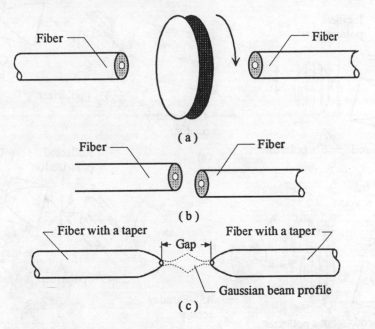

(a)

(b)

(c)

Fig. 8.20: Optical attenuators (a) absorption, (b) axial and longitudinal gap and (c) longitudinal gap types

The attenuator shown in Fig. 8.20a uses materials to absorb light. Since the absorption generally depends on the wavelength, the value of attenuation also depends on the wavelength. Variable attenuation is obtained by rotating the absorption disk as attenuation is different for different disk placement. Backward reflections from the absorption material may affect the source stability. To decrease the reflection intensity, a thin-film coating is used at the surface of the absorption materials. Obliquely aligned or obliquely polished surfaces are also used to avoid reflection. In Fig. 8.20b, variable attenuation is obtained by changing axial offset and longitudinal gap, while in Fig. 8.20c by changing longitudinal gap only.

8.4.2 ISOLATORS

The performance of certain fiber optic communication systems can be severely degraded if part of the launched power is reflected back towards the source. To prevent such undesired reflections, optical isolators are placed immediately after the sensitive device/source. Commercially available optical isolators make use of nonreciprocal nature of Faraday effect, which changes the state of polarization of the incident fields in the presence of a magnetic field parallel to the direction of propagation. Structure of such an isolator is shown in Fig. 8.21.

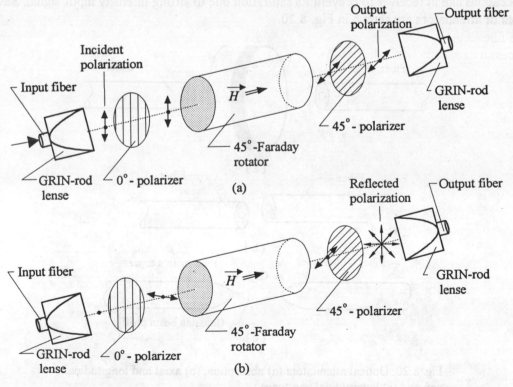

Fig. 8.21: Optical isolator using the Faraday effect (a) forward and (b) backward direction

The light from an input fiber passes through a GRIN-rod lens. A brief explanation on the lens is given in Appendix A2. Let the plane polarized light at the lens output impinges on the input polarizer. Only the vertically polarized component of the input light passes through this polarizer. Then it passes through the Faraday rotator and experience a rotation

$$\theta = VHL \tag{8.19}$$

of its polarization where V, H and L are the Verdet constant, the magnetic field and the length of sample respectively. The parameter V depends on the material (normally a YIG i.e., Yttrium Iron Garnet crystal) and the wavelength of operation. Assume that θ has been adjusted so as to produce a rotation of $\pi/4$ and the output polarizer is rotated $\pi/4$ relative to the input polarizer. It implies that output light from the Faraday rotator will pass through the output polarizer. The backward light coming from undesired reflections will be rotated by another $\pi/4$ when passing once again through the Faraday rotator. This is the result of non-reciprocity of the Faraday effect. The reflected light facing the input polarizer will be orthogonal to the input polarizer's orientation and will thus be blocked by it. Typical insertion loss of this device is 1 dB to 2 dB and isolation of the order of 40 dB.

All these isolators are polarization dependent. In fact, polarization-independent characteristics are required when isolators are not placed in front of a laser diode to prevent reflections. In an ordinary fiber (not a polarization-maintaining fiber), light polarization keeps on changing along the fiber length. Therefore, for an in-line isolator, polarization-independent characteristics are necessary. The configuration of a polarization-independent isolator is given in Fig. 8.22 [50].

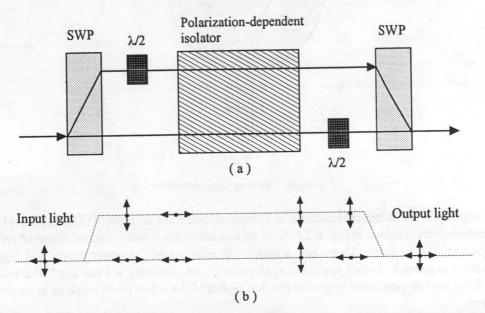

Fig. 8.22: Polarization-independent optical isolator (after [50] copyright 1991, Electronics Letters)

It uses a polarization-dependent isolator, two spatial walk-off polarizers (SWPs) and two half-wave plates (λ/2-wave plates). The input light waves with orthogonal polarizations are separated by an SWP and are set in the same polarization direction by a λ/2-wave plate. These then pass through the polarization-dependent isolator. The polarization direction of one of the two output light is rotated by a λ/2-wave plate and the two light waves are finally recombined by the other SWP. The backward light is blocked by the polarization-dependent isolator. The reported performance is about 55 dB of isolation with 0.38 dB of insertion loss at 1.3 um and about 64 dB of isolation with 0.32 dB of insertion loss at 1.55 μm [50].

8.4.3   CIRCULATORS

A circulator is an n-port optical device. A model of circulator for n = 3 is shown in Fig. 8.23. In an ideal three port circulator, input power from port 1 is output from port 2, input power from port 2 is output from port 3 and input power from port 3 is output from port 1 and there exists no reflection for each port [45].

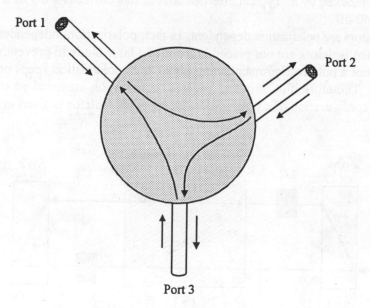

Fig. 8.23: 3-port optical circulator

One application of an optical isolator is in Erbium-doped fiber amplifier (EDFA) [36, 44]. One such configuration is shown in Fig. 8.24. With an optical isolator and a mirror, reuse of reflected pump power can be made. Therefore, a highly efficient use of the pump power and high gain amplification is possible. In this application, all ports in the circulator are not used. For example, a beam from the last port is not output at the first port and the other ports work as in an ordinary circulator.

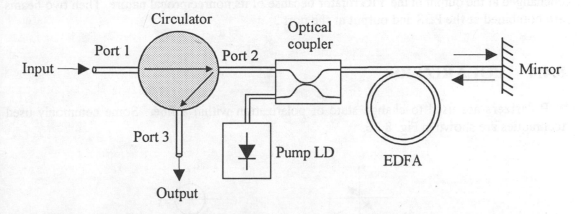

Fig. 8.24: Erbium-doped fiber amplifier using an optical circulator

There are several proposed configurations of optical circulators. As an example, the configuration of a 4-port polarization-independent circulator is shown in Fig. 8.25 [30].

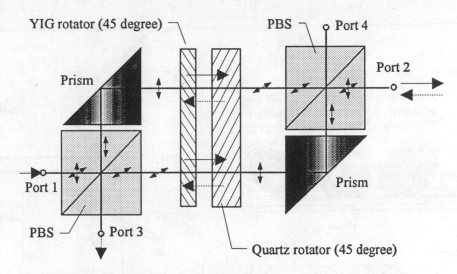

Fig. 8.25: Principle of a polarization-independent circulator (after [30], copyright 1979, Electronics Letters)

It uses a YIG rotator, a quartz rotator, two polarization beam splitters (PBSs) and two right-angle prisms. The incoming beam from port 1 splits at first PBS into two polarized beams. Each beam experiences polarization rotation and polarization is interchanged. The interchanged-polarization beams pass through the second PBS and are output at the port 2 [45]. The input light from port 2 is separated by the PBS. However, the polarization of two separated beams is unchanged at the output of the YIG rotator because of its nonreciprocal nature. Then two beams are combined at the PBS and output at the port 3.

8.4.4 POLARIZERS

Polarizers are used to change state of polarization within a fiber. Some commonly used techniques are shown in Fig. 8.26.

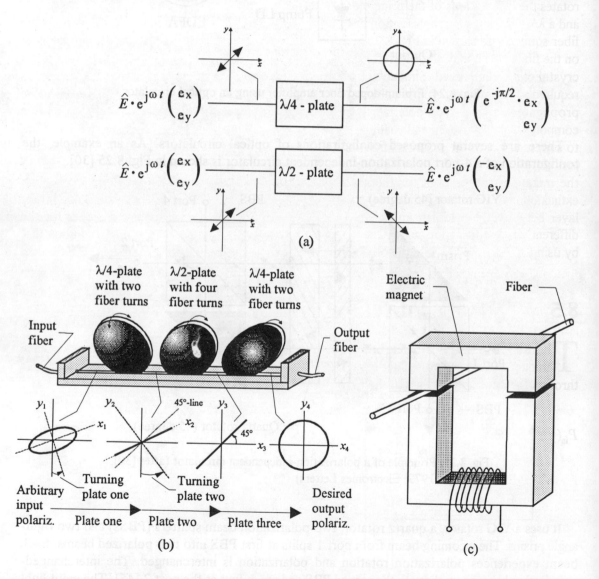

Fig. 8.26: Typical polarizers (a) $\lambda/4$- and $\lambda/2$-plates, (b) Lefevre polarizer and (c) fiber squeezer

The function of λ/4- and λ/2-wave plates as shown in Fig. 8.26a is to shift one of the orthogonal polarizations (i.e., the vertical or the horizontal one) by 90° and 180° respectively. Instead of plates, any other birefringent crystal (BC) known from classical optics can also be used to realize this shift. In the Lefevre polarizer, required birefringence is creating by fiber bending. In this polarizer, as shown in Fig. 8.26b, thin discs with a groove are used to coil a fiber and to achieve the bending. As explained in Chapter five, the bending changes state of polarization. When disc diameter, wavelength of light and number of turns are properly matched, the disc behaviour is analogous to that of a classical λ/4- or λ/2-fractional wave plate. Its rotation directly rotates the orientations of the principle axis of the fiber. By employing a chain of a λ/4-, a λ/2- and a λ/4-disc, any input polarization can be changed into any desired output polarization. In the fiber squeezer as shown in Fig. 8.26c, required birefringence is obtained by mounting pressure on the fiber which may slightly deforms the fiber as well. For this, an electric magnet or a piezo crystal can be employed. Similar to the Lefevre polarizer, a chain of at least three squeezers is required to perform any type of polarization change. When only one polarized light wave is to be propagated, a fiber-type polarizer (not shown in the figure) can be used. These are in-line components which use a metal coating or a BC contacted at one side of a removed cladding fiber to suppress one of the polarized light waves [45]. In the case of metal coating, the propagation loss difference between two modes (whose electric fields are either parallel or perpendicular to the metal surface) is used for the realization of one polarized light wave propagation. High extinction ratio with a low insertion loss can be realized by introducing a properly designed buffer layer between the cladding and metal [23]. In the BC, RI is not constant and differs for the different wave propagation angle. One polarization wave is propagated with a large radiation loss by using a proper BC, while the other is propagated with a small radiation loss [6].

8.5   TUNABLE FILTERS

The tunable filter as a black box is shown in Fig. 8.27. Many signals at different optical frequencies appear at the filter input. The filter is selective enough that only one signal passes through the filter and appears at the output without much of a distortion.

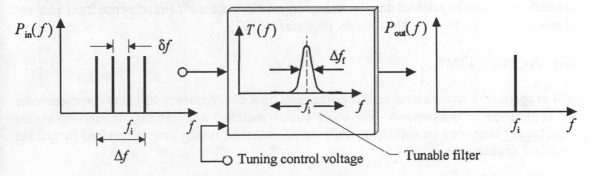

Fig. 8.27: Basic function of a tunable optical filter

The wavelength-selective filters are important in fiber optic systems that carry more than one wavelength. The wavelengths do not interfere with each other in the fiber, but these can interfere at the detector as the detector response is very wide. To prevent this, the different wavelength signals must be separated at the fiber end. This is typically achieved by placing a filter at the fiber end that reflects certain wavelengths and transmit others. Filters are also important to isolate signals from the background noise and to block stray pump light in optical amplifiers [25].

As a wavelength-selective device, one can use a simple prism which works on the principle of chromatic dispersion. Unfortunately, prisms are usually bulky and therefore not used commonly. The other types of filters are based on the interference effects. When such a filter is tuned to light of a desired wavelength, there is a reinforcement by constructive interference, while for the other wavelengths there may be a nonconstructive interference or even completely destructive interference. Some of the filters based on this effect are Mach-Zehnder, electrooptical, acousto-optic tunable filters.

The critical issues involved in the design of tunable filters from the systems point of view are as follows.

(i) NUMBER OF RESOLVABLE CHANNELS

In the Fig. 8.27, Δf is the frequency difference between the lowest and highest frequency channels and δf as the spacing between the adjacent channels (sometimes, it will be more convenient to deal with wavelength and in that case one can use $\Delta \lambda$ and $\delta \lambda$ respectively). If the tuning is to cover the entire Δf of low attenuation window at either 1.3 μm or 1.5 μm, then 200 nm ($\approx$ 25,000 GHz) is probably a reasonable target for tuning range. The minimum interchannel spacing δf depends on the level of crosstalk or crosstalk degradation which can be tolerated and bandwidth Δf_f of the tunable filter. This bandwidth must be smaller than δf. The channel spacing that guarantees < 0.5 dB crosstalk penalty varies practically between 3 to 10 times the required channel bandwidth depending upon the modulation scheme and the optical-filter transfer function. The ratio $N_{max} = .|\Delta f / \delta f|_{max}$ gives the maximum number of equally spaced channels which can be resolved into the tuning range before crosstalk (interference from adjacent channels) severely degrades the receiver performance [22].

(ii) ACCESS TIME

The speed with which a tunable filter can be reset from one frequency to a new one determines its application in the network. For some circuit switched applications, millisecond access time/tuning time may be sufficient, while submicrosecond tuning time is required for packet switched applications.

(iii) ATTENUATION

The signal at the filter output will in general incur some power losses/attenuation because of insertion and internal losses of the filter. This attenuation must be minimized to reduce its impact on the already limited network power budget.

(iv) CONTROLLABILITY

There are two main aspects as far as controllability is concerned. First one is the stability that is once the filter is set to a particular frequency, thermal and mechanical factors should not cause the tuning to drift more than a small fraction of the bandwidth of one channel. Second, the filter must be easily resettable to any designated value. It is most desirable that the filter have a smooth characteristic curve of frequency versus applied signal. If this curve is not smooth, this may add considerable complexity to the control circuitry [22].

(v) POLARIZATION INDEPENDENCE

As polarization-independent filters can work for all possible polarization states of the impinging light, these are preferred over the polarization-dependent filters. Further, the latter type of filters entail the added complexity of polarization control, polarization diversity or polarization scrambling elsewhere in the system [22].

(vi) COST

The filter cost mainly includes the fabrication and pigtailing (fiber attachment) costs. Even if the lithographic technology that has been perfected with electronic integrated circuits is used for the realization, low loss attachment of fibers has often remained as a problem. With others, the pigtailing problem is virtually absent, but due to high cost manufacturing steps there is a little potential for full integration [22].

(vii) SIZE, POWER CONSUMPTION AND OPERATING ENVIRONMENT

Tunable filters and their associated control circuitry must be as compact as possible to meet user requirements. It should be possible to operate the filter from the same power supplies that are available for the other electronic circuitry. Further, the filters must be resistant to shock, vibration, humidity and temperature conditions prevailing in the low cost high usability environments.

In the following Subsections, commonly used tunable filters are discussed in brief.

8.5.1 FABRY-PEROT

A schematic of Fabry-Perot interferometer (FPI) or etalon consisting of a resonant cavity formed by two parallel mirrors is shown in Fig. 8.28.

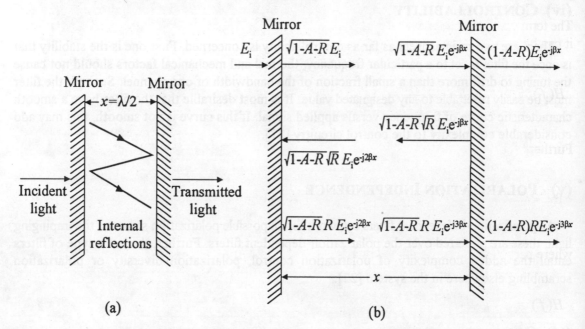

Fig. 8.28: (a) Geometry and (b) electric field strength of successive reflections in a FPI/etalon

Light from an input fiber is collimated (made parallel), passed through the cavity and then refocussed on the output fiber. Let R be the power reflectivity (reflectance) of each of the two mirrors. It implies that on each reflection, the reflected field strength becomes $\sqrt{R}$ times the incident field strength. Let A be the power absorption loss as the light passes through the supporting glass material and reaches the first mirror and let the absorption in passing from the second mirror to the output likewise be A. Light of field strength E_i enters the filter, passes through the first mirror and $\sqrt{(1-A-R)}E_i$ survives and the remaining is lost as heat or reflected back from the mirror. The signal then passes through the resonant cavity presumed to be lossless. The first arrives at the right-hand mirror with field strength $\sqrt{(1-A-R)}\exp(-j\beta x)$, where β represents the propagation constant. A portion $(1-A-R)E_i\exp(-j\beta x)$ passes to the output and another portion $\sqrt{(1-A-R)}\sqrt{R}E_i\exp(-j\beta x)$ is reflected back towards the first mirror. This arrives at the first mirror and is reflected to the second mirror where it emerges $(1-A-R)RE_i\exp(-3j\beta x)$.

Adding up all the successive contributions to the output E_o, the complex transfer function of the field strength is

$$H(f) = \frac{E_o(f)}{E_i(f)} = (1-A-R)\exp(-j\beta x)\left[\sum_{m=0}^{\infty} R^m \exp(-2jm\beta x)\right] \tag{8.20}$$

The term within square bracket is a geometric progression with argument $R\exp(-2j\beta x)$. Therefore, it can be simplified and the above equation can be written as

$$H(f) = \frac{(1-A-R)\exp(-j\beta x)}{1-R\exp(-2j\beta x)} \tag{8.21}$$

Further,

$$\beta x = \frac{2\pi}{\lambda}x = \frac{2\pi f}{c}x = 2\pi f\tau \tag{8.22}$$

where parameter τ represents the one-way propagation time. With this, Eq. (8.21) becomes

$$H(f) = \frac{(1-A-R)\exp(-j2\pi f\tau)}{1-R\exp(-j4\pi f\tau)} \tag{8.23}$$

Therefore, the power transfer function $T(f)$ is given by

$$T(f) = |H(f)|^2 = \frac{(1-A-R)^2}{(1-R)^2 + R[2\sin(2\pi f\tau)]^2} \tag{8.24a}$$

or

$$T(f) = \frac{\left[1-\dfrac{A}{(1-R)}\right]^2}{\left[1+\left(\dfrac{2\sqrt{R}}{(1-R)}\sin(2\pi f\tau)\right)^2\right]} \tag{8.24b}$$

The right hand side of above equation is known as the Airy function and is shown in Fig. 8.29 for three different values of R and $A = 0$.

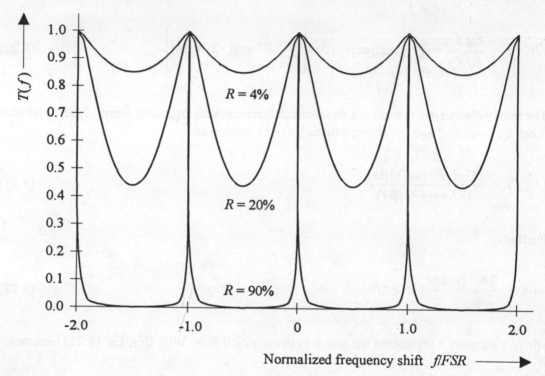

Fig. 8.29: Power transfer function of a FP filter versus normalized frequency shift for various end-face reflectivities R

It is seen from the above figure that $T(f)$ is a periodic function. Its period, called the free spectral range (FSR), is given by

$$FSR = \frac{1}{2\tau} = \frac{c_0}{2nx} \tag{8.25}$$

where c_0 represents the velocity of light in free space and n the RI of the medium between the mirrors. The 3 dB bandwidth of FP filter (full width at half maximum FWHM) is obtained from Eq. (8.24b) and is given by

$$FHWM \approx \frac{c_0}{2nx} \frac{(1-R)}{\pi \sqrt{R}} \tag{8.26}$$

The ratio of the *FSR* to *FWHM* gives an indication of how many channels can be selected by the filter. This ratio is known as finesse F of the filter and is given by

$$F = \frac{FSR}{FWHM} = \frac{\pi\sqrt{R}}{(1-R)} \tag{8.27}$$

Typically, F ranges from 20 to 100. The maximum number of resolvable channels depends on the allowed crosstalk penalty level which in turn depends on the shape of the filter transfer function. For a FP filter, the channel spacing can be as small as three times of FWHM to keep the crosstalk level < 0.5 dB. The maximum number N_{max} of channels that can be selected by a FP filter is therefore given by

$$N_{max} < \frac{F}{3} = \frac{\pi\sqrt{R}}{3(1-R)} \tag{8.28}$$

For $R = 0.97$, N_{max} from the above equation will be 34. To tune the filter from one channel to another, FP cavity length L is to be fine-adjusted. The length needs to be changed by only half the operating wavelength to tune the filter over one entire *FSR*.

One of the most attractive characteristics of etalons is that their characteristics are usually polarization insensitive as long as the mirrors are plane and the intercavity material is not birefringent. When the cavity is filled with material whose velocity of propagation is polarization dependent, the entire device becomes polarization dependent [22].

Among the various design options available, one of the commonly used design is the tunable-fiber FP filter [52]. In this design, low cost as well as low attenuation are achieved by means of reducing or eliminating the pigtailing by forming the mirrors directly on the butt end of the fiber core. The filter is formed by a short gap between two fiber-end faces which are coated with highly reflecting films to create the resonant cavity. The device is tuned by using a piezo-electric control in the configuration as shown in Fig. 8.30. The tuning speed is fairly slow (tuning time ≥ 1 ms) and the filter is therefore only suitable for circuit switched applications [40].

To accommodate more channels, finesse of the filter has to be improved. Two approaches are possible the multipass and the multicavity. In the multipass scheme as shown in Fig. 8.31a, the light passes twice through the same cavity. The composite power transfer function is then the square of the individual transfer function given in Eq. (8.24).

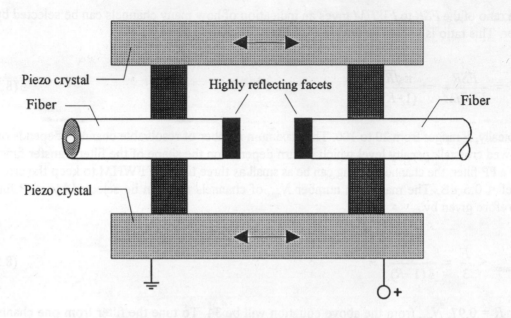

Fig. 8.30: Schematic of a tunable-fiber FP filter

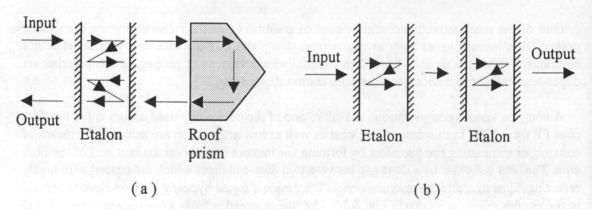

(a)   (b)

Fig. 8.31: Filters with two cascaded etalons (a) two-pass and (b) two-cavity

In the multicavity approach, optical signal passes through two FP filters with an isolator in between as shown in Fig. 8.31b. This approach can bring the finesse to values in excess of 1000 [41]. Further, by sharpening the filter transfer function, the required channel separation is halved relative to the single-stage case. It results in a doubling of the number of resolvable channels over the same *FSR*. With $F = 1000$, over 600 WDM channels can be supported with adequately low level of crosstalk. The disadvantages of this approach are the higher loss than the single-stage because of the need to isolate the two stages and the tuning is more difficult to control.

8.5.2   MACH-ZEHNDER

In a single-stage MZI filter, the multichannel input signal is split into two equal parts by a 3 dB coupler. The two versions of the same signal traverse paths of slightly different lengths and merge together in another 3 dB coupler at the output as shown in Fig. 8.32. Thus MZI filter involves interference of the two versions of the same signal traversing paths of slightly different length. In contrast to this, Fabry-Perot interferometer (FPI) filter involves light interference of many repeated reflections.

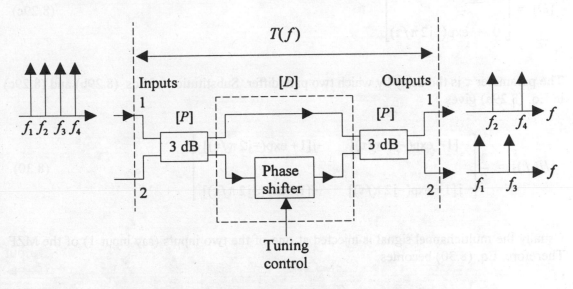

Fig. 8.32: Schematic of a Mach-Zehnder interferometric filter

The overall transfer function $[H(f)]$ of a single-stage Mach-Zehnder filter (MZF) is obtained by multiplying the scattering matrix associated with the 3 dB coupler and that associated with the two different propagation paths

$$[H(f)] = \begin{bmatrix} H_{11}(f) & H_{12}(f) \\ H_{21}(f) & H_{22}(f) \end{bmatrix} = [P][D][P] \tag{8.29a}$$

where

$$[P] = \frac{1}{\sqrt{2}} \begin{bmatrix} 1 & -j \\ & \\ -j & 1 \end{bmatrix} \tag{8.29b}$$

and

$$[D] = \begin{bmatrix} 1 & 0 \\ & \\ 0 & \exp(-j2\pi f\tau) \end{bmatrix} \tag{8.29c}$$

The parameter τ is the delay by which two path differ. Substitution of Eqs. (8.29b) and (8.29c) in Eq. (8.29a) gives

$$[H(f)] = \frac{1}{2} \begin{bmatrix} [1 - \exp(-j2\pi f\tau)] & -j[1 + \exp(-j2\pi f\tau)] \\ & \\ -j[1 + \exp(-j2\pi f\tau)] & -[1 - \exp(-j2\pi f\tau)] \end{bmatrix} \tag{8.30}$$

Usually the multichannel signal is injected at one of the two inputs (say input 1) of the MZF. Therefore, Eq. (8.30) becomes

$$[H(f)] = \begin{bmatrix} H_{11}(f) & 0 \\ & \\ H_{21}(f) & 0 \end{bmatrix} = \frac{1}{2} \begin{bmatrix} [1 - \exp(-j2\pi f\tau)] \\ \\ -j[1 + \exp(-j2\pi f\tau)] \end{bmatrix} \tag{8.31}$$

and the corresponding power transfer function is

$$\begin{bmatrix} |H_{11}(f)|^2 \\ \\ |H_{21}(f)|^2 \end{bmatrix} = \begin{bmatrix} \sin(\pi f\tau)^2 \\ \\ \cos(\pi f\tau)^2 \end{bmatrix} \tag{8.32}$$

It is clear from the above equation that the power transfer function is periodic in frequency with period $1/\tau$. Such filter is often called a periodic filter. To understand the filtering operation, let there are four input channels spaced equally in frequency by $\delta f = 1/2\tau$. Initially, the resonant condition of the MZF is obtained at f_1 such that $\cos^2(\pi f_1 \tau) = 1$. Therefore, the channels at frequency f_1 and f_3 will appear at output 2, while f_2 and f_4 at output 1. Further, filtering of each output can be achieved by using a second MZF whose period is double of the first stage i.e., $2/\tau$.

In case of 16 channels, four stage MZFs will be required. If the outputs at port 1 are considered, the first stage pass only channels 0, 2, 4, 6, 8, 10,...., the second stage to pass only 0, 4, 8 and third stage to pass only 0, 8, 16 etc. This requires that the delay introduced by the successive MZF stages is set according to

$$\tau_n = \frac{1}{2^n \delta f} \qquad (8.33)$$

The overall power transfer function $T(f)$ for output at port 2 from Eqs. (8.32) and (8.33) is

$$T(f) = \prod_{n=1}^{N} \cos(\pi f \tau_n)^2 = \left[\frac{\sin(\pi f/\delta f)}{N \sin(\pi f/N \delta f)} \right]^2 \qquad (8.34)$$

where N is the number of stages. This is shown in Fig. 8.33. It is observed that when device is ideally constructed, infinitely deep nulls are produced at the undesired channel frequencies.

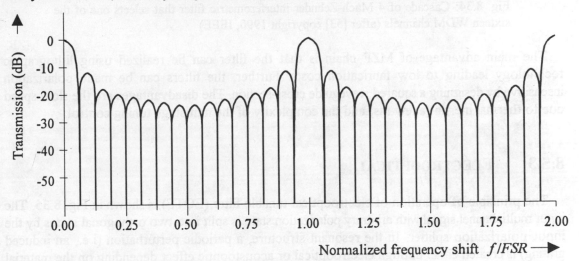

Fig. 8.33: Transmission characteristic of a 4-stage Mach-Zehnder interferometric filter

A four-stage MZF that can multiplex 16 WDM channels is shown in Fig. 8.34 [55]. In this filter, tunability is achieved by thermal variation of the path length difference in each MZF stage. A tunable 128-channel 7-stage MZF has been demonstrated with a crosstalk level of -13 dB and fiber-to-fiber attenuation of 6.7 dB [54].

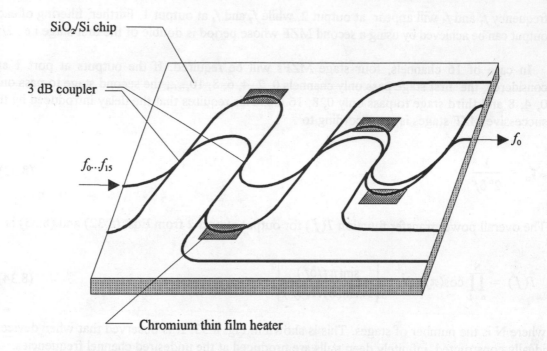

Fig. 8.34: Cascade of 4 Mach-Zehnder interferometric filter that selects one of the sixteen WDM channels (after [53] copyright 1990, IEEE)

The main advantage of MZF chain is that the filter can be realized using lithographic technology leading to low fabrication costs. Further, the filters can be made polarization insensitive by designing a squared waveguide cross-section. The disadvantages are the slow speed due to thermal inertia (a few ms) and the complexity of the mutistage tuning control.

8.5.3   ELECTROOPTICAL

The principle of operation of electrooptical tunable filter (EOTF) is shown in Fig. 8.35. The input multichannel signal with arbitrary polarization state is split into two orthogonal states by the input polarization splitter. In the resonant structure, a periodic perturbation (i.e., an induced grating) is created either through electrooptical or acoustooptic effect depending on the material used. These perturbations change the polarization state of the horizontal and vertical components. Two different effects will be observed. The first to the light that is at or near the phase-matched wavelength and the second to light of other wavelengths. For phase-matched wavelengths, there is a 90° rotation between input and output planes of polarization. The rotation converts a horizontally polarized light at the filter input into a vertically polarized light which is delivered by the second polarization splitter to output 1. Other wavelengths emerge from the grating region still horizontally polarized and pass through to output 2. Similarly, the vertical component of input also has its phase-matched wavelength rotated and passed to output 1, while other wavelengths are transferred by the second polarization splitter to output 2. Irrespective of the state of polarization of the incoming signal, it will reach the output with the same polarization. This is true for both resonance frequencies going to output 1 and for all the off-resonance frequencies going to output 2.

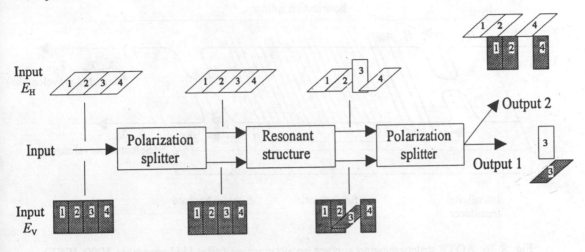

Input E_H

Input

Input E_V

Polarization splitter → Resonant structure → Polarization splitter

Output 2

Output 1

Fig. 8.35: Principle of filtering mechanism in both EOTF and AOTF [22]

Due to electrooptical effect, the tuning speed of EOTF is very fast ($\approx$ns). The tuning range, however, is limited to about 10 nm because of the small electrooptical effect. A 3 dB bandwidth (FWHM) of 1 nm has been achieved. It gives number of resolvable channels to be about 10. A longer perturbation region would improve the filter bandwidth and thereby the number of resolvable channels. But this would be at the expense of increased attenuation.

8.5.4   ACOUSTOOPTIC

The basic principle of acoustooptic tunable filter (AOTF) is same as that of EOTF. As their name implies, in AOTF acoustooptic effect is used while in the EOTF electrooptical effect. The structure of an AOTF is shown in Fig. 8.36. The diffraction grating in this filter is formed by surface acoustic waves (SAWs) and the wavelength tuning is accomplished by varying the SAW frequencies (usually between several tens to several hundreds of megahertz). The periodic perturbation induced by the acoustooptic effect can be viewed as a dynamic grating with a period equal to the acoustic wavelength. Because of this, tuning range of the AOTF can be much broader (can be entire 1.3 μm to 1.6 μm wavelength range) than that of the EOTF [10]. Further, grating created acoustically remains in essentially the same position as light interact with it i.e., there is no Doppler shift of the diffracted light. The reason for this is that light velocity is five orders of magnitude higher than the sound velocity in the medium. Lastly, the AOTF has the powerful and unique capacity to select several wavelength channels simultaneously and independently when multiple acoustic waves are present in the interaction length. This is made possible due to the weak interaction between different acoustic waves. Simultaneous selection up to five wavelengths separated by 2.2 nm has been achieved [10].

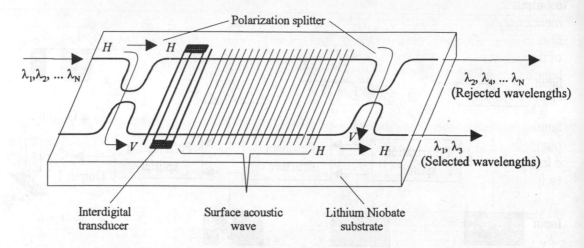

Fig. 8.36: AOTF structure using surface acoustic waves (after [51] copyright 1990, IEEE)

The drawback of the AOTF is its large tuning time ($\approx \mu s$) which is limited by the time for the acoustic wave to fill up the interaction length.

The 3 dB bandwidth of AOTF (*FWHM*) is similar to that of EOTF (≈ 1 nm) as is the insertion loss (≈ 5 dB) [26].

8.5.5   SEMICONDUCTOR-BASED

Single-longitudinal mode (SLM) lasers have built-in gratings (e.g., DFB and DBR lasers) to provide the selection of a single optical frequency. When biased below threshold, these devices can operate as resonant amplifiers amplifying only those input channels whose optical frequencies coincide with those of the semiconductor optical structure. The wavelength tuning is achieved either by current injection (tuning time $\approx$ ns) [34] or temperature variation (tuning time $\approx$ ms).

The main advantage of semiconductor-based tunable filters is their monolithic integration within the receiver as the receivers use the same semiconductor material. Their disadvantages are limited tuning range and strong polarization-dependent characteristics.

8.5.6 FIBER-BRILLOUIN

In the fiber-Brillouin tunable filter, stimulated Brillouin scattering (SBS) in SM fiber is used as a selective amplification mechanism for wavelength-tunable filtering [12]. In this filtering mechanism, an unmodulated optical pump wave is injected from the receiving end of the optical fiber in the direction opposite to that of the incoming WDM signal. Selective amplification of one of the WDM channels takes place provided that the pump power (>10 mW) and the interaction length (several kilometres of fiber) are sufficient.

The filter tunability is achieved by the use of a tunable pump laser. The filter bandwidth is limited by the SBS-gain bandwidth and is about 100 MHz for an unmodulated narrow linewidth pump. It can be increased up to several hundreds of megahertz by broadening the pump spectrum e.g., by modulating the pump. The switching time is determined by the time required for the pump to fill up the fiber. With a 10 km length of fiber, it may be about 5 µs [40].

Various tunable filter characteristics are summarized in Table 8.2 [40].

Table 8.2: Characteristics of tunable filters

Technology	Tuning range (nm)	3 dB band-width (nm)	Number of resolvable channels	Loss (dB)	Tuning speed
FPF-single stage	50	0.5	10s	2	ms
FPF-tandem	50	0.01	100s	5	ms
Mach-Zehnder	5-10	0.01	100s	5	ms
EOTF	10	1	10	≈ 5	ns
AOTF	400	1	10s	≈ 5	µs
Active semiconductors	1-5	0.05	10s	0 (gain possible)	ns
Fiber-Brillouin	10	<0.01	100s	0 (gain possible)	µs

8.6   FIXED FILTERS

An alternative to tunable filters is to use fixed filters or grating devices. The latter devices typically filter out one or more wavelength signals from a WDM signal [7]. Commonly used fixed filters are following.

(i)   GRATING

In a grating filter, a periodic variation of the RI is directly photo-induced in the core of an optical fiber. A Bragg grating will reflect a signal of given wavelength back to the source, while passing the other wavelength signals. Typical spectral bandwidth is of the order of 0.1 nm and reflectivity in excess of 99 % [29]. The drawback of such a filter is that the RI of grating varies with temperature. It increases with temperature resulting in longer wavelengths being reflected. Such grating filters can be used in the implementation of multiplexers, demultiplexers and tunable filters [7].

(ii)   THIN-FILM INTERFERENCE

This type of filter is quite similar to Bragg-grating filter with the exception that such a filter is fabricated by depositing alternate layers of low index and high index materials onto a substrate (see Fig. 8.15 also). Thin-film filter technology suffers from lower thermal stability, high insertion loss and poor spectral profile. Technological advances may overcome the above problems [47].

8.7   MODULATORS

In an optical transmitter, modulation of light can either be performed by direct modulation of the laser's drive current as described in Chapter two or by means of an external optical modulator. The latter one has the advantage of no laser chirping due to modulation since laser source can now operate with a fixed and stabilized drive current.

Depending on the physical mechanisms, external modulators are classified into three types: (i) magnetooptical or Faraday, (ii) acoustooptic (AO) and (iii) electrooptical (EO) modulators. In advanced optical communication systems, the latter one is exclusively used since these modulators can be operated at much higher modulation speeds. The principle structure of an integrated $LiNbO_3$ electrooptical modulator is shown in Fig. 8.37a. A change in the input voltage level U changes the strength of the electrical field across the waveguide and, hence, its refraction index. As a result, velocity of light and phase difference between optical input and output signals are also changed. When electrical input voltage equals U_π, a phase change of 180° is achieved. In this case, the modulator operates as a PSK modulator where phase is shifted from 0° to 180° depending on the digital input signal which is either $U = 0V$ or U_π.

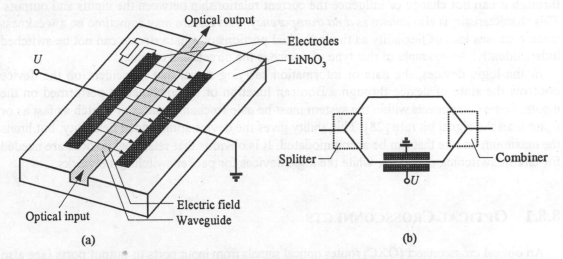

Fig. 8. 37: Integrated electrooptical (a) PSK and (b) ASK waveguide modulators

For amplitude modulation (e.g., ASK), integrated waveguide modulators are often used in a Mach-Zehnder configuration as shown in Fig. 8.37b. In this configuration, one branch is used to modulate the light phase, whereas the other as a simple waveguide delay line. When the output of both the branches are combined, total transmission occurs for equal velocity i.e., for $U = 0V$ drive and no transmission for 180° optical phase difference i.e., for $U = U_\pi$.

Integrated waveguide modulators can be used for digital as well as for analogue signals since these devices show an excellent linearity. For this reason, analogue Mach-Zehnder waveguide modulators, are particularly employed in fiber-based TV-broadcast networks as described in Chapter fourteen. Typical characteristics of a commercially available LiNbO$_3$ waveguide modulator are (i) modulation speed: 10 Gbit/s, (ii) U_π = 5V, (iii) insertion loss: 4 dB. Further, it is a polarization-sensitive device.

8.8 ROUTING AND SWITCHING ELEMENTS

In all optical networks, optical *routing* and *switching components* are used to rout and switch high bandwidth optical data signal without electrooptical conversion. In some devices, switching function is performed electronically whereas the routing function optically i.e., the optical data signal is transparently routed from a given input to a given output. This transparency allows the device to be independent of the data rate and format of the optical signal.

The switching and routing devices can be divided into two types: *relational* and *logical*. In relational devices, a relationship is established between inputs and outputs. This relationship is a function of the control signal applied to the switch and is independent of the contents of input signal. One important characteristic of this device is that the information entering and flowing through it can not change or influence the current relationship between the inputs and outputs. This characteristic is also known as *data transparency*. This feature may sometime be a weakness since it causes loss of flexibility as the individual portions of a data stream can not be switched independently. An example of this type of device is the directional coupler.

In the logic devices, the data or information carrying signal that is incident on the device controls the state of device through a Boolean function or its combination performed on the inputs. Some components within the system must be able to change states or switch as fast as or faster than the signal bit rate [28]. This ability gives the device some added flexibility, but limits the maximum bit rate that can be accommodated. It is obvious that relational devices are needed for circuit switching networks, while the logic devices for packet switching networks.

8.8.1 OPTICAL CROSSCONNECTS

An optical crossconnect (OXC) routes optical signals from input ports to output ports (see also Chapter fourteen). These can be either wavelength insensitive i.e., incapable of demultiplexing different wavelength signals on a given input fiber or wavelength sensitive. A basic crossconnect element is the 2×2 cross-point which routes optical signals from two input ports to two output ports and has two states: cross state and bar state (Fig. 8.38). In the cross state, signal from the upper input port is routed to the lower output port and signal from the lower input port to the upper output port. In the bar state, signal from the upper input port is routed to the upper output port and signal from the lower input port to the lower output port.

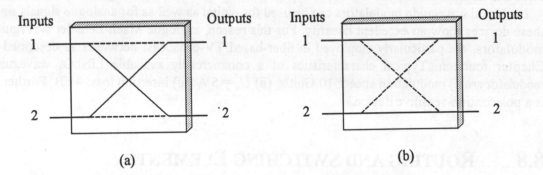

Fig. 8.38: 2×2 crossconnect in the (a) cross and (b) bar states

Optical crossconnects can be implemented using two types of technologies: (i) the generic directive switch, in which light via some structure is physically directed to one of the two outputs and (ii) the gate switch, in which optical amplifier gates are used to select and filter input signals to specific output ports (refer to Fig. 8.39).

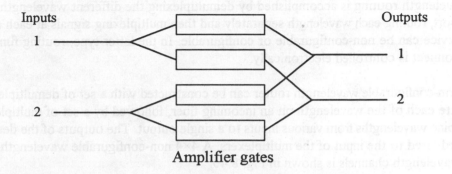

Fig. 8.39: 2×2 amplifier gate switch

8.8.2 NON-CONFIGURABLE WAVELENGTH ROUTER

A wavelength router can route signals arriving at different input fibers (ports) of the device to different output fibers (ports) depending on the wavelength of signals.

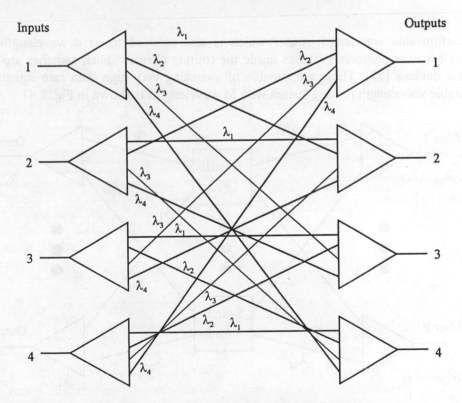

Fig. 8.40: 4×4 non-configurable wavelength router

Wavelength routing is accomplished by demultiplexing the different wavelengths from each input port, routing each wavelength separately and then multiplexing signals at each output port. The device can be non-configurable or configurable. In the latter type, routing function of the crossconnect is controlled electronically.

A non-configurable wavelength router can be constructed with a set of demultiplexers, which separate each of the wavelength on an incoming fiber, followed by a set of multiplexers which recombine wavelengths from various inputs to a single output. The outputs of the demultiplexers are hardwired to the input of the multiplexers. A 4×4 non-configurable wavelength router with four wavelength channels is shown in Fig. 8.40.

The router is non-configurable because the path of a given wavelength channel is fixed after it enters the router on a particular input fiber. The routing matrix which characterizes the router i.e., which wavelength on which input port gets routed to which output port is determined by the internal connections between the demultiplexers and multiplexers [7].

8.8.3   CONFIGURABLE WAVELENGTH ROUTER

A configurable wavelength router which is also referred to as a wavelength-selective crossconnect uses photonic switches inside the routing element. Such switches are based on relational devices [28]. These are capable of switching very high data rate signals. A P×P configurable wavelength routing switch with M wavelengths is shown in Fig. 8.41.

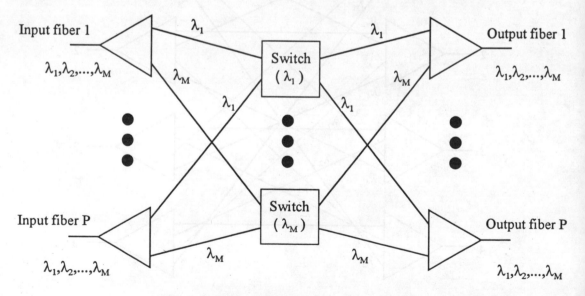

Fig. 8.41: P×P configurable wavelength router

A network built with such switches is more flexible because of the additional control in setting up connections than the passive, non-configurable wavelength routed network. The routing depends on both the wavelength chosen at the source node and the configuration of the switches in the network node [7].

8.8.4   ADD-DROP MULTIPLEXERS

In a WDM network, there is only a small portion of traffic that is destined for a node, but to take out this portion all the traffic must be processed in the node. With the advent of optical add-drop multiplexers (ADMs), the switching node could be relieved of processing load as the wavelength of the beam performs the function of routing. The wavelength of the optical carrier serves as a header that spatially routes itself to the destination. Processing is no longer necessary and hence data rate in the network is not determined by the fastness of the electronics in the node. After dropping a specific wavelength, it can be processed locally. New information can be imposed and eventually the signal can be added to the network. Optical ADM is analogous to adding and retrieving information from a certain time slot in the electrical domain.

Like wavelength sensitive OXCs (see Chapter fourteen), optical ADMs play an important role in WDM networks. The main advantages of the optical ADM are: (i) high tunability, (ii) flexibility to select large number of channels without significant crosstalk, (iii) low attenuation, (iv) insensitivity to polarization fluctuations, (v) high stability and (vi) relatively low cost. Depending on the flexibility, optical ADMs are divided into following three types.

(a) ADM WITH PERMANENT ADD-DROP PATTERN

The simplest approach to realize an optical ADM is the utilization of commercially available state-of-the-art WDM devices (Fig. 8.42).

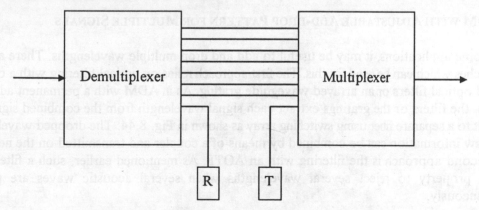

Fig. 8.42: ADM with permanent add-drop pattern

The demultiplexer separates out the combined signal and assigns the different wavelengths to separate fibers. After processing the desired signal, a multiplexer reunites the signals again. This process is quite inefficient. Further, it can not adjust to flexible traffic since only one fixed wavelength can be dropped and added.

(b) ADM WITH ADJUSTABLE ADD-DROP PATTERN FOR ONE SIGNAL

To achieve higher flexibility in the network, tunable optical filter is used in the ADM in place of WDM multiplexer (refer to Fig. 8.43). The filter not only avoids to extract all the signals, but also provide flexibility in the selection of a signal.

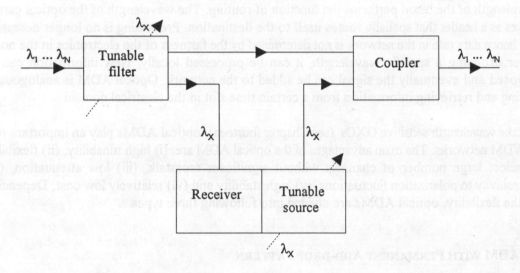

Fig. 8.43: ADM with adjustable add-drop pattern for one signal

(c) ADM WITH ADJUSTABLE ADD-DROP PATTERN FOR MULTIPLE SIGNALS

In some applications, it may be useful to add and drop multiple wavelengths. There are two approaches which can be used for this. The first approach relies on optical filtering with a cascade of fixed optical filters or an arrayed waveguide grating. As in ADM with a permanent add-drop pattern, the filters or the gratings extract each signal wavelength from the combined signal and assign it to a separate fiber using switching array as shown in Fig. 8.44. The dropped wavelengths with new information can be combined by means of a coupler and transmitted on the network. The second approach is the filtering with an AOTF. As mentioned earlier, such a filter has a unique property to reject several wavelengths when several acoustic waves are present simultaneously.

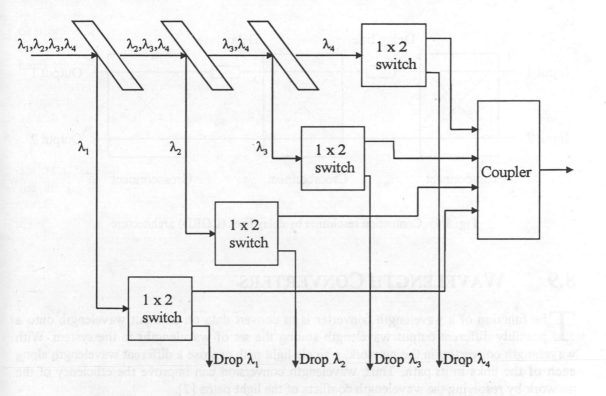

Fig. 8.44: ADM with adjustable add-drop pattern for multiple signals

8.8.5   PHOTONIC PACKET SWITCH

In a packet-switched system, there exists the problem of resource contention when multiple packets contend for a common resource in the switch. In an electronic system, contention may be resolved through the use of buffering, while in the optical systems contention resolution is a more complex issue since it is difficult to implement components that can store optical data. A number of switching architectures that use delay lines to implement optical buffering have been proposed. The contention resolution by delay lines (CORD) architecture [50] consists of a number of 2×2 crossconnect elements and delay lines as shown in Fig. 8.45.

Each delay line functions as a buffer for a single packet when two packets contend for the same output port, one packet may be switched to a delay line while the other packet is switched to the proper output. The packet that was delayed can then be switched to the same output after first packet has been transmitted.

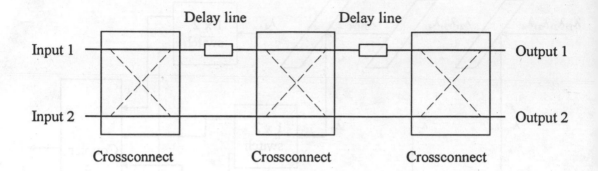

Fig. 8.45: Contention resolution by delay lines (CORD) architecture

8.9   WAVELENGTH CONVERTERS

The function of a wavelength converter is to convert data on an input wavelength onto a possibly different output wavelength among the set of wavelengths in the system. With wavelength converters in the network, a single light path can use a different wavelength along each of the links in its path. Thus, wavelength conversion can improve the efficiency of the network by resolving the wavelength conflicts of the light paths [7].

An ideal wavelength converter should have the following characteristics: (i) transparency to bit rates and signal formats, (ii) fast set up time of output wavelength, (iii) conversion of both shorter and longer wavelengths, (iv) moderate input power levels, (v) possibility for same input and output wavelengths i.e., no conversion, (vi) insensitivity to input signal polarization, (vii) low chirp output signal with high extinction ratio and large signal-to-noise ratio and (viii) simple implementation [18]. The wavelength conversion techniques can be broadly classified into two types: optoelectronic wavelength conversion, in which the optical signal must first be converted into an electronic signal and all-optic wavelength conversion, in which the signal remains in the optical domain. These are discussed below.

8.9.1   OPTOELECTRONIC

In this method, optical signal is first translated into the electronic domain using a photodetector as shown in Fig. 8.46.

The electronic bit stream is stored in the buffer (labelled FIFO for the first-in-first-out queue mechanism). The electronic signal is then used to drive the input of a tunable laser tuned to the desired output wavelength. This method has been demonstrated for bit rates up to 10 Gbit/s [57]. The disadvantages are (i) optoelectronic conversion adversely affects the transparency of the signal requiring optical data to be in a specific modulation format and at a specified bit rate and (ii) all information in the form of phase, frequency and analog amplitude of the optical signal is lost during the conversion process [7]. A less complex optoelectronic converter based on simple photodetector and laser (without FIFO and address decoder) is often used in practice.

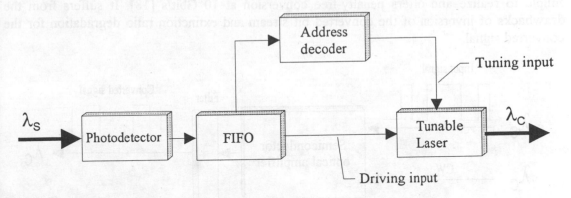

Fig. 8.46: Schematic of an optoelectronic wavelength converter

8.9.2   ALL-OPTICAL

All-optical conversion methods are divided into two types: one employ the coherent effect and the other cross-modulation. Wavelength converters based on coherent effect rely on wave-mixing effect which arises from a non-linear optical response of a medium when more than one waves are present say CW pump and input signal at frequency f_p and f_s respectively. It results in the generation of another wave at frequency $f_c = (n-1)f_p - f_s$ and whose intensity is proportional to the product of the interacting wave intensities. This process preserved both phase and amplitude information, offering strict transparency. Further, it allows simultaneous conversion of a set of multiple input wavelengths to another set of multiple output wavelengths and could potentially accommodate a signal with bit rates exceeding 100 Gbit/s.

The n = 3 corresponds to four-wave mixing (FWM). It is a consequence of third-order non-linearity which causes three optical wave frequencies f_i, f_j and f_k (k≠i, j) to generate a fourth wave of frequency $f_i + f_j - f_k$. Similarly, n = 2 corresponds to difference frequency generation (DFG), which is a consequence of a second-order non-linear interaction of a medium with two optical waves: a pump wave and a signal wave [57]. The DFG technique offers a full range of transparency without adding excess noise to the signal. It is also bidirectional and fast. The disadvantages are low efficiency and high polarization sensitivity. The main difficulties in the implementation of this technique lie in the phase matching of interacting waves and in fabricating a low loss waveguide for high conversion efficiency [7]. The wavelength converters based on cross-modulation utilize active semiconductor optical amplifier (SOA) in the cross-gain modulation (XGM) or in the cross-phase modulation (XPM) mode. In the XGM mode, an intensity modulated input signal modulates the gain in the SOA due to gain saturation. A CW signal at the desired output wavelength λ_c is modulated by the gain variation so that it carries the same information as the original input signal (refer to Fig. 8.47). The CW signal can be launched into the SOA either in the same direction as the input signal (codirectional) or in the opposite direction (counterdirectional). This scheme is simple to realize and offers penalty-free conversion at 10 Gbit/s [18]. It suffers from the drawbacks of inversion of the converted bit stream and extinction ratio degradation for the converted signal.

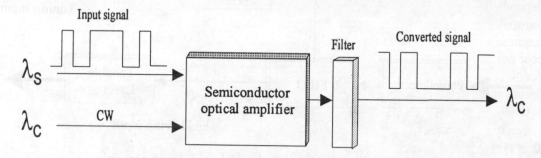

Fig. 8.47: Wavelength converter based on XGM in semiconductor optical amplifier (codirectional)

The other method based on XPM rely on the fact that RI of the SOA is dependent on the carrier density in its active region. An incoming signal that depletes the carrier density will modulate the RI and thereby results in phase modulation of CW signal (wavelength λ_c). As an example, structure of MZI converter is shown in Fig. 8.48.

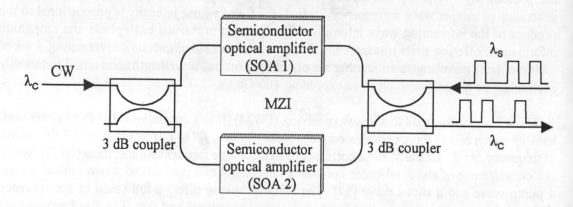

Fig. 8.48: Interferometric wavelength converter based on XPM in semiconductor optical amplifiers

The SOAs are placed in asymmetric configurations so that the phase change in the two amplifiers is different. As a result, the CW light is modulated according to the phase difference.

The asymmetric splitters ensure that an intensity modulated phase difference is achieved between the interferometer arms due to different saturation of SOA1 and SOA2. With the XPM scheme, the converted output signal can be either inverted or non-inverted unlike in XGM scheme where the output signal is always inverted. The XPM scheme is more power efficient as compared to XGM scheme [18].

The actual choice of a technology to be employed for wavelength conversion in a network depends on the requirements. It is clear that optoelectronic converters offer only limited digital transparency. Moreover, deploying multiple optoelectronic converters in a WDM crossconnect requires sophisticated packaging to avoid crosstalk among channels. This leads to an increased cost per converter, making this technology even less attractive than all-optical converters [57]. Among all-optical converters, converters based on SOAs (XGM and XPM) are better suited for practical systems [7].

8.10 OPTICAL BISTABILITY AND DIGITAL OPTICS

An optical bistable device can be regarded as an optical switch. It has two stable states 'on' and 'off'. The process of going from one state to another state is non-linear. It is because of the fact that when the energy supplied passes a certain threshold, the device suddenly jumps to the state where it remains. The threshold for switching up as shown in Fig. 8.49 is higher than the threshold for switching down. Thus, the switch will stay in a given state even if the force that brought it there is removed. The region between these thresholds is the bistable region.

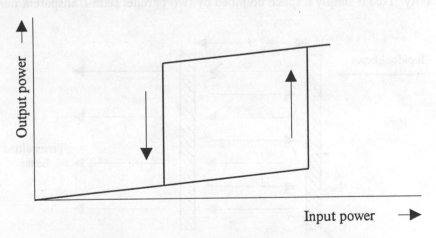

Fig. 8.49: Input-output characteristic of a bistable device

The criterion for evaluating the performance of bistable optical devices (BODs) is as follows [20].

- Switching time: The amount of time required to switch from one state to the other state, to switch back and delay before another switching cycle can start.

- Switching energy: It is the energy needed to switch and how much of it is dissipated in the device?

- Stability: It is the stability of two states and specifically the possibility of spontaneous switching.

The BODs have the potential of very fast switching at low power levels. A BOD exhibiting picosecond switching time using only picojoules of energy would prove far superior to an electronic device which performs the same function. Furthermore, by careful adjustments of the device bias and input level, the BODs can provide logic functions like AND-gate, OR-gate, NOT-gate [56]. These can also be used for memory elements (flip-flop), power limiters and pulse shapers, differential amplifiers and A-D converters [20].

8.10.1   FABRY-PEROT RESONATOR BISTABLE DEVICE

An intrinsic FP resonator bistable device is shown in Fig. 8.50. It basically consists of a resonant cavity. This is simply a space bounded by two parallel semi-transparent mirrors.

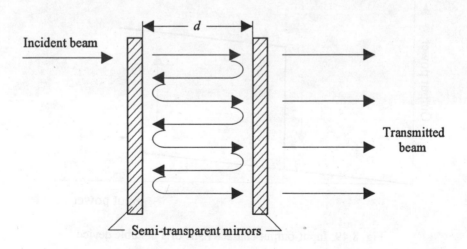

Fig. 8.50: Incident and transmitted beams in a Fabry-Perot resonator

The light which enters through one of the mirrors bounces back and forth between the two mirrors. At each reflection, some of the light is emitted through the mirror. The transmitted beam is therefore composed of rays that have bounced back and forth at different times. If there is a constructive interference between all these rays, the transmitted light intensity will be high. Otherwise, the rays will cancel each other and the transmitted intensity will be low. The condition for constructive interference and high transmission is that path difference for various rays should be an integral multiple of wavelength. Therefore, the resonant condition is

$$2d = m\lambda \tag{8.35}$$

where d is the cavity length and m an integer. Wavelengths (frequencies) that fulfil the above condition are called resonant wavelengths (frequencies). It is obvious that high transmission occurs only at the resonant frequencies. It may be mentioned that high transmitted intensity is coupled with intensity inside the resonator. When constructive interference occurs outside, it also occurs inside [20].

To create a bistable device, a material with an optical non-linearity is placed in the resonant cavity as shown in Fig. 8.51. A typical non-linearity is the change of RI with intensity i.e., the photorefractive effect. When the intensity of light passing through such a material changes, so does its refraction index (RI). When the RI becomes larger, the light passing through it becomes slower. Therefore, the effective path difference between successive beams becomes longer. The light wave must now have a longer wavelength or a lower frequency to fulfil the resonant condition given in Eq. (8.35).

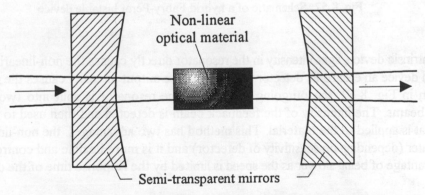

Non-linear optical material

Semi-transparent mirrors

Fig. 8.51: Schematic of a intrinsic Fabry-Perot bistable device

The bistable behaviour of the device can be explained as follows. In the stable state of low transmission, the illuminating beam of light is somewhat off resonance. When its intensity is increased, the characteristic resonant frequency of the cavity changes and becomes nearer to the frequency of the illumination. At a certain intensity level, it is near enough for the interference between the beams to become constructive and an intensity build up occurs inside the resonator. When the intensity is lowered far enough, the intensity in the resonator will not suffice to keep the resonator in tune with the input frequency. As soon as the tuning is lost, there is no more constructive interference inside the resonator. The intensity in the resonator decreases sharply and the device switches back to low transmission state [20].

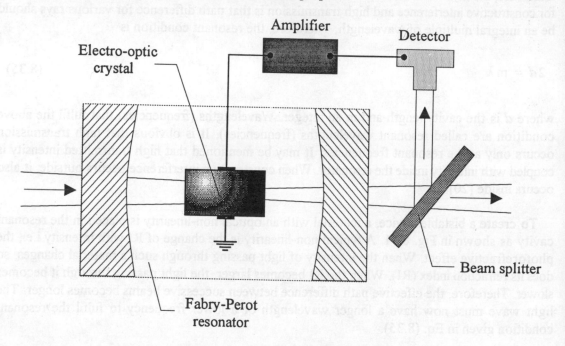

Fig. 8.52: Schematic of a hybrid Fabry-Perot bistable device

In an intrinsic device, light intensity in the resonator directly causes the non-linearity. However, in a hybrid device an electronic detector that detects the output intensity causes the non-linearity. As shown in Fig. 8.52, the outgoing beam from the resonator is split into two: output and feedback beams. The intensity of the feedback beam is detected and then used to modulate the voltage that is applied to the material. This method has two advantages: the non-linearity can be much greater (depending on sensitivity of detector) and it is more flexible and controllable. It has the disadvantage of being slower as the speed is limited by the response time of the detector [20].

8.10.2   SELF-ELECTROOPTICAL EFFECT BISTABLE DEVICE

An alternate hybrid approach is based on the use of materials constructed by alternating thin layers of two different semiconductor materials which exhibit non-linear properties. Combinations used include arsenide and gallium aluminium arsenide; mercury telluride and cadmium telluride;

silicon and indium phosphide. It resulted in the development of the so-called self-electrooptical effect device (SEED) which exhibits hysteresis and bistable characteristics [31].

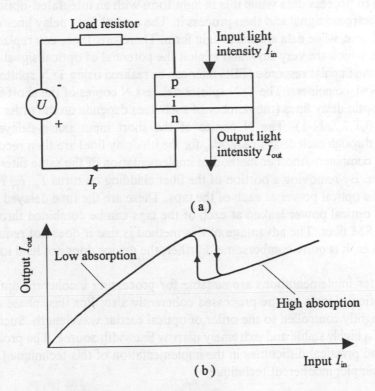

(a)

(b)

Fig. 8.53: (a) Schematic structure and (b) input/output characteristic of a self-electrooptical effect device (SEED) [31]

The schematic of a simple SEED is shown in Fig. 8.53. The operation of the device can be described as follows. Let initially no light is incident on the photodetector. There will not be any photocurrent (except dark current) and the power supply voltage essentially appears across the photodetector. As the input light intensity is increased, the photocurrent causes a voltage drop across the resistor. As a result, voltage across the photodetector decreases. This decrease in voltage causes an increase in the absorption which in turn cause an increase in photocurrent. This increase in photocurrent causes a larger drop across the resistor, further reduction in voltage across the photodetector, further increase in absorption and further increase in photocurrent. This will continue until the photodetector approaches forward bias (nearly 0V). The overall effect is that the device switches abruptly from a high to low voltage state. This switching of high to low voltage state corresponds to the optical output being switched from high to low optical state.

It may be mentioned that in the above SEED, referred to as resistor biased SEED, low inputs are transmitted and high inputs are blocked. In the resonator devices, the opposite is true i.e., low intensities are blocked and high ones are transmitted [20].

8.10.3   FIBER OPTIC DELAY LINE

It is preferable to process data while it is in light form with an integrated-optic circuit than to transform it to electronic signal and then process it. The optical fiber delay lines can be used for processing in real time, while data are still in light form. Therefore, these can replace digital signal processing stages which are very slow and exploit the potential of optical signal processing [2]. For example, a finite impulse response (FIR) filter can be realized using $1 \times N$ splitter, N fiber optic delay lines and $N \times 1$ combiners. The $1 \times N$ splitter delivers N copies of the short input pulse into a set of N fiber optic delay lines (the number of such lines depends on the order of the filter) of lengths L_i (i = 0,1,2,....N-1). The N copies of the short input pulse delayed by different propagation time through each delay line ($L_i n/c_0$ for the ith delay line) are then recombine through a $N \times 1$ fiber optic combiner. Another method of implementation of the same filter is to wind SM fiber onto a drum. By removing a portion of the fiber cladding on turns T_x, T_y, T_z etc., one can tap a portion of the optical power at each of the taps. These are the time delayed versions of the input pulse. The optical power leaked at each of the taps can be combined through a lens and launched into a SM fiber. The advantage of this method is that it does not require splitter and combiner. However, it is quite cumbersome. Further, the device does not lend itself to compact packaging.

The above filter implementations are suitable for processing incoherent optical signals. In principle, the optical signals can be processed coherently also. For this, phase of each tapped signal should be tightly controlled to the order of optical carrier wavelength. Such precision will require the use of a highly stable and extremely narrow linewidth source. The problem of finding such a source and practical difficulties in the implementation of this technique [32] lead to the choice of much simpler incoherent technique.

8.11   REFERENCES

[1]   Andonovic, I.; Uttamchandani, D.: Principles of Modern Optical Systems. Artech House, 1989.

[2]   Anemogiannis, E.; Kenan, R. P.: Integrated optical architectures for tapped delay line. IEEE J. LT-8(1990)8, 1167-1176.

[3]   Arai, H.; Uetsuka, H. et. al.: Fabrication of planar lightwave circuit 1.48μm/1.55μm WDM for EDFA. IEICE Japan National Conf. Rec. on Commun., No.C-213, 1993.

[4]   Arkwright, J. W. et. al.: 7×7 monolithic single-mode star coupler. Electron. Lett. 26(1990)18, 1534-1535.

[5]   Arkwright, J. W. et. al.: Monolithic 1×19 single-mode fused fiber couplers. Electron. Lett. 27(1991)9, 737-738.

[6]   Bergh, R. A.; Lefevre, H. C.; Shaw, H. J.: Single-mode fiber-optic polarizer. Opt. Lett. 5(1980)11, 479-481.

[7]   Borella, Michael S. et. al. : Optical components for lightwave networks. Proc. IEEE, 85(1997)8, 1274-1307.

[8]   Born, M.; Wolf E.: Principles of Optics. Pergamon Press, 1974.

[9]   Boucouvalas, A. C.; Georgiou, G.: A method of beam forming and fabricating optical fiber gap devices. Proc. Euro. Conf. Opt. Commun., Barcelona, Spain, 1986, 361-364.

[10]   Cheung, K. W. et. al.: Electronic wavelength tuning using acousto-optic tunable filter with broad continuous tuning range and narrow channel spacing. IEEE Photon Technology Lett. 1(1989)2, 38-40.

[11]   Chlamtac, I.; Fumagalli, A.; Kazousky, L. G. et. al.: CORD: contention resolution by delay lines. IEEE J. SAC- 14(1996)6, 1014-1029.

[12]   Chraplyvy, A. R. et. al.: Narrowband tunable optical filter for channel selection in densely packed WDM systems. Electron. Lett. 22(1986)20, 1084-1085.

[13]   Cush, R.; Goodwin, M.: Optical logic devices. National Electronics Review, U.K., 1987, 51-55.

[14]   Digonnet, M.; Shaw, H. J.: Wavelength multiplexing in single-mode fiber couplers. Applied Optics 22(1983)3, 484-491.

[15]   Dorn, S. Van.: Fiber optic couplers. Proc. SPIE, 574(1985), 2-8.

[16]   Dragone, C.; Henery, C. H. et. al.: Efficient multichannel integrated optics star coupler on silicon, IEEE Photon. Technol. Lett. 1(1989)8, 241-243.

[17]   Dragone, C.: Efficient N×N star coupler based on Fourier optics. IEEE J. LT-7(1989)3, 479-489.

[18]   Durhuus, T. et. al.: All-optic wavelength conversion by semiconductor optical amplifiers. IEEE J. LT-14(1996)6, 942-954.

[19]   Dyott, R. B.; Schrank, P. F.: Self locating elliptical core fiber with an accessible guiding region. Electron. Lett. 18(1982), 980-981.

[20]   Feitelson, Dror G.: Optical Computing: A Survey for Computer Scientists. MIT Press, 1990.

[21]   Fujii, Y.; Minowa, J.; Tanada, H.: Practical two-wavelength multiplexer and demultiplexer: design and performance. Applied Optics, 22(1983), 3090-3097.

[22]   Green, Paul E. Jr.: Fiber Optic Networks. Prentice-Hall, 1993.

[23]   Gruchmann, D.; Petermann, K. et. al.: Fiber-optic polarizers with high extinction ratio. European Conf. on Optical Communications ECOC, (1983), 305.

[24]   Hanaizumi, O.; Miyagi, M.; Minakata, M.; Kawakami, S.: Antenna coupled Y junctions in 3-dimensional dielectric waveguide. European Conf. on Optical Communications ECOC, (1985), 179.

[25]   Hecht, Jeff: Understanding Fiber Optics. Second Edition, Howard W. Sams&Co., 1997.

[26]   Herrmann, H. et. al.: Integrated optical, TE- and TM-pass, acoustically tunable, double-stage wavelength filters in LiNbO$_3$, Electron. Lett. 28(1992)7, 642-644.

[27]   Hill, A. M. et. al.: Linear crosstalk in wavelength division multiplexed optical-fiber transmission systems. IEEE J. LT-3(1985)3, 643-651.

[28]   Hinton, H. S.: Photonic switching fabrics. IEEE Commun. Magazine, 28(1990)4, 71-89.

[29]   Inoue, A. et. al.: Fabrication and application of fiber Bragg grating-A review. Optoelectron. Devices Technol. 10(1995)3, 119-130.

[30]   Iwamura, H.; Iwasaki, H. et. al.: Simple-polarization-independent optical circulator for optical transmission systems. Electron. Lett. 15(1979), 830.

[31]   Jager, D.; Forsman, F.; Zhai, H. C.: Hybrid optical bistability based on increasing absorption in depletion layer of an Si Schottky SEED device. Electron. Lett. 23(1987)10, 490-491.

[32]   Kar, Subrat.: Code division multiple access in fiber optic networks. IETE J. Education, 38(1997)3&4, 167-173.

[33]   Keiser, G.: Optical Fiber Communications. Second Edition, McGraw-Hill, 1991.

[34]   Kobrinski, H. et. al.: Wavelength selection with nanoseconds switching times using DFB laser amplifiers. Electron. Lett. 24(1988)15, 969-970.

[35] Kominato, T.; Yasu, M. et. al.: Optical WDM circuits with guided wave Mach-Zehnder interferometer configuration. IEICE Japan National Conf. Rec. on Commun., No.C-502, 1989.

[36] Lauridsen, V. R.; Tadayoni, R. et. al.: Gain and noise performance of fiber amplifiers operating in new pump configurations. Electron. Lett. 27(1991)4, 327-329.

[37] Lawson, C. M.; Kopera, P. M. et. al.: In-line single-mode wavelength division multiplexer/demultiplexer. Electron. Lett. 20(1984)23, 963-964.

[38] Mallison, S. R.: Wavelength selective filters for single-mode fiber WDM systems using Fabry-Perot interferometers. Applied Optics, 26(1987)3, 430-436.

[39] Marhic, M. E.: Combined star couplers for single-mode optical fibers. Proc.FOC/LAN 84, Boston, 1984, 175-179.

[40] Mestdagh, Denis J. G.: Fundamental of Multiaccess Optical Fiber Networks. Artech House, 1995.

[41] Miller, C. M. et. al.: Wavelength locked two stages fiber Fabry-Perot fiber for dense wavelength division demultiplexing in an erbium-doped fiber amplifier spectrum. Electron. Lett. 28(1992)3, 216-217.

[42] Miyauchi, E.; Iwama T. et. al.: Compact wavelength multiplexer using optical fiber pieces. Optics Lett. 5(1980)7, 321-325.

[43] Mortimore, D. B.: Wavelength-flattened fused couplers. Electron. Lett. 21(1985)17, 742-743.

[44] Nishi, S.; Aida, K.; Nakagawa, K.: Highly efficient configuration of erbium-doped fiber amplifier. Proc. 16th European Conf. on Optical Commun. ECOC, Vol. I (1990), 99.

[45] Norio Kashima: Passive Optical Components for Optical Fiber Transmission. Artech House, 1995.

[46] Okamoto, K.; Takahashi, H. et. al.: Design and fabrication of integrated optic 8×8 star coupler. Electron. Lett. 27(1991)9, 774-775.

[47] Scobey, M. A.; Spock, D. E.: Passive WDM components using microplasma optical interference filters. OFC'96, Tech. Dig. San Jose, CA, pp.242-243.

[48] Senior, J. M.: Optical Fiber Communication-Principles and Practice. Second Edition, Prentice-Hall, 1992.

[49] Sheem, S. K.; Moeller, R. P.: Single-mode fiber wavelength multiplexer. J. Applied Phys. 51(1980)8, 4050-4052.

[50] Shiraishi, K.: New configuration of polarization-independent isolator using a polarization-dependent one. Electron. Lett. 27(1991)4, 302-303.

[51] Smith, D. A.; Baran, J. E.; Johnson, J. J.; Cheung, K. W.: Integrated-optic acoustically-tunable filters for WDM networks. IEEE J. SAC-8(1990)6, 1151-1159.

[52] Stone, J. et. al.: Pigtailed high-finesse tunable fiber Fabry-Perot interferometers with large, medium and small free spectral range. Electron. Lett. 23(1987)15, 781-783.

[53] Takahashi, H. et. al.: Integrated-optic 1×128 power splitter with multifunnel waveguide. IEEE Photon Technol. Lett. 5(1993)1, 58-60.

[54] Takato, N. et. al.: 128-channel polarization insensitive frequency-selection-switch using high silica waveguide on Si. IEEE Photon Technol. Lett. 2(1990)6, 441-443.

[55] Toba, H.; Nosu, K.O.; Takato, N.: Factors affecting the design of optical FDM information distributing systems. IEEE J. SAC-8(1990)6, 965-972.

[56] Yokohama, I.; Noda, J.; Okamoto, K.: Fiber coupler fabrication with automatic fusion elongation processes for low excess loss and high coupling-ratio accuracy. IEEE J. LT-5(1987)7, 910-915.

[57] Yoo, S. J. B.: Wavelength conversion technologies for WDM network applications. IEEE J. LT-14(1996)6, 955-966.

9 FUNDAMENTALS OF OPTICAL COMMUNICATION SYSTEMS

This Chapter provides the fundamental background required for the study of advanced incoherent optical communication systems with intensity modulation (IM) and direct detection (DD) referred as IM/DD systems as well as coherent systems with more sophisticated modulation and detection schemes. The principle function of incoherent receiver (also referred as direction detection receiver), coherent receiver and an introduction to the basic laws have been described in Section 9.1. In order to assess the performance of coherent receivers, it is useful to compare their features with a reference receiver. For this purpose, direct detection receiver is well-suited since this receiver is well-developed and established in commercial optical communication networks [6, 7, 13] all over the world. Therefore, Section 9.1 begins with a review of conventional direct detection receiver. A more detailed analysis of incoherent and coherent systems is given in Chapters ten and eleven. In coherent systems, there are two different receiver configurations: heterodyne and homodyne. Both concepts are briefly described in Section 9.2 and discussed in detail in Chapter eleven. We then analyse the performance gain of direct detection and coherent receivers in terms of receiver sensitivity and selectivity. Section 9.2 also presents fundamental knowledge of multichannel systems based on wavelength division multiplexing (WDM). Sections 9.3 and 9.4 give an overview of various components and signals in optical fiber systems, in particularly in coherent systems. Finally, Section 9.5 reviews some important aspects of digital communication systems, which are required to analyse optical communication systems. The eye pattern technique and probability of error, also referred as bit error rate (BER), are refreshed as two important and powerful measures for systems comparison.

9.1 PRINCIPLE OF DIRECT AND COHERENT DETECTION

In a *direct detection receiver*, incoming optical light wave electric field E_r at the receiver input is directly converted to an electrical current by means of a PIN-photodiode. When E_r is quite weak, an avalanche photodiode (APD) is often used. The photodiode current in both the cases is directly proportional to the square of E_r and, therefore, proportional to optical power P_r incident upon the photodiode. Let us presume that surface dark current in APD is negligible. With this, dark current I_{dark} becomes equal to bulk dark current I_{db}. The APD current is given by

$$I_{APD} \sim P_r \sim E_r^2$$

$$I_{APD} = M R_0 P_r \tag{9.1}$$

Here, R_0 [A/W] represents the responsivity and M the average gain of photodiode. The responsivity R_0 is given by the well-known expression [13]

$$R_0 = \frac{e\,\eta}{h\,f} \tag{9.2}$$

where e is the electron charge, η the quantum efficiency of photodiode, h the Planck's constant and f the frequency of light.

(a) Direct detection receiver

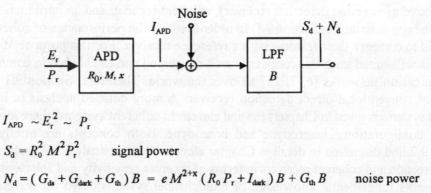

$$I_{APD} \sim E_r^2 \sim P_r$$

$$S_d = R_0^2\,M^2 P_r^2 \qquad \text{signal power}$$

$$N_d = (\,G_{ds} + G_{dark} + G_{th}\,)\,B = e\,M^{2+x}\,(\,R_0\,P_r + I_{dark}\,)\,B + G_{th}\,B \qquad \text{noise power}$$

(b) Coherent detection receiver

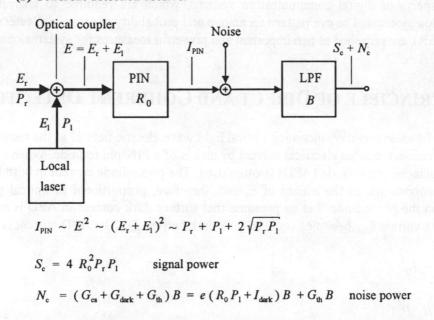

$$I_{PIN} \sim E^2 \sim (\,E_r + E_1\,)^2 \sim P_r + P_1 + 2\sqrt{P_r P_1}$$

$$S_c = 4\,R_0^2 P_r P_1 \qquad \text{signal power}$$

$$N_c = (\,G_{cs} + G_{dark} + G_{th}\,)\,B = e\,(\,R_0\,P_1 + I_{dark}\,)\,B + G_{th}\,B \qquad \text{noise power}$$

Fig. 9.1: Principle block diagram and basic laws of optical (a) direct and (b) coherent detection

It can be seen from Fig. 9.1a that the photodiode current I_{APD} is disturbed by noise with constant power spectral density (psd) $G_d(f) = G_d$. In an optical receiver, this additive noise arises from shot noise (G_{ds}) of the photodiode and electronic or thermal noise of the amplifiers and resistors (G_{th}) frequently referred as circuit noise. For simplicity, amplifiers are not added to the block diagram in Fig. 9.1. The shot noise consists of two components: one component depends on the received optical power level (G_{dp}), while the other component is independent of it (G_{dark}). The latter one is caused by the dark current I_{dark} of photodiode. The overall double-sided psd is given by

$$G_d(f) = G_d = G_{ds} + G_{th} = G_{dp} + G_{dark} + G_{th}$$

$$= e\, M^2\, F(M) \left[R_0\, P_r + I_{dark} \right] + G_{th} \tag{9.3}$$

Here, suffix d represents direct detection and will only be used if there is a difference to coherent detection. In most practical applications, excess noise factor $F(M)$ can be approximated by

$$F(M) \approx M^x \tag{9.4}$$

where x is the excess noise exponent which depends on the photodiode material (refer to Chapter six). If B is the double-sided noise-equivalent bandwidth of the low-pass filter, then signal power, noise power and signal-to-noise ratio (it is abbreviated as SNR or S/N ratio also) will be

$$S_d = \left(R_0\, M\, P_r \right)^2 \tag{9.5}$$

$$N_d = e\, M^{2+x} \left(R_0\, P_r + I_{dark} \right) B + G_{th}\, B \tag{9.6}$$

and

$$SNR_d = \left(\frac{S}{N} \right)_d = \frac{\left(R_0\, M\, P_r \right)^2}{e\, M^{2+x} \left(R_0\, P_r + I_{dark} \right) B + G_{th}\, B} \tag{9.7}$$

Sometimes, N_d and SNR_d are expressed in terms of single-sided psd. In that case, double-sided noise-equivalent bandwidth will be replaced by single-sided bandwidth. It can be seen from Eqs. (9.5) and (9.6) that signal and noise power are related to a reference resistor of 1 Ω. Hence, S_d and N_d are expressed in Amperes squared (A^2) instead of Watts (W). In order to simplify calculation and focus our interest on the principle in this Section, effect of low-pass filter on signal power S_d is not considered here. For this reason, we assume a constant, time-invariant optical power P_r at the input to the receiver and, consequently, a constant shot noise psd G_{ds}. The received optical power P_r is actually a function of time since intensity modulation is normally applied to the transmitter laser source. Therefore, optical systems with intensity modulation and direct detection (IM/DD) are characterized by *signal-dependent noise*.

Next we consider the coherent receiver shown in Fig. 9.1b. Here, incoming optical signal and signal from a local laser are superimposed on the surface of photodiode. This is in contrast to direct detection receiver wherein received signal is directly made to incident on the photodiode. A coherent receiver is called *homodyne receiver* if the frequencies of received and local laser light waves are equal and *heterodyne receiver* if these are different. In Sections 9.3 and 9.4 and in Chapter eleven, heterodyne and homodyne receivers have been considered in more detail. The superposition of light waves is simply described by

$$E = E_r + E_l \tag{9.8}$$

where E_l is the electric field of local laser light wave and P_l the corresponding power level. Like in direct detection receiver, photodiode current is again proportional to optical power at the input of photodiode. Assuming a PIN-photodiode (APD is generally not required in coherent receivers as shown below in Section 9.2.1), current at the output of the photodiode is given by

$$I_{PIN} \sim \left(E_r + E_l\right)^2 \sim \left(P_r + P_l + 2\sqrt{P_r P_l}\right)$$

$$I_{PIN} = R_0 \left(P_r + P_l + 2\sqrt{P_r P_l}\right) \tag{9.9}$$

The photodiode current I_{PIN} has three components, but only the third component

$$I = 2 R_0 \sqrt{P_r P_l} \tag{9.10}$$

is of interest in coherent receivers. With respect to frequency, this mixing produces a baseband signal if homodyne receiver is used (this case is assumed here), whereas it is an intermediate frequency signal in case of heterodyne receiver. The intermediate frequency (IF) itself is equal to the difference in local laser and received optical light wave frequencies. The first and also the second components in Eq. (9.9) can practically be eliminated by means of filtering or by using a balanced receiver with two photodiodes instead of one (Section 9.4). Similar to the constant noise psd G_d of a direct detection receiver, noise psd in coherent optical heterodyne or homodyne receiver is given by

$$G_c(f) = G_c = G_{cs} + G_{th} = G_{cp} + G_{dark} + G_{th}$$
$$= e\left[R_0 P_1 + I_{dark}\right] + G_{th} \tag{9.11}$$

In the strict sense, P_r must also be considered in addition to P_1 to determine noise psd from Eq. (9.11). In most practical applications, however, this is not required because P_r is always some order of magnitudes lower than P_1 i.e.,

$$P_r \ll P_1 \tag{9.12}$$

This important relationship is generally valid due to attenuation of the optical link (for example, fiber attenuation). Let us consider now signal power, noise power and SNR. We obtain

$$S_c = 4\,R_0^2\,P_r\,P_1 \tag{9.13}$$

$$N_c = e\left(R_0\,P_1 + I_{dark}\right)B + G_{th}\,B \tag{9.14}$$

and

$$SNR_c = \left(\frac{S}{N}\right)_c = \frac{4\,R_0^2\,P_r\,P_1}{e\left(R_0\,P_1 + I_{dark}\right)B + G_{th}\,B} \tag{9.15}$$

It may be mentioned that this equation is valid only for the simplified coherent receiver configuration shown in Fig. 9.1b. From Eq. (9.14), it can be seen that noise power N_c at the output of coherent receiver is independent of the received power level P_r and applied modulation. Hence, coherent optical systems are characterized by *signal-independent noise* which is an advantage in comparison to direct detection system. However, in presence of laser phase noise this advantage no longer exists (Chapter eleven).

9.2 RECEIVER PERFORMANCE

In this Section, performance of incoherent and coherent detection receiver is analysed in terms of sensitivity (Section 9.2.1) and selectivity (Section 9.2.2). The latter one is an important issue when realizing multichannel systems based on wavelength division multiplexing (WDM) or optical frequency division multiplexing (OFDM), whereas the first criterion is more important in long-range communication links. With the availability of powerful optical amplifiers (Chapter seven), receiver sensitivity has become usually less important than selectivity in advanced optical fiber networks. In optical free-space communication (Chapter fifteen), however, a high receiver sensitivity still remains an important issue, whereas selectivity is less important.

9.2.1   SENSITIVITY

Comparing the signal power S_d at the output of direct detection receiver (Eq. 9.5) and signal power S_c at the output of coherent receiver (Eq. 9.13), it becomes clear that the avalanche gain M of an APD is merely replaced by the capability of the local laser light power P_l to amplify the signal in the same way: $S_d \sim M^2$ and $S_c \sim P_l$. Difference in noise power N_d and N_c is, however, much more predominant. In direct detection receiver, signal-dependent shot noise (G_{dp}) is an important component of the receiver noise and it increases with M^{2+x}. As x is always greater than zero, M^{2+x} is always more than M^2. For this reason, shot noise in direct detection receiver is always amplified   more than the signal power: $G_{dp} \sim M^{2+x}$ and $S_d \sim M^2$. Hence, there is an optimum avalanche gain M which maximizes SNR_d. It is given by

$$M_{opt} = \left(\frac{2\ G_{th}}{x\ e\ \left(R_0\ P_r\ +\ I_{dark}\right)} \right)^{\frac{1}{2+x}}$$ (9.16)

Unlike direct detection, signal power S_c and signal-independent shot noise (G_{cs}) are amplified in like manner if heterodyne or homodyne detection is used. Therefore, a rise in P_l always improves SNR_c (an exception is explained in [3]). If P_l is sufficiently large, then all sources of noise which are independent of P_l can be neglected. As system performance is now limited by the shot noise of photodiode only, we call it the *shot noise limit*. Here, SNR_c exhibits a saturation and reaches its maximum value which is much higher than SNR_d of direct detection receiver: $SNR_c \gg SNR_d$. It represents a significant advantage of the coherent detection receiver over the direct detection receiver.

Emergence of performance gain is shown in Fig. 9.2 which gives a first qualitative comparison of both types of optical receivers, coherent and direct detection. Fig. 9.2a and 9.2b compare the signal power S_c and S_d and noise power N_c and N_d respectively in terms of ratios

$$V_s = \frac{S_c}{S_d} = 4\ M^{-2}\ \frac{P_l}{P_r}$$ (9.17)

and

$$V_N = \frac{N_c}{N_d} = \frac{R_0 P_1 + I_{dark} + \dfrac{G_{th}}{e}}{M^{2+x}(R_0 P_r + I_{dark}) + \dfrac{G_{th}}{e}} \tag{9.18}$$

as a function of P_1. It is seen from Eqs. (9.17) and (9.18) that V_S and V_N increase proportional to P_1. However, increase in V_N is less than the increase in V_S. This highlights the reason why coherent receivers offer an improvement in the sensitivity over a direct detection receiver.

A powerful measure to assess the performance gain of coherent receivers is defined by the gain in SNR, where

$$GS = 10 \log_{10}(V_{S/N}) \tag{9.19}$$

and

$$V_{S/N} = \frac{\left(\dfrac{S}{N}\right)_c}{\left(\dfrac{S}{N}\right)_d} = V_S\, V_N^{-1} \tag{9.20}$$

The $V_{S/N}$ shown in Fig. 9.2c is obtained by the multiplication of straight V_S (Fig. 9.2a) and hyperbolic $1/V_N$ (dotted line in Fig. 9.2b). With Eqs. (9.17) and (9.18), above equation becomes

$$V_{S/N} = 4\, M^{-2}\, \frac{P_1}{P_r} \cdot \frac{M^{2+x}(R_0 P_r + I_{dark}) + \dfrac{G_{th}}{e}}{R_0 P_1 + I_{dark} + \dfrac{G_{th}}{e}} \tag{9.21}$$

As shown in Fig. 9.2c, $V_{S/N}$ is less than 1 if P_1 is very low. Consequently, GS is negative, which corresponds to a degradation in coherent receiver performance in comparison to direct detection receiver. Within this range of low P_1, direct detection receiver even offers a better performance than the coherent detection receiver. The reason for this unusual result is that the signal amplification by P_1 is too low.

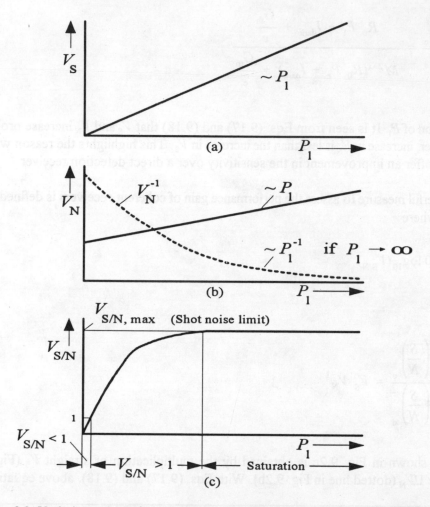

Fig. 9.2: Variations of relative (a) signal power, (b) noise power and (c) signal-to-noise ratio for coherent and direct detection (incoherent) receivers

Example 9.1

Consider that $P_r = 10$ nW ($\triangle$ -50 dBm), $G_{th} = 10^{-23}$ A^2/Hz, $I_{dark} = 10^{-11}$ A, $R_0 = 1$ A/W, $x = 0.9$ and M $= M_{opt} \approx 27$. In order to achieve a ratio $V_{S/N} = 1$ or a gain $GS = 0$ dB, a local laser light power $P_1 \approx 0.56$ μW ($\triangle$ -32.5 dBm) is required. As the optical power level of commercially available laser sources is in general some order of magnitudes higher than -32.5 dBm (normally -10 dBm to 0 dBm), a performance gain by applying coherent detection is always obtainable in practice.

As seen from Fig. 9.2c, performance gain of coherent detection rises first very steeply with the increase in P_1. Finally, shot noise limited receiver operation is achieved when P_1 becomes sufficiently large. The maximum gain GS_{max} in SNR at shot noise limit is

$$GS_{max} = 10 \log_{10} \left(4 M^x \left[1 + \frac{M^{2+x} I_{dark} + \dfrac{G_{th}}{e}}{M^{2+x} R_0 P_r} \right] \right) \tag{9.22}$$

Example 9.2

With the same system parameters as given in example 9.1, it can be deduced from above equation that the performance gain due to coherent detection at high P_1 is $GS_{max} = 20.5$ dB implying $V_{S/N} = V_{S/N,max} \approx 112$. Even for a local laser light power of only $P_1 \approx 0.1$ mW (-10 dBm), $V_{S/N}$ from Eq. (9.21) is 69 implying $GS = 18.4$ dB.

Besides the gain GS in SNR, gain in receiver sensitivity (GR) is another important and an efficient measure to compare systems performance. GR is defined as the logarithm of the ratio of required optical power at the input of both types of receivers (direct and coherent) for SNR_c = SNR_d (i.e., $V_{S/N} = 1$ and $GS = 0$ dB). It is given by

$$GR = 10 \log_{10} \left(\frac{P_{r,d}}{P_{r,c}} \right) \tag{9.23}$$

The required optical input power $P_{r,d}$ and $P_{r,c}$ can be determined from Eqs. (9.7) and (9.15). These are given by

$$P_{r,d} = \frac{M^x e B}{2 R_0} \left(\frac{S}{N} \right)_d \left[1 + \sqrt{1 + \frac{4}{M} \frac{M^{2+x} I_{dark} + \dfrac{G_{th}}{e}}{M^{2+x} e B \left(\dfrac{S}{N} \right)_d}} \right] \tag{9.24}$$

and

$$P_{r,c} = \frac{e B}{4 R_0} \left(\frac{S}{N} \right)_c \frac{R_0 P_l + I_{dark} + \dfrac{G_{th}}{e}}{R_0 P_1} \tag{9.25}$$

From Eqs. (9.23) - (9.25), receiver sensitivity improvement factor V_r from the relationship $GR = 10 \log_{10} (V_r)$ is given by

$$V_r = 2 M^x \frac{R_0 P_1}{R_0 P_1 + I_{dark} + \frac{G_{th}}{e}} \left[1 + \sqrt{1 + \frac{4}{M} \frac{M^{2+x} I_{dark} + \frac{G_{th}}{e}}{M^{2+x} e B \frac{S}{N}}} \right] \quad (9.26)$$

In a coherent receiver, excess noise exponent x, I_{dark} and G_{th}/e do not affect the system performance provided P_1 is high enough. Use of a coherent receiver is more advantageous when the influence of x, I_{dark} and G_{th}/e on the receiver performance is higher. To obtain a high SNR, large optical power $P_{r,d}$ or $P_{r,c}$ is required at the receiver input. Higher the desired SNR, larger the difference in $P_{r,d}$ and $P_{r,c}$ and, therefore, higher the GR. Under shot noise limited receiver operation i.e., $P_1 \to \infty$, maximum gain in receiver sensitivity GR_{max} is achieved and is given by

$$GR_{max} = 10 \log_{10} \left(2 M^x \left[1 + \sqrt{1 + \frac{4}{M} \frac{M^{2+x} I_{dark} + \frac{G_{th}}{e}}{M^{2+x} e B \frac{S}{N}}} \right] \right) \quad (9.27)$$

In the following, implications of both the gain in signal-to-noise ratio and gain in receiver sensitivity are discussed.

(i) GAIN IN SIGNAL-TO-NOISE RATIO, *GS*

With equal power levels $P_{r,d}$ and $P_{r,c}$ at the input of direct and coherent detection receivers, GS represents the performance gain of latter receiver over the former in terms of higher receiver output signal-to-noise ratio. Therefore, GS is a direct measure to assess the improvement in bit error rate obtained by using a coherent instead of a direct detection receiver.

(ii) GAIN IN RECEIVER SENSITIVITY, *GR*

With equal signal-to-noise ratios $(S/N)_d$ and $(S/N)_c$ at the output of direct and coherent detection receivers, GR represents the performance gain of latter receiver over the former in terms of lesser required power level at the receiver input. This performance gain can be used to increase the transmission distance. Therefore, GR is a direct measure to determine the improvement in the transmission distance or the repeater spacing (in km)

$$\Delta l = GR \cdot \alpha^{-1} \qquad (9.28)$$

where α [dB/km] is the attenuation coefficient of the fiber.

Example 9.3

Consider that equal signal-to-noise ratio $(S/N)_d = (S/N)_c = S/N = 40.48$ is required. Receiver parameters are the same as given in the examples 9.1 and 9.2. The coherent receiver is shot noise limited and the double-sided bandwidth of the low-pass filter is $B = 1120$ MHz. From Eqs. (9.27) and (9.28), it can be deduced that GR_{max} is 19.5 dB. With $\alpha = 0.2$ dB/km, Δl from Eq. (9.28) will be 97.5 km. It corresponds to an increase in repeater spacing by about 97 km.

Coherent communication systems are capable to offer a remarkable system performance gain of up to 20 dB in contrast to conventional direct detection system. However, such a high improvement in performance can only be achieved under ideal conditions. A realistic system analysis has to consider the following important aspects as well (Chapter eleven):

- vector and not scalar superposition of electric fields,

- polarization fluctuations,

- laser noise,

- intersymbol interference,

- different degrading effects on binary symbols due to laser phase noise and

- system optimization.

It may be mentioned that such a high gain in sensitivity can also be achieved by using an optical preamplifier in front of direct detection receiver (Chapter ten). Under ideal conditions, it may also achieve shot noise limited operation. In that case, coherent systems no longer give the advantage of better sensitivity. Since optical amplifiers are commercially available, advantage of coherent systems in terms of better sensitivity is no longer evident in many fiber systems.

9.2.2 SELECTIVITY

Receiver selectivity is one of the important issues in multichannel transmission systems such as broadband distribution networks (Chapter fourteen). These systems are based on high-density wavelength division multiplexing (WDM) or optical frequency division multiplexing (OFDM). Actually, WDM and OFDM are different terminologies for the same concept. In both WDM and OFDM, a number of equally spaced optical carriers are used to increase the transmission capacity. The optical carriers can either be used to transmit different information signals or to transmit one ultra-high bit rate signal by dividing the total bit rate in a number of low bit rate signals depending on the number of optical carriers available. Assuming, for example, ten optical carriers each modulated with a 10 Gbit/s signal, then total bit rate equals 100 Gbit/s. Besides the improvement of transmission capacity, WDM is also advantageous to increase the flexibility of a network e.g., in case of an optical overlay network as explained in Chapter fourteen.

Although WDM and OFDM represent same concept, WDM terminology is normally used in conjunction with direct detection systems with relatively large channel spacing e.g., 100 GHz and more, whereas OFDM terminology in conjunction with coherent detection systems with relatively small channel spacing e.g., less than 10 GHz.

In a WDM system, the receiver must be able to select each optical channel properly. Quality of selection depends on the optical channel spacing and the type of receiver. A typical channel allocation in a 32-channel fiber transmission system as defined by ITU-T is given by

$\lambda_1 = 1535.82$ nm	$\lambda_9 = 1542.14$ nm	$\lambda_{17} = 1548.51$ nm	$\lambda_{25} = 1554.94$ nm
$\lambda_2 = 1536.61$ nm	$\lambda_{10} = 1542.94$ nm	$\lambda_{18} = 1549.32$ nm	$\lambda_{26} = 1555.75$ nm
$\lambda_3 = 1537.40$ nm	$\lambda_{11} = 1543.73$ nm	$\lambda_{19} = 1550.12$ nm	$\lambda_{27} = 1556.55$ nm
$\lambda_4 = 1538.19$ nm	$\lambda_{12} = 1544.53$ nm	$\lambda_{20} = 1550.92$ nm	$\lambda_{28} = 1557.36$ nm
$\lambda_5 = 1538.98$ nm	$\lambda_{13} = 1545.32$ nm	$\lambda_{21} = 1551.72$ nm	$\lambda_{29} = 1558.17$ nm
$\lambda_6 = 1539.77$ nm	$\lambda_{14} = 1546.12$ nm	$\lambda_{22} = 1552.52$ nm	$\lambda_{30} = 1558.98$ nm
$\lambda_7 = 1540.56$ nm	$\lambda_{15} = 1546.92$ nm	$\lambda_{23} = 1553.33$ nm	$\lambda_{31} = 1559.79$ nm
$\lambda_8 = 1541.35$ nm	$\lambda_{16} = 1547.72$ nm	$\lambda_{24} = 1554.13$ nm	$\lambda_{32} = 1560.61$ nm

In the above allocation, channel spacing is about 0.8 nm which equals about 100 GHz. In the analysis and design of WDM systems, it is often required to transform a bandwidth $\Delta\lambda = \lambda_{max} - \lambda_{min}$ in terms of wavelength (e.g., 0.8 nm) into a bandwidth $\Delta f = f_{max} - f_{min}$ in terms of frequency (e.g., 100 GHz). This transformation can simply be performed by employing either of the following two relationships:

$$\Delta f = f_{max} - f_{min} = \frac{c}{\lambda_{min}} - \frac{c}{\lambda_{max}} \approx \frac{c}{\lambda_0^2} \cdot \Delta\lambda \tag{9.29a}$$

$$\Delta\lambda = \lambda_{max} - \lambda_{min} = \frac{c}{f_{min}} - \frac{c}{f_{max}} \approx \frac{c}{f_0^2} \cdot \Delta f \tag{9.29b}$$

Here, c represents the velocity of light and $f_0 = 0.5 (f_{min} + f_{max})$ and $\lambda_0 = 0.5 (\lambda_{min} + \lambda_{max})$ are the centre frequency and the wavelength respectively. The approximations given in Eqs. (9.29a) and

(9.29b) above are valid for $\Delta\lambda/\lambda_0 \ll 1$ or $\Delta f/f_0 \ll 1$ (narrowband consideration) which is normally fulfilled in practical optical communication systems.

Example 9.4

For a given 1550 nm semiconductor laser diode with a spectral half width of $\Delta\lambda = 1$ nm, its spectral half width in terms of frequency is $\Delta f = 125$ GHz.

In the following Subsections, selectivity of direct and coherent detection receivers is explained and compared.

(i) DIRECT DETECTION RECEIVER

In a direct detection receiver, channel selection is performed by means of a wavelength-tunable optical filter as shown in Fig. 9.3. Both tunable and non-tunable optical filters are discussed in Chapter eight. In this Section, we restrict our interest on system aspects of these filters which directly influence the selectivity.

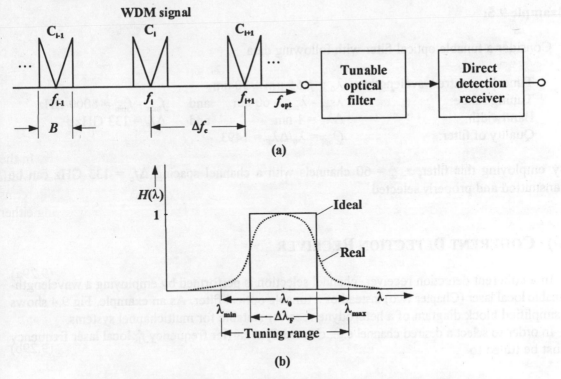

Fig. 9.3: Channel selection by means of a (a) direct detection receiver and (b) tunable optical filter

As shown in Fig. 9.3, within the tuning range of an optical filter

$$z_{max} = \frac{\lambda_{max} - \lambda_{min}}{\Delta\lambda_F} = \frac{f_{max} - f_{min}}{\Delta f_F} \tag{9.30}$$

different information channels can simultaneously be transmitted and selected. Thereby, minimum allowed optical channel spacing is

$$\Delta\lambda_c = \Delta\lambda_F \quad \text{and} \quad \Delta f_c = \Delta f_F \gg B \tag{9.31}$$

where Δf_c and Δf_F represent channel spacing and filter bandwidth in terms of frequency and $\Delta\lambda_c$ and $\Delta\lambda_F$ in terms of wavelength. It is evident that channel spacing (e.g., 133 GHz) must be much larger than bandwidth B of information signal (e.g., 1 GHz). This is in contrast to conventional electrical FDM systems e.g., in a CATV system, where channel spacing is in the order of signal bandwidth.

Example 9.5:

Consider a tunable optical filter with following data:

Tuneable centre wavelength: $\lambda_0 = 1.55 \ \mu m \pm 30$ nm
Tuning range: $\lambda_{max} - \lambda_{min} = 60$ nm   and $f_{max} - f_{min} \approx 8000$ GHz
Bandwidth: $\Delta\lambda_F = 1$ nm   and $\Delta f_F = 133$ GHz
Quality of filter: $Q_{opt} = \lambda_0/\Delta\lambda_F = 1493$

By employing this filter, $z_{max} = 60$ channels with a channel spacing $\Delta f_c = 133$ GHz can be transmitted and properly selected.

(ii) COHERENT DETECTION RECEIVER

In a coherent detection receiver, channel selection is performed by employing a wavelength-tunable local laser (Chapter two) instead of a tunable optical filter. As an example, Fig 9.4 shows a simplified block diagram of a heterodyne receiver suitable for multichannel systems.

In order to select a desired channel e.g., channel C_i at carrier frequency f_i, local laser frequency must be tuned to

$$f_{Li} = f_{IF} + f_i \tag{9.32}$$

where f_{IF} is the fixed centre frequency of the intermediate frequency (IF) filter. Within the tuning range of the local laser, which is approximately same as tuning range of an optical filter,

$$z_{max} = \frac{\lambda_{max} - \lambda_{min}}{\Delta\lambda_{IF}} = \frac{f_{max} - f_{min}}{\Delta f_{IF}} \qquad (9.33)$$

different information channels can be transmitted and selected. Here, $\Delta\lambda_{IF}$ and Δf_{IF} represent IF filter bandwidth in terms of wavelength and frequency. Minimum allowed channel spacing is

$$\Delta\lambda_c = \Delta\lambda_{IF} \quad \text{and} \quad \Delta f_c = \Delta f_{IF} \approx B \qquad (9.34)$$

Thus, optical channel spacing Δf_c can be as small as signal bandwidth B which is same as in conventional electrical FDM systems. However, channel spacing is chosen somewhat larger e.g., $\Delta f_c = 5B$ due to realization considerations.

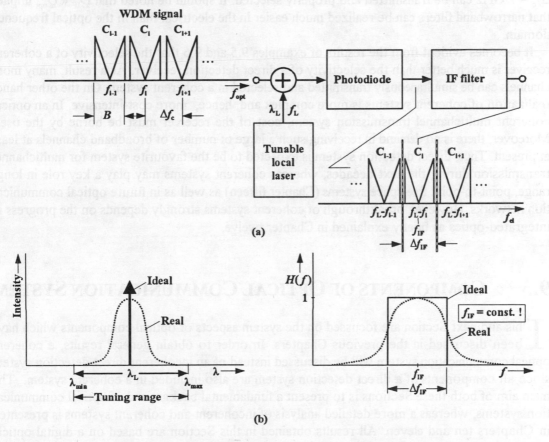

Fig 9.4: Channel selection with (a) heterodyne receiver and (b) tunable local laser and fixed IF filter

Example 9.6:

Consider a signal bandwidth $B = 1$ GHz, a tunable local laser and an IF filter with following typical data:

Local Laser:
Tunable centre wavelength: $\lambda_L = \lambda_0 = 1.55$ μm ± 30 nm
Tuning range: $\lambda_{max} - \lambda_{min} = 60$ nm   and $f_{max} - f_{min} \approx 8000$ GHz
Bandwidth: $\Delta\lambda_F = 1$ nm   and $\Delta f_F = 133$ GHz

IF Filter:
Centre frequency: $f_{IF} = 3B$
Bandwidth: $\Delta f_{IF} = B$
Quality of filter: $Q_{IF} = f_{IF}/\Delta f_{IF} = 3 \ll Q_{opt}$

By employing these components, $z_{max} = 8000$ different broadband channels with a channel spacing $\Delta f_c = 1$ GHz can be transmitted and properly selected. It should be noted that $Q_{IF} \ll Q_{opt}$ implies that narrowband filters can be realized much easier in the electrical than in the optical frequency domain.

It becomes evident from the results of examples 9.5 and 9.6 that the selectivity of a coherent receiver is much better than the selectivity of a direct detection receiver. As a result, many more channels can be simultaneously transmitted and selected in a coherent system. On the other hand, realization of coherent systems is more complex and, hence, more cost-intensive. In an optical coherent multichannel transmission system, cost of the receiver must be borne by the user. Moreover, there is no demand in receiving such a large of number of broadband channels at least at present. Thus, direct detection system is expected to be the favourite system for multichannel transmission during the next decades, whereas coherent systems may play a key role in long-range, point-to-point free-space systems (Chapter fifteen) as well as in future optical communication networks. However, breakthrough of coherent systems strongly depends on the progress in integrated-optics as briefly explained in Chapter twelve.

9.3   COMPONENTS OF OPTICAL COMMUNICATION SYSTEM

This and next Section are focussed on the system aspects of optical components which have been discussed in the previous Chapters. In order to obtain general results, a coherent optical communication system will be discussed instead of an incoherent direct detection system since all components of a direct detection system are also included in a coherent system. The main aim of both these Sections is to present a fundamental background of optical communication systems, whereas a more detailed analysis of incoherent and coherent systems is presented in Chapters ten and eleven. All results obtained in this Section are based on a digital optical communication system with coherent detection shown in Fig. 9.5.

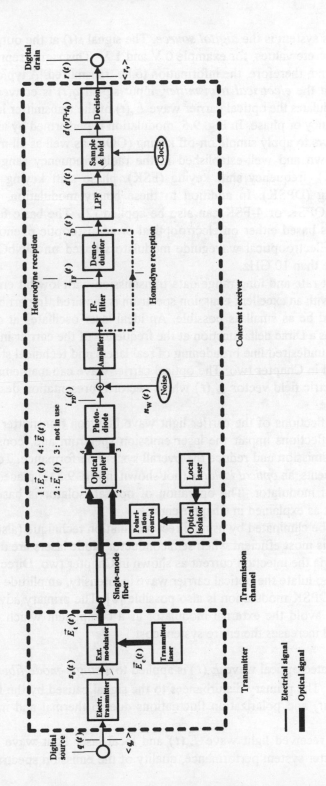

Fig. 9.5: Block diagram of fiber-based coherent optical communication system

The first component of this system is the *digital source*. The signal $s(t)$ at the output of this source is characterized by discrete values, for example 0 V and 1 V. This signal represents the binary symbol sequence $<q_v>$ and, therefore, the information to be transmitted. A typical pattern $<q_v>$ is shown in Fig. 9.6. In the *electrical transmitter*, input signal $q(t)$ is converted to an electrical signal $s_e(t)$ which modulates the optical carrier wave $\vec{E}_c(t)$ of the transmitter laser either in intensity, amplitude, frequency or phase. In Fig. 9.5, modulation is performed by an *external optical modulator* which allows to apply simple on-off keying (OOK) as well as all modulation schemes which are well-known and well-established in the radio frequency range such as amplitude shift keying (ASK), frequency shift keying (FSK), phase shift keying (PSK) or differential phase shift keying (DPSK). In addition to these binary modulation, multilevel modulation schemes such as QPSK or 4-FSK can also be applied [2]. The basic function of external optical modulators is based either on electrooptical or acoustooptic phenomenon as explained in Chapter eight. Electrooptical waveguide modulators based on $LiNbO_3$ allow a modulation frequency of more than 10 GHz.

In order to achieve high bit rate and long range data transmission with low bit error rate, a *single-mode transmitter laser* with an excellent emission spectrum is required. In particular, linewidth of the spectrum should be as small as possible. An ideal laser oscillates at one single frequency only which appears as a Dirac delta function at the frequency of the carrier in the emission spectrum. The reasons for undesired line broadening of real lasers and technical steps for its reduction have been explained in Chapter two. The optical carrier wave can mathematically be expressed by its complex electric field vector $\vec{E}_c(t)$ where vector representation describes the polarization of laser light wave.

There can be undesired reflections of the carrier light wave between transmitter laser and external modulator. These reflections impair the laser emission spectrum and, consequently, deteriorate the quality of transmission and reduce the overall system performance. To avoid or at least to reduce these impairments, an *optical isolator* (not shown in Fig. 9.5) is placed between transmitter laser and external modulator. The operation of optical isolator is based on the magnetooptical Faraday effect as explained in Chapter eight.

The external modulator can be eliminated by using direct modulation technique (also referred as internal modulation) which is most efficient when semiconductor diode lasers are used. Such lasers can easily be modulated via the injection current as shown in Chapter two. Direct modulation of diode laser allows to modulate the optical carrier wave in intensity, amplitude as well as in frequency. Further, direct DPSK modulation is also possible [8]. The primary advantage of using direct modulation is to avoid the external modulator as a component which exhibits a considerable insertion loss and increases the entire system cost.

As shown in Fig. 9.5, modulated optical wave $\vec{E}_t(t)$ is applied to a *single-mode fiber* which is often called monomode fiber. The primary disturbances to the signal caused by the fiber arise from dispersion (Chapter four) and polarization fluctuations due to thermal and mechanical effects (Chapter five).

In the receiver front end, received light wave $\vec{E}_r(t)$ and local laser light wave $\vec{E}_l(t)$ are superimposed. To achieve better system performance, quality of the emission spectrum of the *local laser* must be similar to that of transmitter laser. The superposition of light waves $\vec{E}_r(t)$ and $\vec{E}_l(t)$ is performed by an *optical coupler* for example, a optical waveguide coupler. This coupler normally has two fiber inputs and two fiber outputs called the fiber pigtails. The power splitting ratio with respect to input and output is defined by the coupling efficiency or coupling ratio k as shown in Section 9.4.

The superimposed optical wave $\vec{E}(t)$ at the output of optical coupler is applied to a *photodiode* which generates an electric current $i_{PD}(t)$ proportional to optical input power. For this purpose, a PIN or an APD can be used in principle. As the gain of an APD improves the performance of coherent optical communication systems only insignificantly [3], a PIN-photodiode is normally used in practice. As explained in Section 9.1, optical power at the input of photodiode is proportional to the square of input electric field. Owing to the fact that the superposition of fields $\vec{E}_r(t)$ and $\vec{E}_l(t)$ is linear, photodiode current $i_{PD}(t)$ also contains an intermediate frequency signal $i_{IF}(t)$ referred as IF signal. In coherent heterodyne receivers, we are primarily interested in this IF signal. The IF itself is given by the difference of local laser frequency (f_l) and the frequency (f_r) of received light wave.

Signal and noise are amplified by the *electrical amplifier(s)* in the same way. Therefore, we can assume identical amplification without the loss of generality.

When the frequencies f_l and f_r are equal (i.e., $f_{IF} = 0$), coherent receiver is called *homodyne receiver*. Consequently, the modulated optical signal at the input to receiver is directly converted to baseband. To achieve homodyne detection, great care and accuracy have to be undertaken for the phase and frequency stabilization and tracking of received and local laser light waves. In *heterodyne receiver* ($f_{IF} \neq 0$), IF signal is selected by an *IF filter* and then demodulated with respect to amplitude, frequency or phase in the demodulator or detector circuit. All types of well-known radio frequency demodulators can be used: synchronous demodulator (coherent demodulator), envelope detector or frequency detector. The output signal $i_d(t)$ of the demodulator also contains undesired components such as double-frequency component which can be eliminated usually by a low-pass filter (baseband filter). In a homodyne detection receiver, low-pass filter will reduce additionally the influence of noise by decreasing the noise-equivalent bandwidth. In a heterodyne receiver, noise reduction is primarily performed by the IF filter.

The white Gaussian noise $n_w(t)$ is caused by the shot noise of photodiode and thermal noise of resistors and amplifiers, mainly the first amplifier stage next to photodiode. In *optical free-space communication system* (Chapter fifteen), background noise is also a source of noise in addition to both the above noise sources.

The last components in our coherent system shown in Fig. 9.5 are the sample and hold and the decision circuits which have to recover the digital information $r(t)$. The noisy and disturbed signal at the input of sample and hold circuit is referred as *detected signal $d(t)$* in this book. A data transmission is without any error if transmitted and received sequence of symbols are same i.e., $\langle r_v \rangle = \langle q_v \rangle$. Due to various imperfections and disturbances such as laser noise, shot noise, thermal noise, background noise, dispersion and polarization fluctuations, symbol errors are unavoidable. The most efficient measure to assess the quality of digital communication system is the probability of symbol error, also called the probability of error. This probability is quite important to analyse, optimize and compare digital optical as well as electrical communication systems. Section 9.5 discusses the fundamentals of error probability calculation in more detail. The coherent optical communication systems require great efforts in the design of high-quality *phase- and frequency tracking* circuits. For simplicity, these circuits have not been shown in the block diagram in Fig. 9.5. In order to design and analyse coherent optical communication systems, basic knowledge of the principle of phase- and frequency tracking is unavoidable. Section 11.1.3 reviews the prime features of tracking as far as required in this book.

9.4   SIGNALS IN OPTICAL COMMUNICATION SYSTEM

In this Section, signals in the transmitter, the channel and the receiver of a communication system shown in Fig. 9.5 will be discussed in more detail. In order to obtain a simplified description and straight forward calculation, complex signal representation is preferred. Transfer to the real physical signals, which can actually be measured, is easily accomplished by taking the real part of complex signals.

9.4.1   TRANSMITTER

(i) DIGITAL SOURCE

The digital source in the transmitter generates symbols q_v at equal temporal intervals of duration T. The digital information to be transmitted is represented by the sequence $<q_v>$. We call T the symbol length or symbol period and $1/T$ the symbol rate. The symbols q_v are taken out of a store of M symbols. On the basis of M, we have to distinguish between binary source ($M = 2$) and multilevel source ($M > 2$). In case of binary source with

$$q_v \in \{L, H\} \text{ or } q_v \in \{"0", "1"\} \text{ or } q_v \in \{zero, one\} \text{ or } q_v \in \{space, mark\} \qquad (9.35)$$

symbol period T and symbol rate $1/T$ are frequently called a bit period or duration and bit rate respectively. In this Chapter, we assume that all the symbols q_v are binary and statistically independent. The probabilities of occurrence $p(q_v = H)$ and $p(q_v = L)$ are equal. Hence,

$$p(q_v = H) = p(q_v = L) = 0.5 \qquad (9.36)$$

The physical representation of the binary symbol sequence $<q_v>$ is given by the electrical signal

$$q(t) = \hat{q} \sum_{v=-\infty}^{\infty} a_v \, \text{rect}\left(\frac{t - vT}{T} \right) \tag{9.37}$$

The peak value of this signal is $\hat{q}$ for example, $\hat{q} = 1$ V.

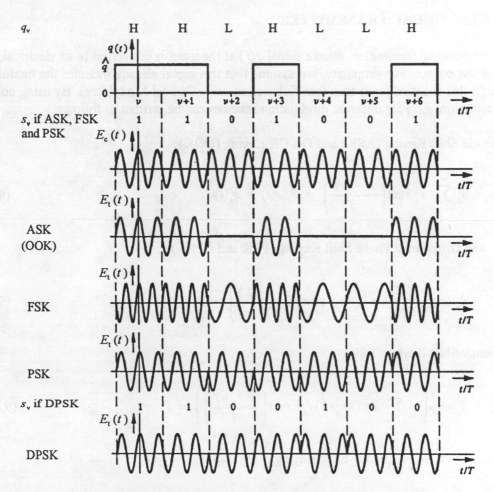

Fig. 9.6: Typical signals in an optical transmitter in absence of noise

In this book, we always use a rectangular source signal $q(t)$ such as shown in Fig. 9.6. This signal is described in terms of the following function:

$$\text{rect}(x) = \begin{cases} 1 & \text{for } |x| \leq 0.5 \\ 0 & \text{otherwise} \end{cases} \tag{9.38}$$

The interrelation between physical source signal $q(t)$ and symbols q_ν is given by the amplitude coefficients

$$a_\nu = 1 \quad \text{for} \quad q_\nu = H \quad \text{and} \quad a_\nu = 0 \quad \text{for} \quad q_\nu = L \tag{9.39}$$

(ii) ELECTRICAL TRANSMITTER

In the electrical transmitter, source signal $q(t)$ at the input is converted to an electrical signal $s_e(t)$ at the output. For simplicity, we assume that this signal already includes the modulation, although this is actually not true (see Subsection iv on Optical Modulator). By using complex representation, $\underline{s}_e(t)$ for different modulation schemes can be written as follows:

Amplitude Shift Keying (ASK) and On-Off Keying (OOK)

$$\underline{s}_e(t) = \hat{s}_e \sum_{\nu=-\infty}^{\infty} s_\nu \ \text{rect}\left(\frac{t-\nu T}{T}\right) = \hat{s}_e \ s(t) = s_e(t) \tag{9.40}$$

Phase and Differential Phase Shift Keying (PSK and DPSK)

$$\underline{s}_e(t) = \hat{s}_e \ \exp\left(j \sum_{\nu=-\infty}^{\infty} \pi(1-s_\nu) \ \text{rect}\left(\frac{t-\nu T}{T}\right) \right) = \hat{s}_e \ \underline{s}(t) \tag{9.41}$$

Frequency Shift Keying (FSK)

$$\underline{s}_e(t) = \hat{s}_e \ \exp\left(j \sum_{\nu=-\infty}^{\infty} 2\pi(2s_\nu-1) f_d t \ \text{rect}\left(\frac{t-\nu T}{T}\right) \right) = \hat{s}_e \tag{9.42a}$$

Continuous Phase Frequency Shift Keying (CPFSK)

$$\underline{s}_e(t) = \hat{s}_e \ \exp\left(j \sum_{\nu=-\infty}^{\infty} \int_{\tau=-\infty}^{t} 2\pi f_d (2s_\nu-1) \ \text{rect}\left(\frac{t-\nu T}{T}\right) \ d\tau \right) = \hat{s}_e \ \underline{s}(t) \tag{9.42b}$$

The peak value $\hat{s}_e$ of electrical transmitter output signal $\underline{s}_e(t)$ is independent of modulation scheme. In case of ASK and OOK, $\underline{s}_e(t)$ is a real signal (Eq. 9.40). Normalizing $\underline{s}_e(t)$ with respect to $\hat{s}_e$, we obtain $\underline{s}(t)$ which is dimensionless. The quantity f_d in Eqs. (9.42a) and (9.42b) is called the frequency shift or frequency deviation. Modulation coefficients s_v and amplitude coefficients a_v are related by

$$s_v = a_v \tag{9.43}$$

for ASK, PSK, FSK and CPFSK, whereas

$$\begin{aligned} s_v &= s_{v-1} \quad \text{if} \quad a_v = 1 \\ s_v &= \bar{s}_{v-1} \quad \text{if} \quad a_v = 0 \end{aligned} \tag{9.44}$$

in case of DPSK. Here, $\bar{s}_v$ represents the inversion of s_v i.e., $\bar{s}_v = 1-s_v$, where $s_v \in \{0, 1\}$. The above relationship between s_v and a_v for ASK, PSK FSK and DPSK is given in Fig. 9.6.

(iii) TRANSMITTER LASER

With a single-mode laser in the transmitter, optical carrier wave can be written as

$$\vec{E}_c(t) = \begin{pmatrix} E_{cx}(t) \\ E_{cy}(t) \end{pmatrix} = E_c(t) \, e^{j2\pi f_c t} \, \vec{e}_c \tag{9.45}$$

Here, normalized unit vector

$$\vec{e}_c = \begin{pmatrix} e_{cx} \\ e_{cy} \end{pmatrix} = \begin{pmatrix} |e_{cx}| \, e^{j\psi_{cx}} \\ |e_{cy}| \, e^{j\psi_{cy}} \end{pmatrix} \quad \text{with} \quad \vec{e}_c \, \vec{e}_c^* = 1 \tag{9.46}$$

describes the *state of polarization* (SOP) of the optical carrier wave (Chapter five). It is often called the polarization unit vector. Laser phase and amplitude noise are included in the complex carrier envelope

$$E_c(t) = |E_c(t)| \, e^{j\Phi_c(t)} \approx \hat{E}_c \, e^{j\Phi_c(t)} \tag{9.47}$$

As an example, Fig. 9.6 shows the real part $\text{Re}\{E_c(t)\}$ of the periodical carrier wave in absence of any noise. In contrast to phase noise $\phi_c(t)$, envelope or amplitude noise $|E_c(t)|$ of the laser is less important in coherent optical communication systems. Its influence on system performance will be considered in detail in Chapter eleven. In this Section, laser phase noise will not be considered i.e., $|E_c(t)| \approx \hat{E}_c$. Optical communication systems with intensity modulation and direct detection are not influenced by phase noise, but they can be very sensitive to laser amplitude noise as shown in Chapter ten.

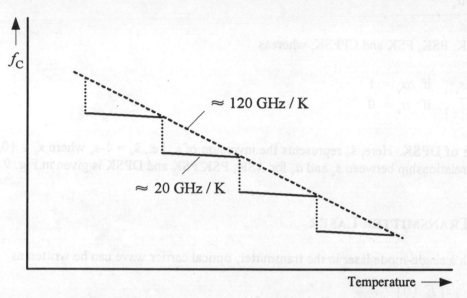

Fig. 9.7: Frequency-temperature characteristic of typical semiconductor laser diode

In a real semiconductor laser, carrier frequency f_c is not constant; rather it changes continuously with time (Chapter two). Fluctuations in temperature are the prime reason for these undesired changes. Even high frequency jumps called mode hopping can occur if temperature is changing continuously (Fig. 9.7). Hence, temperature stabilization and automatic frequency control circuits are unavoidable in coherent optical communication systems.

(iv) OPTICAL MODULATOR

Depending on modulation scheme and information to be transmitted, carrier wave $\underline{E}_c(t)$ of the transmitter laser is changed in amplitude, phase or frequency by the modulator. Thereby, modulated electric baseband signal $\underline{s}_e(t)$ as shown in Fig. 9.6 is shifted to optical frequency. The modulator output signal is represented by $\vec{E}_t(t)$. To describe the modulation process mathematically, an ideal modulator i.e., a simple multiplier is formally considered [4]. Taking into account a modulator coefficient K_m, modulated optical wave at the output of transmitter is given by

$$\vec{E}_t(t) = K_m \; \hat{s}_e \; \underline{s}(t) \; \underline{E}_c(t) \; \vec{e}_c = \underline{E}_t(t) \; \vec{e}_t \quad \text{with} \quad \vec{e}_c = \vec{e}_t \qquad (9.48)$$

The right side of above equation is due to simple formal separation of electric field vector $\vec{E}_t(t)$ in two components: one is the polarization unit vector $\vec{e}_t$ and the other modulated complex envelope $E_t(t) = K_m \hat{s}_e \underline{s}(t) E_c(t)$. To simplify our discussions, we assume that $\vec{e}_t$ and $\vec{e}_c$ are constant and equal. It has been made clear by Fig. 9.6 which shows typical signals obtained by taking the real part of $E_t(t)$.

In order to obtain a simple mathematical representation of the transmitted signals, we have carried out the modulation process in two steps: electrical modulation in the virtual electrical transmitter as considered above (Fig. 9.6) and optical modulation in the real optical modulator. In this model, optical modulation process is described by a simple multiplication of the electrical transmitter output signal $\underline{s}_e(t)$ and transmitter laser signal $E_c(t)$. However, in a real coherent optical transmitter the electric transmitter subsystem is usually a simple booster amplifier. It provides the required electric input power to optical modulator, wherein the modulation is actually performed. Therefore, signals given in Eqs. (9.40) to (9.42) cannot be measured in a real transmitter as they exist only in our model.

9.4.2 TRANSMISSION CHANNEL

In this Chapter, single-mode fiber is always used as transmission channel. Free-space channel will be considered in Chapter fifteen. Due to non-ideal transmission properties of single-mode fibers, transmitted optical signal $E_t(t)$ is disturbed during transmission. Especially the polarization is influenced by various thermal and mechanical strains in the fiber. As a result, state of polarization at the input to the receiver is changing randomly with time. Direct detection receivers are independent of state of polarization, whereas coherent receivers are extremely dependent. Thus, polarization fluctuations are a fundamental source of noise in coherent optical communication systems and have been considered in detail in Chapter five.

A second important imperfection of single-mode fibers is dispersion. Three different types of dispersion have to be considered: modal, chromatic (material and waveguide) and polarization dispersion. The first two types of dispersion are important in every digital optical communication system and the last one, in addition, becomes important in long-range multigigabit systems. Physical mechanism of dispersion is discussed in Chapter four, whereas influence of dispersion is briefly considered in Chapters ten to twelve as well as in various publications e.g., [6, 7, 12]. In this Section, we shall not consider the fiber dispersion. In the absence of dispersion, fiber output signal or the receiver input signal is given by

$$\vec{E}_r(t) = \underline{s}(t) \, E_r(t) \, e^{j2\pi f_r t} \, \vec{e}_r(t) \tag{9.49}$$

where $f_r = f_t$. Unlike the transmitted signal, polarization unit vector

$$\vec{\underline{e}}_r(t) = \begin{pmatrix} \underline{e}_{rx}(t) \\ \underline{e}_{ry}(t) \end{pmatrix} = \begin{pmatrix} |\underline{e}_{rx}(t)| \ e^{j\psi_{rx}(t)} \\ |\underline{e}_{ry}(t)| \ e^{j\psi_{ry}(t)} \end{pmatrix} \quad \text{with} \quad \vec{\underline{e}}_r(t)\,\vec{\underline{e}}_r(t)^* = 1 \qquad (9.50)$$

of received signal is now a function of time and describes the polarization fluctuations at the input of coherent optical receiver. Due to various imperfections and disturbances from the fiber, polarization at the input to the receiver can be regarded as a random process (Chapter five). Owing to the inherent losses of fiber, optical signal will get attenuated during transmission. At the output end of fiber of length L (in km), attenuated amplitude of electric field is given by

$$E_r(t) = \hat{E}_r \ e^{j\phi_r(t)} = \hat{E}_t \ e^{-\alpha_{Np}L} \ e^{j\phi_t(t)} \qquad (9.51a)$$

where α_{Np} represents the fiber attenuation in Np/km. In terms of optical power levels, above equation yields the well-known relation

$$20 \ \log_{10}\left(\frac{\hat{E}_r}{\hat{E}_t}\right) = 10 \ \log_{10}\left(\frac{P_r}{P_t}\right) = -\alpha_{dB} \ L \qquad (9.51b)$$

where P_r and P_t represent the received and transmitted optical power levels respectively. The relationship between α_{Np} (in Np/km) and α_{dB} (in dB/km) is : 1 Np = 8.686 dB.

Without loss of generality, we assume that statistics of phase noise at the transmitter output and receiver input are same. Thus we can take $\phi_t(t) = \phi_r(t)$. As we have neglected fiber dispersion, optical signal $E_t(t)$ at the transmitter output and $E_r(t)$ at the receiver input only differ in amplitude ($\hat{E}_r \ll \hat{E}_t$) and polarization where $\vec{\underline{e}}_t$ is usually constant and $\vec{\underline{e}}_r(t)$ is random.

9.4.3   RECEIVER

(i)   LOCAL LASER

For a local laser with same spectral properties as the transmitter laser, output wave can be expressed as

$$\vec{E}_1(t) = \begin{pmatrix} E_{1x}(t) \\ E_{1y}(t) \end{pmatrix} = E_1(t) \ e^{j2\pi f_1 t} \ \vec{\underline{e}}_1 \qquad (9.52)$$

Similar to the transmitter laser (Eqs. 9.45 and 9.46), we again assume a constant and linear polarization for local laser source. Hence, the local laser polarization unit vector is given by

$$
\vec{e}_1 = \begin{pmatrix} \varrho_{lx} \\ \varrho_{ly} \end{pmatrix} = \begin{pmatrix} |\varrho_{lx}| \ e^{j\psi_{lx}} \\ |\varrho_{ly}| \ e^{j\psi_{ly}} \end{pmatrix} \quad \text{with} \quad \vec{e}_1 \ \vec{e}_1^* = 1
\tag{9.53}
$$

The complex envelope

$$
\underline{E}_1(t) = |\underline{E}_1(t)| \ e^{j\Phi_1(t)} \approx \hat{E}_1 \ e^{j\Phi_1(t)}
\tag{9.54}
$$

again contains laser phase and amplitude noise. Here, also we neglect laser amplitude noise i.e., $|\underline{E}_1(t)| \approx \hat{E}_1$ which is considered in next Chapter.

In comparison to the received optical wave $\underline{E}_r(t)$, local laser wave $\underline{E}_1(t)$ is not attenuated by the transmission channel i.e., by the fiber. Therefore, the following relations are always valid in coherent optical communication systems

$$
\hat{E}_1 \gg \hat{E}_r = \hat{E}_t \ e^{-\frac{\alpha_{NP}L}{1\,Np}} \quad \text{and} \quad P_1 \gg P_r = P_t \ 10^{-\frac{\alpha_{dB}L}{10\,dB}}
\tag{9.55}
$$

where P_1 is the power of local laser.

(ii) OPTICAL COUPLER

The modulated optical wave $\underline{E}_r(t)$ at the input of coherent receiver and local laser wave $\underline{E}_1(t)$ are linearly superimposed in an optical coupler. Usually, this coupler is a four-port coupler with two fiber inputs and two fiber outputs called the pigtails (Chapter eight). As shown in Fig. 9.5, input ports are taken as 1 and 2 and output ports as 3 and 4. An optical coupler is primarily characterized by its coupling ratio k $(0 < k < 1)$ that defines the ratio of the input to the output power. Considering a symmetrical coupler without any internal loss, the coupler can be described by the following matrix:

$$
\begin{pmatrix} \vec{E}_4(t) \\ \vec{E}_3(t) \end{pmatrix} = \begin{pmatrix} \sqrt{1-k} & j\sqrt{k} \\ j\sqrt{k} & \sqrt{1-k} \end{pmatrix} \begin{pmatrix} \vec{E}_1(t) \\ \vec{E}_2(t) \end{pmatrix}
\tag{9.56}
$$

This coupler is exactly same as given in block diagram of Fig. 9.5 except that $\underline{E}_1(t)$ and $\underline{E}_2(t)$ are replaced by $\underline{E}_r(t)$ and $\underline{E}_l(t)$ respectively. Considering the optical power $P_4(t)$ and $P_3(t)$ at the coupler output, following relationships can be obtained:

$$
\begin{aligned}
P_4(t) &\sim \vec{E}_4(t)\vec{E}_4^*(t) = \left|\vec{E}_4(t)\right|^2 \\
&= (1-k)\left|\vec{E}_r(t)\right|^2 + k\left|\vec{E}_l(t)\right|^2 + 2\sqrt{k(1-k)}\,\mathrm{Im}\left\{\vec{E}_r(t)\,\vec{E}_l^*(t)\right\}
\end{aligned}
\tag{9.57}
$$

and

$$
\begin{aligned}
P_3(t) &\sim \vec{E}_3(t)\vec{E}_3^*(t) = \left|\vec{E}_3(t)\right|^2 \\
&= (1-k)\left|\vec{E}_l(t)\right|^2 + k\left|\vec{E}_r(t)\right|^2 - 2\sqrt{k(1-k)}\,\mathrm{Im}\left\{\vec{E}_r(t)\,\vec{E}_l^*(t)\right\}
\end{aligned}
\tag{9.58}
$$

In order to obtain a better signal representation, we now take the advantage of complex description of optical power $P_4(t) = \mathrm{Re}\{\underline{P}_4(t)\}$ and $P_3(t) = \mathrm{Re}\{\underline{P}_3(t)\}$. If we consider that the proportionality between optical power and electric field is same for all the participating electric fields $\underline{E}_r(t)$, $\underline{E}_l(t)$, $\underline{E}_3(t)$ and $\underline{E}_4(t)$, we obtain

$$
\begin{aligned}
\underline{P}_4(t) &= k\,P_1 + (1-k)\left|\underline{s}(t)\right|^2 P_r \\
&+ 2\sqrt{k(1-k)}\,\sqrt{P_r\,P_1}\,\underline{s}(t)\,a_p(t)\,e^{j\phi(t)}\,e^{j\phi_p(t)}\,e^{j2\pi f_{IF}t}
\end{aligned}
\tag{9.59}
$$

and

$$
\begin{aligned}
\underline{P}_3(t) &= (1-k)\,P_1 + k\left|\underline{s}(t)\right|^2 P_r \\
&- 2\sqrt{k(1-k)}\,\sqrt{P_r\,P_1}\,\underline{s}(t)\,a_p(t)\,e^{j\phi(t)}\,e^{j\phi_p(t)}\,e^{j2\pi f_{IF}t}
\end{aligned}
\tag{9.60}
$$

where P_1 and P_r represent the mean optical power level of the local laser light wave and unmodulated received light wave respectively. The intermediate frequency is given by

$$
f_{IF} = f_1 - f_r \quad \text{with} \quad f_r = f_t
\tag{9.61}
$$

and the overall resulting phase noise follows the relation

$$\phi(t) = \phi_l(t) - \phi_r(t) \quad \text{with} \quad \phi_r(t) = \phi_t(t) \tag{9.62}$$

It becomes evident from Eqs. (9.59) and (9.60) that polarization fluctuations caused by fiber imperfections firstly, result in amplitude fluctuations $a_p(t)$ and secondly, phase fluctuations $\phi_p(t)$. These fluctuations are sources of noise and may seriously deteriorate the performance of advanced and coherent communication systems. In the worst-case, detected signal can even become zero as shown in Chapter five. Both sources of noise can be deduced from the following simple relationship:

$$\vec{e}_r(t)\,\vec{e}_l^{\,*}(t) = a_p(t)\,e^{j\phi_p(t)} \tag{9.63}$$

Using Eqs. (9.50) and (9.53), we obtain the following equations for amplitude and phase fluctuations:

$$
\begin{aligned}
a_p(t) &= \left| \vec{e}_r(t)\,\vec{e}_l^{\,*}(t) \right| = \sqrt{\left[\mathrm{Re}\left\{\vec{e}_r(t)\,\vec{e}_l^{\,*}(t)\right\}\right]^2 + \left[\mathrm{Im}\left\{\vec{e}_r(t)\,\vec{e}_l^{\,*}(t)\right\}\right]^2} \\
&= \Big[|e_{lx}|^2\,|e_{rx}(t)|^2 + |e_{ly}|^2\,|e_{ry}(t)|^2 \\
&\quad + 2\,|e_{lx}|\,|e_{rx}(t)|\,|e_{ly}|\,|e_{ry}(t)|\cos\!\left(\psi_{rx}(t) - \psi_{lx} - \psi_{ry}(t) + \psi_{ly}\right) \Big]^{1/2}
\end{aligned}
\tag{9.64}
$$

and

$$
\begin{aligned}
\phi_p(t) &= \arctan\!\left(\frac{\mathrm{Im}\left\{\vec{e}_r(t)\,\vec{e}_l^{\,*}\right\}}{\mathrm{Re}\left\{\vec{e}_r(t)\,\vec{e}_l^{\,*}\right\}} \right) \\
&= \arctan\!\left(\frac{|e_{lx}|\,|e_{rx}(t)|\,\sin\!\left(\psi_{rx}(t) - \psi_{lx}\right) + |e_{ly}|\,|e_{ry}(t)|\,\sin\!\left(\psi_{ry}(t) - \psi_{ly}\right)}{|e_{lx}|\,|e_{rx}(t)|\,\cos\!\left(\psi_{rx}(t) - \psi_{lx}\right) + |e_{ly}|\,|e_{ry}(t)|\,\cos\!\left(\psi_{ry}(t) - \psi_{ly}\right)} \right)
\end{aligned}
\tag{9.65}
$$

The output of coupler is converted to an electric signal by means of a PIN or an APD. Since an optical coupler normally contains two fiber outputs, either of the outputs can be used and the other output remains unused. However, instead of using only one output of the coupler, it is much more efficient to use both the outputs. Both possibilities are discussed below.

(iii) PHOTODIODE

A photodiode generates a current $i_{PD}(t)$ proportional to the absorbed optical power. Consider first a coherent receiver with only one single photodiode. We call it a *single-diode receiver*. Assuming, for example, that port 4 is used, the photodiode current is given by

$$i_{PD4}(t) = k\, R_0\, P_1 + (1-k)\, |\underline{s}(t)|^2\, R_0\, P_r$$
$$+ \, 2\, R_0\, \sqrt{k(1-k)}\, \sqrt{P_1 P_r}\, \underline{s}(t)\, a_p(t)\, e^{j\phi(t)}\, e^{j\phi_p(t)}\, e^{j2\pi f_{IF} t} \qquad (9.66)$$

This current consists of three components: first component $k R_0 P_1$ is a DC without any information. Second component $(1-k)\, |\underline{s}(t)|^2 R_0 P_r$ truly contains information, but is negligibly small as compared to third component since $P_r \ll P_1$. Therefore, only the third component is of practical interest in coherent receiver. This component contains transmitted information $\underline{s}(t)$ which has to be detected and recovered in the demodulator. However, it is severely corrupted by polarization fluctuations in the fiber and laser noise.

Next we consider a coherent receiver with two photodiodes, which is termed as a *balanced receiver*. In this type of receiver, optical power available at both the output ports is fully used. To get maximum signal power, photodiode currents $i_{PD4}(t)$ and $i_{PD3}(t)$ have to be subtracted to obtain the output current $i_{PD}(t) = i_{PD4}(t) - i_{PD3}(t)$. It is given by

$$i_{PD}(t) = (2k-1)\, R_0\, P_1 + (1-2k)\, |\underline{s}(t)|^2\, R_0\, P_r$$
$$+ \, 4\, R_0\, \sqrt{k(1-k)}\, \sqrt{P_1 P_r}\, \underline{s}(t)\, a_p(t)\, e^{j\phi(t)}\, e^{j\phi_p(t)}\, e^{j2\pi f_{IF} t} \qquad (9.67)$$

By comparing Eqs. (9.66) and (9.67), it becomes clear that the amplitude of third component is twice now. It implies a gain of 3 dB in signal power. This gain is an important feature of balanced receiver over single-diode receiver. A second advantage of significant practical interest is obtained when coupling ratio is chosen to be $k = 0.5$. In that case, DC component $(2k-1)R_0 P_1$ as well as the baseband component $(1-2k)\, |\underline{s}(t)|^2 R_0 P_r$ in Eq. (9.67) become zero. Therefore, a balanced receiver configuration becomes more useful particularly when homodyne receivers are used since third component of $i_{PD}(t)$ is already at a baseband frequency ($f_{IF} = 0$).

The phase noise $\phi(t)$ caused by local and transmitter lasers is a fundamental source of noise in coherent optical communication systems. This noise shows a strong influence on BER and, therefore, on the overall transmission quality. Polarization fluctuations due to fiber imperfections also degrade the system performance. Several techniques exist to suppress the effect of polarization fluctuations almost completely as discussed in Chapter five.

With a fixed and stable state of polarization, negligible DC and baseband components, photodiode current can now be written as

$$i_{PD}(t) = \hat{i}_{PD} \; \underline{s}(t) \; e^{j\phi(t)} \; e^{j2\pi f_{IF}t} \tag{9.68}$$

with

$$\hat{i}_{PD} = 2K_R \; \sqrt{k(1-k)} \; R_0 \; \sqrt{P_1 P_r} \tag{9.69}$$

Here, K_R is a new constant which will be 2 for balanced receiver and 1 for single-diode receiver. For a symmetrical balanced receiver with $k = 0.5$, above equation becomes same as Eq. (9.10) derived in Section 9.1 under ideal conditions.

(iv) RECEIVER NOISE

As shown in Fig. 9.5, photodiode current $i_{PD}(t)$ is disturbed by an additive and nearly white Gaussian noise $n_w(t)$. As stated earlier, main sources of this noise are the shot noise of photodiode(s) and the thermal noise of resistors and amplifiers, especially of the first amplifier stage. As the frequency components of both sources of noise are uniformly distributed, this noise is called *white noise*. Its double-sided psd measured in Amperes squared per Hertz (A²/Hz) is given by

$$G_c(f) = G_c = e \left(R_0 \; k \; P_1 + I_{dark} \right) + G_{th} \tag{9.70}$$

Except the coupling ratio k, above equation is same as Eq. (9.11). In the strict sense, both P_r *and* P_1 must be considered to determine the shot noise psd i.e., $eR_0 k(P_1+P_r)$. As $P_r \ll P_1$, this is practically not required.

Let us now consider the balanced receiver shown in Fig. 9.8, which is exclusively used in practice. The shot noise of both photodiodes is absolutely uncorrelated. Therefore, in a symmetrical balanced receiver ($k = 0.5$), noise psd is always twice the psd in a single-diode receiver:

$$G_c(f) = G_c = 2 \; e \left(R_0 \; k \; P_1 + I_{dark} \right) + G_{th} \tag{9.71}$$

With K_R and k as variables, psd of noise in both balanced and single-diode receivers can be expressed by the following common formula:

$$G_c(f) = G_c = e\,K_R\left(R_0\,k\,P_1 + I_{dark}\right) + G_{th} \qquad (9.72)$$

It is infer from this equation that first amplifier stage and, hence, the source of thermal noise is to be placed after the subtraction of photodiode currents. If amplifier is placed immediately after the photodiodes, two amplifiers would be required. In that case, thermal noise G_{th} will get doubled.

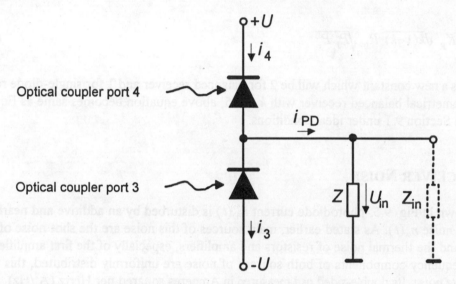

Fig. 9.8: Balanced receiver configuration

In order to reduce the effect of noise, amplified photodiode current must be filtered. It is an IF filter if a heterodyne receiver is used and a low-pass filter for a homodyne receiver. Thereby, the white noise $n_w(t)$ at the filter input will be converted to *band-limited coloured noise* $n(t)$ at the filter output. An important measure of $n(t)$ is its variance. In heterodyne receiver, this variance is given by

$$\sigma_{het}^2 = G_c\int_{-\infty}^{+\infty}|H_{IF}(f)|^2\,df = 2G_c\int_{-\infty}^{+\infty}|H_B(f)|^2\,df \qquad (9.73)$$

where $H_{IF}(f)$ and $H_B(f)$ represent the frequency response (also known as system transfer function) of the IF filter and equivalent virtual baseband filter respectively. In a heterodyne receiver, baseband filter function $H_B(f)$ is used to describe $H_{IF}(f)$ analytically. The IF filter and its baseband representation are related by the following relationships:

$$H_{IF}(f) = H_B(f - f_{IF}) + H_B(f + f_{IF}) \qquad (9.74)$$

and

$$h_{IF}(t) = h_B(t)e^{j2\pi f_{IF}t} + h_B(t)e^{-j2\pi f_{IF}t} \qquad (9.75)$$

Here, $h_{IF}(t)$ and $h_B(t)$ are the impulse response of IF filter and equivalent baseband filter respectively. As mentioned earlier, low-pass filter $H_{LP}(f)$ in a heterodyne receiver (Fig. 9.5) has to reject the double-frequency components caused by the demodulation mechanism. Hence, signal and noise are not influenced by this filter. In a homodyne receiver, $H_{LP}(f)$ is responsible for noise reduction also. At the output of this filter, noise variance is given by

$$\sigma_{hom}^2 = G_c \int_{-\infty}^{+\infty} |H_{LP}(f)|^2 \, df \qquad (9.76)$$

Under the condition that heterodyne IF filter and homodyne low-pass filter are of same type that is $H_{LP}(f) = H_B(f)$, following important relationship is obtained:

$$\sigma_{het}^2 = 2\sigma_{hom}^2 = 2G_c \int_{-\infty}^{+\infty} |H_B(f)|^2 \, df \qquad (9.77)$$

It becomes clear from the above equation that a homodyne receiver always offers a 3 dB higher signal-to-noise ratio or a 3 dB better receiver sensitivity than a heterodyne receiver provided signal power is same. However, it is only valid if ideal laser sources without any laser noise have been used. Further, modulation scheme (ASK or PSK) must be same in both types of systems. When the effect of phase noise is considered, a heterodyne receiver may even offer a better performance than a homodyne receiver (Chapters eleven and twelve).

In order to analyse the performance of coherent communication systems, receiver noise $n(t)$ is considered to be a zero mean Gaussian random process. Its probability density function (pdf) is

$$f_n(n) = \frac{1}{\sqrt{2\pi}\sigma_n} e^{-\left(n^2/2\sigma_n^2\right)} \quad \text{with} \quad \sigma_n = \begin{cases} \sigma_{het} = \sqrt{2}\sigma_{hom} & \text{for heterodyning} \\ \sigma_{hom} & \text{for homodyning} \end{cases} \qquad (9.78)$$

In the analysis of heterodyne receivers, use of the narrow-band representation

$$n(t) = x(t) \cos(2\pi f_{IF} t) + y(t) \sin(2\pi f_{IF} t) \tag{9.79}$$

is very advantageous. In the above representation, use of negative sign instead of plus sign will not affect the following results. The in-phase and quadrature components $x(t)$ and $y(t)$ are statistically independent, zero mean Gaussian random processes with

$$\sigma_n = \sigma_x = \sigma_y \tag{9.80}$$

Narrow-band representation is valid if

$$\frac{B_{IF}}{f_{IF}} \ll 1 \tag{9.81}$$

where B_{IF} is the noise-equivalent bandwidth of IF filter. The complex representation of narrow-band noise is

$$\underline{n}(t) = x(t) \, e^{j2\pi f_{IF} t} - jy(t) \, e^{j2\pi f_{IF} t}$$

$$= \left(x(t) - jy(t) \right) e^{j2\pi f_{IF} t} \quad \text{with} \quad n(t) = \text{Re}\{\underline{n}(t)\} \tag{9.82}$$

In this book, we have primarily considered Gaussian filters. It is represented by

$$H_B(f) = e^{-\pi\left(\frac{f}{2f_g}\right)^2} \quad \bullet\!\!-\!\!\circ \quad h_B(t) = 2f_g e^{-\pi(2f_g t)^2} \tag{9.83}$$

where f_g is the cut-off frequency. The double-sided noise-equivalent bandwidth of this filter is $B = \sqrt{2} f_g$. Thus, the noise variance at the filter output will be

$$\sigma_{het}^2 = 2\sigma_{hom}^2 = 2 \sqrt{2} \, G_c f_g \tag{9.84}$$

(v) Demodulation

In a heterodyne receiver, demodulator has to recover the transmitted information which is included in the IF signal

$$\underline{i}_{IF}(t) = \int_{-\infty}^{+\infty} \underline{i}_{PD}(\tau) \, h_{IF}(t-\tau) \, d\tau + \underline{n}(t)$$

$$= \hat{i}_{PD} \int_{-\infty}^{+\infty} \underline{s}(\tau) \, e^{j\phi(\tau)} \, h_B(t-\tau) \, d\tau \, e^{j2\pi f_{IF}t} \qquad (9.85a)$$

$$+ \hat{i}_{PD} \int_{-\infty}^{+\infty} \underline{s}(\tau) \, e^{j\phi(\tau)} \, h_B(t-\tau) \, e^{j4\pi f_{IF}\tau} d\tau \, e^{-j2\pi f_{IF}t} + \underline{n}(t)$$

When the narrow-band condition (9.81) is satisfied, second integral in the above equation containing the double-frequency term $\exp(j2\pi 2f_{IF}\tau)$ is approximately zero. In that case, Eq. (9.85a) can be approximated as

$$\underline{i}_{IF}(t) \approx \left[\hat{i}_{PD} \int_{-\infty}^{+\infty} \underline{s}(\tau) \, e^{j\phi(\tau)} \, h_B(t-\tau) \, d\tau + x(t) - jy(t) \right] e^{j2\pi f_{IF}t} \qquad (9.85b)$$

Depending on modulation scheme (ASK, FSK, PSK or DPSK), different types of demodulators can be used such as synchronous demodulator (which is often called coherent demodulator), envelope detector, frequency discriminator, one- and two-filter demodulator, autocorrelation demodulator etc. In this book, both heterodyne and homodyne receivers are called coherent receivers. On the basis of demodulation scheme, however, we have to distinguish between incoherent detection (for example, envelope detection) and coherent detection (for example, electrical or optical synchronous detection). In heterodyne receivers, either incoherent or coherent demodulation schemes can be used, whereas homodyne receivers always use coherent demodulation scheme. As mentioned earlier, low-pass filter in heterodyne receiver is responsible only for suppressing the undesired products of demodulation such as signal component of twice the intermediate frequency. Noise reduction and signal shaping are usually not performed by this filter.

In contrast to heterodyne receiver, demodulation process in homodyne receiver is performed by transferring the optical signal to baseband. Homodyne detection and demodulation are actually one single and joint process which cannot be separated. Thus, homodyne receiver can be regarded either as an optical synchronous receiver or as an optical coherent receiver with coherent detection. The low-pass filter in a homodyne receiver is responsible for not only noise reduction but also for undesired signal shaping and, hence, increasing the effect of intersymbol interference due to the restricted low-pass filter bandwidth B. To maximize the system performance, bandwidth and frequency response of this filter must be optimized. In this book, output signal of the low-pass filter is referred as *detected signal* $d(t)$ irrespective of type of receiver (heterodyne or homodyne), modulation and demodulation schemes (coherent or incoherent) used.

(vi) SAMPLE AND HOLD CIRCUIT

In the sample and hold circuit, detected signal $d(t)$ is sampled at equal temporal intervals of duration T (i.e., bit period). Finally, the sampled signal $d(vT + t_0)$ is applied to the decision circuit (Fig. 9.5). The time t_0 exactly defines the sampling point. Thus, t_0 represents an important system parameter which must be optimized to maximize system performance. In case of symmetrical filters, for example a Gaussian filter, optimum decision time is always at the centre of each symbol (bit) i.e., $t_0 = 0$.

(vii) DECISION CIRCUIT

Depending on whether signal $d(vT + t_0)$ is above or below a fixed threshold, symbol r_v = "0" or r_v = "1" is decided by the decision circuit and sent to the data output of digital receiver. In addition to filter transfer function $H_B(f)$ and sampling time t_0, threshold voltage E is a third optimizable system parameter of primary importance (Chapters ten and eleven). In the calculation of bit error probability or BER, shape of the detected signal $d(t)$ and statistical features of its sampled values $d(vT + t_0)$ play a major role.

9.5   EYE PATTERN AND PROBABILITY OF ERROR

This Section presents a brief review of two important and powerful measures to determine the transmission quality of digital optical communication systems. As in digital electrical transmission systems, these measurements are the

- Eye pattern and
- Probability of error or bit error rate (BER).

The general results obtained in this Section will be applied to coherent optical communication systems in Chapter eleven. The readers who are already familiar with this topic may escape this Section and move over to the next Chapter.

9.5.1   INFLUENCE OF NOISE AND INTERSYMBOL INTERFERENCE

The received signal in a digital communication system is significantly perturbed by noise and intersymbol interference (ISI) which is caused by the restricted bandwidth of various components such as filters. The most important signal which directly determines the transmission quality, is the detected signal $d(t)$ at the input of sample and hold circuit. It is given by

$$d(t) = d_0(t) + n(t) \tag{9.86}$$

This can be regarded as a linear superposition of signal $d_0(t)$ and noise $n(t)$. The signal $d_0(t)$ in presence of ISI will look like as in Fig. 9.9.

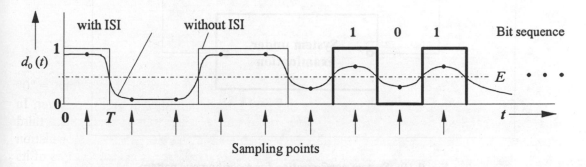

Fig. 9.9: Detected signal $d_0(t)$ with and without ISI

At a given bit rate $R = 1\text{bit}/T$, influence of ISI can be reduced by increasing the available transmission bandwidth of all responsible system components and filters. In case of an infinite bandwidth, signal $d_0(t)$ would be rectangular in shape (i.e., without any ISI). However, large bandwidth will increase the noise. Thus, a trade-off exists between bandwidth and transmission quality. It is clear that the bandwidth is an optimizable system parameter of prime importance.

For the design and development of digital communication systems, some measure to determine the level of ISI is required. For this reason, eye pattern recording represents a very simple and frequently used measurement technique. The eye pattern is briefly described below.

9.5.2 EYE PATTERN

The detected signal $d_0(t)$ when fed to the input of an oscilloscope triggered on bit duration T or multiple of T, a pattern having the shape of an eye is obtained on the screen. The schematic diagram for the measurement is given in Fig. 9.10 and a typical eye pattern in Fig. 9.11. The system under examination in Fig. 9.10 can be either a complete digital communication link or merely a component of it; for example, a single filter. The signal distortion due to ISI can directly be observed on the oscilloscope screen. Therefore, recording of the eye pattern is a measurement technique which operates in the time domain. Depending on the communication system, eye pattern can be asymmetrical in amplitude as well as in time.

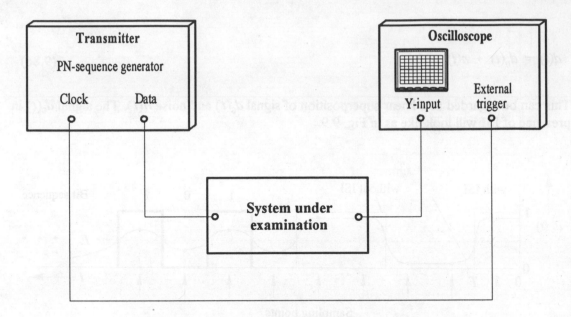

Fig. 9.10: System configuration for recording eye pattern

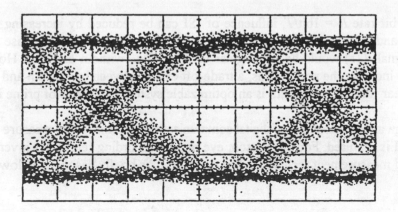

Fig. 9.11: Typical eye pattern

Generation of eye pattern is explained in Figs. 9.12 and 9.13. Here, each bit is disturbed by only one previous and following bit. For a rectangular input bit signal $a(t)$, Fig. 9.12 shows the detected signal $d_0(t)$ as a superposition of impulse response for different bit combinations. Based on Fig. 9.12, generation of the eye pattern in a more comprehensive manner is illustrated in Fig. 9.13.

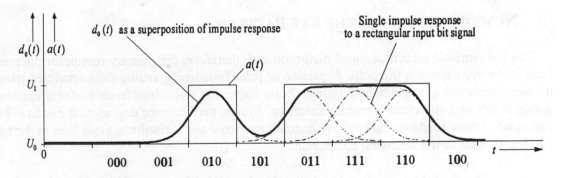

Fig. 9.12: Principle of eye pattern recording (explanation 1)

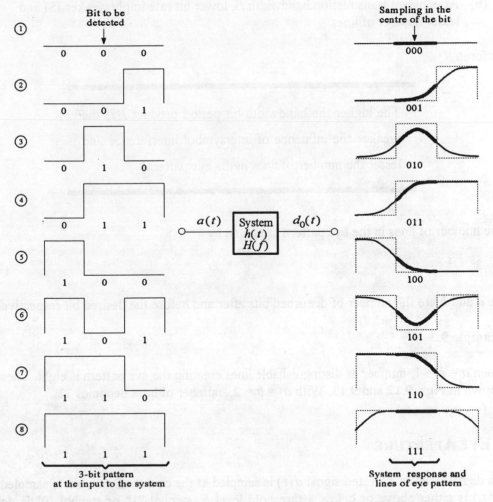

Fig. 9.13: Principle of eye pattern recording (explanation 2)

(i) NUMBER OF LINES IN THE EYE PATTERN

The fundamental reason of signal distortion and, therefore, the primary reason for different lines in the eye pattern is the pulse dispersion or pulse broadening arising from restricted transmission bandwidth (for example, filtering). Pulse dispersion yields interference of neighbouring pulses or bits and, therefore, a mutual distortion. System performance degradation due to ISI is higher when more neighbouring bits have mutually interfered and it results in more lines in the eye pattern. It leads to the following conclusion:

(a) At constant bit rate $R = 1/T$, higher bandwidth imply weaker ISI and lesser number of lines.

(b) At constant transmission bandwidth B, lower bit rate imply weaker ISI and lesser number of lines.

Therefore:

> The higher the bandwidth-bit period product BT, the weaker the influence of intersymbol interference and lesser the number of lines in the eye pattern!

The number of lines in the eye pattern are given by

$$K = 2^{a+b+1}$$

(9.87)

where a and b are the number of disturbed bits after and before the desired bit respectively.

Example 9.3

When $a = b = 1$, number of distinguishable lines creating the eye pattern is eight. These lines are shown in Figs. 9.12 and 9.13. With $a = b = 2$, number of lines becomes 32.

(ii) EYE APERTURE

In a digital receiver, detected signal $d(t)$ is sampled at the centre of each bit. If sampled value $d(vT+t_0)$ is either above or below a threshold level E, symbol "1" or symbol "0" is detected respectively:

$$d(vT + t_0) \geq E \rightarrow 1,$$

$$d(vT + t_0) < E \rightarrow 0.$$

When the sampled value $d_0(vT+t_0)$ is very close to threshold, even a small noise level may be high enough to cause an error. The probability of error becomes maximum for those bit sequences which are responsible for the inner lines in the eye pattern. We call these special sequences the *worst-case pattern*. In practice, the worst-case pattern are often given by

$$\cdots 0\ 0\ 0\ 1\ 0\ 0\ 0\ \cdots \text{ (single one, upper worst-case line) and}$$

$$\cdots 1\ 1\ 1\ 0\ 1\ 1\ 1\ \cdots \text{ (single zero, lower worst-case line).}$$

The characteristic parameters of an eye pattern i..e., eye aperture and worst-case sample values d_{1w} and d_{0w} for the above worst-case pattern are shown in Fig. 9.14.

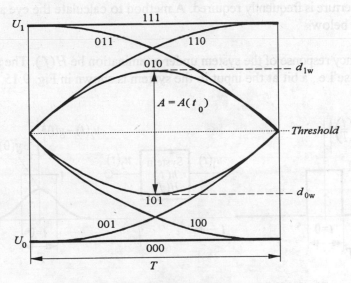

Fig. 9.14: Eye aperture in the eye pattern

As shown in the above figure, worst-case pattern define the inner open part of the eye. The absolute and normalized eye aperture A and a are

$$A = d_{1w} - d_{0w} = (U_1 - U_0) - 2d_{0w} = 2d_{1w} - (U_1 - U_0) \tag{9.88}$$

and

$$a = \frac{A}{U_1 - U_0} \, 100\% \tag{9.89}$$

The eye aperture A is a well-suited and powerful measure to assess the system performance and to determine the probability of error (discussed in Section 9.5.3). In general:

> The smaller the eye aperture A, the
> higher the probability of error!

Normally, the eye pattern and eye aperture are measured by the set up given in Fig. 9.10. To analyse and optimize digital communication systems, analytical calculation or at least an estimation of the eye aperture is frequently required. A method to calculate the eye aperture A has been briefly discussed below.

Let the frequency response of the system under examination be $H(f)$. The system response to a rectangular pulse i.e., a bit at the input to the system is shown in Fig. 9.15.

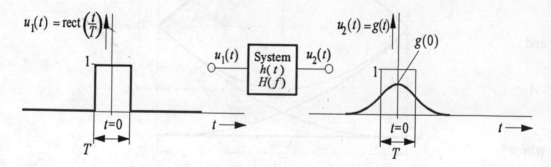

Fig. 9.15: System response to rectangular pulse (bit) signal

The eye aperture A can be determined by the following general expression

$$A = 2(U_1 - U_0)\left[g(t_0) - \sum_{\nu=0}^{\nu=+\infty} |g(t_0 - \nu T)| - \sum_{\mu=0}^{\mu=+\infty} |g(t_0 + \mu T)| \right] \tag{9.90}$$

where $g(t)$ represents the system response to a rectangular pulse as shown in Fig. 9.15 and U_0 and U_1 are the input signal levels corresponding to bit "0" and "1" respectively. When the worst-case pattern of the system are known, calculation of the eye aperture is simplified. With worst-case pattern of "single one" and "single zero", we obtain

$$A = (U_1 - U_0)\left[2g(t_0) - H(0)\right] \tag{9.91}$$

where $H(0)$ represents the DC frequency response of the system which is often equal to unity and $g(t)$ can be easily determined from the convolution $u_1(t) \star h(t)$.

When the frequency response $H(f)$ of the system is same as that of a Gaussian low-pass or baseband filter as given in Eq. (9.83) i.e., $H(f) = H_B(f)$, the characteristic parameters of the eye pattern as shown in Fig. 9.14 are obtained as

$$d_{1w} = (U_1 - U_0) \int_{-T/2}^{+T/2} h_B(-t)\, dt = (U_1 - U_0)\left[1 - 2Q\left(\sqrt{2\pi}f_g T\right)\right] \tag{9.92}$$

$$d_{0w} = (U_1 - U_0)(1 - d_{1w}) = 2(U_1 - U_0)Q\left(\sqrt{2\pi}f_g T\right) \tag{9.93}$$

and

$$A = (U_1 - U_0)\left[1 - 4Q\left(\sqrt{2\pi}f_g T\right)\right] \tag{9.94}$$

where

$$Q(x) = \frac{1}{\sqrt{2\pi}} \int_{x}^{+\infty} e^{-u^2/2}\, du = \frac{1}{2}\, \mathrm{erfc}\left(\frac{x}{\sqrt{2}}\right) \tag{9.95}$$

is the complementary Gaussian error function and erfc the complementary error function [1, 9]. Eye aperture A as a function of system bandwidth $B = \sqrt{2}f_g$ is shown in Fig. 9.16.

When the frequency response of the system equals the frequency response of a matched filter, the eye pattern is opened by 100% i.e., $A = U_1 - U_0$, $d_{1w} = U_1$ and $d_{0w} = U_0$.

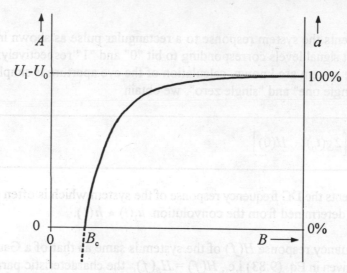

Fig. 9.16: Eye aperture A as function of system bandwidth B

It is clear from Fig. 9.16 that no eye can be observed when bandwidth B is below the critical bandwidth B_c. In this case, eye aperture A mathematically becomes negative. The simple physical reason for this is that worst-case lines single one and single zero do no longer cross each other since there is not enough time to level off.

9.5.3 PROBABILITY OF ERROR AND BIT ERROR RATE

Due to ISI, noise and other inherent system imperfections, bit sequence at the output of a digital receiver is usually different from the transmitted bit sequence i.e., $\langle q_v \rangle \neq \langle r_v \rangle$. Each error in detection is called a *bit error*. Higher the number of errors within a given time, lower is the quality of a digital transmission system. Therefore, probability of bit error or simply probability of error represents an important measure to quantify system performance and assess system quality.

In practice, probability of error is frequently called the *Bit Error Rate* (BER). Actually, this is a misnomer, because the term "rate" is always related to certain interval of time; for example, the bit *rate* of digital systems is in unit of bit/s. Therefore, the unit of bit error *rate* must also be bit/s or error/s, whereas the probability of error p is a dimensionless quantity. As an example, consider a bit rate $R = 10$ Gbit/s (i.e., $T = 0.1$ ns) and a probability of error $p = 10^{-9}$. This means that only one single bit out of 10^9 bits (1 Gbit) is detected incorrectly. In other words, ten bit errors occur during a time interval of 1 second. Hence, the rate of bit errors or *BER* is 10 bit/s.

Generally, the relationship between bit rate R, probability of error p and *BER* is given by the simple expression

$$BER = pR \tag{9.96}$$

In practice, both p and BER are used interchangeably. In this book, we also follow the same trend.

As discussed in the previous Section, probability of error depends upon the transmitted bit sequence. Therefore, each transmitted bit pattern has its own characteristic probability of error. For this reason, calculation of the average BER (p_a) is very comprehensive. In practice, it is much more convenient to determine the worst-case BER (p_w) which is given by

$$p_w = \frac{1}{2}(p_1 + p_0) \geq p_a \tag{9.97}$$

Here, p_1 and p_0 are the probabilities of error for equiprobable binary symbols "1" and "0" respectively in the worst-case pattern. The above equation can be taken as a very good and well-suited approximation i.e., $p_w \approx p_a$. It can also be taken as an upper limit on the probability of error i.e., $p_a \leq p_w$. As discussed in Section 9.5.2, worst-case pattern are frequently given by "single one" and "single zero". Therefore,

$$p_1 = p(1 \rightarrow 0) \quad \text{in case of pattern} \quad \cdots 0001000 \cdots \text{ and}$$

$$p_0 = p(0 \rightarrow 1) \quad \text{in case of pattern} \quad \cdots 1110111 \cdots.$$

The worst-case pattern correspond to the worst-case sample values d_{1w} and d_{0w} which are much closer to threshold than those of all other bit pattern (Section 9.5.2). As a consequence, worst-case pattern are most sensitive to noise and yield much higher probability of error than the other pattern. Hence, worst-case pattern and worst-case BER are of great interest in practical communication system design.

(i)   CALCULATION OF WORST-CASE BER

As shown in Fig. 9.17, bit error occurs with a probability determined by the probability density function (pdf) $f_d(d)$ of the sampled detected signal

$$d(t_0) = d_0(t_0) + n(t_0) \tag{9.98}$$

at the input to the decision circuit. Here, $d_0(t_0)$ represents the signal component. Under worst conditions, it will be

$$d_0(t_0) = \begin{cases} d_{1w} & \text{in case of pattern } \cdots 0001000 \cdots \\ d_{0w} & \text{in case of pattern } \cdots 1110111 \cdots \end{cases} \qquad (9.99)$$

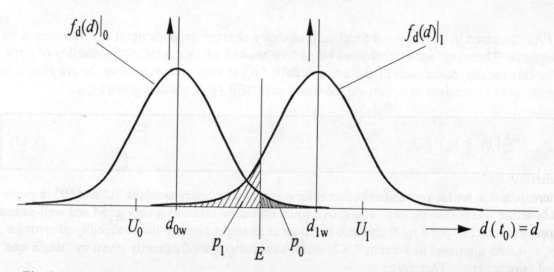

Fig. 9.17: Probability density functions of detected signals and error regions for bit "0" and "1"

In practice, fast estimation of BER is often required. For this purpose, d_{1w} and d_{0w} can be approximated as

$$d_{1w} \approx U_1 \quad \text{and} \quad d_{0w} \approx U_0 \qquad (9.100)$$

which means that ISI has been neglected. In the absence of phase noise, noise in coherent optical communication systems can be regarded as *A*dditive *W*hite *G*aussian *N*oise (AWGN). Hence, the sampled noise $n(t_0)$ will be Gaussian distributed with zero mean and variance σ_n^2 as given by Eq. (9.78). In that case, BER can be calculated as follows:

$$p = \frac{1}{2}(p_1 + p_0) = \frac{1}{2}\int_E^\infty f_d(d)\big|_0 \; dd + \frac{1}{2}\int_{-\infty}^E f_d(d)\big|_1 \; dd$$

$$= \frac{1}{2}\frac{1}{\sqrt{2\pi}\sigma_n}\int_E^\infty e^{-(d-d_{0w})^2/2\sigma_n^2}\,dd \;+\; \frac{1}{2}\frac{1}{\sqrt{2\pi}\sigma_n}\int_{-\infty}^E e^{-(d-d_{1w})^2/2\sigma_n^2}\,dd \qquad (9.101)$$

Here, $f_d(d)|_0$ and $f_d(d)|_1$ represent the pdf of detected signal for the transmitted symbol "0" and symbol "1" respectively. Substitution of $x = (d - d_{0w})/\sigma_n$ and $y = (d - d_{1w})/\sigma_n$ in the above equation yields

$$p = \frac{1}{2} \frac{1}{\sqrt{2\pi}} \int_{\frac{E - d_{0w}}{\sigma_n}}^{\infty} e^{-x^2/2} dx + \frac{1}{2} \frac{1}{\sqrt{2\pi}} \int_{-\infty}^{\frac{E - d_{1w}}{\sigma_n}} e^{-y^2/2} dy \qquad (9.102)$$

$$= \frac{1}{2} Q\left(\frac{E - d_{0w}}{\sigma_n}\right) + \frac{1}{2} Q\left(\frac{d_{1w} - E}{\sigma_n}\right)$$

where Q is again the complementary Gaussian error function as given in Eq. (9.95). It may be mentioned that $f_d(d)|_0$ and $f_d(d)|_1$ are Gaussian distributed only in the absence of phase noise. It becomes clear from Fig. 9.17 that BER reaches minimum if threshold level E is located at the cross point where pdf $f_d(d)|_0$ and pdf $f_d(d)|_1$ meet i.e.,

$$E_{opt} = \frac{d_{1w} + d_{0w}}{2} \qquad (9.103)$$

With the above optimum threshold E_{opt} and eye aperture $A = d_{1w} - d_{0w}$ (Section 9.5.2), we finally get the simple and well-known expression

$$p = Q\left(\frac{A}{2\sigma_n}\right) = Q\left(\frac{1}{2}\sqrt{\frac{S}{N}}\right) \qquad (9.104)$$

It may be mentioned that the above equation is valid only for a binary digital baseband system. For multilevel systems or carrier systems with modulation schemes such as FSK, PSK or DPSK, it has to be modified appropriately (Chapters eleven and twelve).

In most digital systems with AWGN, there is a steep fall in p with the increase in S/N ratio (Fig. 9.18). When phase noise disturb the system in addition, this behaviour is no longer valid. In such cases, error rate curve exhibits a characteristic error rate floor (Chapters eleven and twelve) and is, therefore, different from the curve shown in Fig. 9.18.

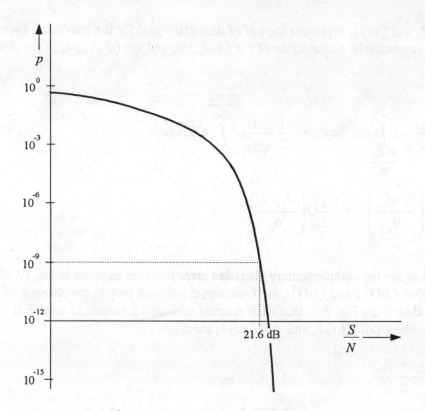

Fig. 9.18: Variations of probability of error with signal-to-noise ratio in digital system with AWGN

In practice, $p = 10^{-9}$ is a frequently demanded BER. For this BER, required S/N ratio can be determined with a high accuracy using the following approximation:

$$Q(6) \approx 10^{-9} \tag{9.105}$$

Finally, some typical eye pattern and their characteristic BERs are shown in Fig. 9.19. All eye pattern shown in this figure have been measured in a realized digital optical communication system [14]. It becomes evident from this figure that there is a strong relationship between eye pattern, eye aperture and bit error rate. With some training and practical experience, it is possible to estimate the bit error rate only by considering the eye pattern. Since a BER measuring equipment is usually more expensive than the set up required for the recordings of eye pattern particularly when the bit rate is high, this method of estimation represents a powerful as well as a low-cost technique.

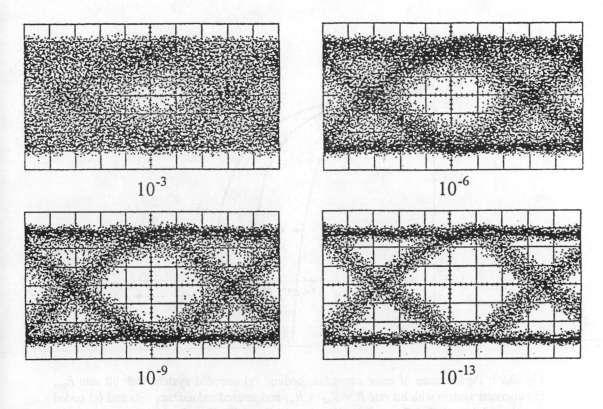

Fig. 9.19: Correlation between eye pattern and BER

BER can be decreased by employing error correction coding techniques [5, 10]. Most suitable are block coding such as the Reed-Solomon code and BCH code and convolutional coding such as Viterbi coding. All these coding techniques use redundancy bits to recognize and correct bit errors. Due to the redundancy, total bit rate (i.e., channel bit rate) increases. A higher bit rate, however, requires a larger bandwidth which, first of all, yields a higher BER since noise power increases (process one). If redundancy bits are used to correct the bit errors, then BER decreases (process two). The process one degrades the system performance, whereas the process two improves it. A performance in terms of a lower BER is obtained when second process is dominant in comparison to the first process. This is normally true in practical systems. Instead of decreasing BER, it is more advantageous in fiber optics to use coding to reduce required S/N ratio or the required received input optical power level for a given BER e.g., 10^{-10}. The performance of error correcting coding, also referred as channel coding, is illustrated in Fig. 9.20. In this figure, BER curves are shown for two uncoded systems, one with a lower and other with a higher bit rate, and one coded system. In the coded system, bit rate is same as in the high bit rate uncoded system. Thus, uncoded high bit rate system can be regarded as a system with redundancy bits which are, however, not used for error correction.

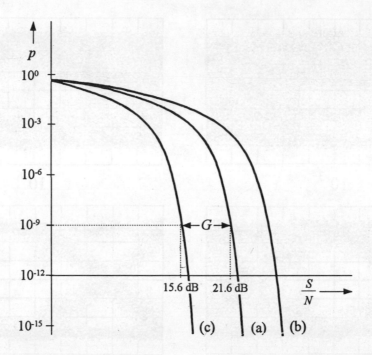

Fig. 9.20: Performance of error correction coding: (a) uncoded system with bit rate R_{inf}, (b) uncoded system with bit rate $R = R_{inf} + R_{red}$ and unused redundancy bits and (c) coded system with bit rate $R = R_{inf} + R_{red}$ and code gain $G = 6$ dB. Here R_{inf} and R_{red} represent the bit rates due to information and redundancy bits respectively.

It becomes clear from this figure that S/N ratio required for a BER of 10^{-10} may be lower in the system employing coding than in the uncoded system. This gain in S/N ratio is known as the *coding gain* which is typically 6 dB. Coding gain of more than 6 dB is still possible. It depends on the type of coding technique and the code rate R_c which is defined as $R_c = R_{inf}/(R_{inf} + R_{red})$.

(ii) INFLUENCE OF FILTERING

Eye pattern and bit error rate are strongly influenced by filtering. As discussed in Section 9.5.2, eye pattern and, consequently, the eye aperture A are functions of the receiver bandwidth. Noise power (N) increases linearly with bandwidth. It means

(a)   large bandwidth B imply large eye aperture A and low BER (without noise) and

(b)   large bandwidth B imply more noise power and high BER (without ISI).

Both the statements above are contradictory. For this reason, there exists an optimum filter bandwidth B - actually an optimum filter frequency response $H(f)$ instead of only an optimum bandwidth - for which S/N ratio is maximum and BER is minimum (Fig. 9.21).

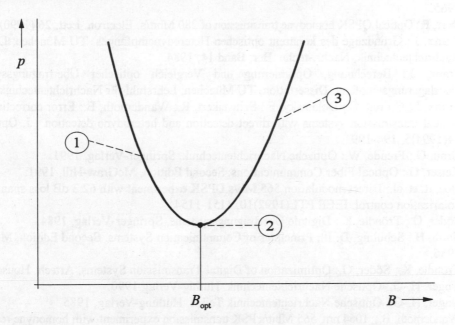

Fig. 9.21: BER p as function of receiver bandwidth B for (1) less noise, but small eye aperture (2) optimum bandwidth and (3) more noise, but large eye aperture

It is well-known from communication engineering that the optimum filter which provides a maximum S/N ratio and a minimum bit error rate at the output of the filter can either be realized by the matched filter or correlation technique. Indeed, both types of optimum filters are different in realization, but result in the same S/N ratio and bit error rate [9, 11]. Therefore, the choice between matched filter and correlator merely depends on realization aspect and not on communication. It may be mentioned that if the type of filter is given, then the optimization is restricted on filter parameters only like the filter bandwidth $B = \sqrt{2}f_g$ for a Gaussian filter.

In digital transmission systems with AWGN, influence of restricted bandwidth can easily be evaluated since a Gaussian noise process always remains Gaussian. However, as soon as phase noise such as laser phase noise is added, evaluation becomes complicated and very comprehensive as the filter input signal is now a non-Gaussian process. Moreover, this non-Gaussian process will additionally be changed by filtering. This problem of coherent optical communication systems is discussed in more detail in Chapter eleven.

9.6   REFERENCES

[1]   Abramowitz, M.; Stegun, I. A.: Handbook of Mathematical Functions. Dover Publications Inc., 1965.

[2]   Derr, F.: Optical QPSK homodyne transmission of 280 Mbit/s. Electron. Lett. 26 (1990)6, 401-403.

[3]   Franz, J.: Grundzüge des kohärent optischen Heterodynempfanges. TU München, Lehrstuhl für Nachrichtentechnik, Nachr.-techn. Ber. Band 14, 1984.

[4]   Franz, J.: Berechnung, Optimierung und Vergleich optischer Übertragungssysteme mit Überlagerungsempfang. Dissertation, TU München, Lehrstuhl für Nachrichtentechnik, 1987.

[5]   Franz, J.; Correll, C.; Dolainsky, F.; Schweikert, R.; Wandernoth, B.: Error correcting coding in optical transmission systems with direct detection and heterodyne detection. J. Opt. Commun. 14(1994)5, 194-199.

[6]   Grau, G.; Freude, W.: Optische Nachrichtentechnik. Springer-Verlag, 1991.

[7]   Keiser, G.: Optical Fiber Communications. Second Edition, McGraw-Hill, 1991.

[8]   Noé, R. et. al.: Direct modulation 565 Mb/s DPSK experiment with 62.3 dB loss span and endless polarization control. IEEE PTL(1992)10, 1151-1154.

[9]   Söder, G.; Tröndle, K.: Digitale Übertragungssysteme. Springer-Verlag, 1984.

[10]   Taub, H.; Schilling, D. L.: Principles of Communication Systems. Second Edition, McGraw-Hill, 1986.

[11]   Tröndle, K.; Söder, G.: Optimization of Digital Transmission Systems. Artech House, 1987.

[12]   Unger, H.-G.: Optische Nachrichtentechnik. Hüthig-Verlag 1990.

[13]   Unger, H.-G.: Optische Nachrichtentechnik Teil II. Hüthing-Verlag, 1985.

[14]   Wandernoth, B.: 1064 nm, 565 Mbit/s PSK transmission experiment with homodyne receiver using synchronization bits. Electron. Lett. 27(1991)19, 1692-1693.

10 INCOHERENT SYSTEMS: ANALYSIS AND OPTIMIZATION

This Chapter reviews the basics of the conventional optical transmission systems with intensity modulation and direct detection (IM/DD) also referred as incoherent detection. The principle operation of optical direct detection system has already been discussed in Chapter nine. In this Chapter, we focus our interest on BER calculation (Section 10.1) and system optimization (Section 10.2). In the analysis, degradation due to intersymbol interferences and additive Gaussian noise i.e., shot noise of photodiode and thermal noise of the amplifiers, in particular of the electrical front-end preamplifier has been taken into account. As already explained in Chapter nine, intersymbol interference is caused by the limited bandwidth of the system components, primarily the receiver baseband filter which is required for noise reduction. For the analysis of the receiver, the detected signal $d(t)$ and its sample values $d(vT+t_0)$ are of fundamental importance. Evaluation of BER, for example, requires the probability density function (pdf) of these sample values.

Performance of advanced optical communication systems with incoherent detection is also degraded by the noise of transmitter laser. The influence of laser noise is described in Section 10.3 and an useful approximation to assess the magnitude of this noise as compared to other sources of noise is given. For a practical receiver, calculation of signal power, noise power and signal-to-noise ratio is made in Section 10.4. Finally, in Section 10.5 performance evaluation of IM/DD system in terms of signal-to-noise ratio has been carried out with optical amplifier in different configurations.

10.1 ANALYSIS

As described in Chapter nine, detected signal $d(t)$ at the input to the sample and hold circuit is of prime importance since it directly influences the BER and, hence, the performance of digital communication systems.

(i) DETECTED SIGNAL $d(t)$ AND SAMPLE VALUES $d(vT+t_0)$

In an optical direct detection receiver as shown in Fig. 10.1, detected signal $d(t)$ is the signal at the output of the baseband filter which is a low-pass filter. Using the results of Chapter nine, we can write

$$d(t) = MR_0 P_r \int_{-\infty}^{+\infty} s(\tau) h_B(t-\tau) \, d\tau + n(t)$$ (10.1)

where $h_B(t)$ represents the impulse response of baseband or low-pass filter, $n(t)$ the additive Gaussian noise at the output of this filter and

$$s(t) = \sum_{\upsilon=-\infty}^{+\infty} s_\upsilon \operatorname{rect}\left(\frac{t-\upsilon T}{T}\right) \quad \text{with} \quad s_\upsilon = \begin{cases} 0 & \text{if } q_\upsilon = "0" \\ 1 & \text{if } q_\upsilon = "1" \end{cases}$$ (10.2)

the transmitted binary sequence. The product $s(t)P_r$ in Eq. (10.1) represents the *intensity modulation* which is usually applied in a direct detection system.

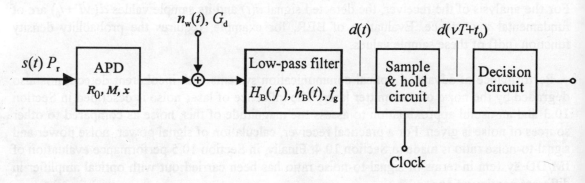

Fig. 10.1: Simplified block diagram of an optical direct detection (incoherent) receiver

For a low-pass filter with Gaussian frequency response and cut-off frequency f_g (Eq. 9.83), variance of the zero mean Gaussian noise $n(t)$ is given by

$$\sigma_n^2 = G_d \int_{-\infty}^{+\infty} |H_B(f)|^2 \, df = \sqrt{2} G_d f_g$$ (10.3)

In an IM/DD system, noise power spectral density G_d is actually not a constant since the shot noise which represents a dominant part of G_d depends on the modulated optical power $s(t)P_r$ at the input to the receiver. Hence, such a system belongs to a class of systems which are characterized by *signal-dependent noise*. For this reason, both the binary symbols "0" and "1" are normally disturbed differently. The transmitted symbol "0" is not or even less disturbed by the shot noise, whereas symbol "1" is always disturbed more.

As well-known from standard digital communication systems, detected signal $d(t)$ is applied to a sample & hold and decision circuits. Due to the additive noise $n(t)$, sample values $d(\upsilon T + t_0)$ are random variables. As the statistical features of $n(t)$ are time-invariant, it is sufficient to determine the statistical parameters of the random variables $d(\upsilon T + t_0)$ at an arbitrary, but fixed point of time, for example, $t = t_0$. The results obtained are valid in general i.e., at each sampling point of time $\upsilon T + t_0$. With the following normalisations and substitutions

$$d := \frac{d(t_0)}{MR_0P_r} \qquad n := \frac{n(t_0)}{MR_0P_r} \qquad \rightarrow \qquad \sigma := \frac{\sigma_n}{MR_0P_r}$$

$$(10.4)$$

$$a := a(t_0) = \int_{-\infty}^{+\infty} s(\tau)\, h_B(t_0-\tau)\, d\tau \quad \text{with} \quad 0 \le a(t_0) \le 1$$

the sampled and normalized detected signal d can be simply expressed as

$$d = a + n \qquad\qquad (10.5)$$

Here d, a and n have the following physical meanings:

- d: normalized sample value of the detected signal,
- a: noiseless part of d disturbed by intersymbol interference (including the transmitted information) and
- n: receiver additive Gaussian noise.

(ii) PROBABILITY DENSITY FUNCTION (pdf) $f_d(d)$

As the additive receiver noise is Gaussian distributed, sampled detected signal d is Gaussian distributed too. As seen from Eq. (10.4), expected value $\eta_d = a$ in this Gaussian pdf is a function of the impulse response $h_B(t)$ of the low-pass filter and transmitted information $s(t)$ i.e., transmitted bit sequence $<q_\upsilon>$. For this reason, normalized standard deviation $\sigma_d = \sigma$ of the sample value d is also a function of the transmitted bit pattern. Depending on whether symbol "0" or symbol "1" is transmitted, following normalized variances are obtained (compare with Eq. 9.3):

$$\sigma_0^2 = \sigma_d^2\big|_{q_\upsilon="0"} = \frac{1}{(M R_0 P_r)^2} \left(eM^{2+x}I_{dark} + G_{th}\right)\sqrt{2}\, f_g \qquad (10.6a)$$

and

$$\sigma_1^2 = \sigma_d^2\big|_{q_v="1"} = \frac{1}{\left(M\,R_0\,P_r\right)^2}\left(eM^{2+x}(R_0P_r + I_{dark}) + G_{th}\right)\sqrt{2}\,f_g \tag{10.6b}$$

Here, $\sqrt{2}f_g$ is the noise-equivalent bandwidth of the low-pass filter and $1/(MR_0P_r)^2$ the normalization factor as used in Eq. (10.4). The Gaussian pdf of sampled detected signal is

$$f_d(d) = \frac{1}{\sqrt{2\pi}\sigma_d}\exp\left(-\frac{(d-a)^2}{2\sigma_d^2}\right) \quad \text{where} \quad \begin{cases} \sigma_d=\sigma_0, \; a=a_{0i} & \text{if} \quad q_v="0" \\ \sigma_d=\sigma_1, \; a=a_{1i} & \text{if} \quad q_v="1" \end{cases} \tag{10.7}$$

where the suffix "i" distinguishes symbol pattern $\langle q_v\rangle_{0i}$ with q_{0i} = "0" and $\langle q_v\rangle_{1i}$ with q_{1i} = "1" and noiseless sample values a_{0i} and a_{1i}. In the strict sense, standard deviation σ_d in the pdf $f_d(d)$ is actually a function of transmitted symbol sequence $\langle q_v\rangle_i$. As the signal-independent additive Gaussian noise usually dominates the signal-dependent shot noise, it is sufficient to consider σ_0 and σ_1. Thus, more accurate division into σ_{0i} and σ_{1i} on the basis of symbol pattern $\langle q_v\rangle_{0i}$ and $\langle q_v\rangle_{1i}$ is normally not required.

(iii) PROBABILITY OF ERROR

As mentioned in Chapter nine, worst-case probability of error p_w is normally used (i) to estimate the average probability of error $p_a \le p_w$ and (ii) as an upper limit. Hence, we again focus our interest on the calculation of p_w.

The worst-case probability of error called the worst-case BER is determined by considering both the worst-case bit pattern $\langle q_v\rangle_{0w}$ and $\langle q_v\rangle_{1w}$. If a Gaussian low-pass filter is employed, then noiseless sample values for these special pattern are given by

$$\langle q_v\rangle_{1w} = \langle\cdots 0\,0\,0\,1\,0\,0\,0\cdots\rangle \;\Rightarrow\; a_{1w} = 1 - 2Q\left(\sqrt{2\pi}f_g T\right)$$

$$\langle q_v\rangle_{0w} = \langle\cdots 1\,1\,1\,0\,1\,1\,1\cdots\rangle \;\Rightarrow\; a_{0w} = 1 - a_{1w} \tag{10.8}$$

where letter Q represents the Q-function as defined in Eq. (9.95). With a_{1w} and a_{0w}, normalized eye aperture A_d of direct detection system can be determined. It is a well-suited and a powerful measure to assess the influence of intersymbol interference (see Chapter nine). We obtain

$$A_d = a_{1w} - a_{0w} = 1 - 2a_{0w} = 2a_{1w} - 1 = 1 - 4Q\left(\sqrt{2\pi}f_g T\right) \tag{10.9}$$

It should be remembered that the constant of normalization is again the amplitude $\hat{\imath}_{PD}$ of the photodiode current. Thus, absolute eye aperture is obtained by multiplying A_d from Eq. (10.9) with $\hat{\imath}_{PD}$.

With the above results, we are now able to calculate the bit error rate. For this, we first have to determine the worst-case probability density functions $f_{d1w}(d)$ and $f_{d0w}(d)$. Considering equal occurrence probabilities for binary symbols "0" and "1", we obtain

$$p_w = \frac{1}{2}\left[\int\limits_{E}^{+\infty} f_{d0w}(d)\ dd + \int\limits_{-\infty}^{E} f_{d1w}(d)\ dd \right] \tag{10.10}$$

where E is the threshold level. Since both the probability density functions $f_{d1w}(d)$ and $f_{d0w}(d)$ are Gaussian, above expression can be rewritten as follows:

$$P_w = \frac{1}{2}\left[Q\left(\frac{E - a_{0w}}{\sigma_0} \right) + Q\left(\frac{a_{1w} - E}{\sigma_1} \right) \right] \tag{10.11}$$

By using Eq. (10.9), the above equation becomes

$$P_w = \frac{1}{2}\left[Q\left(\frac{2E - (1-A_d)}{2\sigma_0} \right) + Q\left(\frac{(1+A_d) - 2E}{2\sigma_1} \right) \right] \tag{10.12}$$

It should again be remembered that both standard deviations σ_0 and σ_1 as well as eye aperture A_d and threshold level E are normalized with respect to the amplitude $\hat{\imath}_{PD} = MR_0P_r$ of the photodiode current.

10.2   OPTIMIZATION

First step in the optimization of a receiver performance is to decide which parameters are optimizable and which are not. In this book, we call a system parameter to be optimizable when there is an optimum value existent which is neither zero nor infinite. In this sense, psd G_{th}, dark current I_{dark}, input power P_r and responsivity R_0 are non-optimizable since their theoretical optimum is located at zero (G_{th} and I_{dark}) or at infinite (P_r and R_0). It becomes clear from Eq. (10.12) that BER can be minimized by optimizing the following *optimizable system parameters*:

- cut-off frequency f_g of the low-pass filter since $\sigma_0 = \sigma_0(f_g)$, $\sigma_1 = \sigma_1(f_g)$ and $A_d = A_d(f_g)$,
- threshold level E and
- avalanche gain M of the photodiode since $\sigma_0 = \sigma_0(M)$ and $\sigma_1 = \sigma_1(M)$.

(i) THRESHOLD E

As mentioned in Chapter nine, optimum threshold E_{opt} for discriminating symbol "0" and symbol "1" is located at the crosspoint where the probability density functions $f_{d0w}(d)$ and $f_{d1w}(d)$ meet. Therefore, the equation for determining the optimum threshold level is given by

$$f_{d1w}(E_{opt}) = f_{d0w}(E_{opt}) \tag{10.13}$$

Since both the probability density functions $f_{d0w}(d)$ and $f_{d1w}(d)$ are Gaussian, it is possible to determine E_{opt} analytically from the above equation. We finally get

$$E_{opt} = a_{0w} + A_d \frac{1 - \dfrac{\sigma_1}{\sigma_0}\sqrt{1 + 2\dfrac{\sigma_1^2 - \sigma_0^2}{A_d^2}\ln\left(\dfrac{\sigma_1}{\sigma_0}\right)}}{\left(1 + \dfrac{\sigma_1}{\sigma_0}\right)\left(1 - \dfrac{\sigma_1}{\sigma_0}\right)} \tag{10.14}$$

When the BER is minimized, following simple relationship between the probabilities of error p_{0w} (symbol "0" is detected wrong) and p_{1w} (symbol "1" is detected wrong) is always valid:

$$p_{0w} \leq p_{1w} \tag{10.15}$$

This inequality is due to asymmetrical and unequal distortion for binary symbol "0" and symbol "1". As mentioned earlier, symbol "1" is always disturbed by shot noise more than the transmitted symbol "0". It is important to note that the optimum threshold E_{opt} and, hence, the minimum BER is normally not obtained when p_{0w} and p_{1w} are chosen to be equal. The equality sign in the above equation is only valid for signal-independent noise i.e., for $\sigma_0 = \sigma_1$.

Now, let us consider the Eq. (10.14) for optimum threshold E_{opt}. It becomes clear from this equation, that the ratio in the square root term is dependent upon the reciprocal of signal-to-noise ratio (SNR). With practical system parameters (for example, $p_w \leq 10^{-9}$), required SNR is high and therefore

$$2 \frac{\sigma_1^2 - \sigma_0^2}{A_d^2} \ln\left(\frac{\sigma_1}{\sigma_0}\right) \ll 1 \tag{10.16}$$

Under this condition, Eq. (10.14) simplifies to

$$E_{opt} \approx a_{0w} + A_d \frac{\sigma_0}{\sigma_0 + \sigma_1} = \frac{a_{0w}\sigma_1 + a_{1w}\sigma_0}{\sigma_0 + \sigma_1} \tag{10.17a}$$

When shot noise is negligibly small as compared to the thermal noise of amplifier, σ_0 and σ_1 become nearly equal. In that case, optimum threshold is exactly located at the centre of the noiseless sample values a_{0w} and a_{1w} as expected.

$$E_{opt} \approx \frac{a_{0w} + a_{1w}}{2} = a_{0w} + \frac{A_d}{2} = 0.5 \tag{10.17b}$$

Again, it should be remembered that this threshold level is normalized with respect to the amplitude $\hat{i}_{PD} = MR_0P_r$ of the photodiode current.

(ii) CUT-OFF FREQUENCY f_g OF THE LOW-PASS FILTER

It can be seen from Eq. (10.12) that minimization of BER requires the maximization of arguments of both Q-functions or the maximization of both the signal-to-noise ratios A_d/σ_1 and A_d/σ_0. As variances σ_0^2 and σ_1^2 are directly proportional to the cut-off frequency f_g, both signal-to-noise ratios reach their maximum value at the same cut-off frequency $f_g = f_{g,opt}$. Therefore, the optimum normalized cut-off frequency $f_{g,opt}T$ is determined by the conditional equation

$$\frac{A_d^2}{\sigma_i^2} \sim \frac{\left|1 - 4Q\left(\sqrt{2\pi} f_g T\right)\right|^2}{f_g T} \rightarrow \max \rightarrow f_{g,opt}T \tag{10.18}$$

where i is equal to 0 or 1, T the bit or symbol period and $1/T$ the bit or symbol rate. By means of a simple computer program, we obtain

$$2f_{\text{g,opt}} T \approx 1.58 \tag{10.19a}$$

which is a well-known result when a Gaussian low-pass filter is used in the digital receiver. It should be noted that irrespective of filter transfer function used, optimum rectangular-equivalent filter bandwidth normalized with bit frequency $1/T$ is always in the range of

$$1.0 < 2f_g T < 2.0 \tag{10.19b}$$

(iii) AVALANCHE GAIN M

In contrast to the optimization of f_g and E performed above, optimization of the avalanche gain M is much more comprehensive since both the Q-functions given in Eq. (10.12) do not reach their maximum value at the same avalanche gain. In practical systems, it is sufficient to determine the optimum avalanche gain from the equation

$$M_{\text{opt}} = \left(\frac{2G_{\text{th}}}{xe\left(R_0 P_r + I_{\text{dark}}\right)} \right)^{\frac{1}{2+x}} \tag{10.20}$$

which has already been derived in Chapter nine. However, this equation only represents an approximation since this semi-optimum avalanche gain only maximizes the second Q-function in Eq. (10.12), but not the first Q-function. For practical system parameters, this is a very good and frequently used approximation of excellent accuracy.

(iv) EVALUATION AND DISCUSSION

After carrying out the optimization, minimum worst-case probability of error p_w by using Eqs. (10.9), (10.11) and (10.17) is obtained as

$$p_w = Q\left(\frac{A_d}{\sigma_0 + \sigma_1} \right) = Q\left(\frac{1}{2}\sqrt{\frac{S}{N}} \right) \tag{10.21}$$

This simple equation represents a well-suited practical tool to estimate the BER of an optical transmission system with intensity modulation and direct detection. As $S = A_d^2$ is a measure of the signal power and $N = \sigma^2$ with $\sigma = (\sigma_0 + \sigma_1)/2$ the noise power, Eq. (10.21) is also a very useful tool to estimate the required signal-to-noise ratio for a given BER p_w.

Example

For a worst-case BER of $p_w = 10^{-9} \approx Q(6)$ following solution is obtained by using Eq. (10.21): $S/N = 144$ or $10 \log_{10}(S/N) = 24.6$ dB.

Let us consider an IM/DD system with the following parameters:

Symbol or bit rate:	$R = 1\text{bit}/T = 560$ Mbit/s,
Bit frequency:	$f_b = 1/T = 560$ MHz,
Frequency of transmitter laser:	$f_t = 200$ THz,
Quantum efficiency of photodiode:	$\eta = 0.83$,
Responsivity of photodiode:	$R_0 = (e\eta)/(hf) = 1$ A/W,
Dark current of photodiode:	$I_{dark} = 10^{-11}$ A,
psd of thermal noise:	$G_{th} = 10^{-23}$ A²/Hz,
Equivalent baseband filter:	$H_B(f) = 1$ for $|f| \leq 1/(2T)$ and
	$= 0$ otherwise
Excess noise exponent:	$x = 0.9$.

The avalanche gain M of the photodiode is chosen to be optimum (Eq. 10.20).

For this system, the BER p_w as a function of received optical power P_r at the input to the direct detection receiver obtained from Eq. (10.12) is shown in Fig. 10.2.

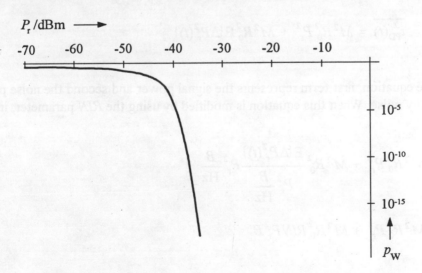

Fig. 10.2: Worst-case probability of error p_w as a function of received optical power level P_r in IM/DD system

It becomes clear from the above figure that the BER decreases rapidly with the increase in the received optical signal power P_r. It implies that even a very small increase in the signal power P_r improves BER by some order of magnitudes. Thus, a direct detection system is characterized by a very steep BER curve, which is true for standard electrical digital communication systems [2, 3, 5] as well.

10.3 INFLUENCE OF LASER NOISE

Performance of incoherent optical communication systems can seriously be degraded by the noise of transmitter laser, in particularly by the laser intensity noise. Intensity fluctuations of the laser output are due to undesired spontaneous emission. As mentioned in Chapter three, the relative intensity noise (RIN) parameter is used to describe and analyse the influence of laser intensity noise. The RIN parameter can normally be taken from data sheet of the laser diode. In the receiver, laser intensity noise get directly transferred to fluctuations in the photodiode current. If there is no modulation, photodiode current is given by

$$i_{PD}(t) = MR_0P_r(t) = MR_0\big(P_r + \Delta P_r(t)\big)$$

(10.22)

Here, P_r represents the received average optical power and $\Delta P_r(t)$ the time-variant optical power due to laser intensity fluctuations. The mean value of $\Delta P_r(t)$ is always zero. Electrical output power is proportional to the mean square of $i_{PD}(t)$ which is

$$\mathrm{E}\{i_{PD}^2(t)\} = \overline{i_{PD}^2(t)} = M^2R_0^2P_r^2 + M^2R_0^2\,\mathrm{E}\{\Delta P_r^2(t)\}$$

(10.23)

In the above equation, first term represents the signal power and second the noise power due to laser intensity noise. When this equation is modified by using the RIN parameter, it becomes

$$\overline{i_{PD}^2(t)} = M^2R_0^2P_r^2 + M^2R_0^2\frac{\mathrm{E}\{\Delta P_r^2(t)\}}{P_r^2\dfrac{B}{\mathrm{Hz}}}P_r^2\frac{B}{\mathrm{Hz}}$$

(10.24)

$$= M^2R_0^2P_r^2 + M^2R_0^2\,RINP_r^2B$$

with

$$RIN = \frac{E\{\Delta P_r^2(t)\}}{P_r^2 B} \tag{10.25}$$

or

$$RIN = \frac{1}{Hz} 10 \log_{10}\left(\frac{E\{\Delta P_r^2(t)\}}{P_r^2 \dfrac{B}{Hz}} \right) \quad \text{in dB/Hz} \tag{10.26}$$

As mentioned in Chapter three, RIN is in the order of -150 dB/Hz or 10^{-15}/Hz. The latter value must be used in Eq. (10.24) above. In order to decide whether laser intensity noise can be neglected or not, it is useful to compare power of intensity noise (second term in Eq. 10.24) with the power of shot noise (see Chapter nine). Their ratio is given by

$$r = \frac{\dfrac{1}{2} M^2 R_0^2 RIN P_r^2 B}{e M^{2+x} R_0 P_r B} = RIN \frac{\eta P_r}{2 M^x hf} \tag{10.27}$$

where B represents the double-sided noise-equivalent bandwidth. In case of a Gaussian baseband filter, this bandwidth equals $\sqrt{2} f_g$ as derived in the previous Chapter. The factor ½ must be used since shot noise was derived in Chapter nine by using double-sided bandwidth (i.e., mathematical bandwidth) whereas RIN parameter taken from a conventional data sheet is normally based on single-sided bandwidth (i.e., physical bandwidth). Laser intensity noise can practically be neglected when $r \ll 1$ i.e.,

$$RIN \ll \frac{2 M^x hf}{\eta P_r} \quad \text{or} \quad P_r \ll \frac{2 M^x hf}{\eta RIN} \tag{10.28}$$

Thus, higher the optical input power P_r at the receiver input, stronger the influence of intensity noise. Assuming worst-case conditions with PIN receiver ($M = 1$), large quantum efficiency η e.g., $\eta = 0.9$, low optical frequency f e.g., $f = c/\lambda = c/1.55\mu m \approx 200$ THz and large RIN e.g., $RIN = -120$ dB/Hz which equals 10^{-12}/Hz, laser intensity noise can be neglected when $P_r \ll 0.3\ \mu W$ or $P_r \ll -35$ dBm e.g., $P_r = -40$ dBm. If RIN is small e.g., $RIN = 10^{-15}$ /Hz, then intensity noise can be neglected when received optical power P_r is much less than -5 dBm e.g., $P_r = -10$ dBm which is usually fulfilled.

10.4   CALCULATIONS FOR A RECEIVER: AN EXAMPLE

In this Section, a practical receiver based on typical data is considered. The receiver configuration is shown in Fig. 10.3 below. It includes a PIN photodiode, a conventional 50-Ohm amplifier and a matched filter which is matched to the rectangular input signal. This signal is disturbed by the approximately white noise of the photodiode and amplifier.

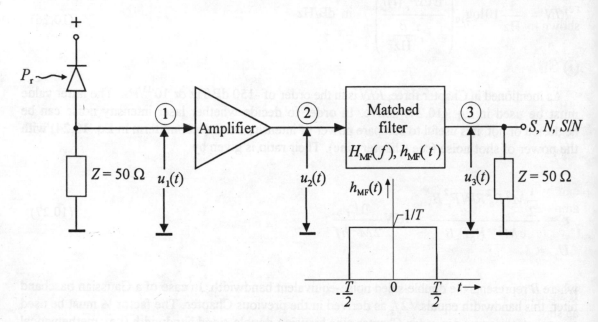

Fig. 10.3: Incoherent receiver with matched filter

Following data are considered for signal and different subsystems.

Optical input signal
Power: $P_r \triangleq -27$ dBm for symbol "1" and $P_r \triangleq 0$ dBm for "0"
Shape: rectangular
Bit rate: $R = 1$bit$/T = 622.08$ Mbit/s (STM-4)

Photodiode
Responsivity: $R_0 = 1.0$ A/W
Dark current: $I_{dark} = 10$ pA

50-Ohm amplifier
Gain: $g_A \triangleq 20$ dB i.e., $g_P = 100$ and $g_V = 10$
Bandwidth: $B_A \gg 1/T$
Noise figure: $F_A \triangleq 3$ dB

Matched filter
System function: $H_{MF}(f) = \sin(\pi f T)/(\pi f T)$
Impulse response: $h_{MF}(t) = 1/T$ for $|t| \leq 0.5\,T$ and $h(t) = 0$ for $|t| > 0.5\,T$
Noise-equivalent bandwidth: $B_{MF} = 1/T$

Here, T is the bit duration and g_P and g_V represent the amplifier gain in terms of power and voltage respectively. In order to determine S/N ratio at the receiver output and to highlight the receiver operation, we separately consider signal and noise power at the measurement points shown in Fig. 10.3.

(i) SIGNAL POWER

At the input of the electrical preamplifier i.e., at measurement point ①, a peak voltage

$$U_1 = R_0\,P_r\,Z \approx 100\,\mu V \qquad (10.29)$$

can be measured for binary symbol "1" whereas $U_1 = 0$ for symbol "0". At the output of the amplifier i.e., at measurement point ②, peak voltage for symbol "1" is

$$U_2 = U_1\,g_V \approx 1\,mV \qquad (10.30)$$

If signal at measurement point ② is measured by means of an oscilloscope, then a typical signal as shown in Fig. 10.4a can be observed.

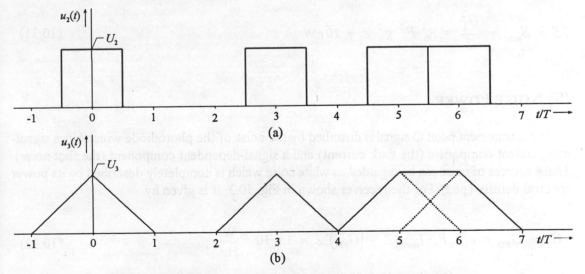

Fig. 10. 4: Signals at the (a) input and (b) output of matched filter

At the output of matched filter i.e., at measurement point ③, the rectangular input signal becomes triangular as shown in Fig. 10.4b. Its peak voltage remains unchanged ($U_3 = U_2$) whereas absolute bit duration increases by a factor of two. As a result, eye pattern is also triangular. As an example, the eye pattern measured in a realized incoherent optical receiver is shown in Fig. 10.5.

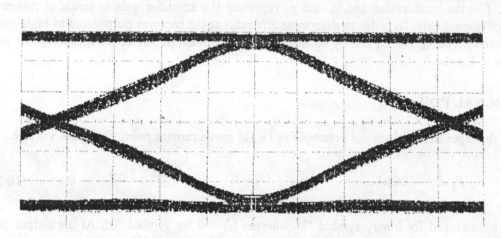

Fig. 10.5: Measured eye pattern at the output of the matched filter

The higher the peak power of signal $u_3(t)$ at the centre of each symbol "1", the better the receiver performance measured in terms of BER or S/N ratio. In should be remembered that each bit is sampled at its centre to decide the transmitted symbol "0" or symbol "1". Thus peak signal power directly influences the BER. Taking into account the above measured voltage, we obtain

$$S = S_{max} = \frac{U_3^2}{Z} = R_0^2 \, P_r^2 \, g_V^2 \, Z \approx 20 \, \text{nW} \tag{10.31}$$

(ii) NOISE POWER

At measurement point ① signal is disturbed by the noise of the photodiode which has a signal-independent component (the dark current) and a signal-dependent component (the shot noise). These sources of noise can be regarded as white noise which is completely described by its power spectral density (psd). For the receiver shown in Fig. 10.3, it is given by

$$G_1 = G_{PD} = e\left(R_0 P_r + I_{dark}\right)Z = \left(G_{\widetilde{PD}}\right)^2 Z \approx 1.6 \cdot 10^{-23} \frac{\text{W}}{\text{Hz}} \tag{10.32}$$

The dark current I_{dark} is normally negligible in comparison to the current $R_0 P_r$. The psd is frequently expressed in terms of pA/$\sqrt{Hz}$ instead of W/Hz. In that case, normalized psd $G_{PD}^{\sim}$ instead of absolute psd G_{PD} is given. With the above data, normalized psd is

$$G_{PD}^{\sim} = \sqrt{G_{PD}} = \sqrt{e(R_0 P_r + I_{dark})} \approx 0.57 \frac{pA}{\sqrt{Hz}} \qquad (10.33)$$

At measurement point ②, noise can still be regarded as white since bandwidth of the amplifier is chosen to be much larger than the bit frequency i.e., $B_A \gg 1/T$. Only in this case, following filter actually acts as a matched filter. The psd at measurement point ② is given by

$$G_2 = G_1 g_V^2 + G_A \qquad (10.34)$$

where G_A is the psd of the amplifier noise. In the data sheet of a commercial 50-Ohm amplifier, noise is normally given in terms of noise figure F_A. Thus, it is required to transform noise figure F_A to psd G_A of amplifier noise or normalized psd $G_A^{\sim}$. This can easily be performed by [4]

$$G_A = \frac{1}{2} k_B T_0 (F_A - 1) g_V^2 \approx 2.1 \cdot 10^{-19} \frac{W}{Hz} \qquad (10.35)$$

and

$$G_A^{\sim} = \sqrt{G_A} = \sqrt{\frac{1}{2} k_B T_0 (F_A - 1) Z^{-1}} \, g_V \approx 64.3 \frac{pA}{\sqrt{Hz}} \qquad (10.36)$$

where k_B is the Boltzmann's constant and T_0 the temperature in degree Kelvin. In case of room temperature, product $k_B T_0$ equals $1.38 \cdot 10^{-23}$ Ws/K·300 K = $4.14 \cdot 10^{-21}$ Ws. In order to calculate the numerical values given on the right sides of Eqs. (10.35) and (10.36), a noise figure $F_A = 2$ (i.e., 3 dB) is taken. Comparing normalized psd of amplifier and photodiode, it becomes clear that in incoherent receivers amplifier noise is normally dominant. This is especially valid for long-range systems where optical power at the receiver input is low. Total noise power at the receiver output i.e., at measurement point ③ is now given by

$$N = \left[\left(G_{PD}^{\sim}\right)^2 g_V^2 + \left(G_A^{\sim}\right)^2 \right] Z B_{MF} = \left[e(R_0 P_r + I_{dark}) Z + \frac{1}{2} k_B T_0 (F_A - 1) \right] g_V^2 \frac{1}{T} \approx 130 \, pW \quad (10.37)$$

(iii) SIGNAL-TO-NOISE RATIO

Using Eqs. (10.31) and (10.37), signal-to-noise ratio is given by

$$\frac{S}{N} = \frac{R_0^2 P_r^2 Z T}{e\left(R_0 P_r + I_{dark}\right)Z + \frac{1}{2}k_B T_0 (F_A - 1)} \approx 154 \quad \text{i.e.,} \quad 21.9\,\text{dB} \tag{10.38}$$

It results in a BER somewhat better than 10^{-9} which requires $S/N = 144$ (i.e., 21.6 dB) as shown in Section 10.1. If the filter used is not a matched filter, then S/N ratio decreases. With a matched filter, eye aperture at the output is 100 %. When a Gaussian filter with an optimum bandwidth for minimum BER is used i.e., $B = 1.58/T$, eye aperture is 90.46%. Thus, S/N ratio is about 1.35 dB less than the S/N ratio for a matched filter receiver.

10.5   INCOHERENT SYSTEMS WITH OPTICAL AMPLIFIERS

In this Section, evaluation of signal-to-noise S/N ratio for an IM/DD system has been made when an optical amplifier is used as preamplifier, postamplifier and in-line amplifier as shown in Fig. 10.6.

(i) PREAMPLIFIER

In this configuration, as shown in Fig. 10.6a, optical signal is first amplified and then detected by a photodetector. In order to include the effect of circuit noise, equivalent noise current due to circuit noise is taken as

$$\overline{i_c^2} = \frac{4k_B T_0 B}{R_1} \tag{10.39}$$

where T_0 is the noise temperature in degree Kelvin and R_1 the photodetector load resistance. Therefore, S/N ratio is given by [1]

$$\frac{S}{N} = \frac{e^2 G^2 \langle n_i \rangle^2}{2e^2 \sigma_{n0}^2 B + 4k_B T_0 B/R_1} \tag{10.40}$$

With the increase in G, beat noise become predominant over the shot noise in the denominator of Eq. (10.40) as the latter noise has higher order dependence on G. Further, beat noise also become predominant over the circuit noise. If b is sufficiently small, S/N ratio approaches the beat noise limited value as given in Eq. (7.19), which under ideal conditions i.e., when $n_{sp} = 1$ is 3 dB worse than the shot noise limited value.

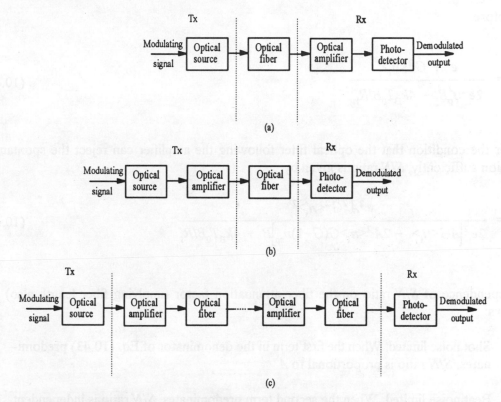

Fig. 10.6: Direct detection system with (a) optical preamplifier, (b) optical postamplifier and (c) cascade amplification using k optical in-line amplifiers

(ii) POSTAMPLIFIER

In this application (refer to Fig.10.6b), input optical signal is first amplified and then made to propagate through the fiber. Finally, the signal is detected by a photodetector.

Let A be the attenuation factor by which the signal will get attenuated while propagating from one end of the fiber to other. At the photodetector output, average number of photons and their variance are given by [1]

$$<n_o> = AG<n_i> + A(G-1)n_{sp}b \qquad (10.41a)$$

and

$$\sigma_p^2 = AG\langle n_i \rangle + A(G-1)n_{sp}b + 2A^2\langle n_i \rangle G(G-1)n_{sp} + A^2(G-1)^2n_{sp}^2b \qquad (10.41b)$$

Therefore,

$$\frac{S}{N} = \frac{e^2A^2G^2\langle n_i \rangle^2}{2e^2\sigma_p^2B + 4k_BT_0B/R_l} \qquad (10.42)$$

Under the condition that the optical filter following the amplifier can reject the spontaneous emission sufficiently, S/N ratio is given by

$$\frac{S}{N} = \frac{e^2A^2G^2\langle n_i \rangle^2}{2e^2\left[AG\langle n_i \rangle + 2A^2\langle n_i \rangle G(G-1)n_{sp}\right]B + 4k_BT_0B/R_l} \qquad (10.43)$$

Dependence of S/N ratio on the fiber attenuation factor (implying fiber length also) is as follows:

- Shot noise limited: When the first term in the denominator of Eq. (10.43) predominates, S/N ratio is proportional to A.

- Beat noise limited: When the second term predominates, S/N ratio is independent of A. Thus, under beat noise limited conditions, noise characteristics of amplifier are similar to an electrical system.

- Circuit noise limited: When the third term predominates, S/N ratio is proportional to A^2.

It is evident from Eqs. (10.40) and (10.43) that under shot noise and circuit noise limited conditions, postamplifier will give lower S/N ratio as compared to preamplifier.

(iii) IN-LINE AMPLIFIER

Let us consider a cascaded amplification system using k optical in-line amplifiers as intermediate repeaters (refer to Fig. 10.6c). For simplicity, let us presume that each amplifier has same characteristics and each section will reduce the signal by the same factor A. If the gain of amplifier is adjusted so that it compensates loss i.e., $GA = 1$, average and variance of photon number at the output can be obtained from Eq. (10.41) by replacing n_{sp} by kn_{sp} and GA by unity and are given by

$$\langle n_o \rangle = \langle n_i \rangle + A(G-1)n_{sp}kb \tag{10.44a}$$

and

$$\sigma_c^2 = \langle n_i \rangle + A(G-1)n_{sp}kb + 2A\langle n_i \rangle(G-1)n_{sp}k + A^2(G-1)^2 n_{sp}^2 k^2 b \tag{10.44b}$$

Therefore,

$$\frac{S}{N} = \frac{e^2 \langle n_i \rangle^2}{2e^2 \sigma_c^2 B + 4k_B T_0 B/R_1} \tag{10.45}$$

Under beat noise limited condition

$$\frac{S}{N} = \frac{\langle n_i \rangle^2}{2\left[2\langle n_i \rangle n_{sp}k + n_{sp}^2 k^2 b\right]B} \tag{10.46}$$

A comparison of Eqs. (7.18) and (10.46) shows that the S/N ratio of the repeatered system is deteriorated at least by a factor of k and ASE-ASE beat noise becomes dominant due to multiplication by k^2. Consequently, in-line amplifiers used for cascaded amplification system should have smaller bandwidth filter. When the filter bandwidth b is sufficiently small, approximate S/N ratio from Eq. (10.46) is given by

$$\frac{S}{N} \approx \frac{\langle n_i \rangle}{4 n_{sp} k B} \tag{10.47}$$

If the amplifiers are not used in the system, S/N ratio in the shot noise limited condition is given by

$$\frac{S}{N} = \frac{A^k \langle n_i \rangle}{2B} = \frac{\langle n_i \rangle}{G^k 2B} \tag{10.48}$$

It is seen from Eqs. (10.47) and (10.48) that S/N ratio gets improved by a factor of $G^k/2kn_{sp}$ with the use of k in-line amplifiers. This, of course, is possible only when ASE-ASE beat noise is sufficiently suppressed. Otherwise as the signal passes through the chain of amplifiers, ASE noise power grows and signal power decreases since amplifier saturation power level is constant for a given amplifier. Eventually, with the increase in the number of amplifiers, signal will get completely lost in the noise. In order to avoid this, either the number of amplifiers have to be restricted or repeaters at the appropriate location have to be used.

10.6 REFERENCES

[1]   Okoshi, T.; Kikuchi, K.: Coherent Optical Fiber Communications. Kluwer Academic Publisher, 1988.
[2]   Söder, G.; Tröndle, K.: Digitale Übertragungssysteme. Springer-Verlag, 1984.
[3]   Stein, S.; Jones, J. J.: Modern Communication Principles. McGraw-Hill, 1967.
[4]   Taub, H.; Schilling, D. L.: Principles of Communication Systems, Second Edition. McGraw-Hill, 1986.
[5]   Tröndle, K.; Söder, G.: Optimization of Digital Transmission Systems. Artech House, 1987.

11 COHERENT SYSTEMS: ANALYSIS AND OPTIMIZATION

Based on the fundamentals of Chapter nine, analysis of various coherent optical communication systems and their optimization is discussed in this Chapter. Different systems are distinguished and classified on the basis of modulation and demodulation schemes used. The following systems will be studied in this Chapter:

- ASK and PSK homodyne systems (Section 11.1),
- ASK, FSK and PSK heterodyne systems with coherent detection (Section 11.2),
- ASK, FSK and DPSK heterodyne systems with incoherent detection (Section 11.3).

In order to highlight characteristic features and performance of coherent systems mentioned above, they are compared with the conventional optical direct detection system with intensity modulation which has been described in the previous Chapter and which represents a well-suited reference for system comparison.

It should be noted that in this Chapter, the terms "coherent" and "incoherent" are exclusively used in conjunction with the demodulation technique applied in the receiver. As it is well-known, each communication system is disturbed by various sources of noise. This Chapter takes into account the shot noise of the photodiode(s), thermal (electronic) noise of the resistors and amplifiers (mainly of the front-end amplifier) as well as the noise of both transmitter and local lasers. To discuss system degradation by noise and other imperfections, this Chapter also considers the influence of filters such as IF and baseband filters on noise and signal (intersymbol interference).

In the *system analysis*, main task is to calculate the probability of error as it represents a powerful measure to assess the system performance and the transmission quality and also as a well-suited parameter for systems comparison (Chapter twelve). To evaluate the probability of error or the bit error rate (BER), signals at different points in the receiver are to be determined which will depend on modulation scheme and noise (see Chapter nine). The detected signal $d(t)$ at the input to the sample and hold circuit is of prime importance since the quality of this signal directly influences the BER and, hence, the performance of a digital communication system.

System analysis requires the examination of probability density function (pdf) of sample values $d(vT+t_0)$ of the detected signal to calculate the BER. In addition, pdf and its statistical characteristics viz., expected value and standard deviation give a clear insight of complex interconnections of various subsystems in a coherent communication system. Probability density function, expected value and standard deviation allow us to identify the important and critical parameters in a system. Furthermore, the different effects of the sources of noise become clear in particular. Results of a change in system parameters can also be assessed very fast.

Another powerful measure in the system performance analysis is the eye pattern. As explained in Chapter nine, eye pattern can easily be measured by an oscilloscope. In system analysis, eye pattern can be calculated mathematically. It can also be simulated by an appropriate computer simulation program. The eye pattern highlights typical features of a system and especially allows to assess the effects of intersymbol interference in the detected signal $d(t)$.

In this book, *system optimization* is always performed with a goal to minimize the probability of bit error. An exact analytical optimization is, however, often not possible. In such a case, optimization is performed by employing appropriate numerical, iterative methods in conjunction with a computer. To minimize computer processing time, simplifying approximations are taken.

11.1 HOMODYNE SYSTEMS

In homodyne systems, modulated optical signal is converted directly into an electrical baseband signal by means of an optical coupler, a local laser and one or two photodiodes. This conversion is true in frequency and phase. Thus, the electrical baseband signal is simply a frequency translated replica of the original optical incoming signal.

Optical homodyning requires that frequency *and* phase of both received and local laser waves are exactly same. This means that both waves have to be *synchronous* i.e., *coherent*. To realize this strong demand, a highly accurate *optical phase-locked loop* (OPLL) circuit is required. The principle of an OPLL is briefly explained in Section 11.1.3.

The process of converting a modulated high frequency signal (HF signal) into a baseband signal by means of a high frequency synchronous local carrier wave is usually called synchronous or coherent detection. Since both modulated HF signal and synchronous local carrier are optical waves in optical homodyne receiver, we call it *optical synchronous detection* or *optical coherent detection*. Hence, optical communication systems with homodyne receivers are strictly to be called as coherent optical communication systems with optical coherent detection. Here, the first "coherent" is related to a coherent laser source, whereas the second "coherent" corresponds to the synchronous demodulation process. It should be noted that this terminology is not an accepted standard. Indeed, heterodyne and homodyne systems are also frequently called as coherent systems in general, irrespective of whether the demodulation process is coherent or incoherent. Thus, this simplified terminology is related to the laser source only which should be coherent as far as possible.

Principle block diagram of a homodyne system for ASK and PSK modulation schemes is shown in Fig. 11.1.

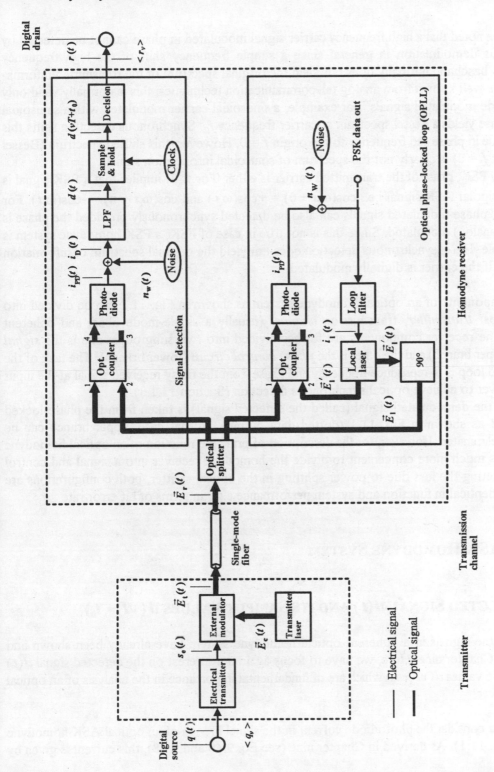

Fig. 11.1: Block diagram of coherent optical communication system with homodyne detection

It must be noted that a high frequency carrier signal modulated in phase cannot be detected by synchronous demodulation in general since a simple frequency shift of the high frequency spectrum to baseband normally does not yield the original spectrum of the transmitted information. As it is well known from analog telecommunication techniques, this is normally valid only for amplitude modulated signals. For example, a sinusoidal carrier modulated with a sinusoidal signal in phase yields a Bessel spectrum at carrier frequency f_C. Synchronous detection shifts this spectrum true in phase and frequency to the origin $f = 0$. However, this shifted spectrum (Bessel spectrum at $f = 0$) is clearly not the spectrum of sinusoidal information signal.

In binary PSK, phase of the transmitted carrier is either 0 or π. It implies that a PSK signal is same as a bipolar ASK signal i.e., $\cos(\omega_c t + 0) = +\cos(\omega_c t)$ and $\cos(\omega_c t + \pi) = -\cos(\omega_c t)$. For this reason, phase modulated signals can also be detected synchronously provided the phase is binary (not analog) modulated. Since this is not true in case of FSK, a FSK homodyne system is not realizable. Here, synchronous detection does not yield the original spectrum of information signal even if the carrier is digitally modulated.

The components of an optical homodyne system as shown in Fig. 11.1 can be divided into three groups: *transmitter*, transmission *channel* (usually a single-mode fiber) and coherent *receiver*. The receiver components are further divided into two subgroups: one is the *signal detector* (upper branch) and the other is the *phase control circuit* (lower branch). The task of the phase-locked loop is to reproduce frequency and phase from the noisy received signal at the input to the receiver to achieve optical synchronous detection (Section 11.1.3).

Usually, the demodulated signal (called the detected signal) is taken from the phase-locked loop circuit as shown in Fig. 11.1 (dotted data output). In that case, upper branch can be completely eliminated. However, for the description of principle function of an optical homodyne system, it is much more convenient to divide the homodyne receiver into a signal and control part. Neglecting the loss due to power splitting in the optical splitter, both configurations are absolutely identical in function and system performance e.g., in terms of bit error rate.

11.1.1   ASK HOMODYNE SYSTEM

(i)   DETECTED SIGNAL $d(t)$ AND ITS SAMPLE VALUES $d(vT + t_0)$

Most of the signals in a coherent optical homodyne system have already been shown and discussed in Chapter nine. Here, we have to focus again our interest on the detected signal $d(t)$ and its sample values $d(vT + t_0)$ which are of fundamental importance in the analysis of an optical digital receiver.

Let us first consider the photodiode current in the signal branch of an optical ASK homodyne receiver (Fig. 11.1). As derived in Chapter nine (see Eq. 9.40 and 9.68), this current is given by

$$i_{PD}(t) = \hat{i}_{PD}\ \underline{s}(t)\ e^{j\phi(t)}$$

$$= \hat{i}_{PD} \sum_{\nu=-\infty}^{\infty} s_\nu\ \text{rect}\left(\frac{t - \nu T}{T}\right)\ e^{j\phi(t)} \tag{11.1}$$

where $s_\nu \in \{0, 1\}$ represents the transmitted binary information. The amplitude $\hat{i}_{PD}$ of the photo-diode current is given by Eq. (9.69). The photodiode current is disturbed by the phase noise

$$\phi(t) = \left[\phi_t(t) - \phi_l(t)\right] - \phi_{PLL}(t) \tag{11.2}$$

where $\phi_t(t)$ and $\phi_l(t)$ are the phase noise of transmitter and local lasers respectively. In addition, phase $\phi(t)$ also includes a noiseless part $\phi_{PLL}(t)$ of the local laser phase which is controlled by the OPLL circuit. In the ideal case which cannot be achieved in practice, control phase $\phi_{PLL}(t)$ equals the resulting phase noise $\phi_t(t) - \phi_l(t)$. As a consequence, the residual phase noise $\phi(t)$ also called the rest-phase noise becomes zero. However, in practical systems a residual phase noise $\phi(t) \neq 0$ is unavoidable.

In order to analyse homodyne systems and especially to calculate the bit error rate, statistical properties of the residual phase noise $\phi(t)$ have to be determined first. As will be seen in Section 11.1.3, residual phase noise $\phi(t)$ is a *Gaussian stationary* random process with *zero mean*. Its standard deviation σ_ϕ is a function of both the laser linewidths Δf_t and Δf_l of transmitter and local lasers and constant noise power spectral density G_c of the shot noise and the thermal noise. Further, σ_ϕ is also a function of the characteristic parameters of the OPLL circuit (see Eq. 11.57).

The detected signal at the output of the low-pass filter (Fig. 11.1) is

$$d(t) = i_{PD}(t) \star h_B(t) + n(t)$$

$$= \hat{i}_{PD} \int_{-\infty}^{+\infty} s(\tau)\ \cos(\phi(\tau))\ h_B(t-\tau)\ d\tau + n(t) \tag{11.3}$$

Here, $h_B(t)$ is the impulse response of the low-pass filter (baseband filter), $s(t)$ the information signal (Eq. 9.40) and $n(t)$ the additive, band-limited and Gaussian distributed receiver noise due to shot noise of photodiodes and thermal noise. This noise is determined by its standard deviation $\sigma_n = \sigma_{hom}$ given in Eq. (9.84).

As $n(t)$ and $\cos(\phi(t))$ are random, sample values $d(vT + t_0)$ of the detected signal are random too. It should be noted that both random processes can be regarded as stationary (Section 3.2.5). Hence, all their statistical properties are independent of sampling time $vT + t_0$. For this reason, it is sufficient to consider a certain, fixed sampling time; for example, $t = t_0$. The results obtained are valid in general i.e., for each sampling time $vT + t_0$.

Comparing the rectangular equivalent pulse duration Δt_B of the low-pass filter impulse response $h_B(t)$ and the rectangular equivalent correlation width Δt_w of the autocorrelation function $R_w(\tau)$ of the random process $w(t) = \cos(\phi(t))$, following important relationship can be observed in most practical systems [5]:

$$\Delta t_B = \frac{1}{h_B(0)} \int_{-\infty}^{+\infty} h_B(t)\,dt \quad \ll \quad \Delta t_w = \frac{1}{R_w(0)} \int_{-\infty}^{+\infty} R_w(\tau)\,d\tau \tag{11.4}$$

Because of this inequality, random process $w(t) = \cos(\phi(t))$ is strongly correlated during the pulse duration Δt_B of the low-pass filter. Therefore, this process can be considered as time-invariant during the time interval Δt_B i.e., $\phi(t) \rightarrow \phi(t_0)$. For this reason, term $\cos(\phi(t_0))$ can be taken outside the integral in Eq. (11.3). However, this constant is changing from one time interval to the other and $\cos(\phi(t_0))$ is actually a random variable. Considering the sampled detected signal $d(t_0)$, we now obtain the following simple expression:

$$d(t_0) = \hat{\imath}_{PD}\, a(t_0)\, \cos\big(\phi(t_0)\big) + n(t_0) \tag{11.5}$$

where

$$a(t_0) = \int_{-\infty}^{+\infty} s(\tau)\, h_B(t_0-\tau)\, d\tau \quad \text{with} \quad 0 \le a(t_0) \le 1 \tag{11.6}$$

Thus, $a(t_0)$ is same as sampled value $d(t_0)$ normalized with respect to $\hat{\imath}_{PD}$ and in absence of any noise i.e., $n(t) = 0$ and $\phi(t) = 0$. Since $a(t_0)$ is a function of impulse response $h_B(t)$, transmitted information $s(t)$ and, consequently, transmitted symbol sequence $<q_v>$, the noiseless sample value $a(t_0)$ is impaired by the intersymbol interference. Thus, the sampled value $a(t_0)$ at the current symbol $q_v = q_0$ is essentially influenced by the neighbouring symbols (compare Section 9.5.1). The limited range of $a(t_0)$ given in the above equation is related to a Gaussian baseband filter $H_B(f)$ as given in Eq. (9.83).

In order to simplify the calculation, it is useful to perform the following substitutions and normalisations:

$$d := \frac{d(t_0)}{\hat{i}_{PD}} \qquad a := a(t_0) \qquad \phi := \phi(t_0)$$

$$n := \frac{n(t_0)}{\hat{i}_{PD}} \quad \rightarrow \quad \sigma := \frac{\sigma_n}{\hat{i}_{PD}} = \frac{\sigma_{hom}}{\hat{i}_{PD}}$$

Now Eq. (11.5) can be written in the following simple form:

$$d = a \cos(\phi) + n = aw + n \quad \text{with} \quad |w| \le 1 \quad \text{and} \tag{11.7}$$

As this fundamental equation will be used frequently in our subsequent considerations, the physical meaning of its different quantities is given below again:

  d: normalized sample value of detected signal,

  a: noiseless part of d disturbed by intersymbol interference (a includes the

    transmitted information),

 $\cos(\phi)$: phase noise factor,

  n: additive Gaussian receiver noise.

It becomes clear from the above equation that the sample values of the detected signal are influenced by two random variables: one is the phase noise term $w = \cos(\phi)$ and the other is additive Gaussian noise n. In addition, the detected sampled signal d is also a function of transmitted information which is included in the term a.

(ii) PROBABILITY DENSITY FUNCTION $f_d(d)$

Since both the random variables w (phase noise term) and n (additive Gaussian receiver noise) are statistically independent, pdf of detected sample value d can easily be obtained by the following convolution:

$$f_d(d) = \frac{1}{|a|} f_w\left(\frac{d}{a}\right) \star f_n(d)$$

$$= \int_{-a}^{+a} \frac{1}{|a|} f_w\left(\frac{w}{a}\right) f_n(d-w) \ dw \ = \ \int_{-1}^{+1} f_w(w) f_n(d-aw) \ dw \qquad (11.8)$$

Here, the last integral takes into account that the pdf $f_w(w)$ is limited on $|w| \leq 1$ as shown in Fig. 3.11. If $a = 0$, then the normalized pdf $(1/|a|) \cdot f(w/a)$ which is included in the first integral changes to the Dirac delta function $\delta(w)$. In this case, pdf $f_d(d)$ is same as the Gaussian pdf $f_n(n)$ of the sampled additive noise n. Substituting the pdfs $f_w(w)$ and $f_n(n)$ from Eqs. (3.47) and (9.78) respectively and after some further simplifications, we obtain [8]

$$f_d(d) = \frac{1}{\pi \sigma_\phi \sigma} \int_0^{+\infty} \exp\left(-\frac{\Psi^2}{2\sigma_\phi^2}\right) \exp\left(-\frac{a^2}{2\sigma^2}\left[\frac{d}{a} - \cos(\Psi)\right]^2\right) d\Psi \qquad (11.9)$$

In this equation, σ_ϕ represents the standard deviation of the stationary rest-phase noise $\phi(t)$ and σ the normalized standard deviation of the additive, stationary and Gaussian circuit noise $n(t)$. The constant of normalization is given by the amplitude $\hat{\imath}_{PD}$ of the photodiode current $i_{PD}(t)$ as mentioned earlier.

The pdf $f_d(d)$ for all "0" (i.e., $s_v = 0$ for all v and, thus, $a = 0$) and all "1" (i.e., $s_v = 1$ for all v and, thus, $a = 1$) are shown in Figs. 11.2a and 11.2b respectively. In the ideal case, where noise and intersymbol interference do not disturb the system, the sampled detected signal d is either zero ($d = 0$) when all "0" are transmitted or one ($d = 1$) when all "1" are transmitted.

Let us now consider the Fig. 11.2a which shows the pdf $f_d(d)$ when a sequence of only "0" (all "0") is transmitted. In this case, the transmitted binary symbol sequence $<q_v>$ is given by the special sequence $<q_v> = \cdots 000 \cdots$. As a result, noiseless sample value a equals zero ($a = 0$) and the phase noise term w is permanently multiplied by zero. It means that laser phase noise does not impair the system. Thus, the detected signal d is only disturbed by the additive Gaussian noise and, consequently, the pdf $f_d(d)$ is Gaussian.

Consider next Fig. 11.2b which gives $f_d(d)$ for all "1". As the transmitted symbol sequence is $<q_v> = \cdots 111 \cdots$, it represents a simple DC signal. Let us first assume that laser phase noise does not impair the system (i.e., $\sigma_\phi = 0$). In such a case, pdf $f_d(d)$ of the detected signal d is again Gaussian, but now shifted to the expected value $d = 1$. However, as soon as laser phase noise influence the system in addition (i.e., $\sigma_\phi \neq 0$), pdf $f_d(d)$ changes its shape and remains no longer

Gaussian. As this influence increases, pdf $f_d(d)$ becomes more and more flat and broad, but only in the direction of the lower levels of d. However, in the opposite direction pdf $f_d(d)$ changes much less. This important result is a typical feature of ASK homodyne and ASK heterodyne systems, which results in some decisive effects; for example, a required shift of the threshold in the decision circuit. As the effects of phase noise are of fundamental importance in homodyne systems, these will be discussed in more detail now.

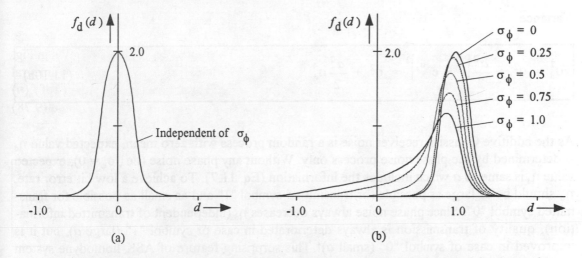

Fig. 11.2: Probability density function $f_d(d)$ of normalized sample values d of the detected signal $d(t)$ in ASK homodyne receiver. The normalized standard deviation of the additive Gaussian receiver noise is fixed at $\sigma = 0.2$. The transmitted information is: (a) all "0" and (b) all "1".

The coherent optical ASK homodyne system is based on an *unipolar modulation scheme*, where signal s_v is either zero ($s_v = 0$) or one ($s_v = 1$). The pdfs of the detected signal for symbols "0" and "1" are usually asymmetrical. This characteristic feature of an ASK system becomes clear in Fig. 11.2. Due to phase noise, symbol "1" is always disturbed more than symbol "0". The mathematical reason for this feature is given by the phase noise term $w = \cos(\phi)$ which is a multiplier to the information signal (Eq. 11.7). In contrast to shot and thermal noise which are *additive* sources of noise, phase noise is a *multiplicative* source of noise. As a result, influence of phase noise rapidly increases with the increase in the amplitude of information signal or the optical power at the input to the receiver. On the other hand, influence of phase noise is less when the incoming optical power is low. Now, performance degradation due to additive Gaussian receiver noise becomes dominant. For this reason, ASK homodyne system belongs to the class of optical communication systems which are characterized by *signal-dependent noise*. It should be remembered that conventional optical direct detection system is also included in this class.

A well-suited measure to assess the different influence of phase noise and additive Gaussian noise is the standard deviation σ_d or variance σ_d^2 and expected value η_d of the sample value d of the detected signal. These can be determined as follows:

Expected value:

$$\eta_d = a \, e^{-\frac{\sigma_\phi^2}{2}}$$

(11.10a)

Variance:

$$\sigma_d^2 = \sigma^2 + \frac{a^2}{2}\left[1 - e^{-\sigma_\phi^2}\right]^2 \approx \sigma^2 + \frac{a^2}{2}\sigma_\phi^4$$

(11.10b)

As the additive Gaussian receiver noise is a random process with zero mean, expected value η_d is determined by the phase noise process only. Without any phase noise (i.e., $\sigma_\phi = 0$), expected value η_d is same as a which includes the information (Eq. 11.7). To achieve a low bit error rate, η_d should be as large as possible for transmitted symbol "1" and as small as possible for transmitted symbol "0". Since phase noise always decreases η_d (independent of transmitted information), quality of transmission is always deteriorated in case of symbol "1" (large a), but it is improved in case of symbol "0" (small a)! This surprising feature of ASK homodyne system becomes more clear when the eye pattern of the detected signal is considered in Chapter twelve.

The variance σ_d^2 of the sampled detected signal d is determined by the variance of the additive Gaussian noise and phase noise. However, the influence of phase noise is much more serious as σ_d^2 depends on the square of the phase noise variance σ_ϕ^2 in contrast to the direct relationship with the variance σ^2 of additive circuit noise (Eq. 11.10b). It may be mentioned that the approximation $\exp(x) \approx 1 + x$ employed in the above expression is valid if x is much less than 1.

(iii) INTERSYMBOL INTERFERENCE AND BIT ERROR RATE (BER)

So far we have considered how the pdf $f_d(d)$ of the detected sample values d is influenced by phase noise and additive Gaussian noise. However, the pdf $f_d(d)$ is also influenced by the transmitted symbol or bit sequence $<q_v>$ which is included in the noiseless part of d i.e., in a (Eq. 11.7). The probability of an errorless detection of a transmitted symbol "1" or "0" is essentially determined by the previous and following symbols.

As explained in Section 9.5.1, mutual distortion of neighbouring symbols is caused by the restricted bandwidth of the system; for example, filters. Each restriction of bandwidth, which, of course, is always required to reduce the influence of shot and thermal noise, yields undesired pulse broadening. It gives rise to intersymbol interference (ISI). As shown in Section 9.5.1, influence of ISI can easily be assessed from the eye pattern measured by means of an oscilloscope. As an example, Fig. 11.3 shows a typical eye pattern.

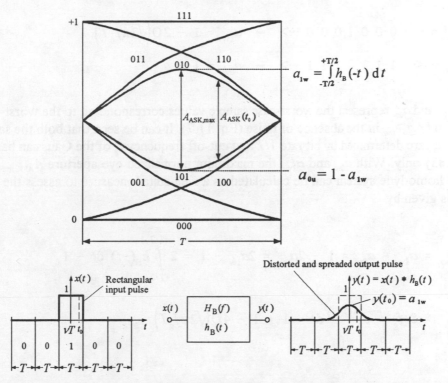

$$a_{1w} = \int_{-T/2}^{+T/2} h_B(-t)\, dt$$

$$a_{0u} = 1 - a_{1w}$$

Fig. 11.3: Eye pattern and aperture in ASK homodyne receiver with Gaussian baseband filter in absence of noise

In order to distinguish the different symbol sequences of a transmitter, an index i is used. Hence, the totality of all symbol sequences is $<q_v>_i = \cdots q_{-2}, q_{-1}, q_0, q_1, q_2 \cdots$ where $i = 0, 1, \cdots \infty$. For determining BER, a second distinction with respect to the present symbol q_0 at the current sampling point t_0 is required. Since this symbol can either be "1" or "0", the following distinction can be made for the sampled detected signal and symbol sequence:

$$d = \begin{cases} d_{1i} = a_{1i} \cos(\phi) + n & \text{if } <q_v>_{1i} = <\cdots\ q_{-2}, q_{-1}, 1 \\ d_{0i} = a_{0i} \cos(\phi) + n & \text{if } <q_v>_{0i} = <\cdots\ \overline{q}_{-2}, \overline{q}_{-1}, 0 \end{cases} \tag{11.11a}$$

It may be mentioned that symbol pattern $<q_v>_{1i}$ and $<q_v>_{0i}$ are complementary to each other. Thus, a symbol $q_v = $ "1" of sequence $<q_v>_{1i}$ is related to a symbol $q_v = $ "0" in the sequence $<q_v>_{0i}$.

In the all possible symbol sequences, the worst-case pattern $<q_v>_w$ are of primary importance since these pattern directly affect the aperture of the eye pattern and, thus, the worst-case BER as shown in Section 9.5.

With a *Gaussian baseband filter* (Eq. 9.83), the worst-case pattern are given by the symbol sequences "single one" and "single zero" i.e.,

$$\langle q_v \rangle_{1w} = \langle \cdots 0\ 0\ 0\ 1\ 0\ 0\ 0\ \cdots \rangle \quad \rightarrow \quad a_{1w} = 1 - 2Q\left(\sqrt{2\pi}f_g T\right)$$

$$\langle q_v \rangle_{0w} = \langle \cdots 1\ 1\ 1\ 0\ 1\ 1\ 1\ \cdots \rangle \quad \rightarrow \quad a_{0w} = 1 - a_{1w}$$

$$(11.11b)$$

Here, a_{1w} and a_{0w} represent the worst-case sample values corresponding to the worst-case pattern $\langle q_v \rangle_{1w}$ and $\langle q_v \rangle_{0w}$ in the absence of noise (Eq. 11.7). It can be seen that both the sample values a_{1w} and a_{0w} are determined by bit rate $1/T$ and cut-off frequency f_g of the Gaussian baseband filter (Eq. 9.83) only. With a_{0w} and a_{1w}, the maximum normalized eye aperture $A_{ASK,max}$ of an ASK optical homodyne system can be calculated as a well-suited measure to assess the influence of ISI. It is given by

$$A_{ASK,max} = a_{1w} - a_{0w} = 1 - 2a_{0w} = 2a_{1w} - 1 = 2\int_{-T/2}^{+T/2} h_B(-t)\ dt - 1$$

$$(11.12)$$

$$= 8 f_g \int_0^{+T/2} e^{-\pi(2f_g t)^2} dt - 1 = 1 - 4Q\left(\sqrt{2\pi}f_g T\right)$$

It should be remembered that the constant of normalization is again the amplitude $\hat{r}_{PD}$ of the photodiode current. The Q-function has already been defined in Chapter nine (Eq. 9.95). In Fig. 11.3, maximum aperture of the eye pattern is located at the centre of the bit provided pulses at the output of the filter are symmetrical. Hence, optimum detection times are given by $t_{0,opt} = \upsilon T$. Comparing the above equation and Eq. (10.9) of Chapter ten, it becomes immediately clear that normalized eye apertures A_{ASK} and A_d for coherent optical ASK homodyne system and incoherent direct detection system are in principle same. However, on comparing the absolute eye apertures, a gain of about 16 dB can be observed i.e., eye aperture in ASK homodyne system is about 6.3 times larger than that of a direct detection system.

With the above results, we are now able to make bit error rate evaluation. For this, we first consider an arbitrary symbol sequence $\langle q_v \rangle_{0i}$ or $\langle q_v \rangle_{1i}$. When the symbol sequence $\langle q_v \rangle_{0i}$ is transmitted, probability of error for the present symbol q_0 = "0" is given by

$$P_{0i} = p(d_{0i} > E) = \int_E^{+\infty} f_{d0i}(d)\ dd = \int_E^{+\infty} \int_{-1}^{+1} f_n(d - a_{0i}w)\ f_w(w)\ dw\ dd \qquad (11.13)$$

which has been derived using Eq. (11.8). In this equation, E is the threshold level normalized to amplitude $\hat{i}_{PD}$ of the photodiode current and $f_{d0i}(d)$ the pdf of the sample value d_{0i}. Hence, this pdf provides that $q_0 = "0"$ or $<q_v>_i = <q_v>_{0i}$. By means of the Q-function or erfc-function

$$Q(x) = \frac{1}{\sqrt{2\pi}} \int\limits_{x}^{+\infty} e^{-\frac{u^2}{2}} du = \frac{1}{2} \mathrm{erfc}\left(\frac{x}{\sqrt{2}}\right) \tag{11.14}$$

the Eq. (11.13) can be simplified as follows [8]:

$$p_{0i} = \int\limits_{-1}^{+1} Q\left(\frac{E - a_{0i} w}{\sigma}\right) f_w(w) \, dw = \int\limits_{-1}^{+1} \tilde{p}_{0i}(w) \, f_w(w) \, dw \tag{11.15}$$

It becomes clear from this equation that calculation of error probability has to be performed in two steps: First, we have to determine the probability of error $\tilde{p}_{0i}(w)$ by neglecting phase noise. Thereby, the random variable $w = \cos(\phi)$ is assumed to be a constant. To show that phase noise is neglected, the sign "~" on the top of the probability $\tilde{p}_{0i}$ has been chosen. The result of the first step is same as obtained in the case of conventional digital communication systems i.e., the well-known dependence of bit error rate on Q-function or erfc-function [e.g., 35, 38]. In the second step, random nature of $w = \cos(\phi)$ has to be considered. For this, pdf $f_w(w)$ determined in Section 3.2.5 will be taken into consideration. Finally, the actual probability of error is obtained by evaluating the expected value of $\tilde{p}_{0i}(w)$, which has been performed in second part of Eq. (11.15).

Next we shall calculate the probability of error p_{1i} for the other symbol $q_0 = "1"$ in the symbol sequence $<q_v>_{1i}$. This calculation can be performed in the same way as for symbol "0". Accordingly, we obtain

$$p_{1i} = p(d_{1i} < E) = \int\limits_{-\infty}^{E} f_{d1i}(d) \, dd = \int\limits_{-\infty}^{E} \int\limits_{-1}^{+1} f_n(d - a_{1i}w) \, f_w(w) \, dw \, dd \tag{11.16}$$

This expression can also be simplified by means of the Q-function (Eq. 11.14). We obtain:

$$p_{1i} = \int_{-1}^{+1} Q\left(\frac{a_{1i} \, w - E}{\sigma}\right) f_w(w) \; dw = \int_{-1}^{+1} \tilde{p}_{1i}(w) \, f_w(w) \; dw \tag{11.17}$$

(a) AVERAGE PROBABILITY OF ERROR p_a

An important measure to assess the quality of transmission and to determine the overall performance of a digital communication system is the average probability of error p_a. As already mentioned in Section 9.5.3, probability of error is frequently called as bit error rate (BER) although it is actually not a rate.

In a realized communication system, BER is usually measured by means of an appropriate measurement equipment which includes at least a digital transmitter (normally a PN-generator) and a digital receiver with a bit error counter. For a new communication system to be designed, the probability of error in the system must be calculated or estimated first. For this, all possible bit pattern $<q_v>_i$ with $i = 1, 2, \cdots \infty$ and their related probabilities of error have to be considered. This is, however, a rather comprehensive task. Normally, a limited number of bits which are in close neighboured influence each other in practical systems. Therefore, only a limited number of symbol sequences need to be considered.

As briefly explained in Chapter nine, number of symbol pattern which must actually be used for the BER calculation equals the number of different lines in the eye pattern of detected signal $d(t)$. If, for example, each bit is only disturbed by n previous bits and v following bits, then the number of distinguishable lines in the eye pattern equals 2^{n+v+1} (see Eq. 9.87). In that case, probability of error is given by

$$p_a = \sum_{i=1}^{2^{n+v}} p\left(<q_v>_i\right) \left(p_{0i} + p_{1i}\right) = 2^{-(n+v+1)} \sum_{i=1}^{2^{n+v}} \left(p_{0i} + p_{1i}\right) \tag{11.18}$$

where $p(<q_v>_i)$ represents the occurrence probability of the symbol sequence $<q_v>_i$ or line number i in the eye pattern. If all the symbols q_v are statistically independent and both the occurrence probabilities $p(q_v = 0)$ and $p(q_v = 1)$ are equal, then the occurrence probabilities $p(<q_v>_i)$ of the symbol sequences are also equal. Let us consider, for example, that only $2^{n+v+1} = 8$ different symbol pattern have to be taken into account. In that case, eye pattern exhibits $2^{n+v} = 4$ distinguishable lines with $q_v = 0$ and 4 distinguishable lines with $q_v = 1$. The occurrence probability $p(<q_v>_i)$ equals $2^{-(n+v+1)} = 1/8$.

(b) WORST-CASE PROBABILITY OF ERROR p_w

Even for a low and limited number of different symbol pattern (for example, $2^{n+v+1} = 8$), calculation of the average probability of error p_a becomes rather comprehensive. In contrast to this, worst-case probability of error p_w can be evaluated much faster. As mentioned earlier, two worst-case pattern $<q_v>_{0w}$ and $<q_v>_{1w}$ are at least existent among all the possible symbol sequences $<q_v>_i$. For this reason, it is useful to define the probability

$$p_w = \frac{1}{2}\left(p_{0w} + p_{1w}\right) \tag{11.19}$$

as the worst-case probability of error or briefly the worst-case BER. In the above expression, p_{0w} and p_{1w} correspond to the worst-case symbol pattern $<q_v>_{0w}$ and $<q_v>_{1w}$ respectively. In the case of an ASK homodyne system, both probabilities p_{0w} and p_{1w} are normally different since an ASK system is disturbed by signal-dependent noise. Remembering that the influence of phase noise is different for symbol "1" and symbol "0". When both threshold level E and cut-off frequency f_g of the Gaussian baseband filter are optimized for a minimum BER (see Subsection iv below), following relationship is valid:

$$p_{0w} \leq p_{1w} \tag{11.20}$$

The equality sign is used for ASK homodyne system without any phase noise. It should be noted that p_{0w} and p_{1w} are always same if an optimized PSK homodyne system instead of an ASK system is used (see Section 11.1.2).

In coherent optical digital communication systems, the product $2f_g T$ of filter bandwidth $2f_g$ and bit duration T is usually more than one i.e., $2f_g T > 1$. For this practical reason, only the adjacent symbols q_{-1} and q_{+1} normally influence the present symbol q_0 at the sampling point t_0. Hence, the eye pattern exactly includes eight distinguishable lines (Fig. 11.3) which are related to eight distinguishable symbol sequences. When all these sequences occur with same probability $p(<q>_i)$ = 1/8, the following practical-based inequality can be given:

$$\frac{1}{4}p_w \leq p_a \leq p_w \tag{11.21}$$

In this relationship, lower bound on the average probability p_a assumes that p_a is determined by both the worst-case pattern only. The contribution of all other (six) symbol sequences is neglected. Thus, their probability of error is assumed to be zero. In contrast to this, upper bound provides an equal and maximum probability of error (i.e., the worst-case BER of the worst-case pattern) for all the eight symbol sequences. Since accuracy of one order of magnitude is mostly sufficient in practice, above estimation represents an useful tool for a fast and practically-based calculation of probability of error or BER. With Eq. (11.15) and Eq. (11.17), worst-case probability of error can now be expressed as:

$$
P_w = \frac{1}{2} \int_{-1}^{+1} \left[Q\left(\frac{E - a_{0w}\, w}{\sigma} \right) + Q\left(\frac{a_{1w}\, w - E}{\sigma} \right) \right] f_w(w)\ dw
\tag{11.22}
$$

Substituting the argument of Q-functions by using the maximum aperture $A_{ASK} := A_{ASK,max}$ of the eye pattern given in Eq. (11.12), we obtain

$$
P_w = \frac{1}{2} \int_{-1}^{+1} \left[Q\left(\frac{2E - (1-A_{ASK})\, w}{2\sigma} \right) + Q\left(\frac{(1+A_{ASK})\, w - 2E}{2\sigma} \right) \right] f_w(w)\ dw
\tag{11.23}
$$

Finally, the pdf $f_w(w)$ of the random process $w = \cos(\phi)$ can be replaced by using Eq. (3.47). After some mathematical operations, following result is obtained:

$$
\begin{aligned}
P_w = \frac{1}{\sqrt{2\pi}\sigma_\phi} \int_{0}^{+\infty} &\left[Q\left(\frac{2E - (1-A_{ASK})\cos(\phi)}{2\sigma} \right) + Q\left(\frac{(1+A_{ASK})\cos(\phi) - 2E}{2\sigma} \right) \right] \\
&\cdot \exp\left(-\frac{\phi^2}{2\sigma_\phi^2} \right)\ d\phi
\end{aligned}
\tag{11.24}
$$

This equation shows that the worst-case BER p_w (and, of course, the average BER p_a also) is a function of several system parameters: threshold E, aperture $A_{ASK,max}(f_g, T, t_0)$ of the eye pattern, standard deviation $\sigma_\phi(\Delta f)$ of the OPLL rest-phase noise (Eq. 11.57), and standard deviation $\sigma(f_g, G_c)$ of the additive Gaussian receiver noise i.e., shot noise and thermal noise. It should be remembered that threshold E and standard deviation σ have been normalized with respect to the amplitude $\hat{\imath}_{PD} = \hat{\imath}_{PD}(P_r, P_l)$ of the photodiode current. Therefore, the worst-case BER p_w is also influenced by the optical power levels P_r and P_l of received and local laser light waves respectively. As these parameters are again functions of other system parameters, the worst-case BER also depends on symbol or bit duration T, cut-off frequency f_g of Gaussian baseband filter, resulting laser linewidth Δf of both transmitter and local laser and power spectral density G_c of the additive Gaussian receiver noise. Finally, the sampling point t_0 is a system parameter of primary importance since it determines the vertical aperture A_{ASK} of the eye pattern and, hence, the BER. In the above expressions, sampling point t_0 is not included since an optimum sampling point t_0 for maximum eye aperture A_{ASK} has already been taken.

(iv) OPTIMIZATION

The aim of this Subsection is to optimize the system parameters for the minimum worst-case BER. To simplify calculation, a Gaussian baseband filter with cut-off frequency f_g is considered (Eq. 9.83). The first step in the process of optimization is to decide which of the system parameters can be optimized and which parameters cannot be optimized.

The characteristic features of a *non-optimizable system parameter* is first a fixed value and second a virtual optimum. The BER could only be minimized when the numerical value of this parameter is chosen to be either infinite or zero and both cannot be realized normally. A typical non-optimizable system parameter is, for example, the emission linewidth Δf of the transmitter laser. This linewidth is a fixed parameter determined by the type of laser which has been selected. Moreover, the optimum linewidth is virtual since the optimum linewidth is zero ($\Delta f = 0$) and, thus, not realizable. Other non-optimizable system parameters are the received light power P_r (the virtual optimum is $P_r \rightarrow \infty$), the bit rate $1/T$ and the power spectral density G_c of the additive Gaussian receiver noise.

In contrast to this, an *optimizable system parameter* exhibits a real, mathematical optimum value. For an ASK coherent optical communication system, three optimizable system parameters are at least existent: threshold level E, cut-off frequency f_g of the Gaussian baseband filter and sampling point t_0. Since we have already selected a Gaussian baseband filter, optimization of this filter is restricted to the cut-off frequency only. In general, entire filter transfer function $H_B(f)$ of the baseband filter has to be optimized. As it is well-known from conventional digital communication techniques, the result of this optimization yields the so-called matched filter which is a very common technique in practice [36]. In view of above and to avoid generality, we have decided to optimize the bandwidth $2f_g$ only and focus our interest on the principle.

(a) SAMPLING POINT t_0

By changing sampling point t_0, effective aperture A_{ASK} of the eye pattern is changed simultaneously. If sampling point is optimized, then the effective eye aperture reaches its maximum value. Considering Fig. 11.3, optimum sampling point $t_{0,opt} = \upsilon T$ is located at the centre of the bit (in the strict sense at the centre of each bit) since this eye pattern is determined on the basis of Gaussian baseband filter which exhibits a symmetrical impulse response $h_B(t) = h_B(-t)$. If a baseband filter with an asymmetrical impulse response $h_B(t) \neq h_B(-t)$ is chosen, then the optimum sampling point is shifted from the centre; for example, to the end of each bit in case of a simple RC filter [35]. In this book, we restrict to symmetrical impulse responses only.

(b) THRESHOLD E

The optimum and normalized threshold E_{opt} is located at the cross point of probability density functions $f_{d0w}(d)$ and $f_{d1w}(d)$ where $f_{d0w}(d)$ and $f_{d1w}(d)$ are related to both the symbols $q_\upsilon = $ "0" and $q_\upsilon = $ "1" in the worst-case pattern $<q_\upsilon>_{0w}$ and $<q_\upsilon>_{1w}$ respectively. Therefore, the equation for determining the optimum threshold level is given by

$$f_{d1w}(E_{opt}) = f_{d0w}(E_{opt}) \qquad (11.25)$$

Proof

$$\frac{dp_w}{dE} = \frac{1}{2} \frac{d}{dE} \left[\int_E^{+\infty} f_{d0w}(d) \; dd + \int_{-\infty}^E f_{d1w}(d) \; dd \right]$$

$$= \frac{1}{2} \left[f_{d1w}(E) - f_{d0w}(E) \right] = 0 \; \rightarrow \; f_{d1w}(E) = f_{d0w}(E) \; \rightarrow \; E = E_{opt}$$

Unfortunately, an analytical solution of Eq. (11.25) is not possible for optimum threshold E_{opt}. Hence, a numerical method of iteration must be used on computer, which also yields the desired solution very fast. The result will be shown and discussed in Subsection (v) below.

An optical ASK homodyne system is disturbed by signal-dependent noise. As a higher signal level is more disturbed than a lower signal level, following inequality can be given for this system:

$$E_{opt} \leq \frac{1}{2}\left(a_{0w} + a_{1w}\right) = 0.5 \qquad (11.26)$$

Here, equality sign applies for an ASK homodyne system without any phase noise i.e., $\sigma_\phi = 0$. In this case, both probability density functions $f_{d0w}(d)$ and $f_{d1w}(d)$ are Gaussian and symmetrical with respect to the optimum and normalized threshold $E_{opt} = 0.5$. Thus, these probability density functions are determined only by the normalized standard deviation $\sigma = \sigma(f_g)$ of the additive Gaussian receiver noise (i.e., shot noise and thermal noise). As the optimum threshold E_{opt} is a function of the standard deviation $\sigma(f_g)$, optimum threshold indirectly is also a function of the optimizable cut-off frequency f_g of the Gaussian baseband filter.

(c) CUT-OFF FREQUENCY f_g OF THE LOW-PASS FILTER

Due to the complexity of both the Eqs. (11.23) and (11.24), an analytical evaluation of the optimum cut-off frequency $f_{g,opt}$ of the low-pass (baseband) filter is also not possible. Therefore, numerical methods of iteration are again to be used. The results are shown and discussed in

Subsection (v) below. The accuracy of numerical results (termed as method one) largely depends on the number of iterative steps and, therefore, on the computer processing time. To get a fast solution, appropriate approximation methods are more convenient.

A rather rough method of approximation (method two) can be obtained by neglecting the phase noise. In that case, optimum cut-off frequency $f_{g,opt}$ is obtained by

$$\frac{A_{ASK}(f_g)}{\sigma(f_g)} \rightarrow \text{maximum provided that } \sigma_\phi = 0 \qquad (11.27)$$

With simple numerical evaluation, this operation of maximization yields

$$f_{g,opt} T \approx 0.79$$

where T is the bit period. This simple approximation is well-suited when the optical power level P_r at the input to the receiver is less than -50 dBm (Fig. 11.4b). Since an ASK homodyne receiver exhibits a high sensitivity, this requirement is normally fulfilled. If a better approximation is needed, then the expected value

$$E\left\{\frac{A_{ASK}(f_g)}{\sqrt{\sigma_d^2(f_g)\big|_{a0w} + \sigma_d^2(f_g)\big|_{a1w}}}\right\} \rightarrow \text{maximum} \rightarrow f_{g,opt} T \qquad (11.28)$$

can be used to perform the maximization and estimate the optimum product $f_{g,opt} T$. We call this method of approximation as method three. In contrast to approximation given in Eq. (11.27), expected value above also takes into account the phase noise and its different influence on both the symbols "0" and "1". As given in Eq. (11.10b), variances $\sigma_d^2(f_g)\big|_{a0w}$ and $\sigma_d^2(f_g)\big|_{a1w}$ of the detected samples d include the influence of additive Gaussian noise (described by $\sigma(f_g)$) as well as the influence of phase noise (described by σ_ϕ). In case of $\sigma_\phi = 0$, which means that phase noise does not disturb the system, both Eqs. (11.27) and (11.28) yield same result for $f_{g,opt} T$.

All three methods discussed above, require a computer or at least an electronic calculator to perform the numerical iterations. While method one allows to determine $f_{g,opt} T$ exactly or at least with desired accuracy (only depending on the number of iterative steps), methods two and three are approximations. Each single iterative step in method one requires the numerical evaluation of an integral which finally requires a relatively large computer processing time. In comparison to that, methods two and three can be performed much faster since no evaluation of integrals is involved. Even when a non-Gaussian baseband filter is used, methods two and three represent a fast and well-suited tool to determine the optimum value of $f_{g,opt}T$.

(v) EVALUATION AND DISCUSSION

The purpose of this last Subsection is to evaluate the results obtained in previous Subsections to achieve a better insight of partial complex interrelations in ASK coherent optical communication system.

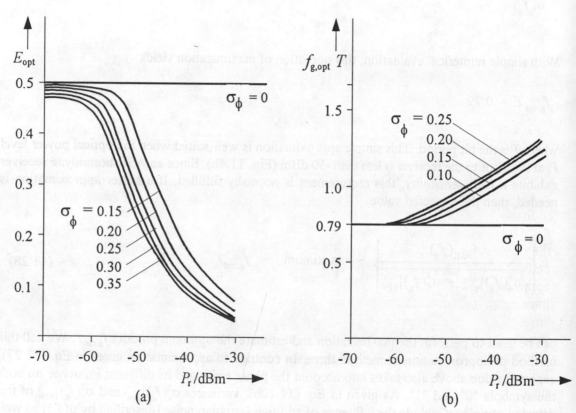

Fig. 11.4: (a) Normalized and optimized threshold E_{opt} and (b) normalized and optimized cut-off frequency $f_{g,opt}T$ of low-pass filter in ASK homodyne receiver

To begin with, Fig. 11.4a shows the normalized and optimized threshold level E_{opt} as a function of the received optical power P_r expressed in dBm. Another parameter in Fig. 11.4a is the standard deviation σ_ϕ of the OPLL rest-phase noise as given in Eq. (11.57). Evaluation of optimum threshold E_{opt} requires the cut-off frequency f_g of the baseband filter. As optimum value of f_g is a function of the received optical power P_r also i.e., $f_{g,opt} = f_{g,opt}(P_r)$, cut-off frequency f_g must be optimized afresh for each P_r to obtain Fig. 11.4a. Thus, each change in P_r requires an optimization of both f_g and E again. For this, method one has been used for calculation (see optimization of cut-off frequency f_g). All other system parameters required to obtain Fig. 11.4a are as follows (Table 11.1):

Table 11.1: Typical system parameters

Transmitted symbol sequence:	$\langle q_v \rangle = \cdots 111 \cdots$ (permanent-"1") and
	$\langle q_v \rangle = \cdots 000 \cdots$ (permanent-"0"),
Symbol or bit rate:	$1/T = 560$ Mbit/s,
Frequency of transmitter laser:	$f_t = 200$ THz,
Quantum efficiency of photodiode:	$\eta = 0.83$,
Responsivity of PIN-photodiode:	$R_0 = (e\eta)/(hf) = 1$ A/W,
Dark current of PIN-photodiode:	$I_{dark} = 10^{-11}$ A,
Thermal noise:	$G_{th} = 10^{-23}$ A^2/Hz,
Equivalent baseband filter:	$H_B(f) = \exp[-\pi(f/2f_g)^2]$,
	$f_g = f_{g,opt}$,
Threshold:	$E = E_{opt}$,
Local laser power:	$P_1 \triangleq -10$ dBm,
Balanced receiver:	$K_R = 2$,
Coupling ratio of optical coupler:	$k = 0.5$ (i.e., 3 dB coupler).

In the absence of phase noise (i.e., $\sigma_\phi = 0$), normalized and optimized threshold always equals $E_{opt} = 0.5$ irrespective of received optical power level P_r. If phase noise impairs the system, then the optimum threshold E_{opt} becomes a function of P_r. Since phase noise always degrades the detection of a transmitted symbol "1", but improves the detection of symbol "0", normalized optimum threshold E_{opt} is always less than 0.5 as shown in Fig. 11.4a. Thereby, the deviation from $E_{opt} = 0.5$ increases with a growing distortion for symbol "1". For this reason, deviation is large when standard deviation σ_ϕ of the OPLL rest-phase noise is large. This deviation also increases with the increase in the optical power P_r at the receiver input. It can be explained as follows: high optical signal power P_r decreases the influence of the additive Gaussian noise and indirectly increases the influence of phase noise factor $\cos(\phi)$. Therefore, it leads to increase in the deviation.

The optimum cut-off frequency $f_{g,opt}$ as a function of the received optical power P_r and standard deviation σ_ϕ of the phase noise is shown in Fig. 11.4b. Analogous to Fig. 11.4a, this figure is also based on method one (exact calculation). Thereby, the threshold E has been optimized afresh for each new value of P_r. It becomes clear from Fig. 11.4b that bandwidth $2f_g$ of the baseband filter must be increased when phase noise disturbs the system. The physical reason for this is a broaden signal spectrum (due to phase noise) which requires a broader filter to give appropriate high energy at the filter output.

An important measure to assess the system performance is the probability of error which should, of course, be as low as possible. The worst-case probability of error p_w as a function of received optical power P_r at the receiver input and standard deviation σ_ϕ of the OPLL rest-phase noise is shown in Fig. 11.5.

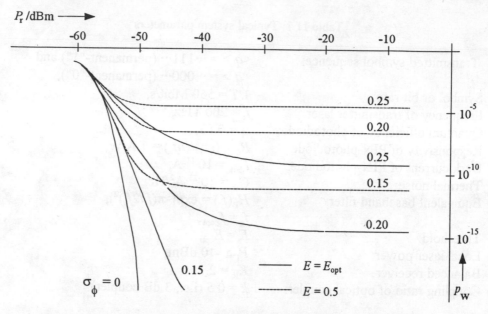

Fig. 11.5: Worst-case probability of error p_w in ASK homodyne receiver

In the above figure, it is considered that the cut-off frequency f_g of the low-pass filter is already optimized. The curves shown in Fig. 11.5 are briefly called the *BER curves* or *performance curves*. It can be seen that these curves can be divided into three characteristic regions:

First region: In this region where received optical power P_r is low, all the curves show a rather steep slope since the additive Gaussian receiver noise dominates the phase noise. This part of the BER curves is in agreement with the well-known BER curves of an optical transmission system with direct detection (see Fig. 10.2) as well as all standard electrical digital transmission systems [35, 36]. Here, even a small increase in the optical power yields a considerable improvement in BER.

Second region: As optical power level P_r increases, influence of the additive Gaussian noise decreases while the influence of the phase noise indirectly increases. As a result, the BER curves exhibit a bend.

Third region: When the received optical power P_r is very high, the system is practically disturbed by phase noise alone. In this case, influence of the additive Gaussian noise is negligibly small. It may be mentioned that the influence of an additive noise can always be reduced by raising the optical power P_r, while the influence of phase noise remains unchanged. For this reason, BER curves reach a saturation level called the *error rate floor* or the *BER floor*.

It becomes evident from the above figure that a desired probability of bit error, for example; $p_w = 10^{-10}$ cannot be achieved when standard deviation σ_ϕ of phase noise exceeds a certain value (for example; $\sigma_\phi \approx 0.25$). Therefore, the error rate floor is of great practical interest. As maximum allowed standard deviation is related to maximum allowed laser linewidth Δf_{max} (Section 11.1.3), Fig. 11.5 directly enables to decide whether a selected laser satisfy the system requirements or not. It should be noted that an error rate floor normally exists in all types of coherent communication systems independent of modulation and demodulation schemes used.

In order to emphasize the influence of a non-optimized threshold E, Fig. 11.5 additionally shows the BER curves for a threshold $E = 0.5$. This normalized threshold only represents optimum threshold for ASK homodyne system without any phase noise. It becomes clear from Fig. 11.5 that a small change in threshold yields a considerable change in system performance.

A comparison of the BER curves of coherent and direct detection systems which are discussed in previous Chapter ten (see Fig. 10.2), yields the following results: (1) BER curve of coherent systems exhibits an error rate floor due to laser noise whereas no such saturation is given when direct detection is employed and (2) BER curve of a direct system is shifted to the right, where received signal power is high. This becomes clear by comparing Fig. 10.2 and Fig. 11.5 above. An ASK homodyne system only requires an optical input power of about -52 dBm to achieve a BER of 10^{-10} (provided laser noise does not disturb the system), whereas the direct detection system requires about -39 dBm. It implies a sensitivity gain of 13 dB for an ASK system.

As mentioned in the previous Sections, worst-case bit error rate is a well-suited and a powerful measure to approximate the average error rate which is the actual error rate of a digital communication system. To verify the accuracy of this worst-case approximation, Fig. 11.6 shows both worst-case probability of error p_w and average probability of error p_a.

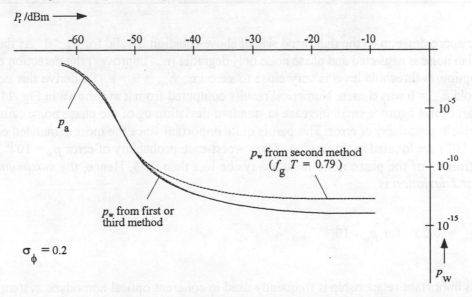

Fig. 11.6: Comparison of average probability of error p_a and worst-case probability of error p_w

It becomes evident from this figure that both the probabilities of error do agree very well. It is important to know that for the calculation of p_a in Fig. 11.6 only eight different symbol sequences are taken. Thus, one previous and one following neighbouring symbols have been considered. As mentioned earlier, only eight symbol sequences are sufficient in case of Gaussian baseband filter with optimum bandwidth $2f_{g,opt}$. This bandwidth must always be optimized afresh for each power level P_r for both minimum worst-case probability of error p_w and minimum average probability of error p_a. As the optimum bandwidth increases with the increase in optical power P_r (Fig. 11.4b), influence of ISI simultaneously decreases. As a result, both probabilities of error become equal at high levels of P_r provided first or second method has been chosen to calculate p_w. It becomes clear from Fig. 11.6 that second and third methods represent very good approximations to predict the average probability of error which can be measured in ASK homodyne system. Moreover, both these methods require very less computer run time in comparison to first method.

In the design of coherent optical communication systems, maximum standard deviation σ_ϕ of the OPLL rest-phase noise to achieve a given probability of error is an important parameter and is required to be known. For this, let us consider that the system is neither disturbed by additive Gaussian receiver noise ($\sigma = 0$) nor by ISI. Thus, the system is only disturbed by phase noise of both transmitter and local lasers. In this case, worst-case probability of error can be easily calculated by

$$p_w = \frac{1}{2} \int_{-1}^{0} f_w(w) \, dw \approx Q\left(\frac{\pi}{2\sigma_\phi}\right) \quad \text{provided } \sigma = 0 \qquad (11.29)$$

The approximation on the right-hand side of above equation is valid for $\sigma_\phi < 1$. As the additive Gaussian noise is neglected and phase noise only degrades (not improves) the detection of symbol "1", optimum threshold level is very close to zero i.e., $E_{opt} = 0 + \epsilon$. To derive this equation, a threshold $E_{opt} = 0$ was chosen. Numerical results computed from it are shown in Fig. 11.7. In the first part of this figure, a small increase in standard deviation σ_ϕ of the phase noise causes a very steep rise in probability of error. This part is quite important since the most demanded error rates ($p_w < 10^{-6}$) are located in this region. For a worst-case probability of error $p_w = 10^{-10}$, standard deviation σ_ϕ of the phase noise must always be less than 0.25. Hence, the *maximum allowed standard deviation* is

$$\sigma_{\phi,max} \approx 0.25 \quad \text{for } p_w = 10^{-10} \qquad (11.30)$$

This important relationship is frequently used in coherent optical homodyne systems. As will be discussed in Section 11.1.3, influence of rest-phase noise primarily depends on the architecture of optical phase-locked loop (OPLL) and, of course, on the laser linewidth of both transmitter laser and local laser. By using Eq. (11.57) discussed in Section 11.1.3, maximum allowed laser linewidth $\Delta f_{max} = \Delta f_{T,max} + \Delta f_{L,max}$ can be calculated. This linewidth represents an important parameter for every coherent optical communication system. A summarized discussion on the phase noise problem and the linewidth requirements is given in Chapter twelve.

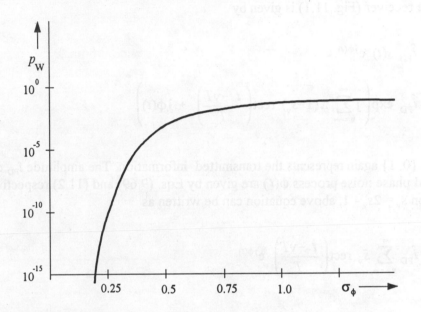

Fig. 11.7: Probability of error for ASK homodyne system when only phase noise disturbs the system

Some important results obtained in this Section on coherent optical ASK homodyne systems are summarized below.

- ASK homodyne systems are very sensitive to the laser phase noise of both transmitter and local lasers.

- Performance degradation due to laser phase noise is different for symbols "0" and "1".

- ASK homodyne systems require an optical phase-locked loop (Section 11.1.3).

- Due to phase noise, BER exhibits an error rate floor.

- Optimum threshold level in presence of laser phase noise is always less than the optimum level when phase noise is absent.

11.1.2   PSK HOMODYNE SYSTEM

(i)   DETECTED SIGNAL $d(t)$ and DETECTED SAMPLE VALUES $d(vT + t_0)$

Like in ASK homodyne system, photodiode current in the signal branch of optical PSK homodyne receiver (Fig. 11.1) is given by

$$i_{PD}(t) = \hat{i}_{PD} \; \underline{s}(t) \; e^{j\phi(t)}$$

$$= \hat{i}_{PD} \; \exp\left(j \sum_{v=-\infty}^{+\infty} \pi(1-s_v) \; rect\left(\frac{t - vT}{T} \right) + j\phi(t) \right) \qquad (11.31a)$$

where $s_v \in \{0, 1\}$ again represents the transmitted information. The amplitude $\hat{i}_{PD}$ of photodiode current and phase noise process $\phi(t)$ are given by Eqs. (9.69) and (11.2) respectively. With the substitution $\tilde{s}_v = 2s_v - 1$, above equation can be written as

$$i_{PD}(t) = \hat{i}_{PD} \sum_{v=-\infty}^{\infty} \tilde{s}_v \; rect\left(\frac{t - vT}{T} \right) e^{j\phi(t)} \qquad (11.31b)$$

Comparing the photodiode currents for ASK system (Eq. 11.1) and PSK system, a close relationship of both homodyne systems is evident. In an ASK system, s_v is either 0 or 1, whereas in a PSK system $\tilde{s}_v$ is either -1 or +1. Thus, a PSK homodyne system can be regarded as an ASK homodyne system with a *bipolar modulation* i.e., $\tilde{s}_v \in \{-1, +1\}$. For this reason, these systems exhibit many similarities. To prevent repetitions, this Section is primarily focussed on the differences only. For other description, Section 11.1.1 can be referred.

According to an optical ASK homodyne system, statistical features of the sampled detected signal

$$d(t_0) = \hat{i}_{PD} \; a(t_0) \; \cos\big(\phi(t_0)\big) + n(t_0) \qquad (11.32)$$

are of fundamental importance to assess the system performance. Difference in the sampled detected signals of both the homodyne systems is only given by the range of values of

$$a(t_0) = \int_{-\infty}^{+\infty} \underline{s}(\tau)\ h_B(t_0-\tau)\ d\tau \quad \text{with} \quad -1 \leq a(t_0) \leq +1 \tag{11.33}$$

It should be remembered that $a(t_0)$ equals the sample value $d(t_0)$ in the absence of any noise i.e., $n(t_0) = 0$ and $\phi(t_0) = 0$ (Eq. 11.6). The range of values given in the above equation presumes a Gaussian baseband filter as given in Eq. (9.83). With the same normalisations and substitutions as in the previous Section, normalized and sampled detected signal $d = d(t_0)/\hat{i}_{PD}$ of PSK homodyne system can be expressed as

$$d = a\ \cos(\phi) + n = aw + n \quad \text{with} \quad |w| \leq 1 \quad \text{and} \tag{11.34}$$

Except the range of noiseless sample value a, this expression is same as Eq. (11.7) for ASK homodyne system. As Eq. (11.34) is of significant importance, physical meaning of various parameters is summarized below.

  d: normalized detected sample value,

  a: noiseless part of d disturbed by intersymbol interference (a includes the transmitted information),

  $\cos(\phi)$: phase noise factor,

  n: additive Gaussian noise.

It becomes clear from the above equation that the sample values of the detected signal are again dependent on two random variables: one is the phase noise term $w = \cos(\phi)$ and the other is additive Gaussian receiver noise n.

(ii) PROBABILITY DENSITY FUNCTION $f_d(d)$

As the sampled detected signals in both ASK and PSK homodyne systems are in priciple equal, probability density functions can be calculated by using the same equation i.e., Eq. (11.9) derived in the previous Section. However, the shape of the probability density functions are different due to the different range of noiseless sample value a.

The pdf $f_d(d)$ in the special case of $a = -1$ and $a = +1$ which means $\tilde{s}_v = -1$ and $\tilde{s}_v = 1$ respectively is shown in Fig. 11.8. In the first case, all "0" are transmitted whereas in the second case all "1". Comparing Figs. 11.8 and 11.2, an important difference between both the homodyne systems becomes clear. The pdf of the detected signal corresponding to symbol "0" and symbol "1" in a PSK homodyne system are now symmetrical around $d = 0$. The reason for this is the bipolar signal transmission scheme i.e., $\tilde{s}_\nu \in \{-1, +1\}$. Thus, both the binary symbols "0" and "1" are equally disturbed by phase noise and additive receiver Gaussian noise. The optimum threshold is always located at zero i.e., $E_{opt} = 0$. In contrast to an ASK homodyne system, a PSK homodyne system belongs to the class of transmission systems which are characterized by *signal-independent noise*.

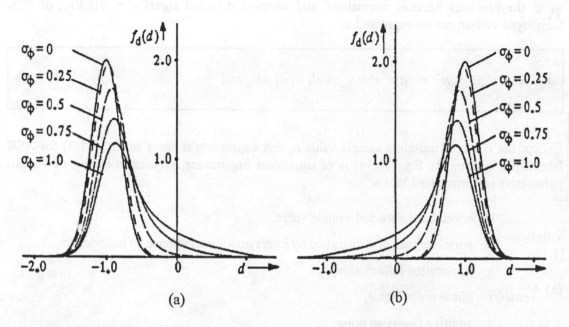

Fig. 11.8: Probability density function $f_d(d)$ of normalized sample value d of detected signal $d(t)$ in PSK homodyne receiver. Normalized standard deviation of additive Gaussian noise is fixed at $\sigma = 0.2$. Transmitted information is: (a) all "0" and (b) all "1" [8].

The pdf of detected signal d is again mainly determined by its expected value $E\{d\} = \eta_d$ and standard deviation σ_d as given in Eqs. (11.10a) and (11.10b) respectively. As both these quantities are functions of the noiseless sample value a, new range of a must be taken into account.

(iii) PROBABILITY OF ERROR OR BIT ERROR RATE (BER)

As the symbols "0" and symbol "1" are equally disturbed in PSK homodyne system, probabilities of error p_{0i} and p_{1i} for symbol pattern $<q_\nu>_{0i}$ and $<q_\nu>_{1i}$ are same. This is in contrast to ASK homodyne system where both probabilities of error are different. Hence, no distinction between the symbol pattern $<q_\nu>_{0i}$ and $<q_\nu>_{1i}$ is required to calculate the probabilities of error if optimum threshold level is used i.e., $E_{opt} = 0$. Hence,

$$a_i := a_{1i} = -a_{0i} \; \rightarrow \; f_{di}(d) := f_{d1i}(d) = f_{d0i}(-d) \; \rightarrow \; p_i := p_{1i} = p_{0i} \qquad (11.35)$$

Like an ASK homodyne system, probability of error for a PSK homodyne system can be calculated as follows (compare with Eqs. 11.13 and 11.16):

$$p_i = \int\limits_{-\infty}^{0} f_{di}(d) \; dd = \int\limits_{-\infty}^{0} \int\limits_{-1}^{+1} f_n(d + a_i w) \, f_w(w) \; dw \; dd \qquad (11.36)$$

By using Q-function given in Eq. (11.14), above equation simplifies to

$$p_i = \int\limits_{-1}^{+1} Q\left(\frac{a_i \, w}{\sigma}\right) f_w(w) \; dw = \int\limits_{-1}^{+1} \tilde{p}_i(w) \, f_w(w) \; dw \qquad (11.37)$$

A detailed explanation of this equation has already been given in the previous Section (Eq. 11.15).

(a) AVERAGE PROBABILITY OF ERROR p_a

Due to symmetrical properties of a PSK homodyne system, average probability of error can be calculated somewhat easier than for an ASK homodyne system (Eq. 11.18):

$$p_a = \sum\limits_{i=1}^{2^{n+v+1}} p(\langle q_v \rangle_i) \, p_i = 2^{-(n+v+1)} \sum\limits_{i=1}^{2^{n+v+1}} p_i \qquad (11.38)$$

It is seen from this equation that a distinction between both the probabilities p_{0i} and p_{1i} is not required since p_{0i} and p_{1i} are equal for PSK homodyne system i.e., $p_{0i} = p_{1i} = p_i$.

(b) WORST-CASE PROBABILITY OF ERROR p_w

In an ASK homodyne system, calculation of worst-case probability of error p_w was made by dividing p_w into two components p_{0w} and p_{1w}. This step was required since ASK is based on unipolar signal transmission with $s_v \in \{0, 1\}$. In contrast, this division is not required when a PSK homodyne system with bipolar signal transmission is considered i.e., $\tilde{s}_v \in \{-1, +1\}$. In this case, we have

$$p_{0w} = p_{1w} = p_w \tag{11.39}$$

By using the above relationship and Eq. (11.37), we obtain

$$p_w = \int_{-1}^{+1} Q\left(\frac{a_w\, w}{\sigma}\right) f_w(w)\ dw = \int_{-1}^{+1} \tilde{p}_w(w)\, f_w(w)\ dw \tag{11.40}$$

where a_w equals the sampled detected signal $d_w = a_w \cos(\phi) + n$ when there is no noise (i.e., $\sigma = 0$ and $\sigma_\phi = 0$) and under worst-case pattern. Assuming a Gaussian baseband filter (Eq. 9.83), normalized worst-case sample value a_w can be determined by using the equation

$$a_w = 1 - 4Q\left(\sqrt{2\pi} f_g T\right) = a_{1w} = -a_{0w} \tag{11.41}$$

As mentioned above, $a_{1w} = +a_w$ corresponds to the worst-case pattern

$$\langle q_v \rangle_{1w} = \langle \cdots\ 0\ 0\ 0\ 1\ 0\ 0\ 0\ \cdots \rangle$$

whereas the inverse pattern

$$\langle q_v \rangle_{0w} = \langle \cdots\ 1\ 1\ 1\ 0\ 1\ 1\ 1\ \cdots \rangle$$

is related to the normalized and noiseless sample value $a_{0w} = -a_w$. Both quantities a_{1w} and a_{0w} determine the *normalized aperture of the eye pattern*

$$A_{PSK} = a_{1w} - a_{0w} = 2a_w = 2\left[1 - 4Q\left(\sqrt{2\pi} f_g T\right)\right] = 2A_{ASK} \tag{11.42}$$

Like in an ASK homodyne system, eye aperture again represents an important measure to determine the system performance, particularly the probability of error. From Eqs. (11.40) and (11.42), we obtain:

$$p_w = \int\limits_{-1}^{+1} Q\left(\frac{A_{PSK}\, w}{2\sigma}\right) f_w(w)\; dw \quad \text{provided} \quad E = E_{opt} = 0 \tag{11.43}$$

The eye aperture A_{PSK} of PSK homodyne system is twice the eye aperture A_{ASK} of ASK homodyne system. For this reason, PSK homodyne system offers a *6 dB higher sensitivity* than the ASK homodyne system provided phase noise does not disturb the system (i.e., $\sigma_\phi = 0$). This results in a lower probability of error. With the same error rate in both the systems, PSK homodyne system requires a *6 dB lower optical power* at the input to the receiver.

Substituting the pdf $f_w(w)$ of random variable $w = \cos(\phi)$ in terms of the pdf $f_\phi(\phi)$ of phase noise, Eq. (11.43) can be written as follows:

$$p_w = \frac{2}{\sqrt{2\pi}\sigma_\phi} \int\limits_{0}^{+\infty} Q\left(\frac{A_{PSK}\,\cos(\phi)}{2\sigma}\right) \cdot \exp\left(-\frac{\phi^2}{2\sigma_\phi^2}\right)\; d\phi \tag{11.44}$$

(iv) OPTIMIZATION

The purpose of this Subsection is to optimize the system parameters for a minimum (worst-case) BER. Since the optimization of coherent optical PSK homodyne system is very similar to that of ASK homodyne system, this Subsection is restricted to the differences of both systems and principles. A more detailed discussion on the optimization of homodyne systems can be found in the previous Section 11.1.1 and also in the following Section 11.1.3, wherein the optical phase-locked loop (OPLL) is considered in more detail.

(a) THRESHOLD E

Due to the symmetrical features of a PSK homodyne system, optimum threshold level is

$$E_{opt} = 0 \tag{11.45}$$

This optimum threshold has already been used to derive the probability of error Eqs. (11.37), (11.43) and (11.44) given above.

(b) CUT-OFF FREQUENCY f_g OF THE LOW-PASS FILTER

As the error rate formulas (11.43) and (11.44) for a PSK homodyne system are much simpler as compared to an ASK homodyne system, optimization of the cut-off frequency f_g becomes easier. It becomes clear from Eq. (11.44) above that the worst-case probability of error reaches its minimum value if the argument of the Q-function is maximum. Therefore, the task of optimization is to maximize the expression

$$\frac{A_{PSK}(f_g) \, \cos(\phi)}{2\sigma(f_g)} = \frac{\left[1 - 4Q\left(\sqrt{2\pi} f_g T\right)\right] \cos(\phi)}{\sigma(f_g)} \rightarrow \text{maxi} \qquad (11.46)$$

with respect to f_g or $f_g T$. It should be noted that the left hand side of this expression is valid in general, whereas the right hand side is restricted on a Gaussian baseband filter. As the random variable $\cos(\phi)$ is not a function of cut-off frequency f_g, this variable (which is actually a constant here) does not influence the result. Hence, this term can be ignored during the maximization process. The normalized and optimized cut-off frequency $f_{g,opt} T$ depends only on the ratio of eye aperture to standard deviation of the additive Gaussian noise, which is same in ordinary digital communication systems [35, 36, 38].

By using appropriate iteration methods, maximization process yields again the following result

$$f_{g,opt} \, T \approx 0.79$$

It must be remembered that above numerical result is valid for a Gaussian baseband filter. In contrast to ASK system, optimum cut-off frequency $f_{g,opt}$ is now independent of the phase noise.

(v) EVALUATION AND DISCUSSION

The worst-case probability of error p_w versus received optical power P_r at the receiver input for different standard deviation σ_ϕ of the OPLL rest-phase noise is shown in Fig. 11.9. Threshold E and cut-off frequency f_g of the Gaussian low-pass filter are already optimized. All other system parameters are summarized in Table 11.1 given in Section 11.1.1.

Except a shift of about 6 dB to the lower optical powers P_r, BER curves shown in Fig. 11.9 exhibit the same characteristic features as for an ASK homodyne system shown in Fig. 11.5. Therefore, the detailed discussion about the Fig. 11.5 in the previous Section is valid here also. The shift of about 6 dB caused by the improved sensitivity of a PSK homodyne system in comparison to an ASK homodyne system represents a significant advantage.

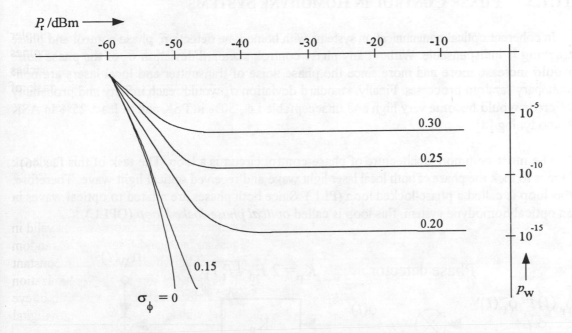

Fig. 11.9: Worst-case probability of error p_w for optical PSK homodyne system

Finally, let us consider again the probability of error in absence of ISI and additive Gaussian noise i.e., $\sigma = 0$. Analogous to Eq. (11.29) in the previous Section, we obtain following worst-case probability of error:

$$p_w = \int_{-1}^{0} f_w(w)\ dw \approx 2Q\left(\frac{\pi}{2\sigma_\phi}\right) \quad \text{provided } \sigma = 0 \tag{11.47}$$

Comparing Eqs. (11.47) and (11.29), it becomes evident that the probability of error is now twice of an ASK homodyne system. The reason for this surprising result is that *both* symbol $q_\upsilon = $ "0" ($s_\upsilon = -1$) and symbol $q_\upsilon = $ "1" ($s_\upsilon = 1$) are disturbed by phase noise in a PSK system, while symbol $q_\upsilon = 0$ ($s_\upsilon = 0$) remains undisturbed in an ASK system. However, this is a rather unrealistic case ($\sigma = 0$) since a sensitivity gain is normally achieved by PSK homodyne system when $\sigma \neq 0$. We have to remember that Eq. (11.47) is indeed a very useful tool to estimate the maximum allowed standard deviation σ_ϕ, but it is not usable to evaluate the true error rate which can be measured in the receiver. For this, Eq. (11.44) must be employed. Assuming, for example, a probability of bit error $p_w = 10^{-10}$, σ_ϕ should not exceed $\sigma_{\phi,max} \approx 0.24$.

11.1.3   PHASE CONTROL IN HOMODYNE SYSTEMS

In coherent optical communication systems with homodyne detection, phase control and phase tracking is indispensable. Without any phase control, standard deviation σ_ϕ of the phase noise would increase more and more since the phase noise of transmitter and local lasers are non-stationary random processes. Finally, standard deviation σ_ϕ would reach infinity and probability of error would become very high and unacceptable i.e., 50% in PSK and at least 25% in ASK homodyning [8].

The most common architecture of phase-control circuit is a loop. The task of this feedback loop is to lock the phase of both local laser light wave and received optical light wave. Therefore, this loop is called a phase-locked loop (PLL). Since both phases are related to optical waves in an optical homodyne system, this loop is called *optical phase-locked loop* (OPLL).

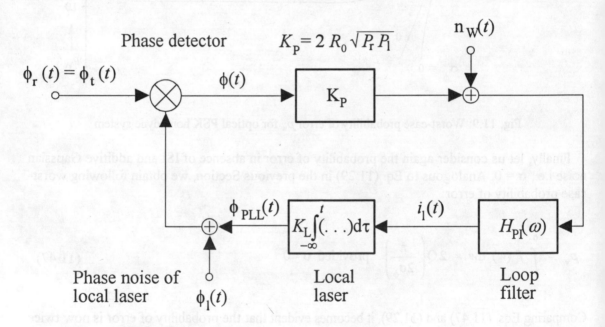

Fig. 11.10: Linearized model of optical phase-locked loop (OPLL)

In order to focus our discussion on the principle, we take advantage of the simplest configuration of optical phase-locked-loop circuit as already shown in Fig. 11.1. A more detailed discussion of phase control circuits is presented, for example, in [2, 3, 12, 16]. Linearized model of OPLL is shown in Fig. 11.10 [3, 16, 32, 33]. We talk about a linearized model, because we take advantage of the approximation $\sin(\phi) \approx \phi$. In contrast to Fig. 11.1, this model uses a balanced receiver with two photodiodes, which means one photodiode at each of the outputs of an optical 3 dB coupler (Chapter nine). In this case, phase detector shown in Fig. 11.10 includes an optical coupler as well as both the photodiodes.

The phase to be controlled in the OPLL, frequently called the leading parameter or the leading phase, is the phase noise of the received optical light wave which equals the phase noise of the transmitter laser i.e., $\phi_r(t) = \phi_t(t)$. Driving parameter is the phase $\phi_{PLL}(t)$ of the local laser located in the OPLL circuit. This phase must continuously be tracked to keep the *phase error* $\phi(t)$ of the OPLL low i.e., the difference between driving and leading phases.

As shown in Fig. 11.10, operation of OPLL is always disturbed by the phase noise $\phi_l(t)$ of the local laser and the receiver's white Gaussian noise $n_w(t)$ i.e., shot noise of photodiodes and thermal noise. As mentioned above, the main operation of OPLL is to keep the phase error

$$\phi(t) = [\phi_t(t) - \phi_l(t)] - \phi_{PLL}(t) = \phi_{tl}(t) - \phi_{PLL}(t) \tag{11.48}$$

low. As the time-varying phase error represents the phase noise which remains uncontrolled, this phase error is also called the *rest-phase noise* or the *residual phase noise*. In the ideal case which can never be practically achieved, $\phi_{PLL}(t)$ and $\phi_{tl}(t) = \phi_t(t) - \phi_l(t)$ are absolutely identical and the rest-phase noise $\phi(t)$ is zero. The phase noise of the received and local laser light waves are statistically independent. Therefore, variance of resulting phase noise $\phi_{tl}(t) = \phi_t(t) - \phi_l(t)$ is simply given by (Section 3.2.2)

$$\sigma^2_{\phi_{tl}} = 2\pi[\Delta f_t + \Delta f_l]t = 2\pi\Delta f t \tag{11.49}$$

where $\Delta f = \Delta f_t + \Delta f_l$ represents the resulting linewidth of both transmitter and local lasers. One important parameter of each phase-locked loop circuit is the phase transmission function $H(\omega)$ which describes the ratio of the spectrum of the leading parameter to the spectrum of the driving parameter. In a second order OPLL, this is given by [14]

$$H(\omega) = \frac{2j\xi\omega_n\omega + \omega_n^2}{(j\omega)^2 + 2j\xi\omega_n\omega + \omega_n^2} \tag{11.50}$$

The phase transmission function includes two characteristic parameters of significant importance: one is the damping constant ξ and the other natural frequency ω_n of the loop. For a PI-loop filter (proportional and integrating-loop filter) with a frequency response

$$H_{PI}(\omega) = \frac{1 + j\omega\tau_2}{j\omega\tau_1} = \frac{\tau_2}{\tau_1}\left[1 + \frac{1}{j\omega\tau_2}\right] \tag{11.51}$$

parameters ξ and ω_n are given by

$$\omega_n = \sqrt{\frac{K}{\tau_1}} \qquad \xi = \frac{1}{2}\omega_n\tau_2 \qquad K = K_P K_L \qquad (11.52)$$

Here, τ_1 and τ_2 are the time constants for PI filter and K the loop gain. It should be noted that a PI-loop filter normally provides sufficient stability to the OPLL. It becomes clear from Fig. 11.10 that the constant K_P is expressed in unit of A (Ampere) and the local laser constant K_L is measured in Hz/A. Hence, the loop gain itself is dimensionless as expected.

Taking now into account the impulse response $h(t) \circ\!\!-\!\!\bullet\, H(\omega)$ of phase transmission function, following equation for the rest-phase noise $\phi(t)$ can be derived directly from the Fig. 11.10:

$$\phi(t) = [\phi_t(t) - \phi_l(t)] - [\phi_t(t) - \phi_l(t)] \star h(t) - K_P^{-1}n_w(t) \star h(t)$$

$$= \phi_{tl}(t) \star [\; \underline{\delta(t) - h(t)}\;] - n_w(t) \star \underline{K_P^{-1}h(t)} \qquad (11.53)$$

$$\frac{1}{2\pi}\big|1 - H(\omega)\big| \qquad\qquad \frac{1}{2\pi}K_P^{-1}H(\omega)$$

Here, the sign $\star$ represents a convolution. The second line of this equation makes use of the sampling properties of the Dirac delta function $\delta(t)$ i.e., $\phi_{tl}(t)\star\delta(t) = \phi_{tl}(t)$. Physical meaning of Eq. (11.53) using well-known components of system theory is illustrated in Fig. 11.11.

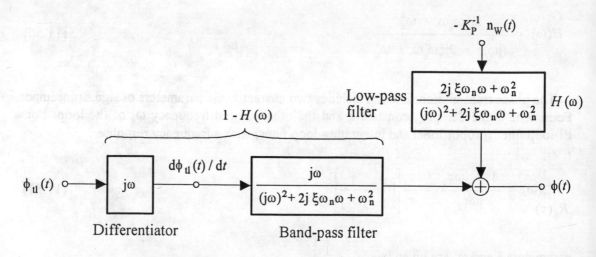

Fig. 11.11: Emergence of OPLL rest-phase noise $\phi(t)$ explained by means of system theory

As shown in this figure, resulting non-stationary laser phase noise $\phi_{tl}(t)$ is first fed to a system block which differentiates its input signal. In accordance with Section 3.2.4, output signal $d\phi_{tl}(t)/dt$ of this block represents a stationary white frequency noise with constant power spectral density $2\pi\Delta f = 2\pi[\Delta f_t + \Delta f_l]$. Thus, OPLL rest-phase noise $\phi(t)$ is composed of two different and independent white sources of noise: one is the white frequency noise $d\phi_{tl}(t)/dt$ and the other receiver white Gaussian noise $n_w(t)$ i.e., shot and thermal noise. The first noise is filtered by a band-pass filter, whereas the second by a low-pass filter. As both sources of noise are stationary and Gaussian with zero mean, rest-phase noise exhibits the same statistical features. Therefore, rest-phase noise is a *stationary Gaussian random process with zero mean*.

For determining the BER, statistical features of OPLL rest-phase noise $\phi(t)$ are of fundamental importance. For this reason, we now derive the following characteristic functions and parameters: noise power spectral density $G_\phi(f)$, autocorrelation function $R_\phi(\tau)$ and standard deviation σ_ϕ.

As both sources of noise are statistically independent, noise power spectral density $G_\phi(f)$ of the OPLL rest-phase noise equals the sum of both power spectral densities: filtered frequency noise and filtered shot and thermal noise. It is given by [14]

$$G_\phi(\omega) = 2\pi\Delta f \left|\frac{1-H(\omega)}{j\omega}\right|^2 + \frac{G_c}{K_P^2}|H(\omega)|^2 \tag{11.54}$$

It may be remembered that G_c represents the power spectral density of the receiver white Gaussian noise $n_w(t)$ (Chapter nine). Replacing the phase transmission function $H(\omega)$ using Eq. (11.50), above equation finally takes the following form:

$$G_\phi(\omega) = \frac{\left[4G_c K_p^{-2}\xi^2\omega_n^2 + 2\pi\Delta f\right]\omega^2 + G_c K_p^{-2}\omega_n^4}{\omega^4 + 2\omega_n^2(2\xi^2-1)\omega^2 + \omega_n^4} \tag{11.55}$$

The autocorrelation function $R_\phi(\tau)$ of the OPLL rest-phase noise $\phi(t)$ is easily obtained by a Fourier transformation of $1/(2\pi)\cdot G_\phi(\omega)$. After some mathematical operations, we obtain

$$R_\phi(\tau) = \begin{cases} \sigma_\phi^2\, e^{-\xi\omega_n|\tau|}\left[(0.5-C_1)\,e^{\omega_n\sqrt{\xi^2-1}|\tau|} + (0.5+C_1)\,e^{-\omega_n\sqrt{\xi^2-1}|\tau|}\right] & \text{if } \xi > 1 \\[2mm] \sigma_\phi^2\, e^{-\xi\omega_n|\tau|}\left[1 - C_2\omega_n|\tau|\right] & \text{if } \xi = 1 \\[2mm] \sigma_\phi^2\, e^{-\xi\omega_n|\tau|}\left[\cos\left(\omega_n\sqrt{1-\xi^2}|\tau|\right) - C_3\sin\left(\omega_n\sqrt{1-\xi^2}|\tau|\right)\right] & \text{if } \xi < 1 \end{cases} \tag{11.56}$$

where

$$\sigma_\phi^2 = R_\phi(0) = \frac{1}{2\pi} \int_{-\infty}^{+\infty} G_\phi(\omega) \, d\omega = \frac{\pi \Delta f}{2\xi \omega_n} + \frac{G_c \omega_n}{K_p^2} \left[\frac{1}{4\xi} + \xi \right] \tag{11.57}$$

represents the variance. Mathematically, this variance is simply obtained by evaluating the area under the noise power spectral density $G_\phi(f)$. The amplitude coefficients C_1, C_2 and C_3 used in Eq. (11.56) are given by [5]

$$C_1 = \frac{1}{2} \frac{a + b(4\xi^2 - 1)}{a + b(4\xi^2 + 1)} \frac{\xi}{\sqrt{\xi^2 - 1}} \tag{11.58}$$

$$C_2 = \frac{a + 3b}{a + 5b} \tag{11.59}$$

$$C_3 = \frac{a + b(4\xi^2 - 1)}{a + b(4\xi^2 + 1)} \frac{\xi}{\sqrt{1 - \xi^2}} \tag{11.60}$$

where

$$a = \frac{2\pi \Delta f}{\omega_n} \tag{11.61}$$

and

$$b = \frac{G_c \omega_n}{K_p^2} \tag{11.62}$$

are used as parameters. As explained in the previous Sections 11.1.1 (ASK homodyne) and 11.1.2 (PSK homodyne), σ_ϕ represents an important system parameter which directly influences the BER and, therefore, the system performance. It was shown that the standard deviation σ_ϕ of the OPLL rest-phase noise should not exceed a certain value to achieve a desired BER; for example, $\sigma_\phi \leq 0.24$ for $p_w \leq 10^{-10}$ in a PSK homodyne receiver. From Eq. (11.57), it can be seen that an optimum natural frequency ω_n must exist for a minimum standard deviation σ_ϕ i.e., a minimum influence of OPLL rest-phase noise. This optimum natural frequency usually called the OPLL-loop bandwidth is given by

$$\omega_{n,opt} = \sqrt{\frac{2\pi \Delta f K_p^2}{G_c(4\xi^2+1)}} \qquad (11.63)$$

When $\omega_n < \omega_{n,opt}$, OPLL is no longer in a position to reduce the influence of the rest-phase noise sufficiently. As a result, influence of rest-phase noise $\phi(t)$ increases and system performance degrades. Under the condition $\omega_n > \omega_{n,opt}$, OPLL would more and more be disturbed by the receiver additive Gaussian noise $n_w(t)$. Again influence of rest-phase noise $\phi(t)$ increases. The optimum loop frequency $\omega_{n,opt}$ given in Eq. (11.63) seems to be the best solution. However, this solution is only valid when the received optical light wave is unmodulated. In the case of modulated optical phase, phase-locked loop is not able to distinguish between the phase noise and the phase information. Therefore, great care has to be taken since a phase-locked loop circuit reduces the signal part of phase in the same way as the noise part of the phase. For this reason, smaller loop bandwidth than given in Eq. (11.63) must be used to prevent signal cancellation. In this book, we consider that loop bandwidth ω_n is fixed at $\omega_n = \omega_{n,max} = 2\pi \cdot 0.001/T$ where T represents the bit duration. This bandwidth provides a well-suited compromise between the reduction of influence of laser phase noise and maintaining the information part of the phase.

In the previous Section, we have derived that the standard deviation of the OPLL rest-phase noise σ_ϕ must be less than 0.24 to achieve a worst-case probability of error $p_w = 10^{-10}$ in an optical PSK homodyne system. If, for example, an OPLL is used with $\xi = 1$ (i.e., non-periodical transient response), then the above requirements yield a *maximum allowed laser linewidth*

$$\Delta f_{max} = \begin{cases} \dfrac{K_p^2}{640 \pi G_c} & \text{if } \omega_n = \omega_{n,opt} \\[3mm] \dfrac{1}{4}\dfrac{\omega_{n,max}}{2\pi}\left[1-20\dfrac{G_c\,\omega_{n,max}}{K_p^2}\right] & \text{if } \omega_n = \omega_{n,max} \end{cases} \qquad (11.64)$$

It should be noted that this linewidth which is derived by using Eqs. (11.57) and (11.63) represents the resulting linewidth of both transmitter and local lasers. For simplicity $\sigma_\phi = 0.24$ has been taken as 1/4.

Example 11.1

Let the parameters of a given homodyne system are: $P_r = -50$ dBm, $P_1 = -10$ dBm, $R_0 = 1$ A/W, $G_c = 2.6 \cdot 10^{-23}$ A^2/Hz, $\xi = 1$ and $T = 6.43$ ns corresponding to a bit rate $R = 155.52$ Mbit/s i.e., STM1 of SDH (Chapter fourteen). In this case, Eq. (11.64) yields a maximum allowed laser linewidth of $\Delta f_{max} = \Delta f_t + \Delta f_1 = 38.88$ kHz. This linewidth corresponds to 0.025% of the bit rate.

We have so far analysed OPLL circuit which is characterized by a simple feedback loop such as shown in Figs. 11.1 and 11.10. Advantages and drawbacks of this special configuration are discussed below. Instead of simple feedback loop, some other approaches are there for matching the phase and frequency of received and local laser waves. Now, we have to focus our interest on these approaches. In this book, a brief discussion is sufficient. The most suitable phase control circuits for coherent detection can be classified into three categories:

- OPLL with pilot carrier,
- Costas loop circuit and
- OPLL with synchronization bit control.

(i) OPLL WITH PILOT CARRIER

The optical phase-control circuit shown in Figs. 11.1 and 11.10 and discussed above belongs to this group. To maintain endless operation i.e., permanent phase tracking without any break, OPLL always requires a permanent signal at the optical carrier frequency. However, an optical wave which is PSK modulated does not normally exhibit a carrier signal a priori. For this , a pilot carrier which requires a certain part of the available transmitter power must be generated [14]. Usually, the pilot carrier is generated by a phase deviation less than ±90°. In the OPLL, phase error signal is separated from the information signal by a simple low-pass filter such as shown in Fig. 11.1 or Fig. 11.10. If an ASK signal is applied to the system, then the OPLL requires a suppression of the amplitude modulation since a permanent DC signal (i.e., $s_u = 0$) would immediately cause a total break of the phase-control function.

An OPLL with pilot carrier needs less number of components and required circuit is simple. Both features represent a significant advantage of using the pilot-carrier technique. Furthermore, this configuration is well-suited for optical integration. However, this technique exhibits some important drawbacks: First, a DC-coupled amplifier is required, which is difficult to realize when bit rate is high. Second, coupling ratio k of the optical coupler must be exactly 50% since any deviation pretends a phase error. Finally, the information signal should not have any spectral component in the OPLL bandwidth which, for example, can be achieved by coding. Since a decreased phase deviation reduces the available signal power, a sensitivity loss of about 0.5 dB can be observed.

(ii) COSTAS LOOP CIRCUIT

The most common OPLL technique is the Costas loop. Here, the phase error signal is obtained by detecting the inphase as well as the quadrature components of the received signal which are multiplied afterwards [15]. The inphase component is same as information signal (launched into the sample and hold circuit). The multiplied signal represents the phase error signal. As the quadrature component requires a certain part of the received optical power, a loss in sensitivity of about 0.5 dB is again there.

The Costas loop technique avoids all the drawbacks of the pilot carrier technique. However, an 90° optical hybrid (90° coupler) is required to separate inphase and quadrature components, which is difficult to realize and integrate. The most convenient way to realize an optical hybrid is to model this device by an optical three-port fiber coupler. In contrast to pilot carrier technique, Costas loop requires at least two complete front ends, which represents a serious drawback.

(iii) OPLL WITH SYNCHRONIZATION BIT CONTROL

A third method of phase tracking is the synchronization bit control [39-41]. This method combines the advantages and avoid the drawbacks of both pilot carrier technique and Costas loop. Requirements of somewhat more sophisticated digital signal processing is the only disadvantage. The synchronization bit approach is based on the principle of Costas loop. The Costas loop technique requires a permanent quadrature component to maintain phase tracking, whereas the synchronization bit technique switches between inphase and quadrature components. Most of the time, received signal is fed into the signal branch of the receiver (inphase component). Thereby, no signal loss occurs. To track the phase, it is periodically required to switch the received signal to the phase-control branch (quadrature component) for a short time. During this time of usually one bit duration, no information is sent. Instead, a synchronization bit is transmitted allowing to generate the phase error signal.

In order to maintain the information bit rate, channel bit rate (information and synchronization bits) must be increased. When one synchronization bit is always transmitted after every eight information bits, the channel rate is 9/8 of the information bit rate. Since an increased channel bit rate requires an increased bandwidth (which increases the influence of the receiver additive Gaussian noise), the sensitivity is reduced by $10 \log_{10}(9/8)$ dB ≈ 0.5 dB. This loss in sensitivity is same as in the Costas loop.

In this book, it is not required to discuss different configurations of phase-control circuits in more detail. Different techniques of PLL circuits are already well-known and well-described in many excellent books and papers for example in [3, 14-16, 20, 21, 32, 33]. In addition, the principle function of a phase-control circuit remains the same irrespective of whether a conventional electrical PLL or an OPLL is considered. Therefore, the reader can see the theory and the design of phase-control circuits in one of the references mentioned above.

11.2   HETERODYNE SYSTEMS WITH COHERENT DETECTION

In a coherent optical heterodyne system, received light wave is first converted to an electrical intermediate frequency (IF) signal by means of an optical coupler, a local laser and one or two photodiodes. The IF is equal to the difference of both transmitter laser and local laser frequencies (Chapter nine). Afterwards, IF signal is demodulated and demodulator structure depends upon the modulation scheme applied in the transmitter. For demodulation, various techniques which are well-known from ordinary digital communication can be employed.

In optical homodyne systems, received optical signal is always detected coherently (homodyne systems with incoherent detection are impossible), whereas in optical heterodyne systems either coherent or incoherent detection of the IF signal can be used in the receiver. This Section is focussed on *coherent detection* or *synchronous detection*.

The principle function of a heterodyne system with coherent detection is very much similar to a homodyne system which already has been discussed in previous Sections 11.1.1 (ASK homodyne) and 11.1.2 (PSK homodyne). Further, heterodyne systems with coherent detection are of little practical interest as compared to heterodyne systems with incoherent detection (Section 11.3) or homodyne systems, in particular PSK homodyne Therefore, system discussion in this Section will be rather restricted.

Due to the twofold meaning of the word "coherent" i.e., coherent laser and coherent (synchronous) detection, the reader may read again the description at the beginning of this Chapter. Like homodyne systems, coherent heterodyne systems can also be realized by using amplitude shift keying (ASK) or phase shift keying (PSK). In addition, frequency shift keying (FSK) can also be used. For this, a two-filter receiver with a synchronous detector in each filter branch is required. It should be remembered that FSK is impossible in homodyne systems.

Synchronous detection of the received IF signal requires the generation of a separate IF carrier signal in the receiver, which must absolutely be synchronous (i.e., coherent) with the received IF carrier signal. Here, synchronous means that both signals must be same in frequency *and* phase. For this, a phase-locked loop circuit is required. In contrast to homodyne system where an OPLL is needed, coherent heterodyne system only requires an electrical PLL which is usually based on standard voltage-controlled oscillator (VCO). The task of this PLL is to match the frequency and phase of the VCO and the received IF carrier signal for coherent detection.

The principle function of OPLL and electrical PLL is the same. Therefore, the model of the optical phase-control circuit shown in Fig. 11.10 (Section 11.1.3) can be used with some simple modifications to describe the electrical PLL in coherent heterodyne systems. The modifications required are summarized in Table 11.2. In addition, following differences must also be taken into account:

- local laser operation is replaced by VCO,

- phase detector only includes electrical components instead of optical components such as photodiode and optical coupler,

- phase noise of IF signal is the leading parameter of the electrical PLL. This is in contrast to OPLL, where this parameter is same as the phase noise of received light wave.

- source of receiver additive noise (shot and thermal noise) is located outside the PLL circuit i.e., at the input to electrical PLL. This is in contrast to OPLL, where this source is located inside the OPLL circuit itself. However, the OPLL model in Fig. 11.10 can still be used if the constant of proportionality K_P is modified appropriately (Table 11.2).

- realization of electrical PLL is much more convenient than the OPLL.

Table 11.2: Phase control in homodyne and coherent heterodyne systems

	Homodyne system	Heterodyne system	Remark
Phase detector	Optical by means of an optical coupler and one or two photodiodes	Electrical	
Local oscillator	Local laser	VCO	
Leading parameter	$\phi_t(t)$	$\phi_t(t) - \phi_l(t)$	$\phi_t(t)$, $\phi_l(t)$: phase noise of transmitter and local lasers
Additional sources of noise in PLL	$n_w(t)$ and $\phi_l(t)$	$n_w(t)$ and $\phi_{VCO}(t)$	$n_w(t)$: white noise (shot and thermal noise) $\phi_{VCO}(t)$: phase noise of VCO (negligible)
Constant K_P of proportionality	$R_0\sqrt{4P_rP_1}$	$R_0\sqrt{2P_rP_1}$	compare with [16]

As the analysis of a coherent heterodyne system is similar to that of a homodyne system, an explicit system calculation will not be carried out in this Section. For the BER calculation, same formulas as derived in Sections 11.1.1 and 11.1.2 can be used. We have only to take into account that the normalized standard deviations σ_ϕ and σ of the PLL rest-phase noise and the additive Gaussian noise have different numerical values in coherent heterodyne receiver (see Table 11.3).

Table 11.3: Characteristic features of coherent heterodyne and homodyne systems

	Homodyne system	Heterodyne system	Remark
Optical input wave $E(t)$	$[\underline{E}_r(t) + \underline{E}_l(t)]\cos(\omega_l t)$	$\underline{E}_r(t)\cos(\omega_r t)$ $+ \underline{E}_l(t)\cos(\omega_l t)$	$\omega_r = \omega_t$ (see Section 9.4.2)
IF signal $i_{IF}(t)$	- - -	$\hat{i}_{PD}\cos(\omega_{IF}t) + n_{IF}(t)$ where $n_{IF}(t) = x(t)\cos(\omega_{IF}t) + y(t)\sin(\omega_{IF}t)$	$(\hat{i}_{PD})^2 = 4\,R_0^2 P_r P_1$ provided a 3 dB coupler and a balanced receiver are used (see Eq. 9.69) $n_{IF}(t)$: noise in IF band
Baseband signal $i_B(t)$	$\hat{i}_{PD} + n_B(t)$	$\frac{1}{2}\hat{i}_{PD} + \frac{1}{2}x(t)$	$(\hat{i}_{PD})^2 = 4\,R_0^2 P_r P_1$ (see remark above) $n_B(t)$: noise in baseband
Power budget in IF domain	- - -	$S_{IF} = 2\,R_0^2 P_r P_1$ $N_{IF} = 2\,G_c B_{IF}$ $= \sigma_x^2 = \sigma_y^2 = \sigma_{het}^2$ $\left(\dfrac{S}{N}\right)_{IF} = 1\dfrac{R_0^2 P_r P_1}{G_c B_{IF}}$	G_c: double-sided noise power spectral density as given in Eq. (9.70) B_{IF}: noise-equivalent bandwidth of IF filter
Power budget in baseband	$S_B = 4\,R_0^2 P_r P_1$ $N_B = 1\,G_c B_B = \sigma_{hom}^2$ $\left(\dfrac{S}{N}\right)_B = 4\dfrac{R_0^2 P_r P_1}{G_c B_B}$	$S_B = 1\,R_0^2 P_r P_1$ $N_B = \frac{1}{2}\,G_c B_B = \frac{1}{4}\sigma_x^2$ $= \frac{1}{4}\sigma_{het}^2$ $\left(\dfrac{S}{N}\right)_B = 2\dfrac{R_0^2 P_r P_1}{G_c B_B}$	B_B: double-sided noise-equivalent bandwidth of the baseband filter $B_B = B_{IF}$ $\dfrac{(S/N)_{B,\,hom}}{(S/N)_{B,\,het}} = 2$
Normalized variance σ^2 of receiver additive Gaussian noise	$\sigma^2 = \dfrac{\sigma_{hom}^2}{\hat{i}_{PD}^2}$ $= \dfrac{1}{4}\dfrac{G_c B_B}{R_0^2 P_r P_1}$	$\sigma^2 = \dfrac{\sigma_{het}^2}{\hat{i}_{PD}^2} = 2\dfrac{\sigma_{hom}^2}{\hat{i}_{PD}^2}$ $= \dfrac{1}{2}\dfrac{G_c B_B}{R_0^2 P_r P_1}$	see Eq. (9.84)
Variance σ_ϕ^2 of rest-phase noise	see Eq. (11.57)	Eq. (11.57) with K_P as given in Table 11.2	

In the absence of phase noise, a homodyne system exhibits a 3 dB sensitivity gain in comparison to a heterodyne system with coherent detection provided that same modulation scheme has been applied (ASK or PSK). Irrespective of whether coherent or incoherent detection is used in the heterodyne receivers, limited bit rate in comparison to homodyne systems is a main disadvantage. A homodyne system can be regarded as a baseband system, whereas a heterodyne system operates in the IF domain. For this reason, photodiodes as well as amplifiers must be able to follow the high frequency IF signal. Assuming, for example, a bit rate of 2 Gbit/s, an IF of about 8 GHz and a cut-off frequency of about 9 GHz for the photodiodes and amplifiers are required. In contrast, both the components only require a cut-off frequency of about 2 GHz if an homodyne system is used. In Chapter twelve, this problem is discussed in more detail.

Main differences of heterodyne systems with coherent detection and homodyne systems are summarized in Table 11.3. All equations in this table are based on the results of Chapter nine. A balanced receiver with two photodiodes i.e., $K_B = 2$ and an optical 3 dB coupler i.e., $k = 0.5$ are presumed. An interesting result is highlighted in this table. In homodyne detection, baseband signal power S_B as well as the baseband noise power N_B are both always higher than that of heterodyne detection. However, homodyne detection offers a 3 dB improvement in the signal-to-noise ratio $(S/N)_B$. The reason for this is a fourfold increase in the signal power, but only a twofold increase in the noise power in comparison to coherent heterodyne system.

Like homodyne systems, performance of heterodyne systems with coherent detection is severely influenced by the phase noise of both transmitter laser and local laser. Thus, requirements for the maximum permissible laser linewidths are very strong. For a PSK heterodyne system, for example, a linewidth-bit duration product of the order of 10^{-4} is required (Chapter twelve). This means that the laser linewidth of both the lasers must be less than 0.01 percent of the bit rate. In contrast to this, an incoherent heterodyne system such as DPSK heterodyne system with autocorrelation detection only requires a linewidth-bit duration product of about 10^{-2}. This corresponds to a maximum permissible laser linewidth of one percent of the bit rate. In incoherent FSK and ASK heterodyne systems with envelope detection, even linewidths same as the bit rate are acceptable (Section 11.3 and Chapter twelve).

11.3 HETERODYNE SYSTEMS WITH INCOHERENT DETECTION

In comparison to systems with coherent detection (i.e., homodyne systems or coherent heterodyne systems), heterodyne systems with incoherent detection do not require a phase tracking circuit, which is a fundamental advantage of employing these systems. To stabilize the intermediate frequency (IF), a standard automatic frequency control (AFC) is quite sufficient. The aim of this Section is to analyse and optimize the following heterodyne systems:

- ASK heterodyne system with envelope detection (Section 11.3.1),

- FSK heterodyne system with one- or two-filter configuration and envelope detection (Section 11.3.2) and

- DPSK heterodyne system with autocorrelation demodulator or two-filter detection (Section 11.3.3).

In the calculation and optimization of the systems listed above, we follow the same procedure as in the previous Sections.

11.3.1 ASK HETERODYNE SYSTEM

This Section describes the system calculation and optimization of an optical ASK heterodyne receiver with standard envelope detection. In particular, the intermediate frequency signal (IF signal) and the signal after envelope detection (called the envelope signal) are derived taking into consideration laser phase noise, receiver additive Gaussian noise, influence of the intermediate frequency filter (IF filter) on both sources of noise and intersymbol interference (ISI). Using the envelope signal which includes the transmitted information, probability density function and the bit error rate are calculated exactly and by approximations. For a minimum bit error rate, optimum IF filter bandwidth is evaluated as a function of linewidths of both transmitter laser and local lasers. To highlight typical features of incoherent ASK heterodyne systems with envelope detection, computer simulated eye pattern are finally presented.

As discussed in the previous Sections, optical homodyne systems (Section 11.1) and optical heterodyne systems with coherent detection (Section 11.2) are very sensitive to the laser phase noise of transmitter and local lasers. The most sensitive system is coherent PSK homodyne, whereas the least sensitive system is incoherent ASK heterodyne with envelope detection. As described in this Section, even in the latter system laser phase noise cannot be neglected.

In an optical ASK heterodyne receiver, phase noise can be regarded as a superposition of an undesired phase- or frequency modulation. In an ideal ASK heterodyne system, this undesired phase- or frequency modulation does not disturb the desired amplitude modulation. In this case, disturbance to the envelope signal is given only by the shot noise of photodiode and thermal noise of the amplifier. Both are additive to the IF signal. The same result can be obtained by neglecting the influence of IF filter on the phase noise. As the analysis of the IF filter influence is rather comprehensive when phase noise disturbs the system, this influence has been usually neglected in the early system calculation [e.g., 7, 23, 42]. In that case, system calculation results in an envelope signal which is again independent of the laser phase noise (Subsection iii below). Therefore, the results obtained are only useful as a rough estimate.

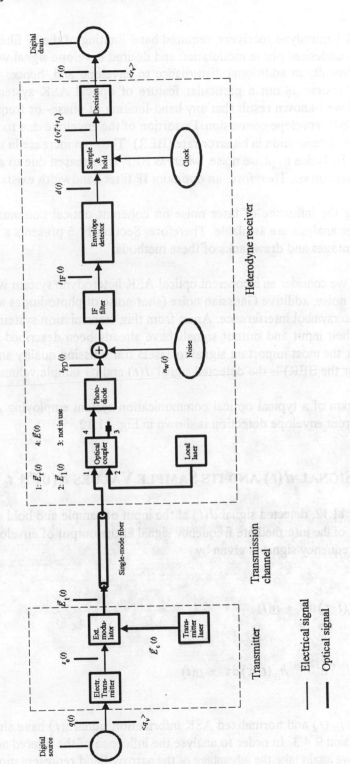

Fig. 11.12: Coherent optical communication system with incoherent ASK heterodyne receiver

—— Electrical signal

—— Optical signal

In practical ASK heterodyne receivers, required band-limiting of the IF filter causes a strong correlation between undesired phase modulation and desired envelope signal which contains the information. As a result, an additional disturbance to envelope and, hence, to information is obtained. This, of course, is not a particular feature of optical ASK systems. It is rather a generally valid and well-known result that any band-limiting of phase- or frequency modulated signal causes a phase-to-envelope conversion. Distortion of the envelope due to laser phase noise leads to an additional deterioration in bit error rate (BER). Thus, an increase in the IF filter band-width decreases the BER due to phase noise, whereas BER is increased due to additive Gaussian receiver noise (Subsection vi). Therefore, an optimum IF filter bandwidth exists (Subsection vii).

For determining the influence of laser noise on coherent optical communication systems, different methods for analysis are available. Therefore, Section 11.5 presents a brief comparison and discusses advantages and drawbacks of these methods.

In the following, we consider an incoherent optical ASK heterodyne system which is disturbed only by laser phase noise, additive Gaussian noise (shot noise of photodiodes and thermal noise of amplifiers) and intersymbol interference. Apart from this, transmission system is ideal. System components and their input and output signals have already been described and explained in Chapter nine. Again, the most important signal to assess transmission quality and system perfor-mance (in particular the BER) is the detected signal $d(t)$ and its sample values $d(\upsilon T + t_0)$.

The block diagram of a typical optical communication system employing ASK modulation scheme and incoherent envelope detection is shown in Fig. 11.12.

(i) DETECTED SIGNAL $d(t)$ AND ITS SAMPLE VALUES $d(\upsilon T + t_0)$

As shown in Fig. 11.12, detected signal $d(t)$ at the input of sample and hold circuit is same as the envelope $|i_{IF}(t)|$ of the intermediate frequency signal at the output of envelope demodulator. The intermediate frequency signal is given by

$$\underline{i}_{IF}(t) = \int_{-\infty}^{+\infty} \underline{i}_{PD}(\tau) h_{IF}(t-\tau) \, d\tau + \underline{n}(t)$$

$$\tag{11.65}$$

$$= \hat{i}_{PD} \int_{-\infty}^{+\infty} s(\tau) e^{j\phi(\tau)} e^{j2\pi f_{IF}t} h_{IF}(t-\tau) \, d\tau + \underline{n}(t)$$

Photodiode current $i_{PD}(t)$ and normalized ASK information signal $s(t)$ have already been given and explained in Section 9.4.3. In order to analyse the influence of the filtered additive Gaussian receiver noise $\underline{n}(t)$, we again take the advantage of the narrow-band representation (see Eq. 9.82)

$$\underline{n}(t) = x(t)\,e^{j2\pi f_{IF}t} - jy(t)\,e^{j2\pi f_{IF}t} = \bigl(x(t) - jy(t)\bigr)\,e^{j2\pi f_{IF}t} \tag{11.66}$$

This equation can be used when the ratio of noise-equivalent IF filter bandwidth B_{IF} to centre frequency f_{IF} of the IF filter is much less than 1.

As explained in Chapter nine, photodiode current $i_{PD}(t)$ at the input to the IF filter normally contains two baseband signal components: one is the DC component kR_0P_1 and the other is baseband component which is proportional to the absolute square of the transmitted information signal $|s(t)|^2$ as given in Eq. (9.66). Both the undesired signal components can be eliminated either by the IF filter itself or by an appropriate balanced receiver configuration (Section 9.4.3).

In order to avoid spectral overlapping of IF spectrum and undesired baseband spectrum in $i_{PD}(t)$, intermediate frequency f_{IF} must be chosen high enough. Because of the quadratic baseband term $|s(t)|^2$, undesired baseband spectrum exhibits a bandwidth of $2f_{s,max}$ i.e., $0 \le f \le 2f_{s,max}$, where $f_{s,max}$ represents the maximum frequency of information signal $s(t)$. For digital modulation, this frequency is approximately $f_{s,max} = 1/(2T)$. The IF spectrum is located in the range of $f_{IF} - f_{s,max}$ to $f_{IF} + f_{s,max}$. Hence, the intermediate frequency should fulfil the following requirement:

$$f_{IF} > 3f_{s,\,max} \tag{11.67}$$

Next we assume an IF filter with frequency response $H_{IF}(f)$ which can be expressed in terms of an equivalent baseband filter $H_B(f)$ i.e., $H_{IF}(f) = H_B(f - f_{IF}) + H_B(f + f_{IF})$. Taking into account the baseband filter impulse response $h_B(t)$ and the narrow-band representation (Eq. 11.66) of the additive noise, IF signal in Eq. (11.65) can be rewritten as follows:

$$\underline{i}_{IF}(t) \approx \left[\hat{i}_{PD} \int_{-\infty}^{+\infty} s(\tau)\,e^{j\phi(\tau)}\,h_B(t-\tau)\,d\tau + x(t) - jy(t) \right] e^{j2\pi f_{IF}t}$$

$$= |\underline{i}_{IF}(t)|\,e^{j\Psi_{IF}(t)}\,e^{j2\pi f_{IF}t} \tag{11.68}$$

It becomes clear from the above equation that the IF signal represents a phase- and amplitude modulated carrier signal with carrier frequency f_{IF}. Eq. (11.68) is an approximation in case of spectral overlapping of the IF filter components $H_B(f - f_{IF})$ and $H_B(f + f_{IF})$ occurs. However, this equation yields the exact result when there is no spectral overlapping (Section 9.4.3). Detected signal $d(t) = |\underline{i}_{IF}(t)|$ and phase $\Psi_{IF}(t)$ are given by

$$
\begin{aligned}
d(t) = |i_{\text{IF}}(t)| &\approx \sqrt{\text{Re}^2\{i_{\text{IF}}(t)\} + \text{Im}^2\{i_{\text{IF}}(t)\}} \\[2mm]
&= \left[\left(\hat{i}_{\text{PD}} \int_{-\infty}^{+\infty} s(\tau)\,\cos(\phi(\tau))\,h_{\text{B}}(t-\tau)\,d\tau + x(t) \right)^2 \right. \\[2mm]
&\quad \left. + \left(\hat{i}_{\text{PD}} \int_{-\infty}^{+\infty} s(\tau)\,\sin(\phi(\tau))\,h_{\text{B}}(t-\tau)\,d\tau - y(t) \right)^2 \right]^{1/2}
\end{aligned}
\tag{11.69}
$$

and

$$
\begin{aligned}
\Psi_{\text{IF}}(t) &= \arctan\left[\frac{\text{Im}\{i_{\text{IF}}(t)\}}{\text{Re}\{i_{\text{IF}}(t)\}} \right] \\[4mm]
&= \arctan\left[\frac{\hat{i}_{\text{PD}} \int_{-\infty}^{+\infty} s(\tau)\,\sin(\phi(\tau))\,h_{\text{B}}(t-\tau)\,d\tau - y(t)}{\hat{i}_{\text{PD}} \int_{-\infty}^{+\infty} s(\tau)\,\cos(\phi(\tau))\,h_{\text{B}}(t-\tau)\,d\tau + x(t)} \right]
\end{aligned}
\tag{11.70}
$$

The detected signal $d(t)$ is sampled at the centre of each bit. If sample value $d(\upsilon T + t_0)$ is either above or below a threshold level, symbol "1" or symbol "0" is detected respectively. The detected signal $d(t)$ can be regarded as a stationary random process. It means that the statistical properties do not change with time. Therefore, it will be sufficient to calculate the statistical properties at the sampling time $t = t_0$. In the strict sense, detected signal is a non-stationary random process since the random processes $\cos(\phi(t))$ and $\sin(\phi(t))$ are also non-stationary. However, both processes become stationary after a very short time $t \gg 1/(2\pi\Delta f)$ as discussed in Section 3.2.5.

In order to determine the statistical properties of the sample value $d(t_0)$, sign of Gaussian random process $y(t)$ with zero mean can be chosen arbitrary without changing the statistical properties of $d(t_0)$. In the following descriptions, a positive sign has been used. Further, following substitutions and abbreviations has also been used to simplify our calculation:

$$d := \frac{d(t_0)}{\hat{i}_{PD}} \qquad n := \frac{n(t_0)}{\hat{i}_{PD}} \rightarrow \sigma := \frac{\sigma_n}{\hat{i}_{PD}} = \frac{\sigma_{het}}{\hat{i}_{PD}}$$

$$x := \frac{x(t_0)}{\hat{i}_{PD}} \qquad y := \frac{y(t_0)}{\hat{i}_{PD}} \rightarrow \sigma_x = \sigma_y = \sigma$$

With the above normalisations of system parameters, local laser power P_1 and received optical power P_r are now only included in the normalized standard deviation σ of the receiver additive Gaussian noise. The normalized detected signal d is given by

$$d = \sqrt{\left[\int_{-\infty}^{+\infty} s(\tau)\,\cos(\phi(\tau))\,h_B(t_0-\tau)\,d\tau + x\right]^2 + \left[\int_{-\infty}^{+\infty} s(\tau)\,\sin(\phi(\tau))\,h_B(t_0-\tau)\,d\tau + y\right]^2} \qquad (11.71)$$

As this expression is of prime importance for our further consideration, physical meaning of the signals included in it is briefly summarized below:

$s(t)$: information signal,
$h_B(t)$: impulse response of the equivalent baseband filter $H_B(f)$ corresponding to real IF filter $H_{IF}(f) = H_B(f - f_{IF}) + H_B(f + f_{IF})$
$\phi(t)$: laser phase noise of both transmitter and local lasers,
$x(t)$: inphase component of the receiver additive Gaussian noise and
$y(t)$: quadrature component of the receiver additive Gaussian noise.

It becomes clear from the above equation that normalized and sampled detected signal d is a function of transmitted information signal $s(t)$ and impulse response $h_B(t)$ of the equivalent baseband filter. This signal is disturbed by three sources of noise: resulting laser phase noise $\phi(t)$, inphase component $x(t)$ and quadrature component $y(t)$ of the receiver additive Gaussian noise.

(ii) PROBABILITY DENSITY FUNCTION $f_d(d)$

The most difficult problem while calculating the probability density function (pdf) and the related BER is the derivation of the statistical properties of the interaction of phase noise and envelope due to restricted bandwidth of IF filter. In particular, the difficulties are due to the non-Gaussian behaviour of signals which are disturbed by laser phase noise (Section 3.4). It is well-known that an additive Gaussian process at a linear filter input always leads to another additive Gaussian process at the filter output. Thus, the shape of the pdf at the filter output remains unchanged when a Gaussian pdf at the filter input is assumed. Examples for such processes are the approximately Gaussian shot noise of the photodiode and the Gaussian thermal noise of the amplifier. However, the shape of the output pdf is different from the input pdf when the input process is non-Gaussian. In such a case, calculation of the pdf at the filter output is generally complicated. The analytical solution of this problem can only be found by comprehensive mathematical and statistical operations as explained in Section 3.4. In the following Subsections, this problem is solved exactly, approximately and by computer simulation.

In order to calculate the pdf $f_d(d)$ of the sampled and normalized envelope $d = |i_{IF}|$ given in Eq. (11.71), we represent the first integral in Eq. (11.71) by A and the second by B i.e.,

$$A = \int_{-\infty}^{+\infty} s(\tau) \, \cos(\phi(\tau)) \, h_B(t_0 - \tau) \, d\tau \tag{11.72}$$

and

$$B = \int_{-\infty}^{+\infty} s(\tau) \, \sin(\phi(\tau)) \, h_B(t_0 - \tau) \, d\tau \tag{11.73}$$

With these parameters, Eq. (11.71) can be written in the following simplified form:

$$d = \sqrt{(A + x)^2 + (B + y)^2} \tag{11.74}$$

It becomes evident from the above equation that the detected signal d is a function of four random variables, namely A, B, x and y. As mentioned in Chapter nine, random variables x and y are statistically independent and each is defined by a Gaussian pdf with zero mean and normalized standard deviation σ. The new random variables A and B are, however, statistically dependent and, in general, non-Gaussian. The non-Gaussian pdf is due to non-Gaussian pdf of the random variables $\cos(\phi)$ and $\sin(\phi)$ as explained in Section 3.2.5. Considering first that the random variables A and B are constant, we evaluate the conditional probability density function $f_{d|A,B}(d \,|A,B)$ by means of the statistical theory of two-dimensional random variables (Section 3.4). We obtain

$$f_{d|A,B}(d,A,B) = \frac{d}{\sigma^2} \, e^{-\frac{d^2 + (A^2 + B^2)}{2\sigma^2}} \, J_0\left(\frac{d\sqrt{A^2 + B^2}}{\sigma^2}\right) \tag{11.75}$$

where J_0 is the Bessel function of order zero [1]. Substitution of

$$C = \sqrt{A^2 + B^2} \tag{11.76}$$

in the above equation yields the well-known Rice probability density function [27]

$$f_{d|C}(d, C) = \frac{d}{\sigma^2}\, e^{-\frac{d^2 + C^2}{2\sigma^2}}\, J_0\!\left(\frac{Cd}{\sigma^2}\right) \approx \frac{1}{\sqrt{2\pi}\sigma}\, e^{-\frac{(d - C)^2}{2\sigma^2}} \tag{11.77}$$

The new random variable C equals the normalized and sampled detected signal d provided variance $\sigma^2 = 0$ i.e., $x = y = 0$. Thus, $C = d(\sigma = 0)$ only includes the influence of phase noise. For a sufficiently high C/σ-ratio corresponding to a high signal-to-noise ratio $(C/\sigma)^2$, the Rice density function can be approximated by a Gaussian density function. Taking into account the above considerations, random variable C can be described by one of the following three alternate forms:

$$C = \left| \int_{-\infty}^{+\infty} s(\tau)\, e^{j\phi(\tau)}\, h_B(t_0 - \tau)\, d\tau \right| \tag{11.78}$$

$$C = \sqrt{ \left[\int_{-\infty}^{+\infty} s(\tau)\, \cos(\phi(\tau))\, h_B(t_0 - \tau)\, d\tau \right]^2 + \left[\int_{-\infty}^{+\infty} s(\tau)\, \sin(\phi(\tau))\, h_B(t_0 - \tau)\, d\tau \right]^2 } \tag{11.79}$$

and

$$C = \sqrt{ \int_{-\infty}^{+\infty}\int_{-\infty}^{+\infty} s(\tau)\, s(t)\, e^{j[\phi(\tau) - \phi(t)]}\, h_B(t_0 - \tau)\, h_B(t_0 - t)\, d\tau\, dt } \tag{11.80}$$

For a randomly varying C with pdf $f_C(C)$, the pdf $f_d(d)$ of the detected signal d is finally given by

$$f_d(d) = \frac{d}{\sigma^2} \int\limits_0^{+\infty} e^{-\frac{d^2 + C^2}{2\sigma^2}} J_0\left(\frac{Cd}{\sigma^2}\right) f_C(C) \, dC \tag{11.81}$$

A particular case of interest is obtained when the influence of IF filter on the phase noise is neglected. In that case, the term $\exp(j\phi(\tau))$ in Eq. (11.78) can be brought outside the convolution integral and, therefore, neglected since $|\exp(j\phi(\tau))| = 1$. This results in $C = C_0$ which is a constant and its pdf $f_C(C) = \delta(C - C_0)$ which is a Dirac delta function. Therefore, Eq. (11.81) yields the following special forms

For $C_0 = 0$, $f_d(d) = \dfrac{d}{\sigma^2} e^{-\frac{d^2}{2\sigma^2}}$ Rayleigh distribution   (11.82)

and

For $C_0 \neq 0$, $f_d(d) = \dfrac{d}{\sigma^2} e^{-\frac{d^2 + C_0^2}{2\sigma^2}} J_0\left(\dfrac{C_0 d}{\sigma^2}\right)$ Rice distribution   (11.83)

which are well-known from standard digital ASK communication theory [36]. It should be noted that a constant C is also obtained in an ideal system i.e., without any phase noise ($\sigma_\phi = 0$) or phase noise-to-amplitude conversion.

(iii) PROBABILITY DENSITY FUNCTION $f_c(C)$

This Subsection describes the exact evaluation of the pdf $f_C(C)$ required to calculate the pdf $f_d(d)$ of the sampled detected signal d and also the BER. As the exact calculation is rather comprehensive, approximations would be much more convenient for a practical system design. For this reason, next Subsection (iv) deals with approximations. As the theoretical considerations in this Subsection are not directly required to understand Subsection (iv), the reader may skip the present Subsection without any discontinuity.

The random variable C defined in Eqs. (11.78) to (11.80) depends upon the random process $\phi(t)$, transmitted information signal $s(t)$ and impulse response $h_B(t)$ of equivalent baseband filter. To calculate the pdf $f_d(d)$ of sampled detected signal d, several analytical methods based on the sampling theorem exists (Section 3.4). One of the methods, namely the method of integral, will representatively be discussed in this Section. By using statistics of two-dimensional random variables [27], pdf $f_C(C)$ of the random variable C can be written as follows:

$$f_{\mathrm{C}}(C) = C\int_{0}^{2\pi} f_{\mathrm{A,B}}\big(C\cos(\phi),\ C\sin(\phi)\big)\ d\phi \tag{11.84}$$

In the above equation, the most difficult part is to evaluate the joint probability density function $f_{\mathrm{A,B}}(A,B)$. As mentioned just above, a particular case of interest is when the influence of the IF filter on the phase noise is neglected. This results in a pdf $f_{\mathrm{C}}(C) = \delta(C - C_0)$ where parameter C_0 depends only on the transmitted symbol pattern $<q_v>$ which includes the information signal $s(t)$. Thus, phase noise is completely extracted from the calculation and pdf $f_{\mathrm{d}}(d)$ and BER become independent of laser phase noise.

In order to obtain a generally valid solution, we make use of the advantage of sampling theorem and describe the random variables A and B by the two sums

$$A = \sum_{n=-\infty}^{+\infty} s_n \cos(\phi_n)\,\alpha_n \tag{11.85}$$

and

$$B = \sum_{n=-\infty}^{+\infty} s_n \sin(\phi_n)\,\alpha_n \tag{11.86}$$

where s_n, α_n and the random variables ϕ_n represent the sample values $s(nT_s)$, $T_s h_{\mathrm{B}}(t_0 - nT_s)$ and $\phi(nT_s)$ respectively. Here, T_s is the sampling period. To fulfil the sampling theorem requirement, sampling period T_s must be chosen small enough. Only in that case, A and B are completely determined from the above sums. The choice of T_s becomes unambiguous when both frequency response $H_{\mathrm{B}}(f)$ *and* frequency spectrum of the IF filter input signal $s(t)\exp(j\phi(t))$ are band-limited, whereas the choice is not definite when these frequency spectra are not band-limited. In such a case, we have to set up an arbitrary band-limit depending upon the required accuracy of calculation. For simplicity, let us presume that impulse response $h_{\mathrm{B}}(t)$ approaches zero within a specified time. In that case, it is sufficient to use summation in the above equations from n = -N to N, where N is finite and appropriately high enough.

By employing the statistical theory of multidimensional random variables (Section 3.4), we now have to design a set of (2N+1) independent equations, where (2N+1) equals the number of random variables included in Eqs. (11.85) and (11.86).

It should be noted that various sets of equations are existent, but not all lead to a straightforward calculation for $f_{\mathrm{C}}(C)$. As explained in Section 3.4, it is better to create a set of equations which can unambiguously be solved i.e., a single solution for the variables on the right hand side of this set. In this sense, the set given below represents an useful set of equations for the calculation of pdf $f_{\mathrm{C}}(C)$.

It can be shown that the calculation of pdf $f_C(C)$ can be simplified further, when two more equations are added to the set of $(2N+1)$ equations. For this, it is required to define the two new random variables u and v. These random variables are statistically independent and will be set to zero later. With above considerations we obtain

$$A = \sum_{n=-N}^{+N} s_n \cos\left(\Delta\phi_{-N} + \cdots + \Delta\phi_n\right)\alpha_n + u$$

$$B = \sum_{n=-N}^{+N} s_n \sin\left(\Delta\phi_{-N} + \cdots + \Delta\phi_n\right)\alpha_n + v$$

$$\phi_{-N} = 0 \quad + \Delta\phi_{-N}$$

$$\begin{matrix} \cdot & & \cdot \\ \cdot & & \cdot \\ \cdot & & \cdot \end{matrix} \qquad (11.87)$$

$$\phi_0 = \phi_{-1} + \Delta\phi_0$$

$$\begin{matrix} \cdot & & \cdot \\ \cdot & & \cdot \\ \cdot & & \cdot \end{matrix}$$

$$\phi_N = \phi_{N-1} + \Delta\phi_N$$

Except the phase ϕ_{-N} at the starting point $t = -NT_s$ all other sampled phases ϕ_n with $-N+1 < n < N$ are gradually circumscribed in Eq. (11.87) by real sampled phase deviations $\Delta\phi_n$ from the starting phase ϕ_{-N}. These real phase deviations are statistically independent and each one is described by a simple Gaussian pdf with zero mean and variance

$$\sigma^2_{\Delta\phi} = 2\pi\left[\Delta f_t + \Delta f_l\right]T_s \qquad (11.88)$$

Here, Δf_t and Δf_l represent the laser linewidth of both transmitter and local lasers (see Eq. 3.20) and T_s the sampling period.

In order to simplify calculation, we further represent the sampled starting phase ϕ_{-N} by a virtual phase deviation $\Delta\phi_{-N}$. Because of the non-stationarity of Gaussian phase itself, variance of the random variable $\phi_{-N} = \Delta\phi_{-N}$ grows linearly with time and is in general infinite (Section 3.2.2). As far as the pdfs of both the non-Gaussian random variables $\cos(\phi_{-N})$ and $\sin(\phi_{-N})$ are concerned, it is immaterial whether the pdf of the starting phase ϕ_{-N} is Gaussian distributed with infinite variance or uniformly distributed between 0 and 2π or $-\pi$ and $+\pi$.

In the following, we consider the random variable ϕ_{-N} to be uniformly distributed. Taking into account the above considerations, multidimensional joint probability density function of all the $(2N+3)$ independent random variables of Eq. (11.87) can be written as

$$f_{A,B,\phi_{-N}\cdots\phi_N}\left(A,\, B,\, \phi_{-N}\,\cdots\,\phi_N\right) =$$

$$\frac{1}{2\pi}f_u\left[A-\sum_{n=-N}^{+N}s_n\cos(\phi_n)\,\alpha_n\right]\cdot f_v\left[B-\sum_{n=-N}^{+N}s_n\,\sin(\phi_n)\,\alpha_n\right]\cdot\prod_{n=-N+1}^{+N}f_{\Delta\phi}\left(\phi_n-\phi_{n-1}\right) \tag{11.89}$$

It should be noted that the generally valid equation for calculating multidimensional joint-density functions is normally much more comprehensive than given in Eq. (11.89) and has been described in Section 3.4. However, due to the proper design of our set of equations (Eq. 11.87), generally valid equation could be replaced by the much easier Eq. (11.89). This equation simply represents a multiple product of independent pdfs. It becomes clear from Eq. (11.89) that the required two-dimensional joint probability density function $f_{A,B}(A,B)$ is obtained by a multiple integration over the random variables ϕ_{-N} to ϕ_N (method of integral). Further, taking into account $u = v = 0$ which corresponds to $f_u(u) = \delta(u)$ and $f_v(v) = \delta(v)$ and inserting $f_{A,B}(A,B)$ into Eq. (11.84), we finally obtain

$$f_C(C) = \frac{C}{(2\pi)^{N+1}\,\sigma_{\Delta\phi}^{2N}}\int_{-\infty}^{+\infty}\cdots\int_{-\infty}^{+\infty}\int_0^{2\pi}\exp\left[-\frac{\sum\limits_{n=-N+1}^{+N}\left(\phi_n-\phi_{n-1}\right)^2}{2\sigma_{\Delta\phi}^2}\right]\cdot \tag{11.90}$$

$$\delta\left[C-\sqrt{\sum_{n=-N}^{+N}\sum_{m=-N}^{+N}s_n\,s_m\,\cos(\phi_n-\phi_m)\,\alpha_n\alpha_m}\right]d\phi_{-N}\cdots d\phi_N$$

Again, it should be remembered that $C = d(\sigma = 0)$ equals the sampled and normalized detected signal d for a zero variance $\sigma = 0$ i.e., $x = y = 0$. This result represents an exact analytical solution for the pdf $f_C(C)$ required to calculate the probability of bit error exactly. However, Eq. (11.90) is only of theoretical interest since evaluation of this multi-integral solution would require too much computer run time. Obviously, the Dirac delta function included in Eq. (11.90) describes a well-defined curve in a (2N+1) dimensional space. This curve is described by setting the argument of the Dirac delta function to zero. For this reason, above equation can also be expressed in terms of a line integral. We obtain

$$f_C(C) = \oint_s \exp\left[-\frac{\sum\limits_{n=-N+1}^{+N}\left(\phi_n-\phi_{n-1}\right)^2}{2\sigma_{\Delta\phi}^2}\right]\left|\,\mathrm{grad}\left(\arg(C,\,\phi_{-N}\,\cdots\,\phi_N)\right)\right|^{-1}ds \tag{11.91}$$

Here, the exponential function pictorially describes a (2N+1) dimensional mountain, s a certain path through this mountain which is defined by the argument of the Dirac delta function given in Eq. (11.90) and ds a small stretch of this path. According to to the above equation, each stretch has to be weighted by the inverse of absolute of the gradient $\text{grad}[\arg(C, \phi_{-N} \cdots \phi_N)]$. In comparison to Eq. (11.90), evaluation of the above equation does not require a multiple integration in (2N+1) dimensional space. Instead an integration along the path s is sufficient. However, this solution is also of theoretical interest as it requires an unacceptable high computer run time.

For the design and realization of coherent optical communication systems with heterodyne detection, a more practical solution is required. For this, approximations and computer simulations represent well-suited and powerful tools. As an example, in the following Gaussian approximation method and some simulation results are explained and discussed.

(iv) GAUSSIAN APPROXIMATION OF PROBABILITY DENSITY FUNCTION $f_C(C)$

To obtain a more practical solution for the pdf $f_C(C)$, we take the advantage of following simple Gaussian approximation

$$f_C(C) \approx \frac{1}{\sqrt{2\pi}\sigma_C} \, e^{-\frac{(c - \eta_C)^2}{2\sigma_C^2}} \qquad (11.92)$$

Some other methods of approximation and a detailed comparison of all these methods will be presented in Section 11.5. The Gaussian approximation is entirely described in terms of mean value η_C (first moment) and variance σ_C^2. Due to the square root dependence of C, it is easy to calculate the second moment $E\{C^2\}$ exactly, whereas the expected value $\eta_C = E\{C\}$ can only be found by approximation (E: expected value). The variance σ_C^2 is given by the well-known statistical relation

$$\sigma_C^2 = E\{C^2\} - \eta_C^2 \qquad (11.93)$$

It may be noted that even a small error in η_C due to approximation could lead to a great error in the variance σ_C^2 when this variance is calculated by using the above equation.

Example 11.2

Let us assume an approximate mean value $\eta_C = 0.999$, a real mean value $\eta_C = 1.000$, and a real second moment $E\{C^2\} = 1.001$. In this particular case, relative error in η_C is only 0.1%, whereas it is 1100% in the variance σ_C^2! Thus, there is no use to calculate the variance σ_C^2 by the equation mentioned above when the second moment $E\{C^2\}$ is given exactly while the first moment η_C is given only approximately. This is valid, in particular, when the predicted variance σ_C^2 is very small.

In order to prevent this large error in variance, we make use of the following substitutions and approximations:

$$C = \sqrt{D} \quad \Rightarrow \quad \begin{cases} \eta_C \approx \sqrt{E\{D\}} = \sqrt{\eta_D} \\ \sigma_C \approx \dfrac{1}{2\sqrt{\eta_D}}\sigma_D \end{cases} \tag{11.94}$$

Here, evaluation of the statistical parameters of C is reduced to the evaluation of the corresponding parameters of $D = C^2 = A^2 + B^2$. Obviously, D equals the argument of the square root given in Eq. (11.80). Thus, D is same as the square of the normalized and sampled detected signal $d(\sigma = 0)$ i.e., $D = C^2 = d^2(\sigma = 0)$. Therefore, the quantity D physically represents a measure of the average power of d including the influence of phase noise. Using Eq. (11.80), we obtain

$$\eta_D = \int_{-\infty}^{+\infty} \int_{-\infty}^{+\infty} s(t_1)\, s(t_2)\, E\left\{e^{j[\phi(t_1)-\phi(t_2)]}\right\} h_B(t_0-t_1)\, h_B(t_0-t_2)\, dt_1\, dt_2 \tag{11.95}$$

$$E\{D^2\} = \int_{-\infty}^{+\infty} \int_{-\infty}^{+\infty} \int_{-\infty}^{+\infty} \int_{-\infty}^{+\infty} s(t_1)\, s(t_2)\, s(t_3)\, s(t_4)\, E\left\{e^{j[\phi(t_1)-\phi(t_2)]+j[\phi(t_3)-\phi(t_4)]}\right\} \tag{11.96}$$

$$\cdot h_B(t_0-t_1)\, h_B(t_0-t_2)\, h_B(t_0-t_3)\, h_B(t_0-t_4)\, dt_1\, dt_2\, dt_3\, dt_4$$

and

$$\sigma_D^2 = E\{D^2\} - \eta_D^2 \tag{11.97}$$

To compute the expected values required to evaluate the integrands of Eqs. (11.95) and (11.96), we have to consider the statistical properties of the Gaussian phase noise, which have already been derived in Chapter three. By using the variables

$$w(\tilde{t}) = e^{j\phi(\tilde{t})} \tag{11.98}$$

and

$$H = 2\pi\Delta f\left(\min(\tilde{t}_4,\ \tilde{t}_2) - \min(\tilde{t}_4,\ \tilde{t}_1) + \min(\tilde{t}_3,\ \tilde{t}_1) - \min(\tilde{t}_3,\ \tilde{t}_2)\right) \tag{11.99}$$

in conjunction with Eq. (3.60), expected values required in Eqs. (11.95) and (11.96) are obtained as follows:

$$E\left\{w(\tilde{t}_1)w^*(\tilde{t}_2)\right\} = E\left\{e^{j\left[\phi(\tilde{t}_1)-\phi(\tilde{t}_2)\right]}\right\} = R_w(\tilde{\tau}) = e^{-\pi\Delta f|\tilde{\tau}|} \tag{11.100}$$

and

$$E\left\{w(\tilde{t}_1)w^*(\tilde{t}_2)w(\tilde{t}_3)w^*(\tilde{t}_4)\right\} = e^{-\pi\Delta f\left[|\tilde{t}_2-\tilde{t}_1| + |\tilde{t}_4-\tilde{t}_3|\right]}e^{H} \tag{11.101}$$

In Eq. (11.100), $R_w(\tilde{\tau}) = R_w(\tilde{t}_2 - \tilde{t}_1)$ represents the autocorrelation function of the complex random process $w(\tilde{t}) = \exp(j\phi(\tilde{t}))$. The variables of time $\tilde{t}_1$ to $\tilde{t}_4$ in Eqs. (11.99) to (11.101) are related to $t = -\infty$ when laser operation and, hence, phase noise process are assumed to have begun. Therefore, $\tilde{t}$ is in principle determined by $\tilde{t} = t + T_\infty$ with $T_\infty \to \infty$ and phase noise variance is already infinite around $t = 0$, which is in the relevant time interval of the integrals given in Eqs. (11.95) and (11.96). It should be noted that the integration which have to be performed in these equations approximately yields zero outside the pulse width Δt_B of the baseband filter impulse response $h_B(t)$.

An alternate equation for the expected value $\eta_D = E\{C^2\}$ is obtained by taking into account the Lorentzian power spectral density $G_w(f)$ of the random process $w(t)$ derived in Chapter three (Eq. 3.67). We obtain

$$\eta_D = \int_{-\infty}^{+\infty} G_w(f)\left|\int_{-\infty}^{+\infty} s(t)\ h_B(t_0-t)\ e^{-j2\pi f(t_0-t)}dt\right|^2 df \tag{11.102}$$

The squared expected value η_C^2 and the variance σ_C^2 as a function of the normalized IF filter bandwidth $B_{IF}T = 2f_gT$ and the normalized laser linewidth ΔfT for the particular bit pattern $\langle q_v \rangle = \cdots0001000\cdots$ (referred as single one) are shown in Figs. 11.13a and 11.13b. The IF filter is assumed to have a Gaussian frequency response $H_B(f) = \exp[-\pi(f/B_{IF})^2]$ in the baseband representation. Physically, the mean square value η_C^2 and variance σ_C^2 represent the signal power (useful power) and the noise power caused by phase noise only. It must be remembered that C equals the sampled detected signal d in absence of receiver additive Gaussian noise i.e., $\sigma = 0$.

As plotted in Fig. 11.13a, signal power $E\{C\}^2 = \eta_C^2$ increases with the increase in the normalized IF filter bandwidth $B_{IF}T$ because the effect of intersymbol interference becomes smaller. On the other hand, signal power will get attenuated when bandwidth is decreased.

Further, Fig. 11.13a indicates that the finite linewidths of real lasers lead to a severe reduction of the signal power.

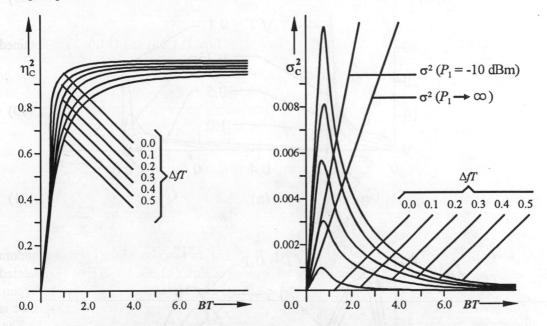

Fig. 11.13: (a) Square of expected value η_C representing the signal power and (b) variance σ_C^2 representing the noise power due to phase noise of random variable $C = d\,(\sigma = 0)$. As a comparison, straight lines in (b) represent the normalized variance σ^2 of additive Gaussian noise.

The normalized noise power σ_C^2 as shown in Fig. 11.13b is zero for $B_{IF}T = 0$ (i.e., the transmission path is broken) and also for $B_{IF}T \to \infty$ (i.e., no influence of the IF filter due to undesired phase noise-to-amplitude conversion). It can be seen from Fig. 11.13b that at certain bandwidths, noise power due to phase noise is maximum. As a comparison, the straight line illustrates the influence of additive Gaussian noise whose normalized power σ^2 increases proportional to the IF filter bandwidth B_{IF}. Finally, it should be noted that the analytical results shown in Figs. 11.13a and 11.13b do agree very well with the respective values obtained by simulation experiments.

(v) COMPUTER SIMULATION OF THE PROBABILITY DENSITY FUNCTION $f_c(C)$

Besides analytical solutions, the computer simulation represents an useful and a powerful alternative. Moreover, simulation gives a clear insight into the typical behaviour of an incoherent ASK heterodyne system. In particular, the effect of a change in system parameters, for example in bit rate can quickly be evaluated and assessed.

The simulated probability density functions $f_C(C)$ and $f_d(d)$ shown in Figs. 11.14a and 11.14b are the typical results of a simulation experiment.

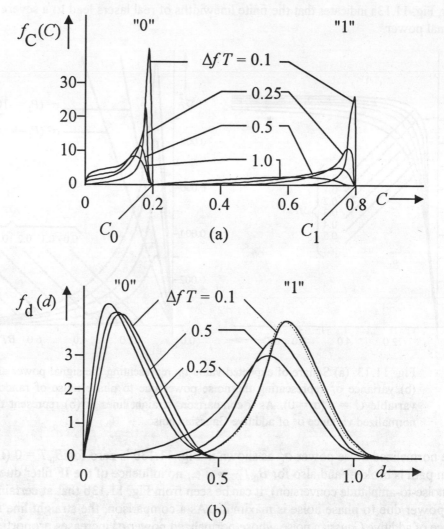

Fig. 11.14: (a) Computer simulated probability density functions $f_C(C)$ and (b) $f_d(d)$ of random variables $C = d\,(\sigma = 0)$ and d (detected signal). The dotted curve makes use of the Gaussian approximation given in Eq. (11.92) and $\Delta f T = 0.1$.

These results are based on 100000 (one hundred thousand) computer simulated random values for each of the random variables C and d. Variable parameter in Fig. 11.14 is the normalized laser linewidth $\Delta f T$. It can be regarded as a direct measure to assess the strength and effect of laser phase noise. In addition, both figures are based on the following system parameters:

- IF filter with a Gaussian frequency response $H_B(f) = \exp[-\pi(f/B_{IF})^2]$ in the baseband representation, $B_{IF}T = 2f_g T = 1$,

- $\langle q_v \rangle_{1i} = \cdots 0001000 \cdots$ (single "1") and $\langle q_v \rangle_{0i} = \cdots 1110111 \cdots$ (single "0")

- receiver additive Gaussian noise with normalized variance $\sigma^2 = 0$ in Fig. 11.14a and $\sigma^2 = 0.01$ in Fig. 11.14b respectively.

It becomes evident from Fig. 11.14a that the laser phase noise give rise to a distortion for transmitted symbol "1", whereas it leads to an improvement for symbol "0"! As already explained in Chapter three, the reason for this very surprising result is that laser phase noise acts as a multiplier with the information signal in ASK heterodyne and homodyne receivers. However, the deterioration is always dominant as compared to improvement. Thus, laser phase noise always yields an overall degradation of system performance as expected.

The values $C_0 = 2Q(\sqrt{\pi/2}B_{IF}T)$ and $C_1 = 1 - C_0$ given in Fig. 11.14a determine the normalized eye opening $A_{ASK} = C_1 - C_0$ of the eye pattern in the absence of both laser phase noise and additive Gaussian circuit noise (see Sections 9.5 and 11.1.1 also).

Asymmetrical effect of laser phase noise becomes clear from Fig. 11.14b. Depending on the normalized laser linewidth $\Delta f T$, optimum threshold level E for a minimum BER could differ from $E_{opt} = 0$ if $\Delta f T$ is large to $E_{opt} = 0.5$ if $\Delta f T$ approaches to zero. It should be remembered that the optimum threshold E_{opt} is always given by the point where the pdfs $f_d(d)$ for symbol "1" and symbol "0" meet. The dotted curve is the Gaussian approximation for $\Delta f T = 0.1$ with mean value $\eta_d = \eta_{C|1}$ and variance $\sigma_d^2 = \sigma_C^2 + \sigma^2$.

The asymmetrical influence of laser phase noise results in an asymmetrical eye pattern of detected signal d. In Chapter twelve (system comparison), this result will be discussed in more detail.

(vi) PROBABILITY OF ERROR

For the calculation of probability of error or BER, we have to distinguish between the bit sequences $<q_v>_{1i}$ with $q_0 = $ "1" (called single one or single "1") and $<q_v>_{0i}$ with $q_0 = $ "0" (called single zero or single "0"). The corresponding probability of errors p_{0i} and p_{1i} are simply obtained by integrating pdfs given in the previous Subsections over the respective areas.

With the same substitutions and abbreviations as used in Section 11.1.1 (ASK homodyne system), we obtain

$$
\begin{aligned}
P_{0i} &= \int_E^{+\infty} f_{d0i}(d) \; \mathrm{d}d = \int_E^{+\infty} \int_0^{+\infty} \frac{d}{\sigma^2} \, e^{-(d^2 + C^2)/2\sigma^2} \, J_0\!\left(\frac{Cd}{\sigma^2}\right) f_{C0i}(C) \; \mathrm{d}C \; \mathrm{d}d \\[2mm]
&\qquad\qquad\qquad\qquad\qquad\qquad\qquad\qquad\qquad\qquad\qquad\qquad (11.103) \\[2mm]
&= \int_0^{+\infty} \tilde{p}_{0i}(C) \, f_{C0i}(C) \; \mathrm{d}C = E\big\{ \tilde{p}_{0i}(C) \big\} \qquad \text{with } C = C_{0i} \geq 0
\end{aligned}
$$

and

$$
\begin{aligned}
p_{1i} &= \int\limits_{0}^{E} f_{d1i}(d)\ \mathrm{d}d = \int\limits_{0}^{E}\int\limits_{0}^{+\infty} \frac{d}{\sigma^2}\, e^{-\frac{d^2+C^2}{2\sigma^2}}\, J_0\!\left(\frac{Cd}{\sigma^2}\right) f_{C1i}(C)\ \mathrm{d}C\ \mathrm{d}d \\
\\
&= \int\limits_{0}^{+\infty} \tilde{p}_{1i}(C)\, f_{C1i}(C)\ \mathrm{d}C = \mathrm{E}\!\left\{\tilde{p}_{1i}(C)\right\} \qquad \text{with } C = C_{1i} \ge 0
\end{aligned}
$$

(11.104)

In the above equations, $C_{0i} = d_{0i}(\sigma = 0)$ and $C_{1i} = d_{1i}(\sigma = 0)$ are the sample values of the envelope $|i_{IF}(t)| = d(t)$ for symbol sequences $\langle q_v\rangle_{0i}$ and $\langle q_v\rangle_{1i}$ respectively provided additive noise does not disturb the system i.e., $\sigma = 0$. Because $C = |i_{IF}(t)| \ge 0$, it is sufficient to integrate from 0 instead from $-\infty$. The last term in both equations follows by interchanging the order of integration of the second term and by representing the new inner integral with $\tilde{p}_{0i}(C_{0i})$ and $\tilde{p}_{1i}(C_{1i})$ respectively. These newly defined probabilities of error depend on amplitudes C_{0i} and C_{1i}, threshold E and normalized standard deviation σ of the additive Gaussian noise, but they are independent of phase noise. Assuming that C_{0i} and C_{1i} are constant, $\tilde{p}_{0i}$ and $\tilde{p}_{1i}$ become the well-known probabilities of error for conventional ASK transmission systems with envelope detection.

It becomes clear from the above equations that the calculation of actual BER can be divided into two steps: First step is the well-known calculation of the BERs $\tilde{p}_{0i}$ and $\tilde{p}_{1i}$ neglecting laser phase noise [36]. In this step, C_{0i} and C_{1i} are constant. In the second step, we consider the randomness of C_{0i} and C_{1i} due to laser phase noise and evaluate the actual BERs p_{0i} and p_{1i} by computing the expected values given in Eqs. (11.103) and (11.104).

(a) AVERAGE PROBABILITY OF ERROR

The average BER p_a is obtained by taking into account occurrence probabilities of all possible symbol sequences $\langle q_v\rangle_i$. Similar to the an ASK homodyne system (Section 11.1.1), we obtain

$$
p_a = \sum_{i=1}^{2^{n+v}} p\big(\langle q_v\rangle_i\big)\big(p_{0i} + p_{1i}\big) = 2^{-(n+v+1)}\sum_{i=1}^{2^{n+v}} \big(p_{0i} + p_{1i}\big)
$$

(11.105)

As mentioned in Section 11.1.1, evaluation of p_a is very comprehensive and not useful for a practical system design. Again, much more convenient is to evaluate the worst-case probability of error as a very good approximation and as an upper bound.

(b) WORST-CASE PROBABILITY OF ERROR

For the calculation of the worst-case probability of error p_w, we have to consider the worst-case sequences $<q_v>_{0w}$ and $<q_v>_{1w}$. Assuming a Gaussian IF filter, worst-case pattern are given by the bit sequences

$$<q_v>_{0w} = \cdots 1110111 \cdots$$

and

$$<q_v>_{1w} = \cdots 0001000 \cdots$$

In order to simplify calculation and to avoid generality, we take advantage of the following approximations and worst-case considerations:

- The pdf $f_{C1w}(C)$ should be calculated as given in Eq. (11.92).

- Instead of using the Rice distribution for calculating the conditional density function $f_{d|C}(d, C)$ with $d = d_{1w}$ and $C = C_{1w}$ we take Gaussian approximation given in Eq. (11.77). This approximation can be employed when signal-to-noise ratio is high enough which is normally fulfilled when symbol "1" is transmitted.

- As mentioned above, laser phase noise always deteriorates the detection of symbol "1", whereas the detection of symbol "0" is always improved. Therefore, under worst-case condition, influence of phase noise can be neglected for symbol "0". In that case, pdf of the detected signal d_{0w} is

$$f_{d0w}(d) = \frac{d}{\sigma^2} e^{-\frac{d^2 + c_{0w}^2}{\sigma^2}} \tag{11.106}$$

Similar to standard ASK transmission systems with envelope detection, pdf $f_{d0w}(d)$ yields a Rice distribution if $C_{0w} \neq 0$ and a Rayleigh distribution if $C_{0w} = 0$.

With these three conditions, worst-case probability of error p_w can now be evaluated as follows:

$$
\begin{aligned}
p_{\mathrm{w}} &= \frac{1}{2}\int_{0}^{E} f_{\mathrm{d1w}}(d)\ \mathrm{d}d \; + \; \frac{1}{2}\int_{E}^{+\infty} f_{\mathrm{d0w}}(d)\ \mathrm{d}d \\[2mm]
&= \frac{1}{2}\,Q\!\left(\frac{\eta_{\mathrm{C1w}} - E}{\sqrt{\sigma_{\mathrm{C1w}}^{2} + \sigma^{2}}}\right) + \frac{1}{2}\int_{E}^{+\infty}\frac{d}{\sigma^{2}}\,\mathrm{e}^{-\frac{d^{2}+C_{\mathrm{0w}}^{2}}{2\sigma^{2}}}\,\mathrm{J}_0\!\left(\frac{C_{\mathrm{0w}}d}{\sigma^{2}}\right)\ \mathrm{d}d
\end{aligned}
\tag{11.107}
$$

In this equation, influence of laser phase noise is considered in terms of mean value η_{C1w} and the standard deviation σ_{C1w} which are determined from Eq. (11.94) under worst-case i.e., $C = C_{\mathrm{1w}}$. As mentioned above, C_{1w} equals the sample value d_{1w} of the detected signal $d(t) = |i_{\mathrm{IF}}(t)|$ under worst-case pattern $\langle q_{\mathrm{v}}\rangle = \langle q_{\mathrm{v}}\rangle_{\mathrm{1w}}$ and in absence of additive Gaussian noise ($\sigma = 0$).

For a minimum BER, threshold level E and IF filter bandwidth $B_{\mathrm{IF}} = 2f_{g}$ which is included in $\sigma = \sigma(f_{g})$, $\eta_{\mathrm{C1w}} = \eta_{\mathrm{C1w}}(f_{g})$ and $\sigma_{\mathrm{C1w}} = \sigma_{\mathrm{C1w}}(f_{g})$ are optimizable system parameters. In contrast to that, laser linewidth Δf included in η_{C1w} and σ_{C1w} and constant power spectral density G_{c} included in σ are non-optimizable system parameters. Moreover, local laser power P_{l} and received light power P_{r} in the normalized standard deviation σ of the Gaussian circuit noise are, of course, also non-optimizable. It must be remembered that σ is normalized with respect to the amplitude $\hat{\imath}_{\mathrm{PD}}$ of photodiode current which is also a function of P_{l} and P_{r}.

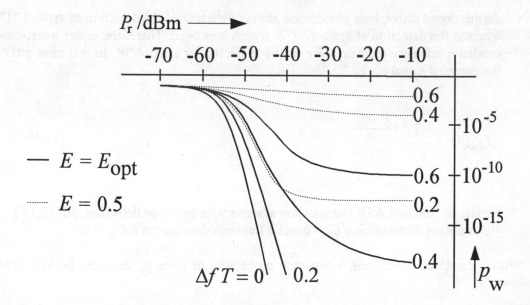

Fig. 11.15: Worst-case probability of error p_{w} for ASK heterodyne system with envelope detection with optimum and non-optimum thresholds

The worst-case BER as a function of the light power P_r at the input to the receiver and normalized resulting laser linewidth $\Delta f\,T$ is shown in Fig. 11.15. The IF filter bandwidth is fixed at $B_{IF} = 1.85/T$ which is the optimum bandwidth in case of $\sigma_\phi = 0$.

The trend of the BER curves shown in Fig. 11.15 are similar to that of coherent optical systems described in the previous Sections. If the received optical power P_r is low, then the additive Gaussian noise dominates the phase noise. Thus, the curves are rather steep. As the optical power P_r increases, influence of additive Gaussian noise decreases, while the influence of the laser phase noise increases. Consequently, BER curves exhibit a bend. As the optical power P_r is increased further beyond this point, performance curves finally reach a saturation called the bit error rate floor (Section 11.1.1).

(vii) OPTIMIZATION

(a) THRESHOLD E

Because of the complexity of Eq. (11.107), optimum threshold level for a minimum BER can only be evaluated by means of numerical methods using computer. Thereby, the relationship

$$E_{opt} \leq 0.5 \tag{11.108}$$

is always valid since both the binary symbols "0" and "1" are influenced differently by the laser phase noise as discussed above. It should be noted that optimum threshold E_{opt} given in Eq. (11.108) is normalized with respect to the amplitude $\hat{\imath}_{PD}$ of photodiode current (Eq. 9.69).

It is seen from Fig. 11.15 that the BER is considerably reduced by taking the optimum threshold E_{opt} instead of the threshold $E = 0.5$ which is optimum only when phase noise is neglected ($\sigma_\phi = 0$).

(b) IF FILTER BANDWIDTH B_{IF}

Similar to the threshold E, optimization of the IF filter bandwidth B_{IF} can only be performed iteratively by means of a computer. One result of prime importance is shown in Fig. 11.16a, where the resulting laser linewidth is fixed at $\Delta f = 0.6/T$. It becomes clear from this figure that optimization of bandwidth $B_{IF} := B = 2f_g$ and threshold E yields a considerable improvement in the system performance i.e., in the BER. For the comparison, Fig. 11.16a also includes the bandwidth $B_{IF} = 1.58/T$ which is the optimum bandwidth when the system is not disturbed by phase noise. It should be noted that the optimization of bandwidth is performed afresh for each received optical power P_r. This is required since the optimum IF filter bandwidth strongly depends on P_r when phase noise disturbs the system (Fig. 11.16b).

An important result of Fig. 11.16a is that the BER curve exhibits *no bit error rate floor* when both bandwidth B_{IF} *and* threshold E are optimized. This means that a bit error rate of 10^{-10} is now generally achievable for any laser linewidth Δf. There exists no limit and no error rate floor in principle. This very surprising behaviour can easily be explained as follows: The system deterioration caused by large laser linewidth can be compensated by a respective broad IF filter bandwidth. This, however, causes a rise in additive Gaussian noise at the IF filter output. On the other hand, effect of this additional additive noise can always be reduced in principle by an appropriate high received light power P_r. The only limitations are the maximum available light power P_r and realizable intermediate frequencies. Nevertheless, to achieve a large repeater spacing, received optical P_r should be allowed to be as low as possible.

Finally, Fig. 11.16b shows the normalized optimum IF filter bandwidth $B_{opt}T$ as a function of the received optical power P_r at the input to receiver and the normalized resulting laser linewidth $\Delta f T$. As in ASK homodyne receiver, a broader filter again reduces the influence of phase noise.

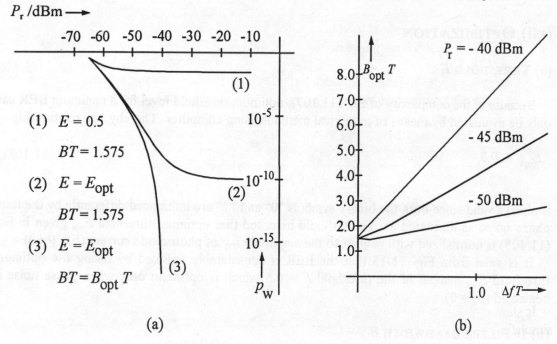

Fig. 11.16: (a) Reduction of probability of error p_w by optimizing IF filter bandwidth B and threshold E for $\Delta f T = 0.6$ and (b) Optimum IF filter bandwidth as a function of normalized resulting laser linewidth $\Delta f T$ and received optical power P_r

The present optimization corresponds to a minimum worst-case BER provided a Gaussian IF filter is used. To obtain a more general solution, an arbitrary IF filter frequency response must be taken into account. Moreover, the system becomes optimum when the system parameters mentioned above are optimized for a minimum average BER instead for a minimum worst-case BER. However, this optimization can only be carried out by very comprehensive and time intensive numerical computer calculation.

Owing to the fact that each bit sequence is related to its own BER, optimum receiver is actually obtained by changing the receiver configuration with only one IF filter to a configuration providing a matched IF filter for each possible bit sequence $<q_v>_i$. It will be the best receiver which can be theoretically obtained. However, it will not be of much practical use due to comprehensive design and realization.

(c) CHARACTERISTIC FEATURES OF INCOHERENT ASK HETERODYNE SYSTEMS

Most important results of our considerations in the previous Subsections on incoherent ASK heterodyne system with envelope detection are summarized below:

- The disturbance to the system caused by laser phase noise is different for transmitting symbol "1" and symbol "0". This results in an asymmetrical eye pattern.

- The optimum threshold level in presence of laser phase noise is always lower than the level when the laser phase noise is absent.

- The optimum IF filter bandwidth increases with the increase in the laser linewidth of both transmitter and local lasers.

- The bit error rate decreases with the decrease in the ratio of laser linewidth to IF filter bandwidth.

- For a fixed IF filter bandwidth e.g., $B_{IF}T = 1$, maximum laser linewidth exists for a desired bit error rate. In this case, the system exhibits a characteristic error rate floor.

- For an optimum IF filter bandwidth, no error rate floor exists in principle. In such a case, any desired bit error rate can be achieved, in principle, by an appropriate optical power level P_r at the input to the receiver.

It should be noted that the first two results become evident in particular when the computer simulated eye pattern is considered (see Chapter twelve).

An ASK heterodyne receiver with incoherent detection (e.g., envelope detection) exhibits a 9 dB lower sensitivity than the most sensitive PSK homodyne receiver and a 3 dB lower sensitivity as compared to incoherent dual-filter FSK heterodyne receiver (Section 11.3.2). In an incoherent ASK heterodyne system, laser linewidths of 20% to 30% of the bit rate $1/T$ can be allowed when the bandwidth of Gaussian IF filter is fixed at $BT = 1.58$. This bandwidth is optimum in the absence of phase noise. When bandwidth is optimized by taking into consideration laser phase noise also, linewidths in the order of bit rate are acceptable. This is in contrast to homodyne systems, where linewidth requirement is very strong e.g., 0.01% of bit rate.

As the linewidth requirements, costs and problems of realization are approximately same in both the incoherent ASK and FSK heterodyne receivers, incoherent ASK heterodyne receiver is of little practical importance. However, most of the results obtained in this Section can be used to analyse incoherent FSK (Section 11.3.2) and DPSK heterodyne system (Section 11.3.3).

11.3.2 FSK HETERODYNE SYSTEM

In conventional analog communication systems such as analog radio and TV broadcasting, frequency modulated (FM) carrier signal is normally demodulated by means of a *frequency discriminator*. In order to obtain a linear discriminator transfer function between the input frequency and output voltage, a proper match between both the discriminator filters is required. Each deviation in linearity directly results in non-linear distortion.

In contrast to this, linear distortion is not as critical in a digital transmission system with FSK modulation. Here, the shape of the received signal is less important, whereas an errorless decision between both binary symbols "0" and "1" is most important. Thus, a linear discriminator is not absolutely required and the centre frequencies of both discriminator filters may be separated by a much larger frequency deviation than in case of a linear discriminator employed in analog FM systems. Since both filters are now no longer matched, this demodulator is usually called *dual-filter demodulator* or *two-filter demodulator*. Owing to the relatively large frequency deviation, these demodulators enable a proper separation and an unambiguous detection of both binary symbols "0" and "1".

The block diagram of a incoherent optical FSK heterodyne transmission system with a two-filter demodulator in the receiver is shown Fig. 11.17. The FSK demodulator contains two branches: one is the "1"-branch or the "1"-channel; other is the "0"-branch or the "0"-channel. Both symbols "0" and "1" are selected by two band-pass filters with centre frequencies f_{IF0} and f_{IF1} respectively. With a frequency deviation f_d and an intermediate frequency (IF) f_{IF}, both centre frequencies are given by

$$f_{IF0} = f_{IF} - f_d \quad \text{and} \quad f_{IF1} = f_{IF} + f_d \qquad (11.109)$$

Hence, lower the frequency deviation f_d, lower is the frequency difference between both the centre frequencies. Thereby, an errorless detection of binary symbols "0" and "1" becomes more and more difficult. In the worst-case i.e., $f_d = 0$, a distinction between both the binary symbols is, of course, no longer possible.

As indicated in Fig. 11.17, both the IF filter output signals $i_{IF0}(t)$ and $i_{IF1}(t)$ which are disturbed by laser phase noise and receiver additive Gaussian noise (i.e., shot noise and thermal noise) are fed to the input of a corresponding envelope detectors. The detector output signals referred as the detected signals $d_0(t)$ and $d_1(t)$ are finally sampled after every T (i.e., symbol- or bit duration) and fed to a maximum decision circuit. Depending on which of the two branches has the higher sample value, decision circuit decides either the symbol "0" or symbol "1".

It becomes clear from Fig. 11.17 that incoherent two-filter demodulator can be regarded as a simple parallel circuit of two incoherent ASK envelope demodulators. Therefore, many results of the previous Section 11.3.1 can be used to analyse an incoherent FSK heterodyne system.

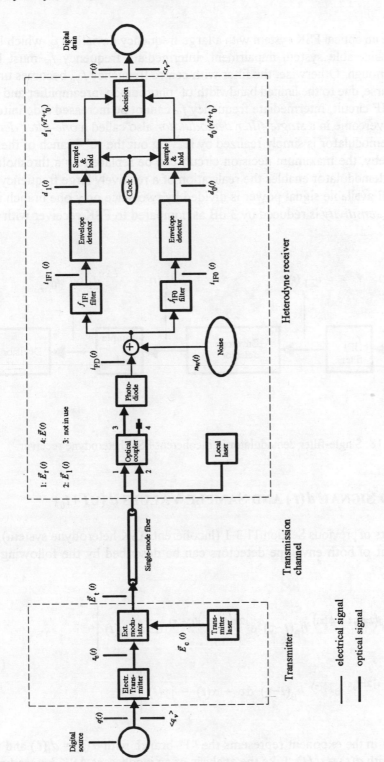

Fig. 11.17: Coherent optical communication system with inncoherent FSK heterodyne receiver

In order to realize an optical FSK system with a large frequency deviation f_d, which is required in presence of considerable system impairment, intermediate frequency f_{IF} must be chosen appropriately high enough. Otherwise, the filter with centre frequency f_{IF0} becomes unrealizable (Eq. 11.109). Of course, due to the limited bandwidth of photodiode, preamplifier and electronic components of the IF circuit, intermediate frequency f_{IF} cannot be increased indefinitely.

This problem is overcome in a *single-filter demodulator* also called a *one-filter demodulator* (Fig. 11.18). This demodulator is simply realized by leaving out the "0"-branch of the two-filter demodulator. Thereby, the maximum decision circuit can be replaced by a threshold decision circuit. Single-filter demodulator enables the realization of a relatively large frequency deviation f_d. However, overall available signal power is divided by two since only one branch is present. Therefore, *receiver sensitivity* is reduced by 3 dB as compared to FSK receiver with dual-filter demodulator.

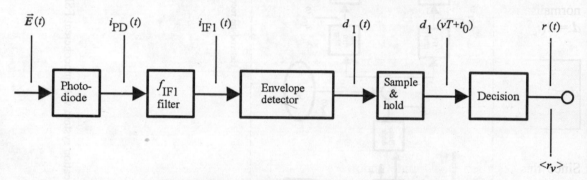

Fig. 11.18: Single-filter demodulator in incoherent FSK heterodyne receiver

(i)   DETECTED SIGNAL $d(t)$ AND SAMPLE VALUES $d(\upsilon T+t_0)$

By using the results of previous Section 11.3.1 (incoherent ASK heterodyne system), detected signals at the output of both envelope detectors can be described by the following common equation

$$d(t) \approx \left| \hat{i}_{PD} \int_{-\infty}^{+\infty} \underline{s}(\tau)\ e^{j\left(2\pi f_{IF}\tau\ +\ \phi(\tau)\right)}\ h_B(t-\tau)\ e^{j2\pi\left(f_{IF}\pm\ f_d\right)(t\ -\ \tau)}\ d\tau\ +\ \underline{n}(t) \right|$$

$$= \left| \hat{i}_{PD} \int_{-\infty}^{+\infty} \underline{s}(\tau)\ e^{\mp j2\pi f_d\tau}\ e^{j\phi(\tau)}\ h_B(t-\tau)\ d\tau\ +\ x(t)\ -\ jy(t) \right|$$

(11.110)

Here, the upper sign in the exponent represents the "1"-branch with $d(t) = d_1(t)$ and the lower sign the "0"-branch with $d(t) = d_0(t)$. Like the analysis of an incoherent ASK heterodyne system

(Section 11.3.1), above equation takes advantage of baseband representation of both the IF filters and also of the narrow-band condition given in Eq. (9.82). Again, $h_B(t)$ represents the impulse response of the equivalent baseband filter $H_B(f)$ as explained in Section 9.4.3. As the modulation scheme is FSK, information signal $\underline{s}(\tau)$ is given by (compare with Eq. 9.42a)

$$\underline{s}(\tau) = \exp\left[j\int_{-\infty}^{\tau}\sum_{\upsilon=-\infty}^{+\infty}2\pi f_d\left(2s_\upsilon-1\right)\mathrm{rect}\left(\frac{t-\upsilon T}{T}\right)\ dt\right] = \exp\left[j\int_{-\infty}^{\tau}\omega_s(t)\ dt\right] = e^{j\left(\phi_s(\tau)\ +\ \phi_0\right)} \quad (11.111)$$

where $\omega_s(t)$ and $\phi_s(t)$ represent the modulated frequency and phase of $\underline{s}(\tau)$. With the same normalisation and substitution as used in the previous Section 11.3.1, normalized sample value $d = d(t_0)/i_{PD}$ of the detected signal can be written as

$$d = \left|\int_{-\infty}^{+\infty}e^{j\phi_0}\ e^{j\left[\phi_s(\tau)\ \mp\ 2\pi f_d\ \tau\right]}\ e^{j\phi(\tau)}\ h_B(t_0-\tau)\ d\tau\ +\ x\ -\ jy\right| \quad (11.112)$$

Since this equation is of prime importance in the following system analysis, meaning of its physical quantities is summarized below:

ϕ_0: arbitrary, but constant phase,
$\phi_s(t)$: signal phase including the transmitted information,
f_d: frequency deviation,
$\phi(t)$: laser phase noise of both transmitter and local lasers,
$h_B(t)$: impulse response of the equivalent baseband filter,
$x(t)$: inphase component of the receiver additive Gaussian noise and
$y(t)$: quadrature component of the receiver additive Gaussian noise.

We have to distinguish "0"-branch and "1"-branch on the basis of term $\phi_s(\tau)\mp2\pi f_d\tau$ in the above equation. In the "1"-branch, this term is either a constant (if q_υ = "1") or proportional to $4\pi f_d\tau$ (if q_υ = "0"). In the "0"-branch, this relationship is reversed i.e., this term is constant for q_υ = "0" and proportional to $4\pi f_d\tau$ for q_υ = "1".

Assuming that no receiver and phase noise exist in the system i.e., $x = y = 0$ and $\phi = 0$, Eq. (11.112) simplifies to

$$d = \left| \int_{-\infty}^{+\infty} e^{j\phi_0}\, e^{j\left[\phi_s(\tau)\, \mp\, 2\pi f_d \tau\right]}\, h_B(t_0 - \tau)\, d\tau \right| \tag{11.113}$$

An alternate equation is obtained by dividing the integral into real and imaginary parts:

$$d = \left[\left(\int_{-\infty}^{+\infty} \cos(\phi_s(\tau)\, \mp\, 2\pi f_d \tau\, +\, \phi_0)\, h_B(t_0 - \tau)\, d\tau \right)^2 \right.$$

$$\left. + \left(\int_{-\infty}^{+\infty} \sin(\phi_s(\tau)\, \mp\, 2\pi f_d \tau\, +\, \phi_0)\, h_B(t_0 - \tau)\, d\tau \right)^2 \right]^{1/2} \tag{11.114}$$

In case of an ideal envelope demodulator, where the detected signal at the output of demodulator equals the absolute of complex IF signal (Eqs. 11.110 and 11.113), a constant phase ϕ_0 does not influence the detection process. Hence, this phase can be either set to zero or chosen in such a way that the imaginary part of the convolution integral (i.e., the part with the sine function) becomes zero. However, this step is only possible in case of an even impulse response $h_B(t) = h_B(-t)$.

Generation of sample value $d = d_1$ in the "1"-branch of a dual-filter FSK demodulator is illustrated in Fig. 11.19. For this, worst-case pattern $\cdots 0001000 \cdots$ (single-"1") and $\cdots 1110111 \cdots$ (single-"0") are assumed to be the transmitted information. The upper curves in Fig. 11.19 first show the corresponding modulated frequency $\omega_s(t)$ and the modulated phase $\phi_s(t)$. In accordance with Eq. (11.114), we then have to subtract the linearly increasing phase $2\pi f_d t$. As a result, we obtain the phase $\phi_1(t) = \phi_s(t) - 2\pi f_d t$ which is also plotted in Fig. 11.19. Because of the fact that an additional constant phase ϕ_0 does not influence the result (i.e., the detected signal at the output of the FSK receiver), we may use ϕ_0 to simplify calculation. It is most convenient to choose ϕ_0 such that the new phase $\phi_2(t) = \phi_1(t) + \phi_0$ is an odd function i.e., $\phi_2(t) = -\phi_2(-t)$.

Next, we have to accomplish the required convolution between the impulse response $h_B(t)$ and signal $\cos(\phi_2(t)) = \cos(\phi_2(-t))$. Finally, the sample value d_1 is obtained by taking the absolute of the convolution result and setting $t_0 = 0$. Owing to our proper choice of the constant phase ϕ_0, imaginary part of the convolution integral yields zero since a sine function is an odd function i.e., $\sin(\phi_2(t)) = -\sin(\phi_2(-t))$.

Comparing ASK and FSK heterodyne receivers, it becomes clear that sample value d_1 (provided single-"1" transmitted) is always higher (i.e., better) when FSK is employed. If a single-"0" is transmitted, then the FSK sample value d_0 can either be above or below the corresponding

ASK sample value. Which of the two cases is actually existent depends on the frequency deviation f_d. If normalized frequency deviation approaches infinity (i.e., $f_d T \to \infty$) or $f_d T = k$ where $k \in \mathbb{N}$, then sample value d_0 is same for both ASK and FSK receivers (see also Fig. 11.20).

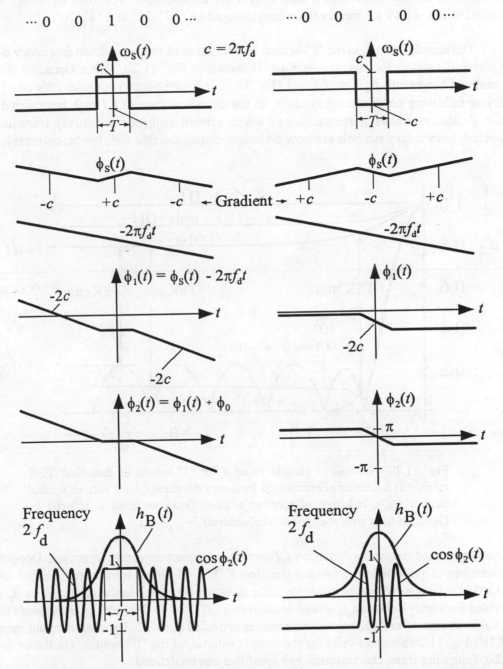

Fig. 11.19: Principle of generation of sample value d_1 in the "1"-branch of dual-filter FSK receiver

It is obvious from Fig. 11.19 that "0"-branch and "1"-branch are absolutely symmetrical. Thus, a symbol "1" in the "0"-branch always yields the same sample value as a symbol "0" in the "1"-branch. Therefore, Fig. 11.19 can also be used to illustrate how the detected signal d_0 is generated in the "0"-branch of a dual-filter FSK demodulator. For this, all binary symbols included in Fig. 11.19 are merely to be interchanged i.e., "0"→"1" and "1"→"0".

The sample value d_1 in the "1"-branch as a function of the normalized frequency deviation $f_d T$ for different symbol sequences $\langle q_v \rangle$ is shown in Fig. 11.20. For a Gaussian filter with normalized bandwidth $B_{IF}T = 2f_g T = 1$ (Eq. 9.83), it is sufficient to consider only one previous and one following neighbouring symbols. In the unrealistic case of $f_d T = 0$, normalized sample value d_1 always equals 1 irrespective of which symbol sequence is actually transmitted. As expected, both binary symbols are now no longer distinguishable and, hence, detectable.

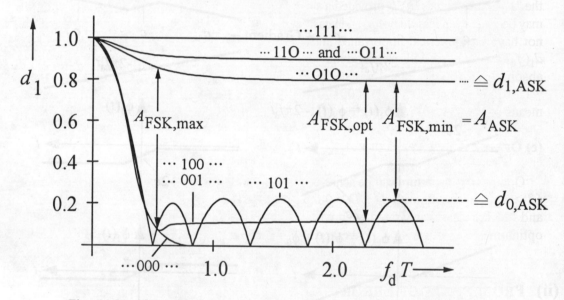

Fig. 11.20: Normalized sample value d_1 of "1"-branch in dual-filter FSK receiver as a function of normalized frequency deviation $f_d T$ for various symbol pattern $\langle q_v \rangle$. Influence of receiver additive Gaussian noise is reduced by Gaussian filter with normalized bandwidth $BT = 2f_g T = 1$.

As normalized frequency deviation $f_d T$ increases, symbol detection improves. Depending on the number of periods in the symbol duration T (Fig. 11.19), a transmitted symbol sequence ···1110111··· (single-"0") yields a sample value d_1 which is located in the range of $0 \leq d_1 \leq d_{0,ASK}$ provided frequency deviation f_d is high enough (e.g., $f_d T > 0.5$). Here, $d_{0,ASK}$ represents the sample value in case of an ASK homodyne receiver provided q_v = "0". By interchanging symbol "0" and "1", Fig. 11.20 becomes valid for the sample value d_0 of the "0"-branch. On the basis of Fig. 11.20, following three characteristic eye apertures can be defined:

(a) MAXIMUM EYE APERTURE $A_{FSK,max}$

This maximum eye aperture is given at a normalized frequency deviation $f_d T \approx 0.5$. When the phase noise does not disturb the system, this aperture is optimum for a minimum BER. However, at $f_d T \approx 0.5$ symbol "0" ($f_{IF} - f_d$) and symbol "1" ($f_{IF} + f_d$) are located closely. As soon as laser phase and frequency noise disturb the system in addition, both symbols can no longer be distinguished properly. It should be noted that laser frequency noise permanently results in a random shift in frequency deviation f_d, which finally increases BER although the eye aperture is large.

(b) MINIMUM EYE APERTURE $A_{FSK,min}$

As shown in Fig. 11.20, eye aperture becomes minimum when sample value $d_1(f_d T)$ of the "1"-branch equals $d_{0,ASK}$ provided a single-"0" is transmitted. Of course, the eye aperture may become even less than $A_{FSK,min}$ in case of $f_d T \ll 0.5$. However, this particular range does not have any practical importance and should not be considered. In the special case of $d_1(f_d T) = d_{0,ASK}$, ASK heterodyne and dual-filter FSK heterodyne systems become absolutely identical. Thus, worst-case BER p_w is defined by both worst-case pattern $\cdots1110111\cdots$ (single-"0") and $\cdots0001000\cdots$ (single-"1"). Moreover, p_w can be calculated by means of Eq. (11.107) irrespective of whether ASK or FSK is used.

(c) OPTIMUM EYE APERTURE $A_{FSK,opt}$

Optimum eye aperture can be achieved by using an appropriate large frequency deviation $f_d T > 5$ in conjunction with $d_1(f_d T) = 0$ for a transmitted single-"0". Here, both symbols "0" and "1" are clearly separated even when laser phase noise disturbs the system. In case of optimum eye aperture $A_{FSK,opt}$, average BER p_a reaches its minimum value.

(ii) PROBABILITY OF ERROR

(a) AVERAGE PROBABILITY OF ERROR p_a

As mentioned in the previous Subsection, average probability of error p_a reaches minimum in case of an infinite frequency deviation provided $d_1(f_d T) = 0$ for a transmitted single-"0" (compare Fig. 11.20). Calculation of p_a is very comprehensive and can only be performed by means of appropriate numerical methods and fast computer program.

As the FSK sample values d are always "better" than the corresponding ASK values (i.e., a symbol "1" always results in a higher and a symbol "0" a lower sample value), average probability of error p_a is somewhat less when FSK is used in the system. In terms of receiver sensitivity, this results in an improvement of about 1 dB in single-filter FSK receiver and about $(1+3)$ dB = 4 dB in dual-filter FSK receiver.

(b) WORST-CASE PROBABILITY OF ERROR p_w

As mentioned earlier, incoherent ASK and FSK systems exhibit same features as far as the worst-case probability of error p_w in concerned provided frequency deviation is sufficiently large ($f_d T > 5$) and $d_1(f_d T) = d_{0,ASK}$ in case of a transmitted symbol sequence $\cdots 1110111 \cdots$ (Fig. 11.20). Therefore, the same formulas can be used to evaluate p_w. Moreover, the BER curves in Fig. 11.15 (incoherent ASK heterodyne system) are also valid for an incoherent FSK heterodyne system with single-filter detection.

In a dual-filter detection, maximum decision must be used instead of simple threshold decision. For this, probability density functions $f_{d1}(d)$ and $f_{d0}(d)$ of the worst-case sample values $d_1 > 0$ and $d_0 > 0$ are to be determined. We obtain

$$p_w = \frac{1}{2}\left[p\left(d_1 > d_0 | q_v = "0"\right) + p\left(d_0 > d_1 | q_0 = "1"\right)\right] = p\left(d_0 > d_1 | q_0 = "1"\right)$$

$$= \int_{d_1=0}^{+\infty} \int_{d_0=d_1}^{+\infty} f_{d1}(d_1)\, f_{d0}(d_0)\, dd_0\, dd_1 \tag{11.115}$$

This equation first assumes same occurrence probabilities for both symbols "0" and "1" and second statistical independence of both sample values d_1 ("1"-branch) and d_0 ("0"-branch). Hence, $f_{d1,d0}(d_1, d_0) = f_{d1}(d_1) \cdot f_{d0}(d_0)$. As noise in "0"-branch and "1"-branch are characterized by different noise spectra, this condition is usually fulfilled provided discriminator filters do not exhibit spectral overlapping. In this case, a dual-filter FSK receiver offers a sensitivity gain of 3 dB in comparison to a single-filter FSK receiver. Evaluation of Eq. (11.115) requires the probability density functions $f_{d1}(d)$ and $f_{d0}(d)$, which can be obtained by applying one of the methods discussed in Section 11.3.1.

Using same approximations as given in Section 11.3.1 for special sequences viz., all "0" and all "1" (i.e., no intersymbol interference), Eq. (11.115) simplifies to

$$p_w = \frac{\sigma_0^2}{\sigma_1^2 + \sigma_0^2}\exp\left(-\frac{C_1^2}{2\left(\sigma_1^2 + \sigma_0^2\right)}\right) \quad \text{where } \sigma_0^2 = \sigma^2 \text{ and } \sigma_1^2 = \sigma^2 + \sigma_{C_1}^2 \tag{11.116}$$

The physical quantities required in the evaluation of Eq. (11.116) can be derived as in Section 11.3.1 by taking all "0" and all "1". The influence of phase noise is included in the amplitude C_1 as well as in the standard deviation σ_{C1}. As described in Section 11.3.1, C_1 represents the sampled detected signal d_1 (i.e., $q_v = "1"$) provided receiver additive Gaussian noise is neglected ($\sigma = 0$).

Assuming that $\sigma_1 = \sigma_0 = \sigma$ which represents a system without any laser phase noise, Eq. (11.115) is simply the well-known BER formula for an electrical FSK transmission system.

(iii) FREQUENCY DEVIATION ESTIMATION

In the realization of incoherent optical FSK heterodyne systems, one question of primary importance arises: What frequency deviation f_d is required to obtain best system performance? To answer this question, a simple estimation is presented and used in this Subsection. Two conditions are, however, required to use this estimation: first we have to provide an ideal frequency discriminator characterized by a linear input-output transfer function (Fig. 11.21) and second we have to neglect receiver additive Gaussian noise i.e., $x = y = 0$ or $\sigma = 0$. Thus, the system is only disturbed by laser phase noise. Influence of additive noise may at least be reduced partly either by an amplitude limiter or appropriate large signal power level.

Transfer function of an ideal frequency discriminator including the corresponding frequencies for both binary symbols "0" and "1" is shown in Fig. 11.21.

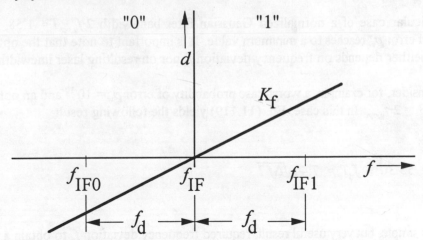

Fig. 11.21: Idealized transfer function of frequency discriminator

At the output of the frequency discriminator, detected signal

$$d(t) = K_\omega\left[\omega_s(t) + \omega(t)\right] = K_\omega\left[\sum_{\upsilon=-\infty}^{+\infty} 2\pi f_d \left(2s_\upsilon - 1\right) \text{rect}\left(\frac{t-\upsilon T}{T}\right) + \omega(t)\right] \qquad (11.117)$$

is obtained. In the above equation, $K_\omega = 2\pi K_f$ represents the gradient of the discriminator transfer function and $\omega_s(t)$ and $\omega(t)$ the angular frequencies of the transmitted information and laser frequency noise respectively. As described in Section 3.2.4, laser frequency noise is a white random process with constant power spectral density

$$G_\omega = 2\pi\Delta f = 2\pi\left(\Delta f_t + \Delta f_l\right) \tag{11.118}$$

Here, Δf_t and Δf_l represent the linewidth of both transmitter and local lasers. When the worst-case pattern $\cdots 0001000\cdots$ and $\cdots 1110111\cdots$ are transmitted and noise is reduced by a Gaussian filter, detected signal given in Eq. (11.117) results in a worst-case probability of error

$$p_w = Q\left[\frac{f_d T\left(1 - 4Q\left(\sqrt{2\pi}f_g\,T\right)\right)}{\sqrt{\Delta f\,T}\sqrt{2}f_g\,T}\right] \tag{11.119}$$

In the particular case of a normalized Gaussian filter bandwidth $2 \cdot f_{g,opt} T \approx 1.58$, worst-case probability of error p_w reaches to a minimum value. It is important to note that the optimum filter bandwidth neither depends on frequency deviation f_d nor on resulting laser linewidth Δf.

Let us consider, for example, a worst-case probability of error $p_w = 10^{-10}$ and an optimum filter bandwidth $B = 2 \cdot f_{g,opt}$. In this case, Eq. (11.119) yields the following result:

$$\frac{\left(f_d T\right)^2}{\Delta f\,T} \approx 55.30 \rightarrow f_d T \approx 7.44\sqrt{\Delta f\,T} \tag{11.120}$$

By using this simple, but very useful result, required frequency deviation f_d to obtain a worst-case probability of error $p_w = 10^{-10}$ can easily be estimated as a function of resulting laser linewidth Δf and bit rate $R = 1\text{bit}/T$.

Example 11.3

For an incoherent optical FSK heterodyne system with bit rate $R = 622.08$ Mbit/s (i.e., STM2 of SDH) and resulting laser linewidth $\Delta f = 56$ MHz, frequency deviation f_d should be at least $f_d = 2.23/T \approx 1.39$ GHz.

It must be noted that Eq. (11.120) has to be handled carefully. This equation suggests that each laser linewidth Δf can be accepted provided that frequency deviation f_d is chosen appropriately high enough. If laser phase noise becomes large, then the undesired interaction between phase noise and amplitude of the IF signal becomes more and more dominant. In this case, phase noise results in amplitude noise in addition.

In the particular case of an ideal frequency detection as considered above, amplitude fluctuations do not influence the detection process at all. If a real detection process is considered; for example, a dual-filter circuit with an envelope demodulator in each branch, amplitude (envelope) fluctuations directly disturb the detection process. This fact, however, is not considered in Eq. (11.120). Nevertheless, the above equation represents a very useful practical tool for a fast estimation of the required frequency deviation f_d, under the conditions and restrictions mentioned above.

When the actual permissible laser linewidth is to be determined, BER curve $p_w(P_r, \Delta f)$ has to be computed as a function of received optical power level P_r and resulting laser linewidth Δf of both transmitter and local lasers. Assuming, for example, an incoherent FSK heterodyne system with a frequency deviation such as given in Eq. (11.120), BER curves are approximately same as the BER curves for an incoherent ASK heterodyne system (Fig. 11.22 and [28]).

(vi) EVALUATION AND DISCUSSION

Worst-case probability of error p_w as a function of received optical power P_r, resulting laser linewidth Δf and frequency deviation f_d is shown in Fig. 11.22. It should be mentioned that this figure is obtained by means of numerical methods and computer program [5].

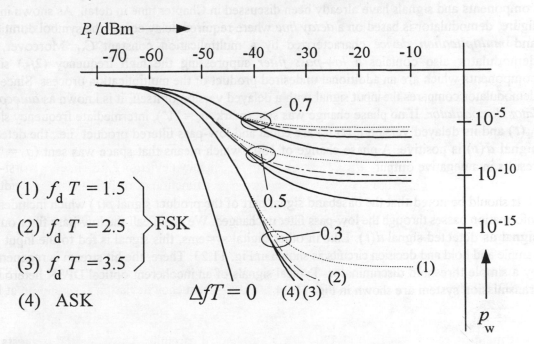

Fig. 11.22: Worst-case probability of error p_w in FSK heterodyne receiver with incoherent single filter detection

It becomes clear from the above figure that ASK heterodyne (curve 4) and FSK heterodyne system (curves 1 to 3) do agree very well provided frequency deviation f_d is chosen large enough (for example, $f_d T = 3.5$). In the special case of an infinite frequency deviation ($f_d \to \infty$) with $d_1(f_d T) = d_{0,ASK}$ (for a transmitted single-"0", see Fig. 11.20), both systems are absolutely identical as far as worst-case BER p_w is concerned.

11.3.3   DPSK HETERODYNE SYSTEM

Differential phase shift keying (DPSK) is a variant of PSK. In a PSK system, information is located in the phase itself, whereas in a DPSK system information is located in the phase difference of two neighbouring symbols. Thereby, symbol "0" is represented by a phase change of 180° and symbol "1" by a phase change of 0° i.e., phase remains unchanged. Each change in phase of 180° (i.e., π) reverse the polarity of received optical signal and, hence, IF signal also. Thus, the transmitted information is included in the change of sign of the received optical signal. Thereby, a change in sign represents a symbol "0", whereas no change in sign represents a symbol "1". Of course, the assignment of symbol and change in sign is arbitrary and can be chosen by the system designer. Therefore, the assignment used above may also be turned over i.e., "0" ≙ 0° and "1" ≙ 180°. In this Section, first assignment will only be used.

The block diagram of a typical DPSK transmission system is shown in Fig. 11.23. The main components and signals have already been discussed in Chapter nine in detail. As shown in this figure, demodulator is based on a *delay line* where required delay equals the symbol duration T and *multiplication device* characterized by a multiplication constant C_M. Moreover, the demodulator also contains a *low-pass filter* suppressing the high frequency ($2f_{IF}$) signal components which are an additional undesired product of the multiplication process. Since this demodulator compares the input signal with a delayed version of itself, it is known as *autocorrelator demodulator*. If no phase change was sent (mark, $q_v = $ "1"), intermediate frequency signal $i_{IF}(t)$ and its delayed version $i_{IF}(t-T)$ are equal and low-pass filtered product i.e., the detected signal $d(t)$ is positive. A phase change of 180°, which means that space was sent ($q_v = $ "0"), results in a negative output.

It should be noted that the baseband signal part of the product signal $p(t)$ which includes the information passes through the low-pass filter unchanged. We again call the low-pass filter output signal as detected signal $d(t)$. Like in other digital systems, this signal is fed to the input of a sample and hold and decision circuits as shown in Fig. 11.23. There, the bit stream is regenerated by a simple threshold discriminator. Typical signals of an incoherent optical DPSK heterodyne transmission system are shown in Fig. 11.24.

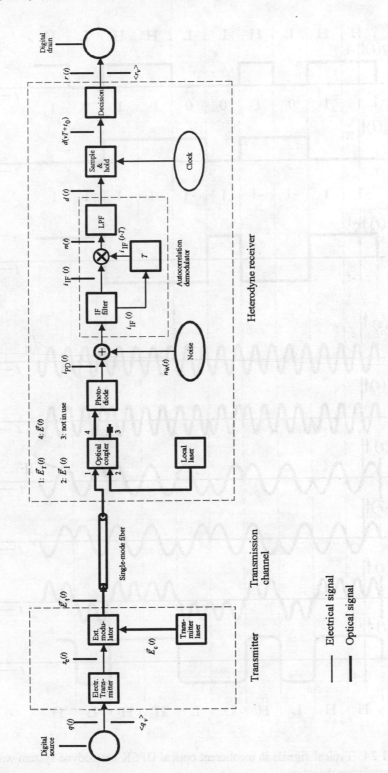

Fig. 11.23: Coherent optical communication system with heterodyne receiver employing autocorrelation detection

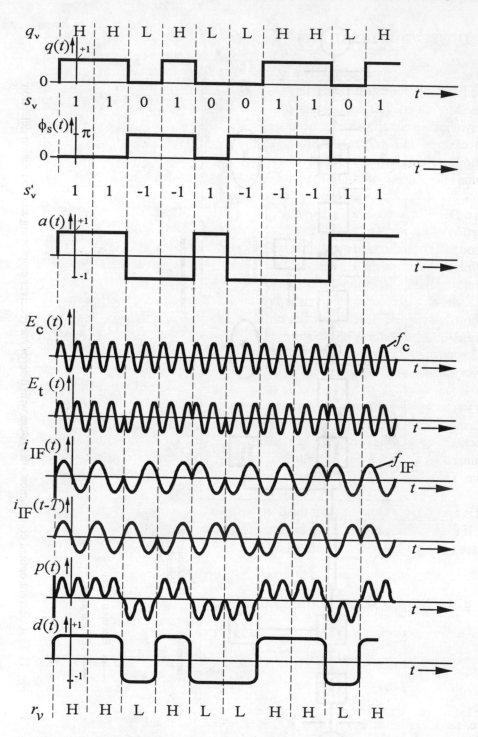

Fig. 11.24: Typical signals in incoherent optical DPSK heterodyne system with autocorrelation detection

(i) DETECTED SIGNAL $d(t)$ AND SAMPLE VALUES $d(\upsilon T + t_o)$

In order to simplify calculation, we consider that the *IF filter* does not influence the useful part of the IF filter input signal. Thus, this filter only reduces the effect of receiver additive Gaussian noise. Moreover, the input signal passes through the filter without any change. Therefore, intersymbol interference do not occur and phase noise remains unchanged.

In case of an incoherent ASK heterodyne receiver, these special conditions result in an envelope or detected signal $d(t)$ which is not disturbed by phase noise. Therefore, BER also become independent of laser phase noise (compare with Section 11.3.1 also).

In a DPSK system, laser phase noise disturbs the detection process twice: First is the *direct distortion* of the signal phase due to additive phase noise and second is *indirect distortion* due to the coupling of phase and amplitude within the IF signal.

Direct distortion is always existent when laser phase is used to transmit information. Therefore, direct distortion can be observed in PSK as well as in DPSK systems. The task of a DPSK demodulator such as given in Fig. 11.23 is to compare the phase of two neighbouring symbols. As DPSK demodulator is not able to distinguish signal phase (which includes the information) and additive phase noise, this detection process is, therefore, directly disturbed.

Like in an ASK heterodyne receiver, *indirect distortion* occurs due to undesired interactions between phase noise and amplitude (envelope) of the IF signal $i_{IF}(t)$. As mentioned in Section 11.3.1, this phase noise-to-amplitude conversion is a result of required bandlimiting of the IF filter to reduce the influence of receiver additive Gaussian noise. With practical system parameters, influence of phase-amplitude coupling is very critical in ASK system with envelope detection, whereas it is negligible in a DPSK system. Here, the direct distortion is absolutely dominant [31]. Therefore, it is allowed to ignore the indirect distortion as already mentioned above. Thus, the task of IF filter is to reduce the additive Gaussian noise only.

It is assumed for simplicity that the IF (intermediate frequency) is an integer multiple of the bit rate. If the ratio is fractional, it can be regarded as a constant phase deviation as discussed below. In case of an integer, detected signal is

$$d(t) = \frac{1}{2}C_M\left[\hat{i}_{PD}\cos\left(\phi_s(t)+\phi(t)\right) + x(t)\right]\left[\hat{i}_{PD}\cos\left(\phi_s(t-T)+\phi(t-T)\right) + x(t-T)\right]$$

$$+ \frac{1}{2}C_M\left[-\hat{i}_{PD}\sin\left(\phi_s(t)+\phi(t)\right) + y(t)\right]\left[-\hat{i}_{PD}\sin\left(\phi_s(t-T)+\phi(t-T)\right) + y(t-T)\right]$$

$$(11.121)$$

Here, random processes $x(t)$ and $y(t)$ are inphase and quadrature components of the receiver additive Gaussian process (i.e., the shot noise of photodiodes and thermal noise of amplifiers and resistors), $\phi(t)$ the resulting laser phase noise of both transmitter and local lasers and

$$\phi_s(t) = \pi\left(1 - s_\upsilon\right)\mathrm{rect}\left[\frac{t - \upsilon T}{T}\right] \qquad s_\upsilon \in \{0,\ 1\} \tag{11.122}$$

the signal phase which includes the transmitted information (compare with Eq. 9.41). The assignment between coefficients s_υ and transmitted binary symbols $q_\upsilon \in \{"0", "1"\}$ becomes clear from Figs. 9.6 and 11.24. Instead of using the so-called phase representation given in Eq. (11.121), detected signal can also be described by making use of the following amplitude representation:

$$d(t) = \frac{1}{2}C_M \left[\hat{i}_{PD}\ a(t)\ \cos(\phi(t)) + x(t) \right]\left[\hat{i}_{PD}\ a(t-T)\ \cos(\phi(t-T)) + x(t-T) \right]$$

$$+ \frac{1}{2}C_M \left[-\hat{i}_{PD}\ a(t)\ \sin(\phi(t)) + y(t) \right]\left[-\hat{i}_{PD}\ a(t-T)\ \sin(\phi(t-T)) + y(t-T) \right] \tag{11.123}$$

where

$$a(t) = \acute{s}_\upsilon \mathrm{rect}\left[\frac{t - \upsilon T}{T}\right] \qquad \acute{s}_\upsilon = 2s_\upsilon - 1 \qquad \acute{s}_\upsilon \in \{-1,\ 1\} \tag{11.124}$$

Without loss of generality, dimensionless product $C_M\hat{i}_{PD}$ can be taken as unity. If we use again same substitutions and abbreviations as in the previous Sections, then the normalized sample value $d := d(\upsilon T + t_0)/\hat{i}_{PD}$ is given by

$$d = \frac{1}{2}\left[a\ \cos(\phi) + x\right]\left[a_T\ \cos(\phi_T) + x_T\right]$$

$$+ \frac{1}{2}\left[-a\ \sin(\phi) + y\right]\left[-a_T\ \sin(\phi_T) + y_T\right] = \frac{1}{2}C_T C + \frac{1}{2}S_T S \tag{11.125}$$

Here, cosine term, delayed cosine term, sine term and delayed sine term are replaced by the abbreviations C, C_T, S and S_T respectively. The suffix T always represents the delayed version. Again, normalization is performed with respect to the amplitude $\hat{i}_{PD}$ of photodiode current. If, for example, all six sources of noise x, x_T, y, y_T, ϕ and ϕ_T are neglected, then Eq. (11.125) simplifies to

$$d = \frac{1}{2}aa_{\mathrm{T}} \tag{11.126}$$

The principle function of a DPSK demodulator i.e., the detection of change in sign of the sampled detected signal d becomes clear from the above simple equation. The decision circuit always generates a received symbol r_{υ} = "1" when d is positive (no change in sign i.e, $a = a_{\mathrm{T}}$) and a symbol r_{υ} = "0" when d is negative (change in sign i.e., $a = -a_{\mathrm{T}}$). This operation is illustrated in Fig. 11.24 above by mean of some typical signals.

Considering again Eq. (11.125), it becomes evident that the sampled detected signal d depends on six (!) sources of noise: x, x_{T}, y, y_{T}, ϕ and ϕ_{T}. Thus, the calculation of pdf $f_{\mathrm{d}}(d)$ requires the evaluation of a sixfold integral. Moreover, calculation of BER which is determined by a certain area under the pdf $f_{\mathrm{d}}(d)$ will require the evaluation of a sevenfold integral! Even with a fast computer program, computer run time becomes extremely large.

Assuming, for example, that only 100 calculation steps are used for one integral (i.e., the effective area below the pdf is divided into 100 single area sections) and each integral may require a computer execution time of 0.01 s, overall computer run time will be $100^6 \cdot 0.01$ s = 10^{10} s = 317 years. Hence, it becomes clear that the BER calculation must really be performed by means of some alternate methods.

(ii) TWO-FILTER REPRESENTATION OF DPSK DEMODULATOR

A different, but functionally equivalent demodulator provided the linear filters are chosen properly is shown in Fig.11.25. It consists of two linear filters followed by envelope detectors. The decision depends on which envelope detector output is high. A very similar system consisting of two band-pass filters is widely used for an incoherent FSK demodulation (Section 11.3.2). It is easier to build this system using current microwave technology since no microwave frequency multiplier is required. In order to make the function of the two-filter representation clear, we first modify Eq. (11.125) by means of some simple mathematical operations. We maintain that

$$d = \frac{1}{8}\left(\left(C+C_{\mathrm{T}}\right)^2 + \left(S+S_{\mathrm{T}}\right)^2 - \left(C-C_{\mathrm{T}}\right)^2 - \left(S-S_{\mathrm{T}}\right)^2\right) = \frac{1}{8}\left(r_{\Sigma}^2 - r_{\Delta}^2\right) \tag{11.127}$$

where

$$r_{\Sigma} = \sqrt{\left(C + C_{\mathrm{T}}\right)^2 + \left(S + S_{\mathrm{T}}\right)^2} \tag{11.128}$$

and

$$r_\Delta = \sqrt{\left(C - C_T\right)^2 + \left(S - S_T\right)^2}$$ (11.129)

are used as parameters which physically represent the normalized and sampled envelopes of IF sum signal $\Sigma(t) = i_{IF}(t) + i_{IF}(t-T)$ and IF difference signal $\Delta(t) = i_{IF}(t) - i_{IF}(t-T)$ respectively.

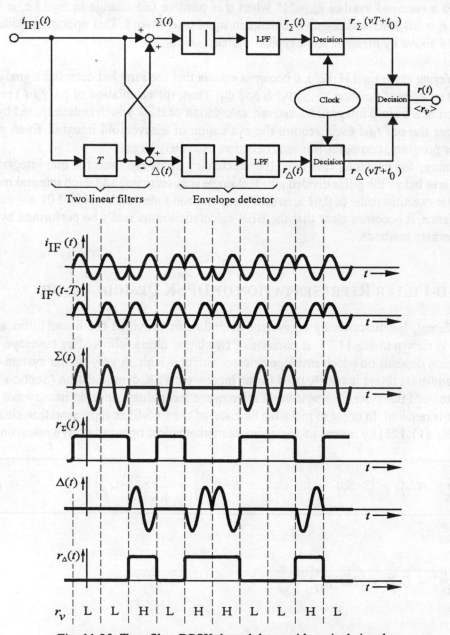

Fig. 11.25: Two-filter DPSK demodulator with typical signals

By using Eq. (11.127), threshold decision $d > 0$ or $d < 0$ is transferred to the maximum decision $r_\Sigma > r_\Delta$ (equal to $d > 0$) or $r_\Sigma < r_\Delta$ (equal to $d < 0$). Since sum term r_Σ and difference term r_Δ are Ricean distributed (due to its mathematical structure), calculation of the BER now becomes much easier as shown below. This is the only and main advantage of using Eq. (11.127).

As mentioned above Eq. (11.127) can be physically described by a two-filter representation of the DPSK autocorrelation demodulator. This representation is not only a powerful mathematical model, but also a realizable alternate circuit for the conventional autocorrelation demodulator. To make the physical operation of the two-filter representation more clear, Fig. 11.25 shows typical signals at different points. Obviously, if a mark ($q_v = $ "1") is sent, the difference $\Delta(t)$ and its envelope $r_\Delta(t)$ are zero and the envelope $r_\Sigma(t)$ of the sum $\Sigma(t)$ is positive, while a space results in the reverse situation (Fig. 11.25).

When we consider the front end of the two-filter representation in detail, it becomes clear that in this front end, sum component and difference component actually represent two linear filters: one is the sum filter $H_\Sigma(f)$ generating the sum signal $\Sigma(t)$ and the other is the difference filter $H_\Delta(f)$ generating the difference signal $\Delta(t)$ as described above. Hence, mathematical representation of filter transfer functions are

$$H_\Sigma(f) = 1 + e^{-j2\pi fT} \tag{11.130}$$

and

$$H_\Delta(f) = 1 - e^{-j2\pi fT} \tag{11.131}$$

Two-filter detection has already been considered in Section 11.3.2 (incoherent FSK heterodyne system), where this detection scheme was referred as dual-filter detection. However, both the filters were band-pass filters matched to possible IF frequencies $f_{IF} - f_d$ for symbol "0" and $f_{IF} + f_d$ for symbol "1." In case of a DPSK, both linear filters are matched in principle to "change in sign" and "no change in sign". This is realized by generating the sum signal and the difference signal from the filter input signal and its delayed version.

Assuming mark ($q_v = $ "1") is sent, envelope $r_\Sigma(t)$ in the sum channel arises from twice the amplitude of the sum signal $\Sigma(t) = 2i_{IF}(t)$. Moreover, the noise power becomes doubled i.e., $2N$ can be measured as compared to the noise power of the IF signal $i_{IF}(t)$. It may be noted that this is only true if the noise $n(t)$ and its delayed version $n(t-T)$ are largely uncorrelated, which requires that the IF filter bandwidth is not less than approximately 78% of the bit rate [11, 31]. The modulation in the difference channel is cancelled ($\Delta(t) = 0$), which results in a Rayleigh distributed sampled envelope r_Δ. Owing to the symmetry of the receiver (see Fig. 11.25), same discussion is valid when a space is sent.

The sampled envelopes r_Σ and r_Δ are statistically independent. If the noise is a stationary Gaussian random process, then this can be proved by transforming a four dimensional pdf of the noise (inphase and quadrature components delayed and undelayed) to r_Σ and r_Δ. This transformed pdf can be separated into the product of two Ricean distributions.

Following table summarizes the relationship of the sample value d (in case of autocorrelation detection) and the envelopes r_Σ and r_Δ (in case of two-filter detection) for transmitted symbol q_υ.

Table 11.4: Relationship of sample values d, r_Σ, r_Δ and transmitted symbol q_υ

Transmitted symbol q_υ	Autocorrelation detection with threshold decision	Two-filter detection with maximum decision
"1"	$d > 0$	$r_\Sigma > r_\Delta$
"0"	$d < 0$	$r_\Sigma < r_\Delta$

(iii) PROBABILITY OF ERROR

When a mark is sent (q_υ = "1"), bit error (i.e., $r_\upsilon \neq q_\upsilon$) will occur if the receiver additive Gaussian noise causes a greater output in the difference channel than in the sum channel i.e., $p = p(r_\Delta > r_\Sigma)$. Instead, if autocorrelation detection is used in the receiver, then a bit error occurs for $d > 0$. For space, calculation can be made in the same way and it yields the same result. If mark and space are sent with equal probability, it is sufficient to determine the BER for only one of them as given below. Neglecting intersymbol interference, probability of error for symbol q_υ is independent of neighbouring symbols. In this particular case, average probability of error is

$$P_a = \frac{1}{2}\left(p_0 + p_1\right) \tag{11.132}$$

Here, p_0 and p_1 represent the probabilities of wrongly detected symbol "0" and "1" respectively. It should be noted that above equation presumes same occurrence probabilities for both binary symbols "0" and "1". The probabilities of error p_0 and p_1 can be calculated as

$$P_0 = p\left(d > 0 | q_\upsilon = "0"\right) = p\left(r_\Sigma > r_\Delta | q_\upsilon = "0"\right) = \int\limits_{0}^{+\infty} \int\limits_{r_\Sigma = r_\Delta}^{+\infty} f_{r_\Sigma, r_\Delta}\left(r_\Sigma, r_\Delta\right) dr_\Sigma \, dr_\Delta \tag{11.133}$$

and

$$p_1 = p\left(d < 0 \,|\, q_\upsilon = "1"\right) = p\left(r_\Sigma < r_\Delta \,|\, q_\upsilon = "1"\right) = \int\limits_0^{+\infty} \int\limits_{r_\Delta = r_\Sigma}^{+\infty} f_{r_\Sigma, r_\Delta}\left(r_\Sigma, r_\Delta\right) \, dr_\Delta \, dr_\Sigma \tag{11.134}$$

The main problem in the calculation of p_0 and p_1 is to determine the two-dimensional joint density function required in the above integrals. Since both the envelopes r_Σ and r_Δ are statistically dependent random variables, calculation of this function becomes rather comprehensive. The problem, however, can be solved by using conditional pdfs. For this, random phases ϕ and ϕ_T will be considered as constants first. In that case, both the random envelopes r_Σ and r_Δ given in Eqs. (11.128) and (11.129) are characterized by a Ricean distribution. This distribution is already known from Section 11.3.1 (incoherent ASK heterodyne system), where the envelope of an ASK signal also had Ricean distribution. As in Section 11.3.1, we obtain

$$f_{r_\Sigma|\Delta\phi}\left(r_\Sigma, \Delta\phi\right) = \frac{r_\Sigma}{2\sigma^2} \exp\left(-\frac{r_\Sigma^2 + 4\sin^2(\Delta\phi/2)}{4\sigma^2}\right) J_0\left(\frac{r_\Sigma \sin(\Delta\phi/2)}{\sigma^2}\right) \tag{11.135}$$

and

$$f_{r_\Delta|\Delta\phi}\left(r_\Delta, \Delta\phi\right) = \frac{r_\Delta}{2\sigma^2} \exp\left(-\frac{r_\Delta^2 + 4\cos^2(\Delta\phi/2)}{4\sigma^2}\right) J_0\left(\frac{r_\Delta \cos(\Delta\phi/2)}{\sigma^2}\right) \tag{11.136}$$

Here,

$$\sigma = \sigma_x = \sigma_y = \frac{\sigma_n}{\hat{i}_{PD}} = \frac{\sigma_{het}}{\hat{i}_{PD}} \tag{11.137}$$

represents the normalized standard deviation of the Gaussian random processes x, x_T, y and y_T with zero mean and

$$\Delta\phi = \phi - \phi_T \tag{11.138}$$

represents the phase difference between current phase ϕ and delayed phase ϕ_T. As $\Delta\phi$ depends on the two random variables ϕ and ϕ_T, it is, of course, a random variable itself (Section 3.2.3).

Considering the conditional probability density functions given in Eqs. (11.135) and (11.136) in detail, it becomes clear that these functions only depend on the phase difference $\Delta\phi$ and not on the absolute phases ϕ and ϕ_T. It is of significant advantage in the calculation of BER. It should be noted that if there is a phase difference $\Delta\phi$ in addition to the $0/\pi$ modulation phase, sum envelope is not twice the transmission amplitude and the difference is no longer zero.

When the Gaussian distributed random variables x, x_T, y and y_T are uncorrelated and, hence, statistically independent, random envelopes r_Σ and r_Δ of sum and difference signals are also statistically independent [31, 36]. The conditions which are required to obtain statistical independence will be discussed later. In case of statistical independence, conditional joint density function

$$f_{r_\Sigma, r_\Delta|\Delta\phi}\left(r_\Sigma, r_\Delta, \Delta\phi\right) = f_{r_\Sigma|\Delta\phi}\left(r_\Sigma, \Delta\phi\right) f_{r_\Delta|\Delta\phi}\left(r_\Delta, \Delta\phi\right) \qquad (11.139)$$

is simply obtained by the product of both the conditional density functions given in Eqs. (11.135) and (11.136). Finally, we have to consider again the statistics of the phase difference $\Delta\phi$. As derived in Section 3.3.3, phase difference $\Delta\phi$ is stationary and Gaussian distributed with zero mean. By taking into account these characteristics, probabilities of error p_0 and p_1 are now given by

$$p_0 = \int_{-\infty}^{+\infty} \int_0^{+\infty} \int_{r_\Sigma = r_\Delta}^{+\infty} f_{r_\Sigma|\Delta\phi}\left(r_\Sigma, \Delta\phi\right) f_{r_\Delta|\Delta\phi}\left(r_\Delta, \Delta\phi\right) f_{\Delta\phi}(\Delta\phi) \; dr_\Sigma \; dr_\Delta \; d\Delta\phi$$

$$\qquad (11.140)$$

$$= \int_{-\infty}^{+\infty} \tilde{p}_0(\Delta\phi) \, f_{\Delta\phi}(\Delta\phi) \; d\Delta\phi$$

and

$$p_1 = \int_{-\infty}^{+\infty} \int_0^{+\infty} \int_{r_\Delta = r_\Sigma}^{+\infty} f_{r_\Sigma|\Delta\phi}\left(r_\Sigma, \Delta\phi\right) f_{r_\Delta|\Delta\phi}\left(r_\Delta, \Delta\phi\right) f_{\Delta\phi}(\Delta\phi) \; dr_\Delta \; dr_\Sigma \; d\Delta\phi$$

$$\qquad (11.141)$$

$$= \int_{-\infty}^{+\infty} \tilde{p}_1(\Delta\phi) \, f_{\Delta\phi}(\Delta\phi) \; d\Delta\phi$$

As in the previous Sections, sign "~" again represents the probability of error when phase noise is neglected. However, a constant phase distortion (i.e., an undesired phase shift) is allowed.

Due to the symmetry of two-filter demodulator, p_0 and p_1 are equal. Therefore, average probability of error in absence of intersymbol interference is given by

$$p_a = \frac{1}{4\sqrt{2\pi}\sigma^4\sigma_{\Delta\phi}} \int\limits_{-\infty}^{+\infty} \int\limits_{0}^{+\infty} \int\limits_{r_\Delta=r_\Sigma}^{+\infty} r_\Sigma r_\Delta \exp\left(-\frac{r_\Sigma^2+r_\Delta^2+4}{\sigma^2}\right) \exp\left(\frac{\Delta\phi^2}{2\sigma_{\Delta\phi}^2}\right)$$

$$\cdot J_0\left(\frac{r_\Delta\cos(\Delta\phi/2)}{\sigma^2}\right) J_0\left(\frac{r_\Sigma\sin(\Delta\phi/2)}{\sigma^2}\right) dr_\Sigma \, dr_\Delta \, d\Delta\phi \qquad (11.142)$$

$$= \int\limits_{-\infty}^{+\infty} \tilde{p}_a(\Delta\phi) \, f_{\Delta\phi}(\Delta\phi) \, d\Delta\phi$$

In the derivation of above equation, Eqs. (11.132), (11.135), (11.136) and (11.140) have been used. To evaluate the average probability of error p_a, variance $\sigma_{\Delta\phi}^2$ of the random phase difference $\Delta\phi$ is required. Using the results of Section 3.2.3, we can write

$$\sigma_{\Delta\phi}^2 = 2\pi\Delta f T \qquad (11.143)$$

where T is the bit duration and Δf the resulting laser linewidth of both transmitter and local lasers. It becomes clear from Eq. (11.142) that the calculation of probability of error is now reduced to a threefold integral instead of a sevenfold integral as described above. An additional simplification is, however, only possible under special conditions. Evaluation of the threefold integral given in Eq. (11.142) by means of computer, run time of about $100^2 \cdot 0.01$ s $= 100$ s is merely required. Here, same numerical values as in Subsection (i) are used. In contrast to 317 years for sevenfold integral, this is truly a very large reduction in the computer run time.

In a system without any phase noise or phase distortion (i.e., $\Delta\phi = 0$), Eq. (11.142) results in the well-known formula of conventional digital DPSK transmission system [36]:

$$p_a\Big|_{\Delta\phi=0} = \tilde{p}_a(0) = \frac{1}{2} e^{-\frac{1}{2\sigma^2}} \qquad (11.144)$$

It should be remembered that the standard deviation σ of the additive Gaussian noise is normalized to the amplitude $\hat{\imath}_{PD}$ of the photodiode current. Therefore, the variance σ^2 actually represents the reciprocal of signal-to-noise ratio.

Considering the conditional probability of error $\tilde{p}_a(\Delta\phi)$ in more detail, some important features of an optical DPSK heterodyne system become evident. For this reason, Fig. 11.26 shows $\tilde{p}_a(\Delta\phi)$ as a function of a constant phase shift $\Delta\phi$.

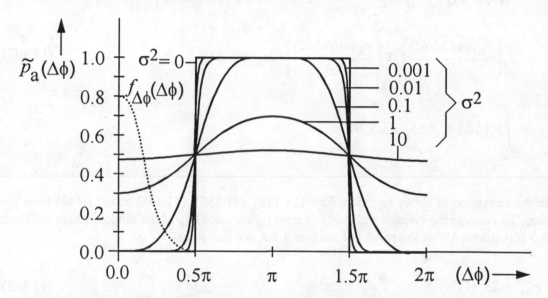

Fig. 11.26: Probability of error $\tilde{p}_a(\Delta\phi)$ in DPSK heterodyne receiver as a function of constant phase shift $\Delta\phi$

When the normalized standard deviation σ of the receiver additive Gaussian noise is low (i.e., the signal-to-noise ratio is high), probability of error is also low provided phase shift $\Delta\phi$ is small enough. As the undesired phase shift $\Delta\phi$ increases, probability of error also increases. In the particular case of $\Delta\phi = \pi/2$ (i.e., 90°), probability of error reaches $\tilde{p}_a(\pi/2) = 0.5$. Thereby, the signal components of two-filter demodulator are completely extinguished and only noise is given to the input of maximum decision circuit. If phase shift is above $\pi/2$, but less than π, then probability of error increases further since both the binary symbols "0" and "1" are interchanged now. In the special case of $\Delta\phi = \pi$ (i.e., 180°), probability of error reaches maximum i.e., $\tilde{p}_a(\pi) = 1$. As each symbol is wrongly detected now, a simple symbol converter ("0"→"1", "1"→"0") yields again an errorless detection with $\tilde{p}_a(\pi) = 0$. Hence, the actual maximum probability of error is 0.5 instead of 1. If $\Delta\phi > \pi$, then $\tilde{p}_a(\Delta\phi)$ again decreases.

It becomes clear from Fig. 11.26 that $\tilde{p}_a(\Delta\phi)$ is symmetrical about $\Delta\phi = 0$ (or multiple of π) and periodical with period 2π. Symmetry and periodicity are caused by both the periodic functions $\sin(\Delta\phi)$ and $\cos(\Delta\phi)$ which are included in the Eq. (11.142). In addition, we have to take into account that the Bessel function $J_0(x)$ is an even function.

In the following, we consider again the statistics of the phase difference $\Delta\phi$ which actually is a random variable due to laser phase noise. This random variable is completely described by a Gaussian pdf $f_{\Delta\phi}(\Delta\phi)$ as shown in Fig. 11.26 (dotted line). According to Eq. (11.142), conditional probability of error $\tilde{p}_a(\Delta\phi)$ must be weighted by $f_{\Delta\phi}(\Delta\phi)$ to obtain the probability of error p_a by integration. In contrast to $\tilde{p}_a(\Delta\phi)$, probability of error p_a now also contains the statistics and, hence, the influence of phase noise. It should be noted that the pdf $f_{\Delta\phi}(\Delta\phi)$ plotted in Fig. 11.26 has been chosen unrealistically broad to obtain a better description. Actually, this pdf is much narrower, especially if a probability of error $p_a = 10^{-10}$ is demanded and would nearly appear as Dirac delta function at $\Delta\phi = 0$.

The probability of error p_a versus received optical power P_r with normalized laser linewidth $\Delta f T$ as a parameter is shown in Fig. 11.27.

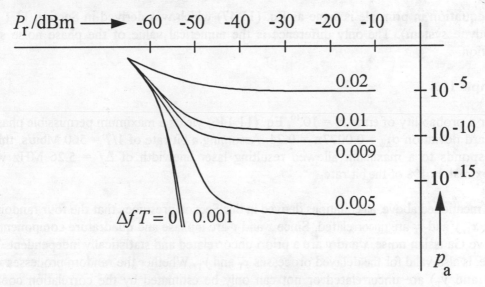

Fig. 11.27: Probability of error $p_a \approx p_w$ in DPSK heterodyne system

It becomes clear from Fig. 11.27 that a DPSK heterodyne system exhibits same characteristic features as ASK and PSK homodyne systems or ASK and FSK heterodyne systems, which all have already been discussed in the previous Sections. Thus, BER curves first fall rapidly, then follow a bend and finally reach to a saturation called the *error rate floor* with increasing optical power levels. Physical description of this typical behaviour of all coherent optical communication systems has been given in Section 11.1.1 (see discussion on Fig. 11.5).

An error rate floor can always be observed when optical input power P_r is high or when standard deviation σ of the additive Gaussian receiver noise is low. If the laser phase noise is low ($\sigma_{\Delta\phi} < 0.5$), which is usually fulfilled in realized systems, then the probability of bit error at the error rate floor becomes

$$p_a = 2 \int_{\pi/2}^{+\infty} f_{\Delta\phi}(\Delta\phi) \, d\Delta\phi \quad \text{for } \sigma = 0 \tag{11.145}$$

By using Q-function given in Eq. (11.14) and Eq. (11.143), above equation yields

$$p_a = 2Q\left(\frac{\pi}{2\sigma_{\Delta\phi}}\right) = 2Q\left(\frac{1}{2}\sqrt{\frac{\pi}{2\Delta fT}}\right) \quad \text{for } \sigma = 0 \tag{11.146}$$

This equation in principle is same as Eq. (11.47) which was derived in Section 11.1.2 (PSK homodyne system). The only difference is the numerical value of the phase noise standard deviation.

Example 11.4

For a probability of error $p_a = 10^{-10}$, Eq. (11.146) yields a maximum permissible phase noise standard deviation $\sigma_{\Delta\phi} = 0.0077\pi = 0.24$. Assuming a bit rate of $1/T = 560$ Mbit/s, this result corresponds to a maximum allowed resulting laser linewidth of $\Delta f = 5.26$ MHz which is approximately 1% of the bit rate.

As mentioned above, all equations derived in this Section presumes that the four random variables x, x_T, y and y_T are uncorrelated. Since x and y are inphase and quadrature components of the additive Gaussian noise, x and y are a priori uncorrelated and statistically independent. This, of course, is also valid for the delayed processes x_T and y_T. Whether the random processes x and x_T (or y and y_T) are uncorrelated or not can only be estimated by the correlation coefficients $\rho_x = \rho_y = \rho_n$. For an IF filter with Gaussian shape and bandwidth $B = 2f_g$, we obtain

$$\rho_x = \rho_y = \frac{E\{xx_T\} - E\{x\}E\{x_T\}}{\sigma_x \sigma_{x_T}} = \frac{E\{xx_T\}}{\sigma^2} = \frac{R_x(T)}{\sigma^2} = e^{-\frac{\pi}{2}(BT)^2} \tag{11.147}$$

where $R_x(\tau) = R_y(\tau) = R_n(\tau)$ represents the autocorrelation function (acf) of the filtered Gaussian noise $n(t)$ at the output of IF filter. After evaluating the above expression, following numerical values are obtained (Table 11.5). It becomes clear that the Gaussian distributed random variables x and x_T (or y and y_T) can be regarded as practically uncorrelated and statistically independent for normalized bandwidths $BT > 1.5$. This result verifies the BER calculation above, where statistical independence was a main precondition.

Table: 11.5: Correlation coefficients ρ_x and ρ_y for several normalized IF filter bandwidths BT

BT	$\rho_x = \rho_y$	Remark
0	1.000	$x = x_T$ and $y = y_T$
0.5	0.675	
1.0	0.208	
1.5	0.029	$BT \approx B_{opt}T$
2.0	0.002	
5.0	≈ 0	uncorrelated

(iv) EFFECTS OF DEVIATION IN INTERMEDIATE FREQUENCY f_{IF} AND TIME DELAY T_d

The phase of undisturbed (i.e., no phase noise and no additive Gaussian noise) intermediate frequency signal $i_{IF}(t)$ is $2\pi f_{IF}t + \phi_s(t)$. Its phase difference with the delayed version $i_{IF}(t-T_d)$ is $2\pi f_{IF}T_d + \phi_s(t) - \phi_s(t-T_d)$, where $\phi_s(t) - \phi_s(t-T_d)$ is either 0 or π. If phase difference is equal to an integer multiple of 2π, the BER approaches minimum. Therefore, $f_{IF}T_d$ should be an integer. If f_{IF} is not chosen properly or implemented delay T_d differs from bit duration T, then phase error

$$\phi_e = 2\pi f_{IF}T_d - k\,2\pi \qquad k \in \{\cdots -1, 0, +1, \cdots\} \tag{11.148}$$

which is additive to the phase difference $\Delta\phi$ cannot be neglected. Its effect is to shift the centre of the pdf $f_{\Delta\phi}(\Delta\phi)$ and, hence, increase the probability of error or the BER. Since the probability of error $\bar{p}(\phi_e)$ is periodical with period 2π and, in addition, independent of the sign of ϕ_e, following alternate equation for the phase error can be used:

$$\phi_e = (2\pi f_{IF}T_d)\bmod(2\pi) \quad \text{with} \quad 0 \le \phi_e \le 2\pi \tag{11.149}$$

Here, the function $a \bmod b = a - \text{Int}(a/b)\cdot b$ decreases a by an integer multiple (namely $\text{Int}(a/b)$) of b.

It can be easily seen from Fig. 11.26 that a rise in ϕ_e increases the BER even to values greater than 0.5 if ϕ_e approaches π. Reversing the assignment of mark and space could again lower the BER, but this is not the solution if the deviation of f_{IF} or T_d is not known. Thus, it is essential to have an integer ratio of f_{IF} and $f_b = 1/T$ which is easily achieved if f_{IF}/f_b is not too large (moderate requirement on a relatively high accuracy).

As the constant phase error ϕ_e can be regarded as the mean of random variable $\Delta\phi$, the probability of error can easily be calculated by

$$p_a = \int_{-\infty}^{+\infty} \tilde{p}_a(\Delta\phi) \, f_{\Delta\phi}(\Delta\phi-\phi_e) \, d\Delta\phi \qquad\qquad (11.150)$$

By setting $\phi_e = 0$, Eq. (11.150) becomes same as Eq. (11.142).

For the realization of an appropriate intermediate frequency (IF), some important points should be kept in mind: First, IF should be high enough since a low IF results in spectral over-lapping of IF and baseband spectrum. Second, IF should not be too large since realization problems increase with increasing IF. Moreover, realizing an integer ratio f_{IF}/f_b is more convenient when f_{IF}/f_b is not too large as already mentioned above. In addition, we have to keep in mind that photodiodes as well as preamplifiers must be able to follow the high frequency IF signal. For this reason, bandwidth of these components should be approximately $0 < f < f_{IF}+f_{s,max}$, where $f_{s,max}$ is the maximum frequency of the information signal.

(v) INFLUENCE OF IF FILTER ON SYSTEM PERFORMANCE

Considering coherent optical heterodyne system, IF filter should first select the modulated IF signal and second reduce the influence of receiver additive Gaussian noise. For this reason, bandwidth of the IF filter should be as small as possible. On the other hand, a small bandwidth leads to intersymbol interference and, in addition, an undesired coupling of phase and amplitude of the IF signal. Hence, for the realization of an appropriate bandwidth, a compromise is always required. If bandwidth is optimum ($B_{IF,opt} = B_{opt}$), then overall distortion and BER are minimum.

In case of a DPSK heterodyne receiver, numerical calculation of the optimum bandwidth B_{opt} is much more comprehensive than that of ASK or FSK heterodyne receivers. This is due to the comprehensive error rate formula given in Eq. (11.142) which requires the evaluation of a threefold integral even if the influence of the IF filter is neglected. An analytical calculation of B_{opt} is impossible. To determine B_{opt} approximately, we consider that phase noise is low. In that case, optimum bandwidth is approximately given by the optimum bandwidth of a DPSK system which is not disturbed by phase noise i.e., $B_{opt}T = 2f_{g,opt}T \approx 1.58$ for a Gaussian IF filter. As phase-to-amplitude conversion decreases with increasing bandwidth, real optimum bandwidth is only insignificantly higher. Influence of the IF filter can be separated in four parts: Firstly, as the filter bandwidth is reduced, there is an useful proportional reduction of the additive noise power. Secondly, if the cut-off frequency is reduced below the bit rate, signal becomes increasingly distorted by the intersymbol interference. This influence is described by the eye pattern shown in Fig. 11.28, which closes if the bandwidth is reduced. Thirdly, the phase noise is converted into additional amplitude modulation, which increases the additive noise. For typical linewidths of 1% of the bit rate, however, this effect (called the indirect distortion) is rather small. The last effect is the reduction of phase noise. It is seen by simulation experiments that the effective linewidth is reduced somewhat by the IF filter, for example $\approx 80\%$ for $B_{opt}T = 1.58$ [31].

Eye pattern can be measured (which is normally the way in practice) or analytically derived. In DPSK heterodyne system, analytical method is, however, much more difficult than in all other heterodyne or homodyne systems. This is primarily due to the multiplication process in the DPSK autocorrelation demodulator. As clear from Eq. (11.126), normalized detected signal $d(t)$ at the output of the demodulator is generated by the weighted product (weight factor 0.5) of both the signals $a(t) \star h_B(t)$ and $a(t-T) \star h_B(t-T)$, which are already disturbed by intersymbol interference. It should be remembered that $h_B(t)$ represents the impulse response of the equivalent baseband filter representation of the IF filter (Section 9.4.3). Physically, convolutional signal $a(t) \star h_B(t)$ represents the envelope of modulated IF filter output signal disturbed by intersymbol interference. To make it more clear, Fig. 11.28a shows the eye pattern of the normalized envelope $a(t) \star h_B(t)$ of IF signal at the input of demodulator and Fig. 11.28b of detected signal $d(t) = 0.5[a(t) \star h_B(t)] \cdot [a(t-T) \star h_B(t-T)]$ at the output of demodulator. Both eye pattern are based on an IF filter with Gaussian frequency response and normalized bandwidth $B = 2f_gT = 1.58$. Phase noise and additive Gaussian receiver noise are neglected ($\sigma_\phi = 0$, $\sigma = 0$).

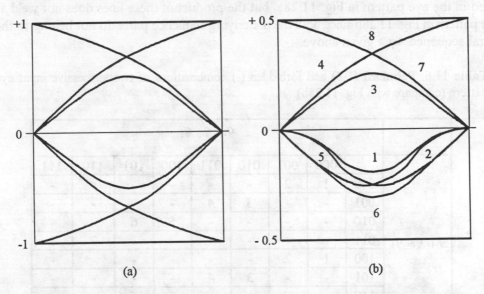

Fig. 11.28: Normalized eye pattern of the (a) envelope $a(t) \star h_B(t)$ of IF signal at the input of DPSK demodulator and (b) detected signal $d(t)$ at the output of demodulator. The numerical values given in (b) are explained in Table 11.6

The eye pattern of the envelope $a(t) \star h_B(t)$ shown in Fig. 11.28a is, of course, same as eye pattern of delayed envelope signal $a(t-T) \star h_B(t-T)$. It should be noted that the eye pattern of Fig. 11.28a is also same as eye pattern of detected signal in a PSK homodyne system (see Fig. 12.4) and, in addition, of detected signal in an ASK heterodyne or homodyne system. For the underlying system parameters, eye pattern in Fig. 11.28a is characterized by eight eye lines only.

Keeping in view some special conditions (discussed below), eye pattern of the DPSK detected signal shown in Fig. 11.28b can be regarded as a product of two eye pattern: one is the eye pattern of the envelope $a(t) \star h_B(t)$ and other is the eye pattern of the delayed envelope $a(t-T) \star h_B(t-T)$. Since both these eye pattern are equal, the eye pattern in Fig. 11.28b corresponds to the square of the eye pattern in Fig. 11.28a. As a result, eye pattern in Fig. 11.28b will be based on $8 \cdot 8 = 64$ eye lines. However, this is actually not true since some special conditions must also be considered. As explained in the following example, only lines which belong to the same symbol sequence $\langle q_v \rangle_i$ are allowed to multiply.

Example 11.5

Let us consider a symbol sequence $\langle q_v \rangle = \cdots q_{-2}, q_{-1}, q_0, q_1, q_2 \cdots = \cdots 01101 \cdots$. Here, it is allowed to multiply sequence part $q_{-1}, q_0, q_1 = 110$ which represents a certain line in the eye pattern of $a(t) \star h_B(t)$ and sequence part $q_{-2}, q_{-1}, q_0 = 011$ which corresponds to the delayed eye pattern of $a(t-T) \star h_B(t-T)$. In contrast to this, it is not allowed to multiply the line which belongs to $q_{-1}, q_0, q_1 = 110$ and the line which belongs to $q_{-2}, q_{-1}, q_0 = 010$. It is true that both lines are included in the eye pattern in Fig. 11.28a, but the product of these lines does not yield a line in the eye pattern in Fig. 11.28b since both the underlying sequence parts do not belong to the same temporal sequence $\langle q_v \rangle$ given above.

Table 11.6: Permitted (1-8) and forbidden (-) combinations of two successive input eye pattern (compare with Fig. 11.28b)

		q_{-1}, q_0, q_1							
		000	001	010	011	100	101	110	111
q_{-1}, q_0, q_1	000	1	2	-	-	-	-	-	-
	001	-	-	3	4	-	-	-	-
	010	-	-	-	-	5	6	-	-
	011	-	-	-	-	-	-	7	8
	100	1	2	-	-	-	-	-	-
	101	-	-	3	4	-	-	-	-
	110	-	-	-	-	5	6	-	-
	111	-	-	-	-	-	-	7	8

It should be noted that $d(t)$ is not symmetrical with respect to time axis because of the non-linear behaviour of demodulator. The "continuous mark", for example, will result in a received signal without any phase change and a steady $d(t)$ i.e., line 8. In contrast, "continuous space" has a rectangular envelope which is shaped sine-like by the IF filter. This yields a negative, but not constant demodulated signal (i.e., line 1). This makes the decision of optimal threshold difficult. No bit sequence $<q_v>$ can be found that leads to a constant normalized signal $d(t) = -0.5$, except for an infinite large IF filter bandwidth. Table 11.6 shows the permitted combinations of the input pattern (symbols q_v) of two successively transmitted bits. Not all 64 combinations are allowed and only 8 of the remaining 16 combinations are distinguishable because DPSK reception is polarity independent.

(vi) OPTIMIZATION

In contrat to the systems discussed in the previous Sections such as ASK and PSK homodyne and ASK and FSK heterodyne systems, optimization in a DPSK system is more comprehensive. Here, optimization is restricted on approximations or computer simulations since an exact analytical optimization is too comprehensive and without practical use.

(a) IF FILTER BANDWIDTH B_{IF}

The optimum IF filter bandwidth for a minimum BER for continuously sent spaces (i.e., ···00000···) is $B_{IF} = B = 1.58 f_b$ (like PSK). In contrast, spectrum of "continuous mark" is a single line at the intermediate frequency that would pass through any filter bandwidth even approaching zero. It should be noted that continuously sent spaces represent the worst-case bit sequence in DPSK. This is in contrast to the systems discussed in the previous Sections, where worst-case pattern was given by single-"0" and single-"1". Therefore, no simple calculation for the actual optimum bandwidth exist and the value obtained for "continuous space" can be used as a worst-case approximation. It must be remembered that $B = 1.58 f_b$ is valid for a Gaussian IF filter. The actual optimum bandwidth is only somewhat above (not below) this approximation.

(b) THRESHOLD E

Due to the asymmetry of the detected signal (see eye pattern in Fig. 11.28b), optimum threshold is no longer located at $E = 0$. It becomes clear from the eye pattern in Fig. 11.28b that the optimum threshold E_{opt} is always located somewhat above zero. The real value of E_{opt} can only be found by numerical methods using a fast computer program [31]. For practical system parameters, however, the optimum threshold remains approximately at zero. Hence, $E_{opt} \approx 0$ can be used as a very good approximation.

(c) CONCLUSION

An optical DPSK heterodyne system with autocorrelation detection can achieve a bit error rate of $p_a = 10^{-10}$ if the resulting laser linewidth is less than $\approx 1\%$ of the bit rate. For such a small linewidth, phase noise is largely unaffected by the IF filter which can, therefore, be optimized considering only the additive noise. The result is $B = 1.58f_b$ for a Gaussian IF filter. The linewidth can be increased due to this intermediate filter to about 1.2% [11, 31].

Similar to the other heterodyne and homodyne systems, a DPSK heterodyne system also exhibits a bit error rate floor. Increasing the received light power P_r even to infinity cannot reduce the BER below a limit that is only determined by the normalized, resulting laser linewidth $\Delta f T = (\Delta f_t + \Delta f_i)T.$

As compared to other optical systems, a DPSK heterodyne system represents a very good compromise. It has higher sensitivity than the incoherent ASK and FSK heterodyne systems and does not require the extremely small linewidth, for example, as in a PSK homodyne system. Moreover, no optical phase-locked loop (OPLL) circuit is required.

11.4   COHERENT SYSTEMS WITH OPTICAL AMPLIFIERS

In this Section, evaluation of signal-to-noise (S/N) ratio has been made when an optical amplifier is used as preamplifier, post amplifier and in-line amplifier. In coherent communication systems, local oscillator (LO) power is much higher than the input optical signal power. As a consequence, shot noise and beat noise will predominate over the circuit noise and the effect of latter noise can be neglected. This is in contrast to direct detection system, wherein the circuit noise may be comparable with the other sources of noise (Section 10.5).

(i)   PREAMPLIFIER

A laser preamplifier followed by heterodyne receiver is shown in Fig. 11.29a. The average output photon number and its variance are given by

$$<n_o> = G<n_i> + (G-1)n_{sp}b + <n_l> + 2\sqrt{G<n_i><n_l>}\cos(2\pi f_{IF}t) \tag{11.151a}$$

and

$$\sigma_h^2 = G<n_i> + (G-1)n_{sp}b + <n_l> + 2<n_i>G(G-1)n_{sp}$$
$$+ 2<n_l>(G-1)n_{sp} + (G-1)^2 n_{sp}^2 b \tag{11.151b}$$

In the above equation, $<n_1>$ is the average photon number from the LO and f_{IF} the intermediate frequency (IF). The fifth term in Eq. (11.151b) represents the ASE-LO beat noise. When the LO power is sufficiently high, this term becomes predominant over other terms. In that case, IF S/N ratio is given by

$$\frac{S}{N} = \frac{2e^2 G <n_i> <n_l>}{2e^2 \left[2<n_l>(G-1) n_{sp} \right] 2B}$$

(11.152)

For $G \gg 1$, it becomes

$$\frac{S}{N} = \frac{<n_i>}{4n_{sp}B}$$

(11.153)

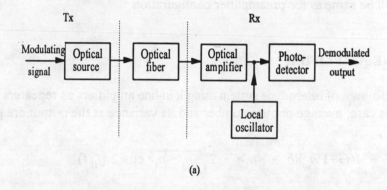

(a)

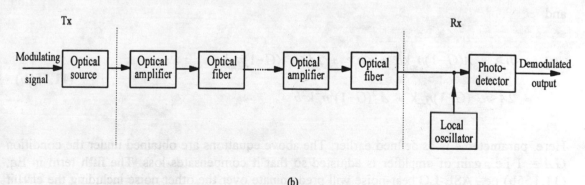

(b)

Fig. 11.29: Heterodyne system with (a) optical preamplifier and (b) cascade amplification using k optical in-line amplifiers

In coherent systems, beat noise limited SNR given by Eq. (11.153) can be achieved more easily than in a direct detection system (Section 10.5). Further, shot noise limited S/N ratio without an optical preamplifier from Eqs. (11.151a) and (11.151b) is given by

$$\left(\frac{S}{N}\right)_{shot} = \frac{<n_i>}{2B} \tag{11.154}$$

It is evident from Eqs. (11.153) and (11.154) that there is no advantage of using optical preamplifier with heterodyne receiver. In fact, it will degrade the performance by a factor of $2n_{sp}$. This, of course, is true only when the LO power is sufficiently high [18, 19].

(ii) POSTAMPLIFIER

Under the condition that LO-ASE beat noise predominates over the other noise, S/N ratio expression will be same as for preamplifier configuration.

(iii) IN-LINE AMPLIFIER

The block diagram of heterodyne system using k in-line amplifiers as repeaters is shown in Fig. 11.29b. In this case, average photon number and its variance at the output are given by

$$<n_o> = <n_i> + A(G-1)n_{sp}kb + <n_l> + 2\sqrt{<n_i><n_l>}\,\cos(2\pi f_{IF}t) \tag{11.155a}$$

and

$$\sigma_a^2 = <n_i> + A(G-1)n_{sp}kb + <n_l> + 2A<n_i>(G-1)n_{sp}k$$
$$+ 2A<n_l>(G-1)n_{sp}k + A^2(G-1)^2 n_{sp}^2 k^2 b \quad . \tag{11.155b}$$

Here, parameter A is as defined earlier. The above equations are obtained under the condition $GA = 1$ i.e., gain of amplifier is adjusted so that it compensates loss. The fifth term in Eq. (11.155b) i.e., ASE-LO beat-noise will predominate over the other noise including the circuit noise if LO power is sufficiently large. In that case

$$\frac{S}{N} = \frac{\langle n_i \rangle}{4 n_{sp} k B} \tag{11.156}$$

This S/N ratio is same as obtained for the direct detection system (Eq. 10.47). In coherent systems also, transmission distance can be increased considerably by using cascaded amplifiers.

11.5 COMPARISON OF CALCULATION METHODS IN HETERODYNE DETECTION

Last Section of this Chapter considers again incoherent optical ASK and FSK heterodyne systems. Several different calculation methods are existing for the evaluation of BER and determining the performance of ASK or FSK heterodyne receivers with incoherent demodulation. In Section 11.3, only one of these methods has been used to analyse incoherent heterodyne systems. This Section compares these methods with regard to accuracy, complexity and required computer processing time.

11.5.1 WHY APPROXIMATION?

Scanning the literature, a large number of publications exist which theoretically examine the incoherent optical heterodyne systems. A small representative sample of these papers is given by [6, 8, 9, 10, 13, 16, 17, 21-26, 28-30, 37, 42]. The primary theoretical problems in analysing optical heterodyne and homodyne systems are solved. However, the available solutions are not same and differ in accuracy, complexity and required computer processing time.

A typical example is given by the receiver performance evaluation of an incoherent ASK or FSK heterodyne system including the effects of laser phase noise and also the influence of IF filter bandwidth. As an incoherent FSK heterodyne receiver with dual- or single-filter detection operates in nearly the same manner as an incoherent ASK heterodyne receiver, in this Section we restrict our consideration to the latter receiver only.

As exact analytical calculation of the system performance is usually not possible, appropriate approximations must be derived. However, the problem is to assess the degree of accuracy of the different approximation methods since the exact solution is unknown. Of course, the best verification of the accuracy is obtained by comparing the approximation results with the measurement results for a realized system; for example, eye pattern, BER and signal shape. On the other hand, an approximate system calculation is usually the first step in the design and realization of an optical communication system. Therefore, a comparison with realized system is not possible since this system is normally not available. In such a case, computer simulation of the system represents a very useful tool to analyse the system and to verify the accuracy of approximations. As the accuracy of simulation results only depend on the available computer run time, it can theoretically be improved arbitrarily. Therefore, simulation results can usually be regarded as the exact solution.

In this Section, we make use of this substantial advantage of computer simulation to compare different approaches of approximate performance evaluation. It should be remembered that simulation results have already been employed and discussed in the previous Chapters several times.

11.5.2   PROBABILITY DENSITY FUNCTION AND BER

This Section first summarizes the basic formulas describing probability density function (pdf) and BER of an optical ASK heterodyne receiver and the results of Section 11.3.1.

In the following, we consider an incoherent optical ASK heterodyne balanced receiver disturbed by laser phase noise, shot noise and thermal noise (Fig. 11.12). The signal of interest in this receiver is the envelope or detected signal $d(t) = |i_{IF}(t)|$ given in Eqs. (11.69) and (11.71). To calculate the BER, pdf $f_d(d)$ of the sampled (and normalized) detected signal d is required. As derived in Section 11.3.1, this pdf can be written as follow

$$f_d(d) = \frac{d}{\sigma^2} \int_0^{+\infty} e^{-\frac{d^2 + c^2}{2\sigma^2}} J_0\left(\frac{Cd}{\sigma^2}\right) f_C(C) \; dC \qquad (11.157)$$

It should be remembered that random variable C equals normalized and sampled envelope d provided standard deviation of the receiver additive Gaussian noise is zero i.e., $\sigma = \sigma_x = \sigma_y = 0$. Thus, shot noise and thermal noise of receiver are excluded.

In order to simplify BER calculation and focus our interest on the topic of this Section, we neglect intersymbol interference. Thus, $C = 0$ if a symbol "0" is transmitted, whereas symbol "1" leads either to $C = 1$ when phase noise $\phi(t)$ is neglected or $C < 1$ when phase noise is included. As derived in Section 11.3.1, worst-case probability of error is given by

$$P_w = \frac{1}{2}\left(p_0 + p_1\right) = \frac{1}{2}\int_E^{+\infty} f_{d0}(d) \; dd + \frac{1}{2}\int_0^E f_{d1}(d) \; dd$$

$$\qquad (11.158)$$

$$= \int_0^1 \tilde{p}(C) \, f_C(C) \; dC$$

where E represents the decision threshold and $\tilde{p}(C)$ the well-known BER formula of conventional ASK heterodyne systems in the absence of phase noise. It should be noted that occurrence probabilities for symbol "0" and "1" are assumed to be equal.

The Eq. (11.158) enables a nearly "exact" calculation of the BER for incoherent ASK heterodyne systems. The only inherent approximations are the narrow-band consideration of the noisy IF signal, no intersymbol interference (which are considered in Section 11.3.1) and the use of an ideal envelope detector with $d(t) = |i_{IF}(t)|$.

As explained in Section 11.3.1, main problem in solving Eq. (11.158) is the determination of pdf $f_C(C)$ of random envelope C which physically represents the envelope of IF signal in the absence of additive Gaussian shot and thermal noise. In principle, it is possible to obtain analytically the required pdf $f_C(C)$ as shown in Section 11.3.1. However, the numerical evaluation on computer is very time consuming, especially for a high laser linewidth-to-bandwidth ratio $\Delta f/B_{IF}$ [6, 9, 10]. For this reason, goal is to find useful methods of approximation for the pdf $f_C(C)$. Approximation problems will be discussed in more detail by considering three representative approaches in the following Section.

11.5.3   METHODS OF APPROXIMATION

(i)   APPROXIMATION METHOD 1
- QUASI-CONSTANT FREQUENCY APPROXIMATION -

This approximation method, first proposed by Garrett and Jacobsen [13, 17] and already introduced in Subsection 3.4 (vi) presumes that the intermediate frequency noise

$$\frac{d\phi(t)}{dt} = \dot{\phi}(t) = \omega(t) = \omega_c \tag{11.159}$$

is approximately constant during the time interval

$$\Delta t_B = \frac{1}{h_B(0)}\int\limits_{-\infty}^{+\infty} h_B(t)\ dt = \frac{H_B(0)}{h_B(0)} = \frac{1}{B} \approx T \tag{11.160}$$

which is in the order of bit duration T. Physically, Δt_B represents the rectangular equivalent pulse width of the filter time response $h_B(t)$ and $B = (\Delta t_B)^{-1}$ the double-sided noise-equivalent bandwidth of the filter frequency response $H_B(f)$. As explained in Chapter nine, $H_B(f)$ is the baseband representation of the intermediate frequency filter i.e., $H_{IF}(f) = H_B(f-f_{IF}) + H_B(f+f_{IF})$. For example, a Gaussian baseband filter with frequency response $H_B(f) = \exp[-\pi(f/B)^2]$ and bandwidth $B = 2f_g = 1/T$ yields $\Delta t_B = T$.

As shown in Fig. 11.30 (reproduced from Chapter three), a constant frequency ω_c during Δt_B is related to a linearly increasing or decreasing (ω_c can also be negative) phase

$$\phi(t) = \omega_c t + \phi_c \qquad\qquad (11.161)$$

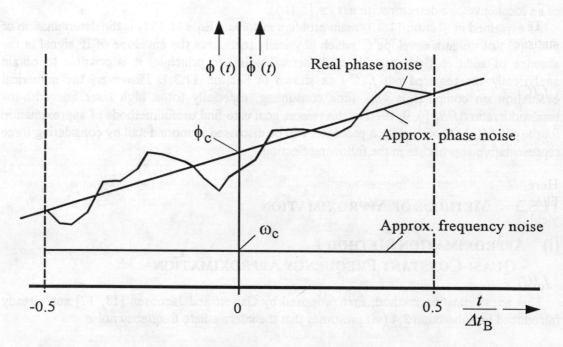

Fig. 11.30: Phase noise and frequency noise approximation related to approximation method 1

It may be mentioned that ω_c is actually a random variable which statistically changes its value from one time interval (or from one bit) to the following time interval (the next bit). The pdf of the zero mean Gaussian random frequency ω_c is given by

$$f_{\omega_c}(\omega_c) = \frac{1}{\sqrt{2\pi}\sigma_{\omega_c}} \exp\left(-\frac{\omega_c^2}{2\sigma_{\omega_c}^2}\right) \qquad\qquad (11.162)$$

Its variance is

$$\sigma_{\omega_c}^2 = 2\pi\Delta fB \qquad\qquad (11.163)$$

where Δf represents the resulting laser linewidth of both transmitter and local lasers. Next, we reconsider the random envelope C for symbol "1" as given in Eq. (11.78). By substituting the phase noise $\phi(t)$ from Eq. (11.161), we obtain

$$C = |H_B(\omega_o)| \tag{11.164}$$

The required pdf $f_c(C)$ can now easily be calculated by using the simple one-dimensional statistical transformation [27]

$$f_C(C) = f_{\omega_c}(\omega_c) \, \frac{1}{\left| \dfrac{dH(\omega_c)}{d\omega_c} \right|_{\omega_c = H^{-1}(C)}} \tag{11.165}$$

Here, H^{-1} represents the inverse function of H. As an example, we consider again a Gaussian filter with frequency response $H_B(f) = \exp[-\pi(f/B)^2]$. In this particular case, above equation yields

$$f_C(C) = \sqrt{\frac{B}{\Delta f}} \; C^{\frac{B}{\Delta f}-1} \; \frac{1}{\sqrt{-\pi \ln(C)}} \tag{11.166}$$

By substituting Eq. (11.166) in Eq. (11.158), we finally obtain

$$P_w = \frac{1}{2}\exp\left(-\frac{E^2}{2\sigma^2}\right) + \frac{1}{2}\sqrt{\frac{B}{\Delta f}} \int_0^1 Q\left(\frac{C-E}{\sigma}\right) \frac{C^{\frac{B}{\Delta f}-1}}{\sqrt{-\pi \ln(C)}} \, dC \tag{11.167}$$

This equation clearly indicates that the BER of incoherent optical heterodyne system is primarily determined by three parameters: First, the normalized standard deviation σ of the additive Gaussian shot and thermal noise; second, the filter bandwidth-to-laser linewidth ratio $B/\Delta f$ and third, the threshold level E. Here, E and $B/\Delta f$ can be optimized to minimize the BER. Eq. (11.167) includes only one integral which, however, cannot be solved analytically. Nevertheless, Eq. (11.167) represents a relatively fast and also accurate method for calculating the BER for incoherent optical ASK heterodyne system.

(ii)   APPROXIMATION METHOD 2
- GAUSSIAN APPROXIMATION -

This method, already proposed and used in Section 11.3.1, takes advantage of the simple Gaussian approximation

$$
f_C(C) \approx \frac{1}{\sqrt{2\pi}\sigma_C} \exp\left(-\frac{\left(C - \eta_C\right)^2}{2\sigma_C^2}\right)
\tag{11.168}
$$

of the pdf $f_C(C)$. The remaining problem is to calculate the mean η_C (first moment) and the standard deviation σ_C. Owing to the square-root dependence of C (see Eq. 11.78 to 11.80), it is easy to calculate exactly the second moment $E\{C^2\} = \sigma_C^2 + \eta_C^2$, whereas the mean value $\eta_C = E\{C\}$ can only be found by approximations (E = expected value). Calculation of η_C and σ_C have been carried out in Section 11.3.1 (incoherent ASK heterodyne system). If an ideal square-law detector with $d(t) = |i_{IF}(t)|^2$ instead of an ideal envelope detector with $d(t) = |i_{IF}(t)|$ is used, then the required moments η_D and σ_D with $D = C^2$ can be calculated exactly.

With the above Gaussian approximation (Eq. 11.168), calculation of the BER becomes very easy. We obtain

$$
P_w = \frac{1}{2}\exp\left(-\frac{E^2}{2\sigma^2}\right) + \frac{1}{2}\, Q\left(\frac{\eta_C - E}{\sqrt{\sigma^2 + \sigma_C^2}}\right)
\tag{11.169}
$$

This simple formula includes no integral provided the Q-function is installed in the computer. Thus, the BER calculation can be performed very fast and with a relatively good accuracy. However, a large computer execution time is necessary to evaluate the statistical values η_C and σ_C or η_D and σ_D. As these values only depend on the linewidth Δf and type of filter used in the receiver, it is possible to store these values in computer data files as a function of Δf for some typical filters normally used in practice.

(iii)   APPROXIMATION METHOD 3
- POWER-ACCOMMODATION APPROXIMATION -

This approximation method proposed by Kazovsky [23] is entirely different from the methods discussed in Subsections (i) and (ii) as it is not based on foregoing BER calculation. This method takes into account the fact that the IF filter bandwidth $B_{IF} = \Delta f_M$ (the index M stands for modulation) must be in the order of three times the bit rate $R = 1bit/T$ to accommodate 95% of the ASK signal power [23]. Considering phase noise also, IF-filter bandwidth must now be increased to [23]

$$B_{IF} = \sqrt{\Delta f_M^2 + \Delta f_{95}^2} = \sqrt{(3R)^2 + \Delta f_{95}^2} \qquad (11.170)$$

where Δf_{95} represents the 95% bandwidth of the photodetector current due to laser phase noise. It is assumed that bandwidth Δf_{95} and linewidth $\Delta f/2$ of a laser are related by $\Delta f_{95} = 12.7\Delta f$. Provided that a

$$penalty = \frac{B_{IF}(\text{with phase noise})}{B_{IF}(\text{without phase noise})} = \sqrt{1 + \left(\frac{1}{3}\Delta f_{95}T\right)^2} \qquad (11.171)$$

of not more than 1 dB is permissible, the following allowable linewidth-bit duration product can be obtained:

$$\Delta f T \leq 0.09 \qquad (11.172)$$

Obviously, this simple relationship enables a very fast (no computer is required), but only a rough estimation of the maximum permissible laser linewidth. Because of the fact that a penalty of 1 dB does not necessarily lead to a total system collapse, a larger Δf than predicted from Eq. (11.172) can be used. A practical example for this is given in [34].

The Eq. (11.172) does not give any information about BER. It only imply that the unknown BER will increase if the penalty is above 1 dB and decrease otherwise. The BER, however, remains unknown. Clearly, this is the drawback of this approximation. However, the permissible linewidth obtained from Eq. (11.172) is in good agreement with those obtained by the much more comprehensive methods 1 and 2 explained above.

11.5.4 COMPARISON

The objective of this Section is to compare the different approximation methods discussed above. The underlying comparison consider the following: accuracy, complexity and required computer processing time of the calculation.

The problem of comparing accuracy is that no exact analytical solution (for example for the BER) exists which can be used as a reference. However, a very good alternative reference is given by computer simulated numerical solutions which has already been used in Section 11.3.1. The accuracy of this solution depends only on the computer run time and can, therefore, theoretically be as high as required. For this reason, the results obtained by a computer simulation can be regarded as the exact solution.

Mean value η_C and standard deviation σ_C obtained by computer simulation and by approximation methods 1 and 2 are compared in Fig. 11.31. It may be noted that $C = d(\sigma = 0)$ physically represents the envelope of IF signal corrupted by phase noise when shot and thermal noise have been neglected ($\sigma = 0$). The corresponding non-Gaussian pdf $f_C(C)$ of the random envelope C is required for an accurate calculation of BER from Eq. (11.158). The mean value η_C and standard deviation σ_C essentially determine location and width of $f_C(C)$.

We can see from Fig. 11.31 below that there is some deviation between both the approximate solutions and the computer simulated (exact) solution. This deviation increases with the increase in the laser linewidth-to-bandwidth ratio $\Delta f/B_{IF}$. For $\Delta f/B_{IF}$ ratio less than about 0.2, deviation in η_C becomes negligibly small, while the deviation in σ_C still remains considerable. As shown in Fig. 11.31a, the exact (computer simulated) solution for η_C is always below the both approximate solutions, whereas the exact solution for σ_C lies between these solutions (Fig. 11.31b).

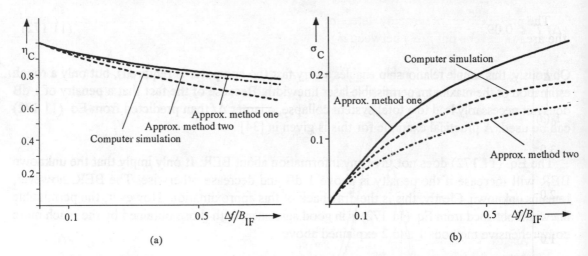

Fig. 11.31: (a) Mean value η_C and (b) standard deviation σ_C of the random envelope C which physically represents the envelope of IF signal $i_{IF}(t)$ corrupted by phase noise (additive Gaussian noise is neglected).

The pdf $f_C(C)$ at $\Delta f/B = 0.4$ is shown in Fig. 11.32. It can be seen that $f_C(C)$ calculated by using approximation method 2 is slightly more close to the computer simulated plot than the approximation method 1. However, the main part of $f_C(C)$ for the BER calculation is the part for $C < 0.5$. The reason for this is that the normalized optimum threshold E is always less than 0.5

in an ASK heterodyne receiver [10]. It becomes clear from Fig. 11.32 that in the range of $C < 0.5$, approximation method 1 represents the more accurate solution.

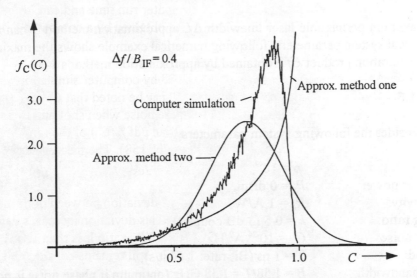

Fig. 11.32: Probability density function $f_C(C)$ of the random envelope C

The probability p_1 of wrongly detecting symbol "1" (as a part of the overall BER) is given by the area under the pdf $f_d(d)$ between $d = 0$ and $d = E < 0.5$.

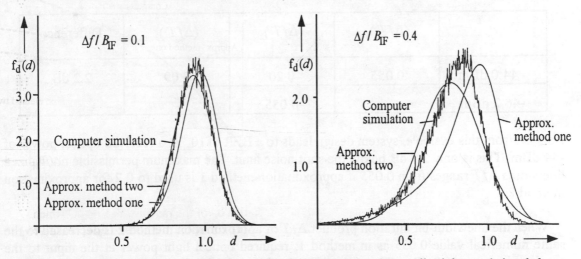

Fig. 11.33: Probability density function $f_d(d)$ of the sampled and normalized detected signal d

It becomes clear from Fig. 11.33 that the deviation between approximation method 1 and 2 is less when $\Delta f/B$ decreases. As $\Delta f/B$ increases, approximation method 1 again represents the more accurate solution. Approximation method 1 always yields a higher BER than method 2. The difference between BERs can reach several orders of magnitude [4]. In practice, where required input light power P_r for a desired BER and a given laser linewidth is of interest, difference between both approximation methods is less (see Example 11.7).

For a maximum permissible laser linewidth Δf, approximation method 3 can also be used. Based on typical system parameters, following numerical example shows the maximum allowed linewidth-bit duration product $\Delta f T$ obtained by approximation methods 1-3.

Example 11.5

Let us consider the following system parameters:

BER:	$p = 10^{-9}$,
Local laser power:	$P_1 = 0$ dBm,
Responsivity:	$R_0 = 1$ A/W,
Coupling ratio:	$k = 0.5$ (3 dB coupler),
Receiver noise:	$G_c = 10^{-24}$ A²/Hz,
Bit period:	$T = 1$ ns (Bit rate: 1 Gbit/s),
IF filter bandwidth:	$B = 1.58/T = 1.58$ GHz (optimum if phase noise is neglected) and
Threshold level:	$E = E_{opt}$.

When the maximum permissible resulting laser linewidth Δf for optical input power levels $P_r = -44$ dBm and -46.2 dBm are calculated, the following results are obtained:

P_r	$(\Delta f T)_1$	$(\Delta f T)_2$	$(\Delta f T)_3$	Difference
- 44.0 dBm	0.055	0.20	0.09	2.2 dB
-46.2 dBm	-	0.055	-	

As seen from this example, system design leads to a BER of 10^{-9} at a received optical power of -44 dBm. This is only 2.9 dB below the shot noise limit. The maximum permissible normalized linewidths $\Delta f T$ ranges from 0.055 if approximation method 1 is used to 0.2 for approximation method 2.

When the linewidth-bit duration product $\Delta f T$ of approximation method 2 is decreased to the same numerical value 0.055 as in method 1, required optical light power at the input to the receiver is now -46.2 dBm. Thus, the difference in received light power is only 2.2 dB for the same BER of 10^{-9} and laser linewidth. However, this difference increases if the linewidth-bit duration product is large; for example, $\Delta f T = 1$.

The computer processing time (CPU-time) for calculating the required optical input power P_r, the BER or the maximum permissible laser linewidth Δf is nearly the same for approximation methods 1 and 2. For example, it will take only a few seconds to calculate the BER from Eq. (11.167) or Eq. (11.169) as a function of P_r. Somewhat more time is needed when the maximum permissible laser linewidth Δf is to be evaluated. If this calculation is performed for a simultaneous optimization of the threshold level E and also the IF filter bandwidth $B_{IF} = B$, then the required computer run time can raise from a few minutes to some hours (depending on the computer). The computing time will continuously increase if intersymbol interference are also taken into account [9]. In contrast to this, no computer run time is needed to calculate the permissible laser linewidth Δf by approximation method 3, because Eq. (11.172) inherently contains the desired solution.

Finally, we shall briefly consider the complexity of the theoretical derivations of approximation methods 1 to 3. The derivation of Eq. (11.172) as the result of approximation method 3 is straightforward and relatively easy. No difficult calculation steps are required in this method. In contrast to this, the derivations of approximation methods 1 and 2 are much more complex and involve some non-trivial calculation steps. Nevertheless, the derivations are also conceptually straightforward.

Table 11.6 Features of three approximate calculation methods for evaluating the performance of incoherent optical heterodyne systems

	Approximation method 1 (quasi-constant frequency)	Approximation method 2 (Gaussian approx.)	Approximation method 3 (power-accommodation)
Accuracy	very good	good	sufficient
CPU-time	medium	medium / low, if η_C and σ_C are stored in data files	zero
Complexity	low	low	very low
Remarks	methods 1 and 2 are similar		only maximum permissible laser linewidth is approximated (not the BER)

To conclude, Table 11.6 above summarizes the results of our comparison performed in this Section. Again, this table considers the accuracy of calculation, required computer processing time and complexity of the underlying mathematical derivations as the most important criteria to compare the approximate calculation methods for incoherent optical heterodyne receivers as discussed above.

It becomes clear from this table and above discussion that the choice of an approximate calculation method depends on its application. If, for example, a fast estimation of the maximum permissible laser linewidth for a given bit rate is required, then approximation method 3 represents the best solution. Unfortunately, no information about the BER can be obtained by this method, which represents a main drawback of this simple method. If a more accurate calculation is required, then both approximation method 1 or method 2 can be used. Here, the required computer execution time is approximately the same. As method 1 is somewhat more accurate, especially for linewidth-to-bandwidth ratios greater than one, it gives the better result. If a solution is required that gives an overall and clear insight into the complex functional behaviour of the system including the interactions of all fundamental system parameters (i.e., BER, bit rate, laser linewidth, variance of shot and thermal noise, threshold, filter bandwidth and optical input power), computer simulation is the best method.

11.6   REFERENCES

[1]   Abramowitz, M.; Stegun, I. A.: Handbook of Mathematical Functions. Dover Publications Inc., 1965.
[2]   Best, R.: Theorie und Anwendungen des Phase-locked Loops. Fachschriftenverband Aargauer Tagblatt AG, Aarau/Schweiz 1976.
[3]   Bonek, E.; Leeb, W. R.; Scholtz, A. L.; Philipp, H. K.: Optical PLLs see the light. Microwaves & RF (1983), 65-70.
[4]   Fischer, E.: Auswirkungen eines ZF-Filters auf phasenverrauschte Eingangssignale. Diplomarbeit, Tech. Univ. München, 1989.
[5]   Fleischmann, M.: Berechnung, Optimierung und Vergleich verschiedener optischer Übertragungssysteme mit Überlagerungsempfang. Diplomarbeit, TU München, Lehrstuhl für Nachrichtentechnik, 1987.
[6]   Foschini, G. J.; Greenstein, L. J.; Vannucci, G.: Noncoherent detection of coherent lightwave signals corrupted by phase noise. IEEE COM-36(1988)3, 306-314.
[7]   Franz, J.: Grundzüge des kohärent optischen Heterodynempfanges. TU München, Lehrstuhl für Nachrichtentechnik, Nachr.-techn. Ber. Band 14, 1984.
[8]   Franz, J.: Evaluation of the probability density function and bit error rate in coherent optical transmission systems including laser phase noise and additive Gaussian noise. J. Opt. Commun. 6(1985)2, 51-57.
[9]   Franz, J.: Receiver analysis of incoherent optical heterodyne systems. J. Opt. Commun. 8(1987)2, 57-66.
[10]   Franz, J.: Optische Übertragungssysteme mit Überlagerungsempfang. Springer-Verlag, 1988.
[11]   Franz, J.; Schaller, N.: Limits of optical heterodyne DPSK transmission systems with envelope detection. J. Opt. Commun. 10(1989)1, 28-32.
[12]   Gardner, F. M.: Phaselock Techniques. John Wiley & Sons Inc., 1979.

[13] Garrett, I.; Jacobsen, G.: Theoretical analysis of heterodyne optical receivers for transmission systems using (semiconductor) lasers with nonneglible linewidth. IEEE J. LT-4(1986)3, 323-334.

[14] Hodgkinson, T. G.: Phase-locked-loop analysis for pilot carrier coherent optical receivers. Electron. Lett. 21(1985)25/26, 1202-1203.

[15] Hodgkinson, T. G.: Costas loop analysis for coherent optical receivers. Electron. Lett. 22(1986)22, 394-396.

[16] Hodgkinson, T. G.: Receiver analysis for synchronous coherent optical fibre transmission systems. IEEE J. LT-5(1987)4, 573-586.

[17] Jacobsen, G.; Garrett, I.: Error-rate floor in optical ASK heterodyne systems caused by nonzero (semiconductor) laser linewidth. Electron. Lett. 21(1985)7, 268-270.

[18] Jain, V. K.: Optical preamplifier in noncoherent space communications. J. Opt. Commun. 14(1993)2, 75-88.

[19] Jain, V. K.: Study of heterodyne and direct detection receivers with an optical pre-amplifier in space communications. J. Opt. Commun. 14(1993)5, 189-193.

[20] Kazovsky, L. G.: Decision-driven phase-locked loop for optical homodyne receivers: performance analysis and laser linewidth requirements. IEEE J. LT-3(1985)6, 1238-1247.

[21] Kazovsky, L. G.: Balanced phase-locked loops for optical homodyne receivers: performance analysis, design considerations, and laser linewidth requirements. IEEE J. LT-4(1986)2, 182-195.

[22] Kazovsky, L. G.: Performance analysis and laser linewidth requirements for optical PSK heterodyne communication systems. IEEE J. LT-4(1986)4, 415-425.

[23] Kazovsky, L. G.: Impact of phase noise on optical heterodyne communication systems. J. Opt. Commun. 7(1986)2, 66-78.

[24] Kikuchi, K.; Okoshi, T.; Nagamatsu, M.; Henmi, N.: Degradation of bit-error rate in coherent optical communications due to spectral spread of the transmitter and the local oscillator. IEEE J. LT-2(1984), 1024-1033.

[25] Okoshi, T.; Emura, K.; Kikuchi, K.; Kersten, R. Th.: Computation of bit-error rate of various heterodyne- and coherent-type optical communication schemes. J. Opt. Commun. 2(1981)3, 89-96.

[26] Okoshi, T.; Kikuchi, K.: Coherent Optical Fiber Communications. Kluwer Academic Publisher, 1988.

[27] Papoulis, A.: Probability, Random Variables and Stochastic Processes. McGraw-Hill, 1985.

[28] Pietzsch, J.: Der Einfluß des Phasenrauschen auf die Fehlerquote bei Übertragung von frequenzmodulierten Signalen. AEÜ 42(1988)2, 132-138.

[29] Saito, S.; Yamamoto, Y.; Kimura, T.: S/N and error rate evaluation for an optical FSK heterodyne detection system using semiconductor lasers. IEEE J. QE-19(1983)2, 180-193.

[30] Salz, J.: Coherent lightwave communications. AT&T Tech. J. 64(1985)10, 2153-2209.

[31] Schaller, H. N.: Berechnung optischer DPSK-Überlagerungssysteme unter Berücksichtigung des Laserphasenrauschens. Diplomarbeit, TU München, Lehrstuhl für Nachrichtentechnik, 1987.

[32] Scholz, A.; Leeb, W. R.; Philipp, H. K.: Detection homodyne pour systemes de communication laser. IOOC-ECOC (1982), 541-546.

[33] Scholz, A.; Philipp, H. K.; Leeb, W. R.: Receiver concepts for data transmission at 10 microns. ESA SP-202(1984), 107-114.

[34] Shikada, M. et. al.: 100-Mbit/s ASK heterodyne detection experiment using 1.3μm DFB laser diodes; OFC-84, Paper No. TUK 6(1984).

[35] Söder, G.; Tröndle, K.: Digitale Übertragungssysteme. Springer-Verlag, 1984.

[36] Stein, S.; Jones, J. J.: Modern Communication Principles. McGraw-Hill, 1967.

[37] Tamburrini, M.; Spano, P.; Piazzolla, S.: Influence of semiconductor-laser phase noise on coherent optical communication systems. Optics. Lett. 8(1983)3, 174-176.

[38] Tröndle, K.; Söder, G.: Optimization of Digital Transmission Systems. Artech House, 1987.

[39] Wandernoth, B.: 1064 nm, 565 Mbit/s PSK transmission experiment with homodyne receiver using synchronization bits. Electron. Lett. 27(1991)19, 1692-1693.

[40] Wandernoth, B.: 5 photon/bit low complexity 2 Mbit/s PSK transmission breadboard experiment with homodyne receiver applying synchronization bits and convolutional coding. ECOC (1994), 59-62.

[41] Wandernoth, B.: Phasensynchronisation für hochempfindliche optische PSK-Homodynempfänger. Dissertation, TU München, Lehrstuhl für Nachrichtentechnik, 1993.

[42] Yamamoto, Y.: Receiver performance evaluation of various digital optical modulation-demodulation systems in the 0.5-10 μm wavelength region. IEEE J. QE-16(1980)11, 1251-1259.

12 SYSTEMS COMPARISON, APPLICATIONS AND PHYSICAL LIMITS

O ne of the aims of this Chapter is to compare the optical communication systems analysed and optimized in previous Chapters ten and eleven. First of all, we have to decide which criterion should be applied to make a fair comparison. Number of different criteria exist depending on the required application. If, for example, an optical undersea communication link with a very large repeater spacing is required, then either an IM/DD system with optical in-line amplifiers or a highly sensitive optical PSK homodyne system represent excellent options. Latter one is also well-suited in optical free-space communications. Here, the criterion of comparison is the repeater spacing or the receiver sensitivity achievable by the different systems. When a short distance and low-cost optical multichannel communication system is needed, an IM/DD system combined with WDM and an incoherent FSK heterodyne system are good options. Now, the criterion of comparison is system cost and inherent problems of realization. In addition to above, other important criteria for a fair comparison are maximum permissible laser linewidth, eye pattern and maximum bit rate.

As an overview, this Chapter starts with systems comparison under ideal conditions (Section 12.1). Subsequently, the comparison is made again by taking into account real conditions (Section 12.2). Some typical applications and problems of realization are described in Section 12.3. In the last Section 12.4, physical limits of optical communications are discussed.

12.1 SYSTEMS COMPARISON UNDER IDEAL CONDITIONS

I n this Section, the term *ideal conditions* means that only the receiver additive Gaussian noise frequently called the circuit noise i.e., shot noise of photodiodes and thermal noise of amplifiers disturb the systems. Other sources of impairments like laser noise and intersymbol interference are neglected. In this case, formulas for the BER calculation become very simple (see Table 12.2) and can, therefore, be evaluated very fast. For this, following physical quantities which have already been derived and discussed in Chapters nine to eleven are required:

$$\hat{i}_d = MR_0P_r \tag{12.1}$$

$$\sigma_{d0}^2 = G_{d0}\int_{-\infty}^{+\infty}|H_B(f)|^2\ df \quad \text{with} \quad G_{d0} = eM^{2+x}I_{dark} + G_{th} \tag{12.2}$$

$$\sigma_{d1}^2 = G_{d1} \int_{-\infty}^{+\infty} |H_B(f)|^2 \, df \quad \text{with} \quad G_{d1} = eM^{2+x}(R_0 P_r + I_{dark}) + G_{th} \tag{12.3}$$

$$\hat{i}_c = 2K_R \sqrt{k(1-k)} R_0 \sqrt{P_r P_1} = \hat{i}_{PD} \quad \text{(compare with Eq. 9.69)} \tag{12.4}$$

and

$$\sigma_c^2 = \sigma_{het}^2 = 2G_c \int_{-\infty}^{+\infty} |H_B(f)|^2 df \quad \text{with} \quad G_c = eK_R(R_0 k P_1 + I_{dark}) + G_{th} \tag{12.5}$$

Typical numerical values of different parameters subsequently required are given in the following table.

Table 12.1: Parameters for systems compared in Table 12.2

Parameters valid for all systems	
Transmitted symbol sequence:	$<q_v> = \cdots 111 \cdots$ (all "1") and
	$<q_v> = \cdots 000 \cdots$ (all "0"),
Symbol or bit rate:	$1/T = 560$ Mbit/s,
Frequency of transmitter laser:	$f_t = 200$ THz,
Quantum efficiency of photodiode:	$\eta = 0.83$,
Responsivity of photodiode:	$R_0 = (e\eta)/(hf) = 1$ A/W,
Dark current of photodiode:	$I_{dark} = 10^{-11}$ A,
psd of thermal noise:	$G_{th} = 10^{-23}$ A²/Hz,
Equivalent baseband filter:	$H_B(f) = 1$ for $|f| \le 1/(2T)$ and
	$= 0$ otherwise
Threshold:	$E = E_{opt}$.
Additional direct detection system parameters	
Avalanche gain of photodiode:	$M = M_{opt}$ (see Eq. 9.16),
Excess noise exponent:	$x = 0.9$.
Additional coherent system parameters	
PIN-photodiode:	$M = 1$,
Local laser power:	$P_1 \triangleq -10$ dBm,
Balanced receiver:	$K_R = 2$,
Coupling ratio of optical coupler:	$k = 0.5$ (i.e., 3 dB coupler).

A comparison of performance of various optical communication systems is given in the following table.

Table 12.2: Comparison of optical communication systems under ideal conditions

Communication system	Probability of error p	Required optical input power P_r	
		$P_1 = -10$ dBm, $\eta = 0.83$	$P_1 \to \infty$, $\eta = 1$
Direct detection	$Q\left(\dfrac{\hat{i}_d}{\sigma_{d0} + \sigma_{d1}}\right)$	-39.4 dBm	-40.2 dBm (1287)
ASK heterodyne with envelope detection	$\dfrac{1}{2}\exp\left(-\dfrac{\hat{i}_c^2}{8\sigma_c^2}\right)$	-48.9 dBm	-51.8 dBm (89)
FSK heterodyne with envelope detection (single-filter)	$\dfrac{1}{2}\exp\left(-\dfrac{\hat{i}_c^2}{8\sigma_c^2}\right)$	-48.9 dBm	-51.8 dBm (89)
FSK heterodyne with synchronous detection (single-filter)	$Q\left(\dfrac{\hat{i}_c}{2\sigma_c}\right)$	-49.3 dBm	-52.2 dBm (81)
FSK heterodyne with envelope detection (dual-filter)	$\dfrac{1}{2}\exp\left(-\dfrac{\hat{i}_c^2}{4\sigma_c^2}\right)$	-51.9 dBm	-54.8 dBm (45)
FSK heterodyne with synchronous detection (dual-filter)	$Q\left(\dfrac{\hat{i}_c}{\sqrt{2}\sigma_c}\right)$	-52.3 dBm	-55.2 dBm (40)
ASK homodyne	$Q\left(\dfrac{\hat{i}_c}{\sqrt{2}\sigma_c}\right)$	-52.3 dBm	-55.2 dBm (40)
DPSK heterodyne	$\dfrac{1}{2}\exp\left(-\dfrac{\hat{i}_c^2}{2\sigma_c^2}\right)$	-54.9 dBm	-57.8 dBm (22)
PSK heterodyne with synchronous detection	$Q\left(\dfrac{\hat{i}_c}{\sigma_c}\right)$	-55.3 dBm	-58.2 dBm (20)
PSK homodyne	$Q\left(\dfrac{\sqrt{2}\hat{i}_c}{\sigma_c}\right)$	-58.3 dBm	-61.2 dBm (10)

This table, in addition to the approximate equations for the BER calculation (column 2), also gives the minimum received light power level P_r required to achieve a BER of 10^{-10}. The column 3 is based on real local laser power of -10 dBm and photodiode quantum efficiency of 83%, whereas the last column presumes a shot noise limited receiver operation with infinite local laser power and 100% quantum efficiency. It should be noted that in the table above, $(i_c/\sigma_c)^2$ represents the signal-to-noise ratio for a coherent system and $(i_d/\sigma_d)^2$ for a direct detection system. Here, σ_d is the average of σ_{d0} and σ_{d1} i.e., $\sigma_d = 0.5 \cdot (\sigma_{d0} + \sigma_{d1})$.

It becomes clear from Table 12.2 that receiver sensitivity decreases from the top of this table to the bottom. Hence, a direct detection system exhibits lowest sensitivity and a PSK homodyne system is characterized by the highest sensitivity. A comparison of columns 3 and 4, shows that the shot noise limited receiver operation and operation at $P_1 \triangleq -10$ dBm in terms of required input power only differ by about 3 dB.

While determining the sensitivity of an optical communication system, two alternate measures can be employed: One is the required optical input power P_r and the other is number n_s of photons per bit. Both measures are related as

$$n_s = \frac{P_r T}{hf} \tag{12.6}$$

Here, the products $P_r T$ and hf represent the *bit energy* and the *photon energy* respectively. The column 4 of Table 12.2 shows both required number n_s of photons per bit and optical power P_r at the input to the receiver. A direct detection system with intensity modulation usually requires more than 1000 photons per bit, whereas coherent systems only require less than 100 photons per bit. With a PSK homodyne system, even 10 photons are sufficient! Based on column 4, Fig. 12.1 indicates the relative sensitivity gains of different optical communication systems over the a direct detection system.

The upper block in this figure represents a conventional direct detection system based on the intensity modulation (IM) of light. All the other blocks represent coherent systems. As shown in the figure, coherent systems can be separated in *homodyne and heterodyne systems*. In addition, coherent systems can also be separated on the basis of *coherent (synchronous) detection* and *incoherent detection*; for example, envelope detection. In this classification, systems using coherent detection (demodulation) require phase-locking (PLL or OPLL), whereas it is not required in systems with incoherent demodulation (see Chapter eleven).

Maximum sensitivity gain of coherent systems under ideal conditions is highlighted in the same figure. As mentioned above, ideal conditions mean no laser noise, no intersymbol interference and infinite local laser power which allows shot noise limited receiver operation. As shown in this figure, improvement in sensitivity increases from the upper blocks to the lower blocks and reaches to a maximum of about (12+3+3+3) dB = 21 dB in a coherent PSK homodyne system. This large improvement in sensitivity can be attributed to a large gain of about 12 dB by changing from direct detection with APD detector to heterodyne detection and three steps of 3 dB each by changing the modulation and demodulation schemes.

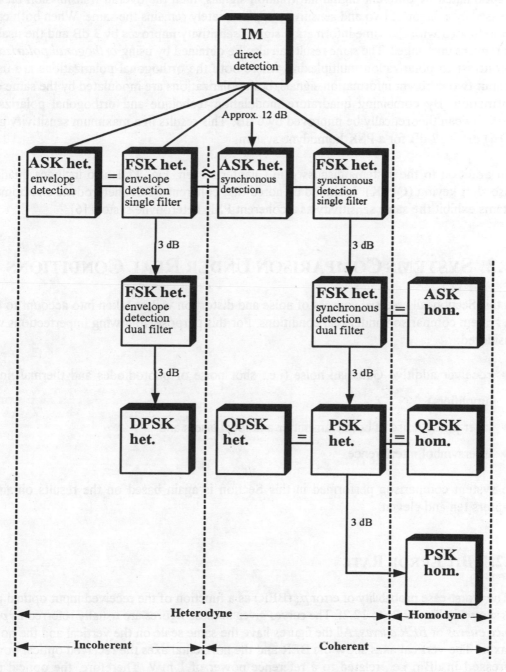

Fig. 12.1: Comparison of sensitivity gain of various optical communication systems under ideal conditions

Theoretically, sensitivity can further be improved by applying *quadrature modulation*. This technique is based on the orthogonality of sine and cosine carrier signals. If both optical carriers are modulated by different digital information signals, then the overall transmission bit rate is increased by a factor of two and sensitivity approximately remains the same. When both carriers are modulated with the same information signal, sensitivity improves by 3 dB and the usable bit rate remains unchanged. The same result can also be obtained by using *orthogonal polarizations*. In contrast to polarization multiplexing where both the orthogonal polarizations are used to transmit two different information signals, both polarizations are modulated by the same digital information. By combining quadrature modulation technique and orthogonal polarizations, sensitivity can theoretically be improved by 6 dB. This results in a maximum sensitivity gain of (21+6) dB = 27 dB for a PSK homodyne system.

In addition to the systems discussed in Chapter eleven, Fig. 12.1 also includes quadrature phase shift keying (QPSK). It should be noted that coherent QPSK heterodyne and homodyne systems exhibit the same sensitivity as a coherent PSK heterodyne system [6].

12.2 SYSTEMS COMPARISON UNDER REAL CONDITIONS

In this Section, all essential sources of noise and distortion will be taken into account to obtain a system comparison under real conditions. For this purpose, following imperfections will be considered:

- receiver additive Gaussian noise (i.e., shot noise of photodiodes and thermal noise of amplifiers),

- laser phase noise of both transmitter and local lasers and

- intersymbol interference.

The system comparison performed in this Section is again based on the results obtained in Chapters ten and eleven.

12.2.1 BIT ERROR RATE

The worst-case probability of error p_w (BER) as a function of the received input optical power P_r is shown in Fig. 12.2a to 12.2f. The curves given in these figures are usually referred as *performance curves* or *BER curves*. All the figures have the same scale on the vertical and the horizontal axes. The vertical axes show the BER and the horizontal axes the received optical power P_r expressed in dBm i.e., related to a reference power of 1 mW. Therefore, the optical power increases from the left side to right side.

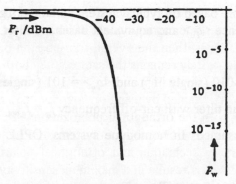

(a) Incoherent direct detection system

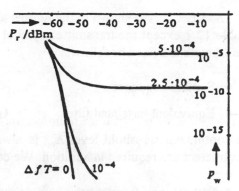

(b) Coherent PSK homodyne system

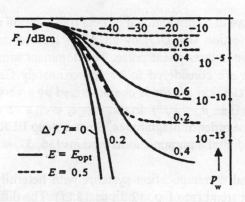

**(c) Incoherent ASK heterodyne system
with envelope detection**

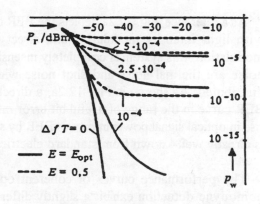

(d) Coherent ASK homodyne system

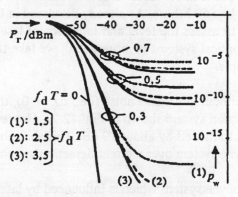

**(e) Incoherent FSK heterodyne system
with envelope detection (single filter)**

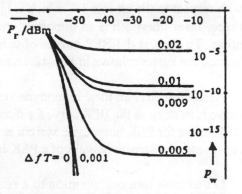

(f) Incoherent DPSK heterodyne system

Fig. 12.2: BER curves for various optical communication systems

The BER curves shown in the figures above are based on typical system parameters given in Table 12.1 except the transmitted symbol sequence $<q_v>$ and equivalent baseband filter. These parameters are fixed as follows:

- Symbol sequences: $<q_v>$ = 010 (single "1") and $<q_v>$ = 101 (single "0"),
- Equivalent baseband filter: Gaussian filter with cut-off frequency $f_g = f_{g,opt}$.

An optimized threshold level E_{opt} is always assumed. In homodyne systems, OPLL system parameters are required in addition. We consider:

- Natural loop frequency: $\omega_n = 0.001 \cdot 2 \cdot \pi \cdot (1/T)$ and
- Damping constant: $\xi = 1$.

As a reference, Fig. 12.2a depicts the BER curve of a conventional optical transmission system with light intensity modulation and direct detection (IM/DD) by an APD. Due to intensity modulation, this system is completely insensitive to laser phase noise. The dominant sources of noise are thermal noise and shot noise which are considered to be approximately Gaussian distributed. As shown in Fig. 12.2a, a direct detection system is characterized by a very steep BER curve in the range of useful bit error rates (i.e., $p_w < 10^{-6}$). In this range, even a very small rise in optical signal power improves BER by some order of magnitudes. Such a steep BER curve is already well-known from standard electrical digital communication systems [36, 37, 43].

The performance curves of coherent optical communication systems with heterodyne or homodyne detection exhibit a slightly different trend (see Fig. 12.2b to 12.2f). The difference arises due to laser phase noise which only influences coherent systems with ASK, FSK, PSK or DPSK modulation, but not an IM/DD system. In coherent systems, resulting laser linewidth of both transmitter and local lasers is a parameter of prime importance. This linewidth is normalized with respect to the bit rate $1/T$ which is taken to be 560 Mbit/s. As already discussed in Chapter three, laser linewidth is an important measure to assess the level and influence of laser phase noise. To discuss the BER curves of coherent optical systems in more detail, we take the PSK homodyne system shown in Fig. 12.2b as a sample.

Assuming first an ideal homodyne system without any phase noise (i.e., $\Delta f = 0$), the BER curve is as steep as the BER curve for direct detection system shown in Fig. 12.2a. However, the BER curve for PSK homodyne system is shifted to the left by about 19 dB. This shift physically represents the sensitivity gain of a PSK homodyne system over a direct detection system.

Let us now turn our attention to a real homodyne system which is influenced by laser phase noise in addition. In this case, BER curves first remain very steep when the optical input power P_r is low. In this range, a coherent system exhibits the same characteristic features as a direct detection system or any other standard electrical digital communication system. The reason for this is that the influence of laser phase noise is negligibly small as compared to additive Gaussian noise of the receiver provided received optical power P_r is low enough. With the increase in P_r, influence of the phase noise increases, while the influence of the receiver additive noise decreases. As a result, a characteristic bend in the BER curve can be observed. When P_r is increased further, laser phase noise finally dominates the other noise. Influence of additive receiver noise can, in principle, be reduced by increasing P_r, whereas the influence of laser phase noise is independent of P_r. Therefore, the BER reaches a characteristic point of saturation termed as *error rate floor* or the *BER floor*.

The fundamental effect of the error rate floor is that certain probabilities of error (for example, $p_w = 10^{-10}$) cannot be achieved when the laser phase noise is too strong. In other words, every laser is not suitable to provide a BER of 10^{-10}. In homodyne systems, requirements for the maximum permissible laser linewidth are strong, in particular. This becomes clear by comparing the numerical values of the normalized linewidth $\Delta f T$ given in the Figs. 12.2b to 12.2f.

The BER curves for an ASK heterodyne system with incoherent detection and an ASK homodyne system are shown in Figs. 12.2c and 12.2d. These curves are similar to the BER curve for a PSK homodyne system in Fig. 12.2b i.e., first a steep course, then a bend and finally a BER floor. When normalized laser linewidths are compared in these figures, it becomes clear that an incoherent heterodyne system is much less sensitive to phase noise than a homodyne system. Even lasers with a relatively poor quality in spectral emission are able to provide a BER of 10^{-10}. Further, above figures also illustrate the strong influence of threshold level. Even a small deviation in threshold level from the optimum threshold E_{opt} deteriorates the BER considerably.

The performance curves of incoherent FSK heterodyne system with a single-filter detection is shown in Fig. 12.2e. For this system, normalized frequency deviation $f_d T$ is another parameter of importance. As explained in Section 11.3.2, this parameter is optimizable for a minimum BER. Finally, Fig. 12.2f shows the BER curves for a DPSK heterodyne system with autocorrelation detection.

In conclusion, the following characteristic features of coherent optical communication systems are highlighted in Figs. 12.2a to 12.2f.

In *quality*, all the BER curves exhibit the same characteristic features irrespective of modulation scheme applied to the system. All BER curves show first a steep course, then a bend and finally a BER floor at high input optical power levels.

In *quantity*, the BER curves are very different. For example, a BER of 10^{-10} is obtained at rather different received optical power levels P_r and laser linewidths Δf. The quantitative shape of the BER curve is essentially determined by various other system parameters. The OPLL parameters (for example, the loop frequency) substantially influence the BER of a homodyne system, whereas the BER of a heterodyne system is fundamentally influenced by the features of IF filter (for example, the IF filter bandwidth) as already discussed in Chapter eleven.

12.2.2 LASER LINEWIDTH REQUIREMENTS

In the design and realization of coherent optical communication systems, two questions of primary importance are: How much maximum laser linewidth can be allowed to achieve a BER of 10^{-10}? What minimum optical power at the input to the receiver is required? The answers to both these questions is available in Fig. 12.3.

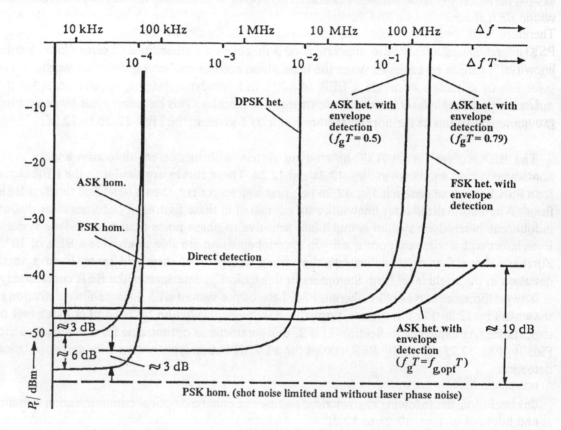

Fig. 12.3: Maximum permissible resulting laser linewidth Δf at a bit rate of $1/T = 560$ Mbit/s and bit error rate 10^{-10}

The horizontal axis in this figure shows the resulting laser linewidth $\Delta f = \Delta f_t + \Delta f_l$ of both transmitter and local lasers. Besides the normalized linewidth $\Delta f T$ (normalized to bit rate), this axis also shows the actual laser linewidth Δf for a bit rate of 560 Mbit/s. The required optical input power P_r expressed in dBm is given on the vertical axis. All other system parameters required to compute the results shown in this figure are taken from the Table 12.1

As seen from the figure above, laser linewidth increases from left to right side, while the required optical power at the input to the receiver from bottom to top. Thus, an optical communication system is more sensitive when it is placed at the bottom of this figure. In addition, influence of phase noise is less when the system is located more on the right. As a reference, a direct detection system has been picked up in Fig. 12.3. This system is insensitive to laser phase noise (i.e., independent of laser linewidth Δf). Therefore, it appears as a horizontal straight line. In order to obtain best system performance, direct detection system has been optimized with respect to threshold level, avalanche gain of APD and bandwidth of low-pass filter as discussed in Chapter ten. As a second reference, the same figure includes an optical PSK homodyne system under ideal conditions i.e., shot noise limited receiver operation without any laser phase noise. Therefore, it represents the theoretical optimum. Since laser phase noise is neglected in this ideal PSK homodyne system, it also appears as a horizontal straight line (i.e., independent of laser linewidth Δf).

On the basis of Fig. 12.3, coherent optical communication systems can be divided into three groups as far as laser linewidth requirements are concerned.

The *first group* contains the coherent optical homodyne systems which exhibit a high sensitivity (i.e., the lines are located far down). However, the requirements for maximum permissible laser linewidth are very strong (i.e., the lines are located on the extreme left side). Heterodyne systems with coherent (i.e., synchronous) detection not shown in the figure are also belong to this group. These systems have somewhat lesser sensitivity.

The *second group* includes only the DPSK heterodyne system based on incoherent autocorrelation detection (see Section 11.3.3). This system represents a good compromise between the sensitivity gain and laser linewidth requirements. A DPSK heterodyne system can tolerate linewidths of about 1 % of the bit rate.

The *third group* contains all incoherent ASK and FSK heterodyne systems. For these systems, laser linewidths of 20% to 30% of the bit rate $1/T$ can be allowed provided normalized bandwidth of the IF filter is $BT = 2f_g T = 1.58$. As discussed in Chapter eleven, this bandwidth is optimum only when there is a Gaussian IF filter with cut-off frequency f_g and no phase noise. If the bandwidth of IF filter is optimized by taking into consideration the influence of laser phase noise also, then lasers with linewidths in the range of bit rate $1/T$ and even more can be employed ($\Delta f T > 1$). As an example, Fig. 12.3 shows an optimized ASK heterodyne system with envelope detection. No error rate floor can be observed. As discussed in Section 11.3.1, optimum bandwidth of IF filter strongly depends on the received optical power P_r when laser phase noise is taken into account. It becomes clear again that heterodyne systems with incoherent detection exhibit the lowest receiver sensitivity.

Comparing the required incoming power levels P_r in the absence of laser phase noise (i.e., $\Delta f T = 0$), a difference of about 1 dB to 2 dB as compared to the optical power levels given in Table 12.2 can be recognized. The reason for this difference is that Table 12.2 is based on the particular symbol sequences all "0" and all "1", whereas Fig. 12.3 is based on single "0" and single "1". Thus, Fig. 12.3 also includes the influence of intersymbol interference. This influence has been excluded in Table 12.2.

It may be mentioned that complexity of realization and cost drastically increase for the systems located on the left side of this figure to those on the right side.

12.2.3 SENSITIVITY GAIN OF COHERENT SYSTEMS

The sensitivity gain of coherent optical communication systems has been shown in Fig. 12.1 under ideal conditions ($\Delta f = 0$). When phase noise disturbs the system, a similar, but modified graphic can be drawn. However, each change in laser linewidth would yield another constellation of the blocks. In this Section, procedure for obtaining the modified graphic will be discussed in brief.

In order to design a new graphic indicating the sensitivity gain of coherent systems corrupted by phase noise, results of Fig. 12.3 must be taken into account. With a fixed laser linewidth Δf, corresponding sensitivity gains can be taken directly from this figure by drawing a vertical straight line at the given linewidth Δf. The difference in received optical power levels P_r in dBm can either be positive or negative for the systems included in this figure. Hence, this difference represents the gain or the loss in receiver sensitivity. Depending on the laser linewidth, location of different blocks in Fig. 12.1 will change. Thereby, systems with a high sensitivity gain may change their position from the bottom to the top. For example, if laser linewidth is very broad, then a PSK homodyne system may even exhibit a lower sensitivity than a conventional direct detection system with intensity modulation. In addition, when the laser linewidth exceeds a certain limit, underlying BER of $p = 10^{-10}$ is no longer achievable.

12.2.4 EYE PATTERN

As already mentioned in the previous Chapters, eye pattern measurement represents a powerful and low-cost tool to assess system performance and to highlight characteristic features of electrical or optical digital communication systems. The eye pattern measurement technique is based on the detected signal at the input to the sample and hold and decision circuits. When this signal is displayed bitwise on the screen of oscilloscope (see Section 9.5), a pattern similar to an eye is created. As explained in Chapters nine to eleven, quality of detected signal is of fundamental importance for the performance of overall transmission system, primarily for the BER.

Let us consider again the coherent optical communication systems discussed in Chapter eleven. These systems can in principle be classified into two groups:

- systems with a horizontal symmetrical eye pattern and
- systems with a horizontal asymmetrical eye pattern.

With the exception of both PSK homodyne and heterodyne systems (see Fig. 12.1), all other systems yield an asymmetrical eye pattern.

As an example, Fig. 12.4 displays the symmetrical and the asymmetrical eye pattern of coherent PSK and ASK systems respectively. Both the noisy eye pattern shown at the right side of Fig. 12.4 are simulated by means of fast computer program. In the simulation, receiver additive Gaussian noise (i.e., shot noise and thermal noise) have been neglected. This means that these eye pattern are obtained when the detected signal is disturbed only by laser phase noise. This figure also illustrates the corresponding noiseless eye pattern which are similar in shape, but different in amplitude level. All four eye pattern shown in Fig. 12.4 are based on a Gaussian filter of bandwidth $B = 2f_g = 1/T$ (see Eq. 9.83).

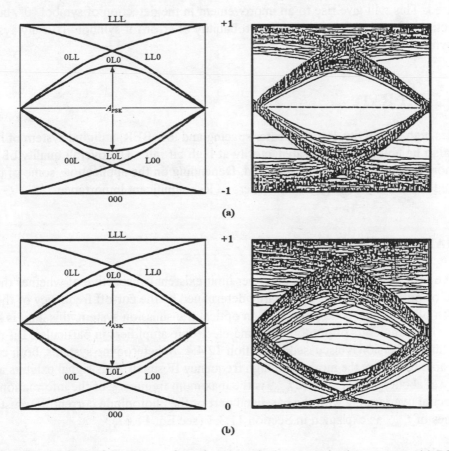

Fig 12.4: Comparison of eye pattern for coherent optical communication systems: (a) PSK system and (b) ASK system

Let us first consider the *eye pattern of PSK system* shown in Fig. 12.4a. This eye pattern is characterized by normalized amplitude levels +1 (symbol "1") and -1 (symbol "0"). Thus, PSK system is a bipolar digital transmission system and both the transmitted symbols "0" and "1" are likewise disturbed by phase noise (see Section 11.1.2). As a result, we observe a symmetrical eye pattern with respect to the horizontal axis at amplitude level 0. The optimum threshold level is always located at the centre line of the eye pattern i.e., $E_{opt} = 0$. In contrast to this, *eye pattern of ASK system* shown Fig. 12.4b is characterized by normalized amplitude levels +1 (symbol "1") and 0 (symbol "0"). Hence, ASK system is an unipolar digital transmission system (see Sections 11.1.1 and 11.3.1). It becomes clear from Fig. 12.4b that a transmitted symbol "1" is always disturbed by phase noise much more seriously than a transmitted symbol "0". This results in an asymmetrical eye pattern with an optimum threshold at $E_{opt} < 0.5$ which is always below the optimum threshold given in absence of laser phase noise (i.e., $E_{opt} = 0.5$). Further, Fig. 12.4b clearly indicates that the detection of symbol "1" is always deteriorated, while the detection of symbol "0" is always improved by phase noise! It becomes evident that the signal level (amplitude level) is always decreased by laser phase noise irrespective of whether a symbol "0" or "1" is transmitted. This will give rise to an improvement in the detection of symbol "0", but degrades the detection of symbol "1". Hence, the probability of error for symbol "1" is always more than that of symbol "0".

12.2.5   BIT RATE

Besides high sensitivity, large repeater spacing and low BER, a digital system of high quality should also be able to operate satisfactorily at high bit rate. The overall quality of a system is better, more the above features are satisfied. Depending on the application, some of the features mentioned above may be of great and others of less significant importance.

(i)   MAXIMUM BIT RATE

For the bit rate, there is always an upper limit existent irrespective of whether the system is electrical or optical. This limit is primarily determined by the cut-off frequency or the restricted bandwidth of employed components. In an optical transmission system, this limit is specified in terms of the parameters of photodiodes and electronic amplifiers in particular. Bit rate limit in terms of fiber dispersion is discussed in Section 12.4.4. In heterodyne systems, both components must be able to follow the modulated high frequency IF signal. This signal exhibits a maximum frequency of about $f_{IF} + f_{s,max}$, where $f_{s,max}$ is the maximum frequency of the information signal. To avoid spectral overlapping with the baseband part of the photodiode current, IF must be at least three times of $f_{s,max}$ as explained in Section 11.3.1 (see Eq. 11.67).

In the following, maximum permissible bit rate $R_{max} = 1\text{bit}/T_{min}$ (in bit/s) or the minimum permissible bit period T_{min} will be estimated. For this, we presume that photodiodes and amplifiers are characterized by a common cut-off frequency f_{pa}. We also presume that information signal is bandlimited to $0 < f_s < f_{s,max} = 1/(2T_{min})$. This means that only the sinusoidal wave with period $2T_{min}$ is transmitted when the periodic symbol sequence $\cdots01010101\cdots$ is applied to the system (see Section 12.4.4). It should be noted that this periodic symbol pattern represents the worst-case pattern with respect to maximum transmission speed and, therefore, maximum signal frequency $f_{s,max}$. Obviously, the sinusoidal signal still enables an errorless decision irrespective of whether symbol "0" or symbol "1" has been transmitted.

Summarizing the results of the above discussion, maximum permissible bit rate can be determined as

$$R_{max} = \frac{1\,bit}{T_{min}} = \frac{1\,bit}{2}\,f_{pa} \quad \text{for heterodyne systems}$$

$$(12.7)$$

$$R_{max} = \frac{1\,bit}{T_{min}} = 2\,bit\,f_{pa} \quad \text{for homodyne systems}$$

Thus, in an optical homodyne system ($f_{IF} = 0$), maximum allowable bit rate is at least four times the maximum bit rate of an optical heterodyne system ($f_{IF} \neq 0$) under ideal conditions. If real conditions are considered such as non-ideal filter with gradual instead of a sharp cut-off frequency response, then difference in bit rate is even somewhat higher.

The coherent optical communication systems are able to transmit very high bit rates. In the development of coherent systems, transmission speed has progressed from some Mbit/s to more than 10 Gbit/s. In combination with coherent multichannel communication (CMC) or wavelength division multiplexing (WDM) and optical time division multiplexing (OTDM), coherent optical communication systems even enable ultrahigh-speed transmission. In such systems, each optical carrier (up to hundred and more) is modulated by a high bit rate signal of up to 100 Gbit/s by means of OTDM (see Section 12.3).

Example 12.1

For ten optical carriers (channels) at different optical wavelengths and a bit rate of 100 Gbit/s at each optical carrier obtained by CMC and OTDM, the overall bit rate in the fiber link is 1000 Gbit/s = 1 Tbit/s.

It becomes clear by this simple example that coherent optical communication provides the basis for highly sophisticated Tbit/s transmission systems which offer many new interesting applications; for example, three-dimensional (holographic) multipoint-to-multipoint video transmission. Three-dimensional high-resolution video transmission is very promising for remote-controlled medical help or video conference with "telepresenting" speakers.

(ii) Interrelation between Bit Rate and Laser Linewidth

For a fixed laser linewidth Δf, the allowable bit rate $R = 1$ bit/T or bit frequency $f_B = 1/T$ is of significant importance in highly sophisticated transmission systems and networks; for example, data highways. As discussed in Chapter eleven, a high bit rate transmission system is relatively more tolerant to phase noise of the laser sources than a low bit rate system. Performance degradation due to laser phase noise is lower, better the following relationship is fulfilled:

$$\Delta f\, T \ll 1 \quad \text{or} \quad \Delta f \ll f_B \qquad (12.8)$$

Considering, for example, an incoherent DPSK heterodyne system, this feature of coherent optical communication systems becomes clear in particular. In this system, variance of phase noise (in the strict sense phase noise difference) increases proportional to the bit duration T and laser linewidth Δf. As the bit rate increases (implying T decreases), variance of phase noise difference decreases and hence the influence of laser phase noise. Similar results can be observed by considering ASK and PSK homodyne or ASK, PSK and FSK heterodyne systems.

As explained in Section 11.3.1, each IF filter give rise to an undesired coupling of phase and amplitude. Hence, phase noise results in additional amplitude noise. As the IF filter bandwidth decreases, efficiency of the undesired phase noise-to-amplitude conversion increases. When the bit rate is high, distortion by phase-to-amplitude conversion is low since high bit rates always require a large bandwidth. On the other hand, lower the bit rate, stronger is the influence of the undesired interaction of phase noise and amplitude. If bit rate reaches a certain lower limit, then system performance again improves by further decreasing the bit rate. This surprising result can be explained as follows: When the bit rate is very low and IF filter bandwidth is small, filtering of the phase noise becomes dominant (which yields an improvement in system performance) as compared to the undesired phase-to-amplitude conversion (which deteriorates the system performance). This, however, can only be observed when intersymbol interference is neglected. At a fixed bit rate and laser linewidth, optimum IF filter bandwidth exists for a minimum BER (see Fig. 11.16). To make it more realistic, effect of intersymbol interference must also be taken into account.

Performance of homodyne systems can also be improved by providing $\Delta f\, T \ll 1$. The reason for this is the operation principle of OPLL. As explained in Section 11.1.3, an OPLL cannot distinguish between phase noise and phase information. Therefore, noise and information signal must be separated a priori in order to guarantee OPLL operation i.e., permanent phase tracking with low rest-phase error. Best separation is obtained when phase noise and information are characterized by non-overlapping frequency spectra. If the bit rate is low, then these spectra are located in the same frequency range and a spectral separation becomes nearly impossible. It should be remembered that the degrading effect of laser phase noise is more in the low frequency range (see Section 3.2.2). When bit rate is high enough, both spectra can be easily separated.

12.2.6 SENSITIVITY GAIN BY OPTICAL AMPLIFICATION

Optical amplifiers are used as booster amplifier in the transmitter, in-line amplifiers in the fiber link or as preamplifier in the receiver (see Chapter seven). Optical preamplifier when employed to improve the sensitivity of optical receiver may provide performance close to shot noise limit. Hence, optical preamplifier is well-suited in combination with a direct detection receiver which usually exhibits a difference of about 20 dB and more from the shot noise limit. A comparison of preamplified optical direct detection receiver and heterodyne receiver without an optical amplifier shows that both can have equivalent performance if the optical filters can be made with a bandwidth comparable to electronic IF filters.

As a heterodyne receiver already operates close to shot noise limit, sensitivity gain due to optical preamplification is negligible. In the strict sense, sensitivity even deteriorates by about 3 dB if the receiver is shot noise limited without preamplifier. The reason for this is given by the inherent noise of optical amplifier due to spontaneous emission (see Chapter seven).

The sensitivity gain of direct detection and heterodyne receivers as a function of amplifier gain G and local laser power P_1 is compared in Fig. 12.5.

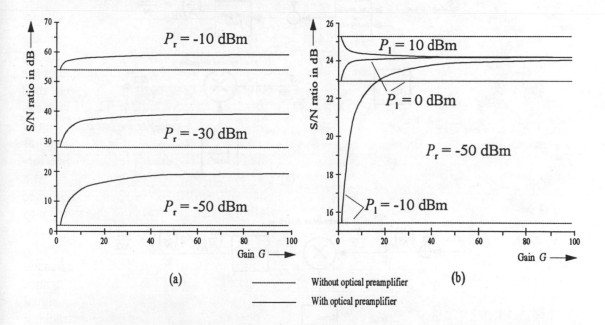

Fig. 12.5: Sensitivity gain of (a) direct detection and (b) coherent detection receivers with optical preamplifier

It becomes clear from this figure that sensitivity always reaches a saturation irrespective of whether a direct or a coherent detection is used. In a direct detection receiver, maximum sensitivity is achieved at the point of saturation. A coherent receiver may even exhibit a sensitivity loss (see curve for $P_1 = 10$ dBm in Fig 12.5b). As shown in this figure, performance of a coherent receiver can only be improved when local laser power is low, which means that a poor coherent receiver has been employed.

In addition, Fig. 12.5 also illustrates that a preamplified direct detection receiver attains a sensitivity close to the high sensitivity of a coherent receiver. For this reason, a direct detection system with optical amplifier represent a well-suited, technical and economical alternative to high sensitivity coherent receivers.

12.2.7 MICROWAVE, OPTICAL HETERODYNING AND HOMODYNING

In this Section, we briefly discuss the difference between coherent microwave and coherent optical detection (Fig. 12.6).

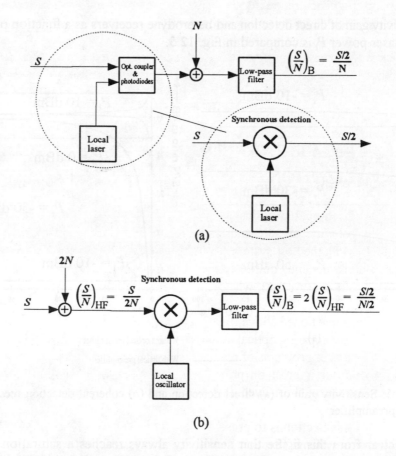

(a)

(b)

Fig. 12.6: Comparison of coherent (a) optical and (b) microwave detection receivers

When a microwave signal disturbed by additive Gaussian noise is detected by means of a heterodyne or homodyne receiver, sensitivity of heterodyne receiver normally equals the sensitivity of homodyne receiver. In comparison, a difference of 3 dB in sensitivity will exist when a fiber-based optical heterodyne or homodyne system is considered. Microwave heterodyne and homodyne system give nearly the same performance as optical heterodyne system, but their performance is 3 dB poorer as compared to an optical homodyne system. This interesting result can be explained as follows: Let us consider microwave homodyne and optical homodyne systems. In microwave systems, received signal is normally disturbed by additive noise while in optical system it is not so. Hence, the key difference between coherent microwave and coherent optical detection systems is the location of source of noise. As shown in Fig. 12.6, noise in an optical receiver is primarily generated in the mixer i.e., within the photodiode. On the other hand, noise in a microwave receiver is already existing in front of the mixer.

It should be noted that in Fig. 12.6, same noise power spectral density in microwave and optical receivers has been considered. In a microwave receiver, noise is in the HF range, while it is in the baseband frequency range in case of an optical receiver. As a result, noise power in the former case will be twice of noise power in the latter case. This difference in microwave and optical systems is at least partly reduced when optical space communication is considered in Chapter fifteen. Due to background noise which is a fundamental source of noise in optical free-space communication systems, we now have one source of noise at the input to the optical receiver (background noise) and the other in the mixer (shot noise and amplifier noise).

12.3   APPLICATIONS AND SPECIAL PROBLEMS OF REALIZATION

We have so far compared optical systems by considering different criteria such as BER, receiver sensitivity, bit rate and laser linewidth requirements. In order to answer the question "which is really the best system", these features alone are not sufficient. The answer to this question primarily depends on the application. Each application is usually characterized by its own optimum solution (system), which represents a trade-off instead of a real optimum.

Besides the criteria mentioned above, the complexity of realization as well as cost are another important features for comparing optical systems. One aim of this Section is to perform a system comparison on the basis of problems of realization and typical applications (Section 12.3.1). For this, we shall focus our interest on principle problems since special problems of realization primarily depend on recent short-lived developments.

A second aim of this Section is to present some typical applications such as high-speed long-haul data transmission systems. (Section 12.3.2). These systems, in particular, are most important in transoceanic data transmission without regenerators. Here, bit rate-distance products in excess of 100 Tbit/s·km seem to be realizable [40]. Two other important applications of optical communications which are optical networks and optical space communications are discussed in detail in Chapters fourteen and fifteen.

12.3.1 SYSTEM COMPARISON BASED ON REALIZATION ASPECTS

As far as complexity of realization and cost are concerned, substantial differences can be observed between homodyne, coherent and incoherent heterodyne as well as direct detection systems. As compared to the realization of coherent optical communication systems, realization of a *direct detection system* is very economical since local laser, optical coupler, PLL or OPLL, AFC and polarization control are not required. This represents an important advantage of employing optical transmission system with intensity modulation and direct detection (IM/DD). Because of this, direct detection systems are used for different applications. In particular, optical direct detection systems are most advantageous for single-wavelength data transmission over short and medium distances; for example, data transmission in local area networks (LANs), plants, cars, aircrafts and so on. IM/DD systems are commercially available for data rates up to 10 Gbit/s and more. In combination with WDM and optical amplifiers such as in-line, booster and preamplifiers, direct detection systems offer a performance which is comparable with coherent optical communication systems.

In *optical homodyne systems*, complexity of realization and cost are maximum. The main reason for this is the OPLL which is required to match optical phase and frequency of both received and local laser light waves. In contrast to a direct detection system, homodyne systems are, in addition, much more sensitive to change in the environmental conditions such as change in temperature or mechanical and acoustic vibrations. Moreover, optical homodyning requires high-quality single-mode laser sources with extremely narrow linewidths of about 0.01% of the bit rate. Hence, solid-state lasers (e.g., Nd:YAG laser) are frequently used in homodyne systems.

Optical homodyne systems, because of their large complexity of realization and high costs, are useful for a very limited number of applications only. It should be remembered that these limitations have been overcome to some extent by using a synchronization bit controlled OPLL instead of a Costas loop or an OPLL employing a pilot carrier (see Section 11.1.3). Since a PSK homodyne system exhibits best receiver sensitivity, this system is most suitable for *repeaterless long-haul point-to-point data transmission*; for example, transisland or transoceanic transmission links (Fig. 12.7a) and intersatellite links (Chapter fifteen). Here, repeaterless means that *neither* an electric repeater *nor* an optical in-line amplifier is used and transmission channel is simply a passive fiber or free space i.e., without any active components.

The long-haul point-to-point transmission links can also be realized by using a direct detection system with a chain of optical in-line amplifiers such as Erbium-doped fiber amplifiers (see Chapter seven). Since each optical amplifier requires the supply of electric power to maintain amplification mechanism, this is actually an active instead of a passive fiber link.

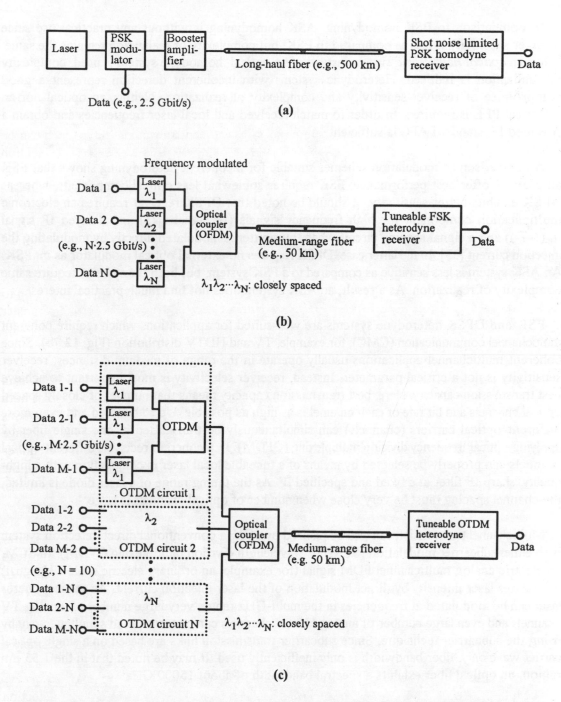

Fig. 12.7: Typical applications of coherent optical communication systems: (a) repeaterless, highly sensitive, long-haul point-to-point transmission using PSK, (b) coherent multichannel communication (CMC) with optical frequency division multiplexing (OFDM) and (c) multigigabit data transmission (e.g., 1 Tbit/s) with OFDM and optical time division multiplexing (OTDM)

In comparison to PSK homodyning, ASK homodyning is without any practical use since receiver sensitivity is less as compared to PSK, but complexity of realization remains the same. When an *optical heterodyne system* instead of an optical homodyne system is used, complexity of realization is reduced. Heterodyne systems with incoherent detection represent a good compromise of receiver sensitivity and complexity of realization. Neither an optical nor an electrical PLL is required. In order to match received and local laser frequencies and obtain a specified IF, standard AFC is sufficient.

A comparison of modulation schemes suitable for incoherent heterodyning shows that FSK and DPSK offer best performance. FSK requires somewhat less critical components, whereas DPSK exhibits better sensitivity. It should be noted that DPSK receiver requires an electronic multiplication component for high frequency signals (IF signal $i_{IF}(t)$ and delayed IF signal $i_{IF}(t-T)$) and a signal shaping circuit when transmitter is modulated directly by modulating the injection current [33]. In the latter case, DPSK requires no external optical modulator as in PSK. An ASK system is less sensitive as compared to a FSK system, but both the systems require same complexity of realization. As a result, an ASK system does not find much practical interest.

FSK and DPSK heterodyne systems are well-suited for applications which require coherent multichannel communication (CMC); for example, TV and HDTV distribution (Fig. 12.7b). Since coherent multichannel applications usually operate in the range of medium distances, receiver sensitivity is not a critical parameter. Instead, receiver selectivity is most important to achieve best transmission capacity. Here, best transmission capacity means that number of closely spaced optical channels and bit rate of each channel is as high as possible. Up to hundred and even more different optical carriers (channels) can simultaneously be transmitted in one single fiber by applying optical frequency division multiplexing (OFDM). In a coherent receiver, different optical channels can properly be selected by means of a tuneable local laser in conjunction with a high-quality, sharp IF filter at a fixed and specified IF. As the tuning range of a laser diode is limited, the channel spacing must be very close when number of optical channels is large.

Multichannel transmission can also be realized by using a conventional direct detection system with either subcarrier modulation (SCM) or wavelength division multiplexing (WDM). In SCM, an electric analog multichannel FDM signal (for example, an ordinary electric cable TV signal) modulates laser intensity by direct modulation of the laser injection current. As semiconductor laser can be modulated at frequencies in the multi-GHz range, very large number of analog TV channels and even large number of analog HDTV channels can be transmitted simultaneously by using the subcarrier technique. Since subcarrier transmission links are based on a single optical carrier wave only, fiber bandwidth is only inefficiently used. It may be noted that in the 1.55 μm region, an optical fiber exhibits a spectral bandwidth of about 15000 GHz!

In WDM, channel selection is performed by a set of optical filters or a single tunable optical filter as described in Section 9.2.2. As the bandwidth of an optical filter is very broad as compared to a microwave filter, only a limited number in the order of 8 to 32 optical carriers with relatively large channel spacing can be transmitted simultaneously. Therefore, WDM also does not enable to use the overall fiber bandwidth when direct detection is used. In contrast to this, CMC systems with closely-spaced OFDM are able to use the large fiber bandwidth most efficiently. In CMC systems, each optical carrier is again modulated by an analog microwave multichannel FDM signal or a digital signal of ultrahigh bit rate. By employing optical time division multiplexing in addition, even Tbit/s communication systems becomes possible (Fig. 12.7c).

Table 12.3: Comparison of optical communication systems in terms of bit rate, repeater spacing, complexity of realization and typical applications

	Homodyne system	Coherent heterodyne system	Incoherent heterodyne system	Direct system
Typical bit rate R	high e.g., 1 Gbit/s to 10 Gbit/s	relatively low e.g., 0.5 Gbit/s to 1 Gbit/s	relatively low e.g., 0.5 Gbit/s to 1 Gbit/s	high e.g., 1 Gbit/s to 10 Gbit/s
Typical repeater spacing at $R = 560$ Mbit/s and $\alpha = 0.2$ dB/km.	ASK: 212 km PSK: 242 km	ASK: 197 km FSK: 212 km PSK: 227 km	ASK: 195 km FSK: 210 km DPSK: 225 km	147 km
Complexity of realization	very large	large	relatively small	small
Phase and frequency tracking	OPLL required	PLL required	no PLL, but AFC required	no PLL and AFC required
Typical applications	Long-range transmission Transoceanic transmission Optical space Communications		Multichannel FDM BISDN HDTV distribution Multimedia services	Short range transmission e.g., LANs, remote control in plants, cars, aircrafts Long range transmission with optical in-line amplifiers

In the realization of OFDM systems with a large number of optical channels, optical transmitter becomes very comprehensive since each optical channel requires a single-mode laser. To achieve a fixed frequency spacing, all lasers must be stabilized and synchronized. This can be achieved either absolutely or relatively by means of a multichannel stabilization unit. For this, all frequencies present in the CMC transmitter comb are periodically measured, which can, for example, be performed by using a scanning heterodyne spectrometer.

Coherent multichannel communication (CMC) systems with FSK or DPSK modulation represent the primary application of coherent optical communications. Because of their flexibility, they are also well-suited for multipoint-to-multipoint networks (multiple-access networks) such as ultrahigh bit rate broadband ISDN (BISDN) or HDTV distribution networks including also video-on-demand. CMC systems will play a major role in highly sophisticated broadband networks based on information superhighways offering a broad spectrum of new services; for example interactive broadband multimedia services.

Fundamental differences and applications of various optical communication systems are summarized in Table 12.3 above. It becomes clear from the above discussion and Table 12.3 that only four optical communication systems are of significant practical interest:

- optical systems with intensity modulation and direct detection,

- optical FSK heterodyne systems with incoherent detection,

- optical DPSK heterodyne systems with incoherent detection and

- optical PSK homodyne system.

With the realization of low-cost coherent optical communication systems, application field of a direct detection system may be more and more taken over by coherent systems. However, the technical and economical breakthrough of coherent optical communication systems is closely related with the development of *integrated optics*.

Optoelectronic integrated circuits (OEICs) represent the key components of low-cost economical and commercial coherent optical multichannel transmission systems. OEIC allows to integrate all required lasers along with the synchronization circuit and modulation electronics on one single substrate in the transmitter of an optical multichannel transmission system. In the optical receiver, OEIC combines optical components (for example, the optical coupler), optoelectronic components (tuneable local laser and photodiodes) as well as the electrical components (low noise front-end amplifier) on one single substrate. In the realization of OEIC, know-how of realizing standard electronic integrated circuits (IC) can be used; for example, the clean-room technology. However, OEIC and ordinary electronic IC exhibit some essential differences which are summarized in Table 12.4 below.

Table 12.4: Comparison of optoelectronic and conventional microelectronic integrated circuit

	Microelectronic integrated circuit (IC)	Optoelectronic integrated circuit (OEIC)
Wavelength	electrical	electrical *and* optical
Guide	metallic	metallic *and* optical waveguide
Components	only electrical	electrical, optical and optoelectronic
Length-side ratio	about 1 : 1	about 1 : 1000 (e.g., laser diode)
Technology	planar	epitaxial
Material	Si, GaAs	GaAs, InP
Remark	many similar components (e.g., many thousand transistors)	less, but very different components (e.g., optical coupler, PIN diode, electrical amplifier)

In the following, some important practical requirements are briefly discussed. We focus our interest on comparison only.

(a) TRANSMITTER

The main requirements for a transmitter laser are narrow linewidth, uniform FM response in case of direct modulation and high output power. Moreover, frequency stabilization is of primary importance. In CMC transmission systems, in addition of frequency stabilization, synchronization of all optical carriers must also be performed.

(b) CHANNEL

In order to achieve a long-haul high bit rate transmission, single-mode fiber should have a low attenuation of less than 0.2 dB/km and a low dispersion, which can be achieved by dispersion equalization at the optical level or by dispersion-shifted fiber. Free-space channels should have low background noise in addition.

(c) COHERENT RECEIVER

Irrespective of whether a homodyne receiver or heterodyne receiver is used, states of polarization of the optical incoming signal and the local laser must be matched. As explained in Chapter five, various polarization-handling schemes exist which primarily differ in loss and speed. For the local laser, narrow linewidth and high output power are again the most important requirements. In addition, wavelength tuning is especially required when a coherent multichannel OFDM system has to be realized.

(d) HETERODYNE RECEIVER

In coherent optical communication systems with heterodyne detection, stable IF is of fundamental importance since temperature, mechanical and electromagnetic disturbances as well as low-frequency laser phase noise may cause fluctuations. The IF fluctuations are usually compensated with an AFC which tracks the frequency of the local laser with respect to the frequency of the received input signal. These fluctuations have to be kept to less than ±10 MHz to keep the sensitivity loss below 0.1 dB. Imperfect demodulation circuit and non-ideal IF signal processing are primarily responsible for a deviation from shot noise limited operation in practical receivers.

(e) HOMODYNE RECEIVER

The most serious problem in the realization of a homodyne receiver is the OPLL. Here, great care must be taken to separate phase noise and phase information and to keep the residual phase error low.

(f) INTRADYNE RECEIVER

An optical intradyne receiver is characterized by an intermediate frequency that is very close to zero. Like in a homodyne receiver, received optical signal is converted into a baseband signal. It offers the possibility of transmitting very high bit rates. On the other hand, no optical phase matching (i.e., no OPLL) is required. Instead, this receiver requires either a symmetrical six-port coupler or an optical 90^0 hybrid as critical component. In addition, the electronic part of receiver is more comprehensive as compared to heterodyne and homodyne receivers.

12.3.2   HIGH-SPEED LONG-HAUL DATA LINKS

Advanced communication networks as described in Chapter fourteen require transmission of high bit rates over large distances [1, 4]. For this, high-speed long-haul data links based on advanced fiber optics are needed [11]. Typical applications are transoceanic links, interoffice links and network backbones.

> High-speed long-haul data links are key components of information superhighways.

At present, fiber loss no longer imposes a serious limitation on high-capacity light wave transmission systems due to the advent of highly reliable and highly efficient Erbium-doped fiber amplifiers (EDFAs) operating in the 1550 nm wavelength region (Chapter seven). The important obstacles remain to be overcome in order to enhance single wavelength system capacity are fiber dispersion, fiber non-linearities and *electronic bottlenecks* at very high speed.

In conventional fibers being used in current networks, dispersion is about 17 ps/(km·nm) at 1550 nm. Thus, the maximum, practical transmission distance in these vastly deployed fibers is about 60 km at a bit rate of 10 Gbit/s and only about 4 km at 40 Gbit/s [1]. This limit is referred as dispersion limit. An explanation of this limit is given in Section 12.4.4. If dispersion is lowered by selecting a dispersion-compensation technique, self-phase modulation (SPM) may play a critical role in such systems. Even if zero dispersion is chosen, SPM can induce spectral broadening resulting in substantial pulse distortion. In addition, polarization-mode dispersion (PMD) caused by residual fiber birefringence will impose a limit on the transmission distance. Nevertheless, impressive results have been achieved in terms of error-free transmission over 9000 km fiber at 10 Gbit/s [38] and error-free transmission over 200 km at 100 Gbit/s [21] by choosing to operate at the averaged zero-dispersion wavelength.

It is difficult to realize high-speed optical transmission systems by means of electrical circuits if bit rate exceeds 10 Gbit/s. However, there exist several techniques to upgrade the existing transmission systems without replacing the installed fiber. At present, high-speed long-haul data transmission systems can be realized by means of

- dispersion compensation,
- wavelength division multiplexing (WDM),
- solitons and
- spectral inversion.

In future, optical time division multiplexing (OTDM) will be an additional option to increase bit rate as discussed in the previous Section and in e.g., [8, 47]. By combining OTDM and WDM (or OFDM, CMC), very high bit rates can be achieved [21, 24]. Even bit rates in the order of some Tbit/s become realistic. In addition, code division multiplexing (CDM) may be a possible option to increase the transmission capacity further [44].

Dispersion compensation will be required in optical amplified or non-amplified long-haul 10 Gbit/s systems using conventional fibers (see Subsection (i) below). Now-a-days, commercial systems employing single-channel transmission in conventional fibers offer, for example, a repeaterless transmission distance as long as 300 km at 622 Mbit/s. This transmission capacity can be increased by using dispersion-compensation techniques such as dispersion-shifted fibers (DSFs) which are also referred as dispersion-compensation fibers (DCFs). Another well-suited approach to increase the capacity of a fiber or an optical network is to use *WDM techniques* as described in Subsection (ii). In principle, WDM and compensation techniques can also be combined. However, this is difficult to realize due to the narrow wavelength range of sufficiently small dispersion. Ultimately, the use of *solitons* in a DSF appears to be the best choice, not only in transoceanic systems, but also in medium-haul, ultra-high speed systems as discussed in Subsection (iii). At present, the mid-span *spectral inversion* (MSSI) technique has the best figure of merit [1].

High-speed long-haul data links can be realized without and with optical amplifiers which can be employed as booster amplifiers, in-line amplifiers (repeaters), and preamplifiers. Depending on whether repeaters are used or not transmission systems are referred as non-repeaterless systems or repeaterless systems respectively. A typical repeaterless system consists of a low chirp transmitter, an optical power booster also referred as postamplifier and a highly sensitive optical preamplifier receiver.

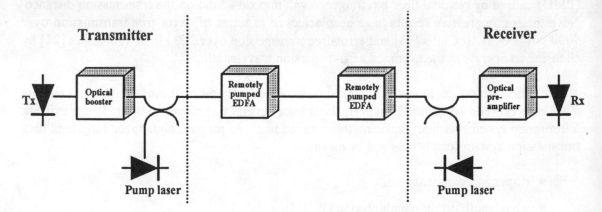

Fig. 12.8: Use of remotely pumped EDFAs to increase power budget

The recent developments of optical amplifiers have led to substantial increase in the span of repeaterless systems. An optical postamplifier boosts the output power of a transmitter by about

20 dB and an optical preamplifier improves the receiver sensitivity by about 10 dB. The use of so-called remotely pumped amplification, which consists in sending the pump beam from the terminals (transmitter and receiver) towards an optical amplifier located several tens of kilometres away, allows further increase of power budgets. A typical example of remotely pumped data transmission system is shown in Fig. 12.8.

By employing booster amplifier, preamplifier and cascaded in-line amplifiers (repeaters) such as EDFAs, transmission distance can be increased drastically as compared to the distance of a repeaterless transmission system.

(i) DISPERSION COMPENSATION

Fiber dispersion drastically reduces transmission distance when bit rate is high. As shown in Section 12.4.4 below, under ideal conditions

> the dispersion limit of transmission distance for a standard single-mode fiber link at 1.55 μm for 10 Gbit/s is about 100 km.

This limit presumes non-chirped on-off keying (OOK) or amplitude shift keying (ASK) signal and the absence of fiber non-linearities. Thus, dispersion compensation will especially be required in optical amplified long-haul transmission systems (more than about 100 km) using conventional fibers at 10 Gbit/s or more.

At present, a lot of non-dispersion-shifted fiber (i.e., standard single-mode fiber, SSMF) has already been installed world-wide for transmission at bit rates up to 2.5 Gbit/s. Investigations to upgrade the existing systems to 10 Gbit/s have become an important task [19, 45]. Laser chirp and fiber dispersion are the basic limitations of transmission distance in high-speed data transmission links. The influence of chirp can be eliminated by using an external optical modulator instead of modulating the laser injection current. System degradation due to dispersion can either be reduced by a dispersion-shifted fiber (DSF), where the zero dispersion wavelength is shifted from 1.3 μm to $\lambda = 1.55$ μm (Chapter four) or by employing a dispersion-compensation technique. In WDM systems, a dispersion-flattened fiber (DFF) is more advantageous as compared to a DSF. With such fibers, WDM can be used effectively for obtaining very high bit rates. Both the fibers are realized by modifying the refraction index profile [22]. For upgrading an existing network without replacing the installed fibers, DSF and DFF are not feasible.

Among the techniques available for upgradation, the most promising one is to use the negative-dispersion fiber (Chapter four). These fibers have opposite sign of the dispersion as compared to the dispersion of transmission fiber used in the existing network. Other approaches are transmitter compensation techniques which are the most easily implemented but provide a limited amount of compensation and receiver-based compensation techniques. In the first one, dispersion compensation is achieved by pre-chirping the data pulses. To generate chirped pulses, frequency modulation of the laser can be employed [18].

Besides dispersion-shifted fibers, transmitter and receiver dispersion-compensation techniques, optical solitons are the other attractive approach to compensate for the pulse distortion due to fiber dispersion (see Subsection (iii) below). For the oceanic transmission systems, the use of solitons appears to be the best choice. Most advantageous for very high-speed data transmission is, however, a combination of optical solitons and dispersion-shifted fibers.

(ii) WAVELENGTH DIVISION MULTIPLEXING

The most popular approach to realize Tbit/s-transmission links or networks appears to be to divide the optical spectrum into many different channels, each channel corresponds to a different wavelength (or frequency). This approach, called *wavelength division multiplexing* (WDM), allows us to realize ultrahigh-speed communication link using several hundred to thousand channels at different wavelengths, each operated at moderate speeds of say 2.48 Gbit/s. Carrier spacing is about 0.8 nm (see Section 9.2.2), whereas in dense WDM (DWDM) it is less than 0.2 nm. At each wavelength, we could have another level of electronic frequency division multiplexing (FDM) where the channel bandwidth can further be subdivided into many RF channels, each at a different microwave frequency. This is termed *subcarrier multiplexing* or *subcarrier modulation*. Each RF carrier can either be modulated by an analog (e.g., video) or a digital signal (e.g., computer data). Instead of electronic frequency division multiplexing, a wavelength can also be shared among different channels by electronic time division multiplexing (TDM). WDM is more tolerant to dispersion since the bit rate per channel is reduced at the expense of more channels.

WDM systems can be classified into three categories: (i) point-to-point transmission links, (ii) broadcast-and-select networks and (iii) wavelength-routing networks. As far as the evolution of these WDM applications is concerned, two stages can be anticipated: transmission-only structures appearing first (categories i and ii) and networks with optical switching appearing later (category iii). WDM light wave systems expand the transmission capacity by increasing the number of optical channels rather than by increasing the data rate of individual channels. This yields a significant economic advantage as it enables graceful network growth to match traffic growth. WDM systems also enable flexible network management, easier service provision and path set up as well as more reliable network architectures. WDM systems with sixteen 2.5 Gbit/s channels transmitted 1420 km and seventeen 20 Gbit/s channels transmitted 150 km [5] have been demonstrated in system experiments using EDFA repeaters.

When WDM techniques are employed to increase transmission capacity, the practical difficulty in this multi-wavelength operation is channel stabilization and selection. In repeaterless transmission applications, it is absolutely required to use a receiver with the highest possible receiver sensitivity. Transmitter with high quality laser (e.g., multi-quantum well laser) and external modulator offer near information bandwidth limited signal spectrum which is needed to minimize dispersion and to permit a narrow filter bandwidth optical preamplifier receivers (OPRs) [3]. However, their output power is not high, ranging from -5 dBm to +5 dBm. Almost all the repeaterless transmission systems use optical power boosters before the transmitter output is launched into the fiber. EDFAs serve well for this purpose since they can amplify the transmitter output to above +20 dBm. Electro- and acoustooptic tunable filters (EOTFs and AOTFs) are especially suited in multi-wavelength networks. An AOTF offers broad tuning range (hundreds of nm), narrow passband (≈ 1 nm) and relatively fast switching speed (a few μs). It also offers similar switching speed and passband performance, reduced drive complexity (as no RF surface acoustic wave transducers are required), do not introduce any frequency shift into the output wavelength(s). Also, an EOTF can be designed to offer a wide tuning range (over the EDFA wavelength window 1530 nm - 1560 nm) and the capability to select a number of wavelengths simultaneously [34].

Besides channel stabilization and selection, WDM additionally shows the following problems: channel crosstalk, non-uniform amplifier gain, management and architecture as well as cost/performance of multi-wavelength components. On the other hand WDM avoids typical TDM problems such as limits of high speed devices and dispersion at 10 Gbit/s and more.

(iii) SOLITON TRANSMISSION

Normally, non-linear effects are responsible for severe system performance degradation. In analog audio transmission systems, for example, non-linear effects cause signal distortion by generating harmonic waves of higher order. Here, degree of distortion is expressed in terms of harmonic distortion factor. In analog systems, non-linear effects primarily arises from non-linear system components such as non-linear amplifiers. In digital transmission systems, non-linear effects are not as critical as in analog systems since recovery of transmitted information only requires to distinguish binary symbol "0" and "1" by means of a threshold instead of recovery of signal shape completely.

Like an electronic or optical amplifier, a passive optical fiber also exhibits non-linearities i.e., new waves can be observed at fiber output although they are not launched into the fiber input as briefly discussed in Chapter four and e.g., [2, 7, 41, 42]. For fiber non-linearities, we have to distinguish different effects known as Raman effect, Brillouin effect, four-wave mixing and self-phase modulation (SPM) which is based on the Kerr effect (see Chapter four). In this Subsection, we focus our consideration on SPM only since it is most important in high-speed long-haul data transmission systems.

The SPM is a non-linear fiber effect. However, in contrast to the other non-linear effects mentioned above (which all yield a system performance degradation), SPM is able to improve system performance. This is a very surprising and unusual result, which needs to be explained in more detail. The Kerr effect refers to a phenomenon that can be simply described by an intensity-dependent refraction index [22]. This results in an intensity-dependent phase shift in a fiber which is known as SPM. Thus Kerr non-linearities can alter the frequency spectrum of a pulse travelling through a fiber. An interesting phenomenon occurs when one examines the Kerr effect in the negative dispersion region of a fiber.

For a high-intensity pulse, the Kerr non-linearity causes the leading edge of the pulse to develop a red shift (a shift toward lower optical frequencies) and the trailing edge to develop a blue shift (toward higher optical frequencies). In addition, because of the negative group velocity dispersion, the red-shifted light will travel slower than the unshifted centre of the pulse and the blue-shifted light will travel faster. By carefully choosing a proper combination of pulse shape, pulse intensity and pulse width, the effects of the Kerr non-linearity and the negative dispersion will cancel out exactly and the input pulse will travel without any change.

> A pulse which travel without any change through a fiber is referred to as a soliton.

By employing solitons, it is possible to compensate pulse broadening caused by chromatic dispersion completely [15, 16]. Solitons clearly show that non-linear effects, which normally disturb data transmission, can also be very useful. Since the transmission of solitons would extremely advantageous in communication systems, extensive research has been carried out in this area in the eighties and nineties e.g., [4, 13, 25-32].

There are two interesting applications of soliton communications. One is long distance transoceanic communication over 10,000 km, in which the transmission speed is limited to 5 Gbit/s to 40 Gbit/s due to various dispersion and amplifier spacing conditions. The other interesting application area is relatively short distance communication over 1000 km, where the transmission speed is 100 Gbit/s to 1 Tbit/s. Both applications open the way to ultrahigh-speed information superhighways. [9].

At present, soliton-like transmission is very attractive since conventional optical components same as in traditional optical communication systems can be employed. In particularly, such transmission systems use standard DFB laser diodes, standard single-mode fibers and state-of-the-art NRZ signals. Soliton-like transmission is primarily based on (i) periodic dispersion compensation by means of fiber-Bragg gratings or dispersion-flattened fibers and (ii) proper attenuation management. The latter one is performed by optical in-line amplifiers which are spaced more closely than in conventional optical transmission links [10, 35].

12.4 PHYSICAL LIMITS FOR OPTICAL COMMUNICATIONS

In optical communications, the following questions frequently arises: (i) How many photons per bit are at least required to achieve a certain BER; for example, 10^{-10}? (ii) What are the maximum bit rate and the maximum transmission length? To answer these questions, the physical limits for optical communication systems have to be considered. In practice, different physical limits are defined to determine the maximum possible performance of optical communication systems. The aim of this Section is to discuss the physical meaning of these limits and make their comparative study. The following limits will be discussed in details:

- Quantum limit,
- Shot noise limit,
- Shannon limit and
- Dispersion limit.

12.4.1 QUANTUM LIMIT

The quantum limit is related to an ideal direct detection receiver which is a simple photon counter. It is well-known from conventional optical communications that the photons of the incoming light wave are Poisson distributed. The random nature of photons is illustrated in Fig. 12.9. This figure shows a typical intensity modulated optical signal which can be observed at the input of a photon counter.

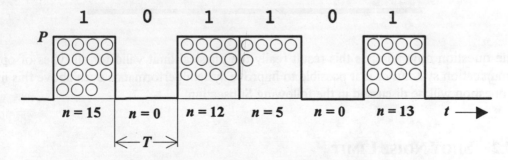

Fig. 12.9 : Poisson distributed photons at the input to optical receiver

It becomes clear from this figure that the number of photons varies from bit to bit when symbol "1" is transmitted, whereas the number of photons is always zero when the transmitted symbol is "0".

The Poisson distribution is described by the following fundamental expression:

$$f(n) = \frac{n_s^n e^{-n_s}}{n!} \qquad (12.10)$$

where n_s represents the average number of photons for bit "1" (see Eq. 12.6). A bit error occurs when absolutely no photon is counted during the transmitted symbol "1". On the other hand, transmitted symbol "0" is always detected correctly. Hence, the average probability of bit error is

$$p = \frac{1}{2} f(n=0) = \frac{1}{2} e^{-n_s} \qquad (12.11)$$

Here, occurrence probabilities for symbol "0" and symbol "1" are assumed to be equal. For a BER of $p = 10^{-9}$, on an average $n_s = 21$ photons are required from Eq. (12.11) above. In conclusion:

> To obtain a BER of 10^{-9}, an ideal optical receiver (photon counter) requires
>
> 21 photons/bit!

The question now arise: Is this result really the physical limit valid for all types of optical communication systems or is it possible to improve system performance even above this limit? This question will be discussed in the following Subsection.

12.4.2   SHOT NOISE LIMIT

Shot noise limited receiver operation can either be obtained by direct detection or coherent detection. The shot noise limit can practically be achieved in a coherent detection receiver provided local laser power is high enough (see Chapter nine), whereas this limit is only of theoretical interest in a direct detection receiver.

(i) DIRECT DETECTION RECEIVER

As explained in Chapter nine, signal-to-noise ratio of an optical direct detection receiver with an APD is given by

$$\left(\frac{S}{N}\right)_d = \frac{(M \, R_0 P_r)^2}{\left(eM^{2+x}R_0 P_r + eI_{dark} + G_{th}\right)B} \tag{12.12}$$

Shot noise limited detection can only be achieved with an infinite or at least an unrealistic high input light power P_r to the receiver. It is due to this reason that shot noise limit is of theoretical interest only in a direct detection receiver. It must be noted that this statement is not valid if an optical preamplifier is used in addition. With sufficient high power P_r, we obtain from Eq. (12.12)

$$\left(\frac{S}{N}\right)_d = \frac{R_0 P_r}{eM^x B} = \frac{\eta P_r}{hfM^x B} \tag{12.13}$$

This shot noise limited signal-to-noise ratio reaches its maximum value

$$\left(\frac{S}{N}\right)_{d,max} = \frac{P_r T}{hf} = n_s \tag{12.14}$$

for $M^x = 1$ (i.e., PIN-diode is used or $x = 0$), $\eta = 1$, and $B = 1/T$ (i.e., the double-sided noise-equivalent bandwidth of a matched filter). As explained in Chapter ten, BER is given by

$$p = Q\left(\frac{1}{2}\sqrt{\left(\frac{S}{N}\right)_d}\right) = Q\left(\frac{A_d}{2\sigma_d}\right) \tag{12.15}$$

where A_d represents the eye aperture and $\sigma_d = 0.5 \cdot (\sigma_0 + \sigma_1)$ the average standard deviation of the additive Gaussian noise. Taking again a BER of $p = 10^{-9}$, on an average $n_s = 144$ photons per bit are required from Eq. (12.15). It should be remembered that $Q(6) \approx 10^{-9}$. In conclusion:

> To obtain a BER of 10^{-9}, a shot noise limited
> optical direct detection receiver requires
>
> 144 photons/bit!

Main features of a shot noise limited direct detection receiver are summarized below:

- Infinite optical power P_r at the input to the receiver,
- Optimum photodiode quantum efficiency $\eta = 1$,
- PIN-photodiode or APD with $x = 0$,
- Matched filter for rectangular input signals with a double-sided noise-equivalent bandwidth $B = 1/T$.

(ii) COHERENT RECEIVER

In a coherent optical receiver, signal-to-noise ratio in the IF domain is given by (see Chapters nine and eleven)

$$\left(\frac{S}{N}\right)_c = \frac{2R_0^2 P_r P_1}{2\left(eR_0 P_1 + eI_{dark} + G_{th}\right)B} \tag{12.16}$$

In contrast to a direct detection receiver, shot noise limited operation can easily be obtained by sufficiently high local laser light power P_1 (compare with Table 12.2). We obtain

$$\left(\frac{S}{N}\right)_{c,IF} := \frac{S}{N} = \frac{P_r T}{hf} = n_s \quad \text{in IF domain} \tag{12.17}$$

and

$$\left(\frac{S}{N}\right)_{c,B} = 2\frac{P_r T}{hf} = 2n_s \quad \text{in baseband frequency domain} \tag{12.18}$$

The bit error rate of a coherent optical receiver depends on the modulation scheme applied to the system. For important modulations schemes used in practice, bit error rates are given below (compare with the formulas given in Table 12.2):

$$p = Q\left(2\sqrt{\frac{S}{N}}\right) \qquad \text{PSK homodyne} \qquad\qquad (12.19)$$

$$p = \exp\left(-\frac{S}{N}\right) \qquad \text{DPSK heterodyne} \qquad\qquad (12.20)$$

and

$$p = \exp\left(-\frac{1}{2}\frac{S}{N}\right) \qquad \text{FSK heterodyne with incoherent dual-filter detection} \qquad (12.21)$$

For a BER $p = 10^{-9}$, we obtain the following result from Eqs. (12.19), (12.20) and (12.21):

> To obtain a BER of 10^{-9}, a shot noise limited ideal heterodyne/homodyne receiver requires
>
> 41 photons/bit for FSK heterodyne,
>
> 21 photons/bit for DPSK heterodyne and
>
> 9 photons/bit for PSK homodyne!

Characteristic features of ideal shot noise limited coherent receivers are summarized below:

- Infinite local laser power P_l,
- PIN-photodiode,
- Maximum photodiode quantum efficiency $\eta = 1$,
- Matched filter for rectangular input signals with a double-sided noise-equivalent bandwidth $B = 1/T$.

It becomes clear from the above results that performance of a PSK homodyne system is actually better than the quantum limit! Instead of 21 photons in case of quantum limit, a PSK homodyne system only requires 9 photons, which means an improvement of 3.5 dB. This very surprising and important result, however, can easily be explained as follows: In a photon counter system, photons are only transmitted in case of symbol "1", whereas in a coherent homodyne system photons are transmitted for symbol "0" also. Thus, overall energy available in the receiver to decide whether symbol "0" or "1" has been transmitted is higher in a PSK homodyne system.

Now, the same question arises again: What is the minimum number of photons required for a demanded BER or what is the real physical limit for optical communications? Is it possible to obtain a BER of 10^{-9} with less than 9 photons/bit?

12.4.3 SHANNON LIMIT

Given a digital source generating a bit rate R and a channel of capacity C (both in bit/s). If

$$R \leq C \tag{12.22}$$

there exists a coding technique such that the output of the source can be transmitted over the channel with a probability of error which may be made arbitrarily small. This important statement is known as the Shannon's theorem. The channel capacity of a white, band-limited Gaussian channel is given by

$$C = B \, \log_2\!\left(1 + \frac{S}{N}\right) \text{ bit/s} \tag{12.23}$$

where B is the single-sided channel bandwidth, S the signal power and N the overall total noise power in the channel bandwidth [12, 13, 17, 23, 39].

For a fixed signal power and in presence of additive white Gaussian noise, channel capacity approaches an upper limit (termed as the Shannon limit) with increasing bandwidth:

$$C_{max} = \frac{1}{\ln(2)} \frac{S}{G_d} \text{ bit} \tag{12.24}$$

In an optical PSK homodyne system, signal power is $S = 4R_0^2 P_r P_1$ and single-sided power density of white shot noise is $G_d = 2eR_0P_1$. Therefore, with $\eta = 1$ and $R_{max} = 1 \text{ bit}/T = C_{max}$ we obtain

$$R_{max} T = \frac{1}{\ln(2)} \frac{2 P_r T}{hf} \text{ bit } = \frac{2 n_s}{\ln(2)} \text{ bit } = 1 \text{ bit} \tag{12.25}$$

This yields

$$n_s = n_{s,min} = \frac{\ln(2)}{2} = 0.347 \text{ photons !} \tag{12.26}$$

This important result actually represents the real physical limit of an optical communication system. Thus, only 0.347 photons per bit are required on an average to obtain errorless data transmission! It may be mentioned that realization of this limit requires advanced highly sophisticated coding techniques. In a direct detection system, an optical preamplifier is needed in addition since coding gain is restricted to some dB only. Sensitivities better than nine photons per bit for a PSK homodyne system has already been demonstrated, where only five photons per bit are required for a BER of 10^{-9} [46].

12.4.4 DISPERSION LIMIT

In Section 12.3.2, high-speed long-haul data transmission links have been discussed as a typical application of optical communications. When analysing such systems, the following question frequently arises: What is the maximum bit rate which can really be transmitted by means of fiber optics? In this Section, maximum bit rate R_{max} of a standard single-mode fiber at 1.55 μm should be estimated. As R_{max} is primarily defined by the dispersion characteristics of the fiber, this bit rate limit is called *dispersion limit*. In order to obtain a general view of transmission capacity of digital fiber links, this Section starts with some basic knowledge of data transmission.

(i) RELATIONSHIP BETWEEN SIGNAL AND CHANNEL BANDWIDTH

Given a digital signal of bandwidth B and a channel (e.g., a fiber) of bandwidth B_C. Then, if

$$B \le B_C \tag{12.27}$$

an errorless transmission is possible in principle. It is a modified version of Eq. (12.22) which was first given by Shannon. Next we have to answer both the question: What is the actual bandwidth B of digital signal and what is the actual bandwidth B_C of transmission channel?

(ii) DIGITAL SIGNAL BANDWIDTH - A PRACTICAL ESTIMATION

To estimate the bandwidth of a digital signal, we restrict our consideration on the *periodic worst-case pattern* ···010101010··· shown in Fig. 12.10. This pattern represents a digital signal of highest speed. If we are able to transmit this pattern properly through the channel (e.g., errorless), all other pattern can be transmitted errorless also.

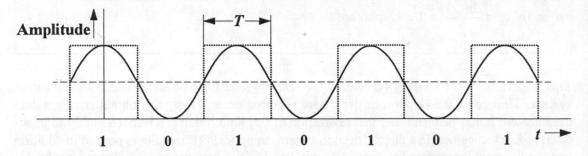

Fig. 12.10: Periodic worst-case pattern (dotted line) and its normalized first harmonic (solid line)

In a digital transmission system, receiver must be able to decide whether a symbol "1" or a symbol "0" has been transmitted. For this, it is not required that receiver input signal is rectangular in shape as the transmitted signal. Errorless transmission of worst-case pattern shown above requires that at least first sinusoidal harmonic of this pattern should be able to pass through the channel (Fig. 12.10). The period of the first harmonic is twice the bit duration i.e., $2T$. All other harmonics are not required for an errorless symbol decision. Thus, the physical bandwidth of the digital signal is

$$B = \frac{1}{2T} \tag{12.28}$$

Here, B represents the single-sided bandwidth whereas $2B = 1/T$ the double-sided bandwidth which is also referred as mathematical bandwidth. The channel must be able to transmit the first harmonic and higher order harmonics need not to be transmitted (Fig. 2. 11). As a result, the upper cut-off frequency (bandwidth) of the channel is given by

$$B_{\text{c}} = f_{\text{c}} \geq \frac{1}{2T} = \frac{f_{\text{b}}}{2} = B \tag{12.29}$$

where f_{b} is called the bit frequency in Hz which is related to the bit rate R in bit/s by $R = 1$ bit·f_{b}. Again, $B_{\text{c}} = f_{\text{c}}$ is the physical bandwidth of the channel and $2B_{\text{c}} = 2f_{\text{c}}$ the double-sided bandwidth..

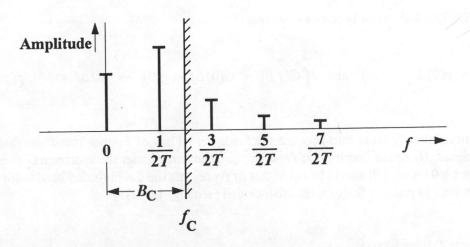

Fig. 12.11: Spectrum of periodic worst-case pattern with cut-off
frequency f_c and bandwidth B_C of channel

(iii) RELATION BETWEEN SIGNAL DURATION AND BANDWIDTH

As it is well known from system theory, duration Δt and bandwidth Δf of a signal are related
by

$$\Delta t \Delta f = 1 \tag{12.30}$$

This fact can easily be proved by considering a signal $g(t)$, its spectrum $G(f)$ and the rectangular
equivalents of signal and spectrum as shown in Fig. 12.12.

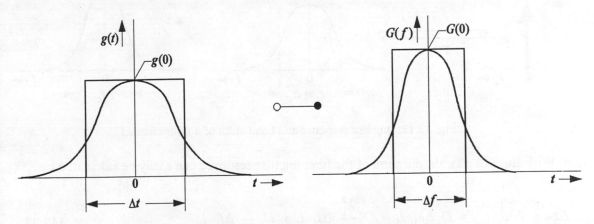

Fig. 12.12: Relationship of signal duration and bandwidth

From Fig. 12.12 above, it becomes clear that

$$\int_{-\infty}^{+\infty} g(t)\,dt = g(0)\Delta t = G(0) \quad \text{and} \quad \int_{-\infty}^{+\infty} G(f)\,df = G(0)\Delta f = g(0) \quad \rightarrow \quad \Delta t \Delta f = 1 \qquad (12.31)$$

This equations already takes into account the fundamental law of Fourier transform that area under the signal $g(t)$ equals spectrum at $f = 0$ i.e., $G(0)$ and area under the spectrum $G(f)$ equals the signal at $t = 0$ i.e, $g(0)$. It should be noted that Δf represents the double-sided bandwidth since negative as well as positive frequencies are included (see Fig. 12.12).

(iv) FIBER DISPERSION

If fiber input signal $g_1(t)$ is of duration Δt_1 then fiber output signal $g_2(t)$ is of duration

$$\Delta t_2 = \sqrt{(\Delta t_1)^2 + (D\,\Delta\lambda_L\,L)^2} \qquad (12.32)$$

where $\Delta\lambda_L$ is the spectral width of transmitter laser, L the fiber length and D the fiber dispersion. If fiber input signal is an infinitesimal pulse with $\Delta t_1 \rightarrow 0$ (i.e, an impulse), then fiber output equals the impulse response of the fiber (Fig. 12.13) which represents an important characteristic feature of a fiber.

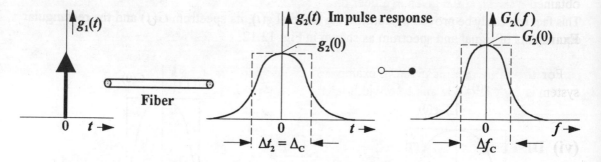

Fig. 12.13: Impulse response and bandwidth of a fiber channel

With Eq. (12.32), the duration of the fiber impulse response can easily be calculated by

$$\Delta t_C = \Delta t_2 \big|_{\Delta t_1 \rightarrow 0} = D\,\Delta\lambda_L\,L \approx D\,\frac{\lambda_L^2}{c}\,\Delta f_L\,L = D\,\frac{c}{f_L^2}\,\Delta f_L\,L \qquad (12.33)$$

Here, $f_L = c/\lambda_L$ and λ_L represent frequency and wavelength of transmitter laser, Δf_L the spectral width (i.e., bandwidth) of modulated transmitter laser in the frequency domain and c the velocity of light in the fiber. The approximation given in Eq. (12.33) is valid for $\Delta\lambda_L/\lambda_L \ll 1$ or $\Delta f_L/f_L \ll 1$ (known as the narrowband condition) which is normally fulfilled in realized optical communication systems.

Example 12.2

Let us consider a typical fiber with dispersion parameter $D = 20$ ps/(nm km) and length $L = 10$ km; a laser with spectral half width (linewidth) $\Delta\lambda_L = 1$ nm and an input signal (e.g., a bit) of duration $\Delta t_1 = 1$ ns. As a result, the output bit duration Δt_2 from Eq. (12.32) is 1.02 ns which represents an increase of 2 %. The duration of the impulse response of this fiber link is $\Delta t_C = 0.2$ ns.

(v) FIBER BANDWIDTH

Using Eqs. (12.30) and (12.33), single-sided physical bandwidth of the fiber channel can easily be calculated by

$$B_C = \frac{\Delta f_C}{2} = \frac{1}{2\Delta t_C} = \frac{1}{2D\Delta\lambda_L L} \approx \frac{c}{2D\lambda_L^2 \Delta f_L L} \tag{12.34}$$

If this bandwidth is multiplied by fiber length, *bandwidth-length product $B_C L$* of the fiber is obtained which is often given in a fiber data sheet.

Example 12.3

For the same data as given in example 12.2 above, physical bandwidth of the above fiber system is $B_C = 2.5$ GHz and bandwidth-length product $B_C L = 25$ GHz·km.

(vi) DISPERSION LIMIT

In order to transmit a binary digital signal with a bit rate R (in bit/s) or a bit frequency $f_b = R/$bit (in Hz), a physical fiber bandwidth

$$B_C \geq \frac{f_b}{2} = \frac{R/\text{bit}}{2} \tag{12.35}$$

is required (see Eq. 12.29). From this and Eq. (12.34), dispersion limit in terms of maximum bit rate can be calculated as follows:

$$R_{max} = \frac{1}{D \, \Delta\lambda_L \, L} \text{bit} \approx \frac{c}{D \, \lambda_L^2 \, \Delta f_L \, L} \text{bit} \tag{12.36a}$$

In the ideal case, bandwidth Δf_L of modulated laser output equals bandwidth of information signal (similar to single-sideband AM) i.e., $\Delta f_L = B = 1/(2T) = f_b/2$. From this and Eq. (12.34) dispersion limit in terms of maximum fiber length is

$$L_{max} = \frac{1}{D\Delta\lambda_L f_b} \approx \frac{c}{D\lambda_L^2 \Delta f_L f_b} = 2\frac{c}{D\lambda_L^2 f_b^2} \tag{12.36b}$$

At a wavelength of $\lambda_L = 1.55$ um, fiber dispersion parameter is $D \approx 17$ ps/(nm km). With $c \approx c_0/1.5 = 2\cdot10^8$ m/s (velocity of light in fiber core), results given in table below are obtained.

Table 12.3: Dispersion limit in terms of maximum fiber length at 1.55 um under ideal conditions

Bit rate R	155.52 Mbit/s (STM-1)	622.08 Mbit/s (STM-4)	2488.32 Mbit/s (STM-16)	10 Gbit/s	40 Gbit/s
Dispersion limit L_{max}	404,925 km	25,307 km	1,581 km	97 km	6 km

It must be mentioned that these results are valid under ideal conditions only i.e.,

- a non-chirped laser modulation,
- a modulated laser linewidth identical to bandwidth of information signal i.e., $\Delta f = f_b/2$ and
- absence of fiber non-linearities.

Under practical conditions, maximum transmission length at e.g., 10 Gbit/s is about 40 km to 60 km instead of 97 km as given in the table above. It becomes clear from Table 12.3 that if bit rate is less than 10 Gbit/s, fiber dispersion can normally be neglected. In this case, fiber attenuation is the limiting factor of the optical transmission system. For bit rates of 10 Gbit/s and above, dispersion compensation is absolutely required.

12. 5 REFERENCES

[1]   Andrekson, P. A. : Ultrahigh speed transmission and multiplexing. ECOC (1994), 109-116.

[2]   Blow, K. J.; Doran, N. J.: Nonlinear effects in optical fibres and fibre devices. Proceeding of IEE Pt. J. 134(1987)3, 138-144.

[3]   Bouley, J.-C.; Destefanis, G.: Multi-quantum well lasers for telecommunications. IEEE Commun. Magazine, 32(1994)7, 54-60.

[4]   Chesnoy, J.; et. al.: Übertragung mit ultrahoher Bitrate für die Jahre nach 2000. Elektrisches Nachrichtenwesen, 3. Quartal 1994, 241-250.

[5]   Clesca, B.; et. al.: 2.5-Gbit/s, 1291-km transmission over non-dispersion-shifted fibre using an integrated electroabsorption modulator/DFB laser module. ECOC (1995), 995-998.

[6]   Derr, F.: System performance of an optical QPSK homodyne transmission system with Costas loop. J. Opt. Commun. 14(1985)2, 42-44.

[7]   Doran, N. J.; Blow, K. J.: Solitons in optical communications. IEEE J. QE-19(1983)12, 1883-1888.

[8]   Ellis, A. D.; et. al.: Ultra-high-speed OTDM networks using semiconductor amplifier-based processing nodes. IEEE J. LT-13(1995)5, 761-770.

[9]   Georges, T.: Recent advances on soliton systems. ECOC (1995), 311- 316.

[10]   Georges, T.; Favre, F.: Solitonic transmission: a review. ECOC (1997), Tutorial.

[11]   Goodfellow, R. C.; et. al.: Optoelectronic components for multigigabit systems. IEEE J. LT-3(1985)6, 1170-1179.

[12]   Gordon, J. P.: Quantum effects in communication systems. IRE Proc., 50(1962)9, 1898-1908.

[13]   Haber, F.: An Introduction to Information and Communication Theory. Addison-Wesley, 1974.

[14]   Hasegawa, A.; Kodama, Y.: Signal transmission by optical solitons in monomode fiber. Proc. IEEE 69(1981)9, 1145-1150.

[15]   Hasegawa, A.: Optical Solitons in Fibers. Springer 1989.

[16]   Haus, H. A.: Molding light into solitons. IEEE Spectrum 30(1993)3, 48-53.

[17]   Jodoin, R.; Mandel, L.: Information rate in an optical communication channel. J. of Optical Society of America, 61(1971)2, 191-198.

[18]   Jopson, B.; Gnauck, A.: Dispersion compensation for lightwave fiber systems. IEEE Commun. Magazine, 33(1995)6, 96-102.

[19]   Jørgensen, B. F.; Pedersen, R. J. S.; Rasmussen, C. J.: Transmission of 10 Gbit/s beyond the dispersion limit of standard single mode fibers. ECOC (1995), 557-564.

[20]   Kawanishi, S.; et. al.: Single polarization completely time-division-multiplexed 100 Gbit/s optical transmission experiment. Techn. Digest of ECOC '93, ThP12.1, Montreux, Sep. 1993.

[21]   Kawanishi, S.; et. al. : 100 Gbit/s, 200 km optical transmission experiment using extremly low jitter PLL timing extraction and all-optical demultiplexing based on polarization insensitive four-wave mixing. Electron. Lett. 30(1994), 800-801.

[22]   Keiser, G.: Optical Fiber Communications, Eecond Edition. McGraw-Hill 1991.

[23]   Mandel, L.: Fundamental limits on information capacity of an optical communication channel. Kinam, Vol.5, Series C (1983), 213-232.

[24]   Matera, F.; et. al.: Study on the implementation of TDM and WDM optical systems up to 40 Gbit/s·km. ECOC (1994), 833-836.

[25]   Mollenauer, L. F.; Stolen, R.H.: Solitons in optical fibers. Laser Focus Magazine, 18(1982)4, 193-198.

[26] Mollenauer, L. F.; Stolen, R.H.: The soliton laser. Optical Society of America, Optics Letters 9(1984)1, 13-15.

[27] Mollenauer, L. F.: Solitons in optical fibres and the soliton laser. Phil. Trans. R. Soc. Lond. A. 315(1985), 437-450.

[28] Mollenauer, L. F.; et. al.: Soliton propagation in long fiber with periodically compensated loss. IEEE J. QE-22(1986)1, 157-173.

[29] Mollenauer, L. F.; et. al.: Demonstration, using sliding frequency guiding filters, of error-free soliton transmission over more than 20 Mkm at 10 Gbit/s, single-channel, and over more than 13 Mkm at 20 Gbit/s in a two channel WDM. Electron. Lett 29 (1993), 910-911.

[30] Nakazawa, M.; et. al.: Straight-line soliton data transmission over 2000 km at 20 Gbit/s and 1000 km at 40 Gbit/s using erbium-doped fibre amplifiers. Electron. Lett. 29(1993), 1474-1475

[31] Nakazawa, M.: Soliton transmission in telecommunication networks. IEEE Commun. Magazine 32(1994)3, 34-41.

[32] Nakazava, M.: Single-polarization 80 Gbit/s soliton data transmission over 500 km unequal amplitude solitons for timing clock extraction. ECOC (1994), post-deadline paper, 41-44.

[33] Noé, R. et. al.: Direct modulation 565 Mb/s DPSK experiment with 62.3 dB loss span and endless polarization control. IEEE PTL(1992)10, 1151-1154.

[34] Nuttall, C. L.; Croston, I. R.; Parsons, N. J.: Electro-optic tuneable filters for multi-wavelength networks. ECOC (1994), 767-774.

[35] Reid, J. J. E.; Liedenbaum, C. T. H. F.: Realisation of 20 Gbit/s long haul soliton transmission at 1300 nm on standard single mode optical fibre. ECOC (1994), post-deadline paper, 61-64.

[36] Söder, G.; Tröndle, K.: Digitale Übertragungssysteme. Springer-Verlag, 1984.

[37] Stein, S.; Jones, J. J.: Modern Communication Principles. McGraw-Hill, 1967.

[38] Taga, H.; et. al.: Recent progress in amplified undersea systems. IEEE J. LT-13(1995)5, 829-840.

[39] Taub, H.; Schilling, D. L.: Principles of Communication Systems. Second Edition, McGraw-Hill, 1986.

[40] Tkach, R. W.: Transcontinental optical network: Opportunities and Obstacies. ECOC (1994), 915-918.

[41] Tomlinson, W. J.; et. al.: Compression of optical pulses chirped by self-phase modulation in fibers. Optical Society of America B. 1(1984)2, 139-149.

[42] Tomlinson, W.J.; Stolen, R. H.: Nonlinear phenomena in optical fibers. IEEE Commun. Magazine, 26(1988)4, 33-44.

[43] Tröndle, K.; Söder, G.: Optimization of Digital Transmission Systems. Artech House, 1987.

[44] Vannucci, G.: Combining frequency-division and code-division multiplexing in a high-capacity optical network. IEE Network, 3(1989), 21-30.

[45] Vodhanel, R. S.; et. al.: Performance of directly modulated DFB lasers in 10 Gb/s ASK, FSK, and DPSK lightwave systems. IEEE J. LT-8(1990)9, 1379-1385.

[46] Wandernoth, B.: 5 photon/bit low complexity 2 Mbit/s PSK transmission breadboard experiment with homodyne receiver applying synchronization bits and convolutional coding. ECOC (1994), 59-62.

[47] Zamkotsian, F.; et. al.: An InP-based optical multiplexer integrated with modulators for 100 Gbit/s transmission. ECOC (1994), post-deadline paper, 105-108.

13   FIBER LINK DESIGN

Over the past several years, optical fiber communication systems are extensively used all over the world for the telecommunication and data transmission purposes. The design of optical links to meet a given requirements has become an important issue. With the communication trend being towards digital, more and more digital based optical communication links are being designed and installed for commercial applications. In view of this, greater emphasis has been placed on the digital system though the same concepts are applicable for analog systems as well.

One of the motivations for developing optical fiber communication is its application to long-haul point-to-point communication systems and optical networks. The point-to-point feature implies a direct fiber route between the transmitter and receiver with no intervening user stations. In long-haul systems, repeaters are often installed if the distance involved is sufficiently large. Applications like transfer of high resolution graphics, development of super computer networks and broadband integrated service digital networks (BISDNs) demand data transfer rates of several gigabits per second. Conventional networks cannot provide the required bandwidth, while the optical offer the most appropriate solutions.

This Chapter describes design procedure of a digital point-to-point link. Two design examples with typical parameters are discussed.

13.1   LONG-HAUL SYSTEMS

The simplest transmission link is a point-to-point having transmitter on one end and receiver on the other. The link length can vary from less than a kilometre to thousands of kilometres depending on the type of application. It can be a digital computer data link connecting computer and terminals in the same building or between two buildings. It can also be an undersea light wave communication system for intercontinental communication over a distance of more than thousands of kilometres. In this application, low attenuation and large bandwidth of the fiber are more important, while in the former other properties of fiber like immunity to electromagnetic interference, low weight, small diameter are of prime importance [1].

In long-haul systems, communication links involved are point-to-point. As such the design of an optical link is a very involved process since it involves many interrelated variables among the fiber, source and photodetector characteristics. Link design is an iterative process and sometimes it may be necessary to relax the system requirements if these cannot be met with the existing technology or are too expensive. The main objective in the link design is not only to meet the system requirements, but it should also be cost effective.

In long distance communication links, where it is not feasible to transmit signals directly from the point of origin to the final destination, the total distance to be covered is divided into several shorter paths in tandem so that the transmission of adequate quality can be achieved. Each shorter path is connected to the next either through an optical amplifier or a regenerative repeater. In designing a multilink system, total system specifications must be reduced to one specification for each link e.g., if the overall system is required to have an error rate of less than 10^{-9} at a given data rate, then the individual link must have better error performance which is directly proportional to the number of links and other link parameters [12].

The key system requirements needed in the link design are: (i) data or bit rate/bandwidth, (ii) bit error rate (BER)/signal-to-noise (S/N) ratio and (iii) transmission distance or link length [4, 5]. Two basic issues involved in the link design are: (i) attenuation which determines the power available at the photodetector input for a given source power and (ii) dispersion which determines the limiting data rate or usable bandwidth. These are often referred as link power and time budgets respectively. Generally, first the link power budget is made and if found unsatisfactory, then some components might be changed. Once established, the designer can prepare time budget to ensure that the desired overall system performance is achieved.

When bit rate is moderate (e.g., $\leq$ 1 Gbit/s), one can operate near 1550 nm to take advantage of the lowest fiber loss occurring in this wavelength region. If bit rate is high (e.g., $\geq$ 10 Gbit/s), one has to operate either at 1300 nm where dispersion is minimum or to use a dispersion-shifted fiber and operate at 1550 nm. In the latter case, both dispersion and loss become minimum at 1550 nm. On the basis of bit rate R and link length L, a tentative choice of operating wavelength, fiber and source can be made by using the RL product as given in Table 13.1 below. The final choice, of course, is to be made on the basis of link power and time budgets. It may be mentioned that maximum modulation speed is about 1 Gbit/s for an LED and 40 Gbit/s for a laser. With a DFB laser source and a dispersion-shifted fiber, even RL products in the order of Tbit/s·km become realistic at 1550 nm. As far as system cost is concerned, it is lowest near 850 nm and increases as the operating wavelength moves to 1300 nm or 1550 nm.

Table 13.1: Typical data rate and distance products in Mbit/s·km of LEDs and lasers (Frequently used combinations are shadowed. GI: graded-index, SI: step index, SM: single-mode and MM: multimode fibers)

	First window 850 nm	Second window 1300 nm		Third window 1550 nm	
	SI-MM	GI-MM	SI-SM	GI-MM	SI-SM
LED	10	500	5,000	50	500
Laser	50	5,000	50,000	1,000	10,000

For optical fiber, choice is to be made between single-mode (SM) and multimode (MM). The MM fiber could be step index (SI) or graded-index (GI). The choice depends on the amount of dispersion that can be tolerated and the power to be coupled into the fiber. LEDs are generally used with MM fibers, although edge emitting LEDs can launch sufficient power into SM fiber at data rates up to 560 Mbit/s over a distance of several kilometres. With a laser source, both SM and MM fibers can be used.

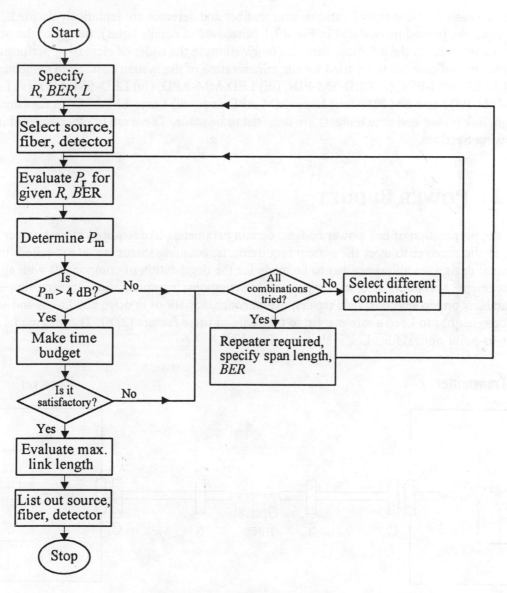

Fig. 13.1: Flow chart for link design

In making a choice of photodetector, minimum optical power required to satisfy the BER requirement at the specified data rate and receiver complexity are to be considered. A PIN-photodiode receiver is simpler and more stable with changes in temperature than an avalanche photodiode (APD) receiver. However, the sensitivity of an APD receiver is much higher (more than 10 dB to 15 dB). Therefore, it is to be used when the received optical power level is quite low.

On the basis of above considerations, source, fiber and detector are tentatively decided. Then following the procedure outlined in Fig. 13.1 (discussed in details latter), design of the optical fiber link is made. In the iteration, various combinations in the order of increasing performance, complexity and cost are to be tried for the minimization of the system cost. Such combinations are: (i) LED-MM-PIN, (ii) LED-SM-PIN, (iii) LED-MM-APD, (iv) LED-SM-APD, (v) Laser-MM-PIN, (vi) Laser-SM-PIN, (vii) Laser-MM-APD and (viii) Laser-SM-APD. In the course of design, link power and time budgets are required to be made. These are briefly described in the following Sections.

13.2   POWER BUDGET

In the preparation of link power budget, certain parameters like required optical power level P_r at the receiver to meet the system requirements, coupling losses etc. are required. In any practical design, an allowance has to be made for the degradation of components with ageing, replacement, variations due to temperature fluctuations, manufacturing spreads, imperfect repeatability on reconnection, field repairs, maintenance, variations in drive conditions and so on. The designer has to keep a safe margin to take care of these factors [2, 8]. The loss model for a point-to-point optical fiber link is shown in Fig. 13.2.

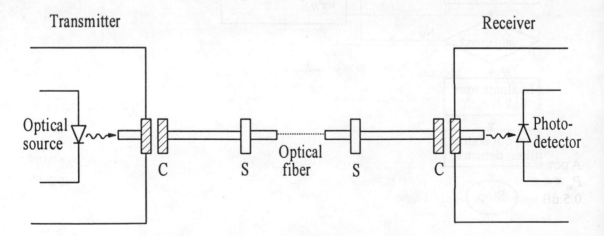

Fig. 13.2: Loss model of point-to-point optical fiber link (C: connector and S: splice)

After computing various losses and fixing safety margin, power budget of the link is made as follows [5]:

Table 13.2 : Calculation of power budget

* L_0 is the factory unit length of fiber and
  $[L/L_o]$ represents the integer part of L/L_o.

$ α is the attenuation coefficient of fiber in dB/km.

Source output power, P_t .. dBm

Minimum receiver power, P_r (min)   .. dBm

Source-to-fiber coupling loss, L_{sf} .. dB

Splice loss:

     - Number of splices, $N = [L/L_0]^*$

     - Fiber-to-fiber coupling loss, L_{ff} .. dB

     - Total splice loss, NL_{ff} .. dB

Fiber loss$, αL .. dB

Fiber-to-detector coupling loss, L_{fd} .. dB

$$\text{Power margin } P_m = \left(P_t - P_r(\text{min}) - L_{sf} - NL_{ff} - \alpha L - L_{fd} \right) \text{ dB} \qquad (13.1)$$

A power margin $P_m \geq 4$ dB is acceptable otherwise some components need to be upgraded. With $P_m < 4$ dB, system will become less reliable. The typical values of L_{sf}, L_{ff} and L_{fd} are about 2 dB, 0.5 dB and 0.2 dB respectively.

13.3   TIME BUDGET

O bjective of time budget is to ensure that the system is able to operate properly at the desired data rate. Sometimes, it may happen that the total system is not able to operate at the desired data rate even if the bandwidth of the individual subsystem exceeds the data rate. If the transmitter, fiber and receiver are considered to have a Gaussian impulse response with rms pulse widths σ_{tx}, σ_f and σ_{rx} respectively, the overall system response will also be Gaussian with a rms pulse width [12]

$$\sigma_{sys} = \sqrt{\sigma_{tx}^2 + \sigma_f^2 + \sigma_{rx}^2} \tag{13.2}$$

As a safety margin, it is usually taken to be 10% higher than the above value. System time budget is considered to be satisfactory if σ_{sys} does not exceed $(1/4R)$ where R is the data rate. The rms pulse widths for different subsystems are determined as follows:

13.3.1   TRANSMITTER RISE TIME

Rise time of transmitter primarily depends on the electronic components of driving circuit and electrical parasitics associated with the optical source. It is few nanoseconds for LED-based transmitter, but can be as short as 0.1 ns for a laser-based transmitter [7]. Presuming an exponential rise and decay, σ_{tx} is nearly equal to half of the rise time [12].

13.3.2   RECEIVER RISE TIME

The receiver rise time t_{rx} is determined by 3 dB electrical bandwidth of the receiver front end. If B is the receiver bandwidth in MHz and the receiver front-end is modelled by a first order low-pass filter having a step response $[1-\exp(2\pi Bt)]u(t)$ where $u(t)$ represents the unit step function, then t_{rx} in ns is approximately given by [6]

$$t_{rx} = \frac{350}{B} \tag{13.3}$$

As for the transmitter, σ_{rx} is determined from the rise time.

13.3.3 FIBER RMS DISPERSION

The rms pulse width of the fiber σ_f includes the contributions of both modal and material dispersion through the relation

$$\sigma_f = \sqrt{\sigma_{mod}^2 + \sigma_{mat}^2} \qquad (13.4)$$

where σ_{mod} and σ_{mat} are the rms pulse widths due to modal and material dispersion of the fiber respectively. For SM fiber, modal dispersion does not contribute and, therefore, σ_f becomes same as σ_{mat}.

The optical fiber link may consist of several concatenated sections and the fiber in each section may have different dispersion characteristics. Further, there may be mode mixing at the splices and the connectors. As a consequence, propagation delay associated with different modes tends to average out. In the absence of mode mixing, σ_{mod} for SI fibers is [10]

$$\sigma_{mod} \approx \frac{n_1 \Delta L}{2\sqrt{3}\,c} \qquad (13.5)$$

where c is the velocity of light and the parameter Δ depends on the core and cladding refraction indices ($\approx (n_1 - n_2)/n_1$). For GI fibers, delay time is a function of g. The minimum intermodal rms pulse broadening with an optimum g is [3, 10]

$$\sigma_{mod} \approx \frac{n_1 \Delta^2 L}{20\sqrt{3}\,c} \qquad (13.6)$$

In both the cases, effect of mode mixing can be included by changing the linear dependence on L by a sublinear dependence L^q, where q is a constant ranging from 0.5 to 1 and its typical value is 0.7.

The σ_{mat} is approximately given by

$$\sigma_{mat} \approx |D_{mat}|\sigma_\lambda L \qquad (13.7)$$

where σ_λ is the rms spectral width of the source. The dispersion parameter D_{mat} may change along the fiber link due to different dispersion characteristics of fibers in the link. Therefore, average value of D_{mat} should be used in Eq. (13.7).

13.4   MAXIMUM LINK LENGTH CALCULATION

When a particular set of components meets the design requirements, one would like to know the maximum distance up to which these components could be used. Further, if the link length is quite large, it will help in determining the repeater location. The maximum link length is determined presuming that link is attenuation limited and there is no dispersion effect. And again when the link is dispersion limited and there is no attenuation effect. Minimum of the two is taken as the maximum practicable link length. Under attenuation limited condition, total fiber attenuation should be at least equal to the difference between power at the input end of fiber and minimum power required at the output end of fiber to meet the receiver power requirement. If there are splices and/or connectors in-between, their attenuation is also to be considered (see Example 13.1). The maximum link length is the total attenuation divided by the fiber attenuation per unit length. For a dispersion limited link, maximum data rate that can be transmitted over an optical fiber system is given by [11, 12]

$$R = \frac{1}{4\sigma_{sys}} \tag{13.8}$$

The maximum allowable fiber dispersion will be

$$\sigma_{al} = \sqrt{\left(\frac{1}{4R}\right)^2 - \sigma_{tx}^2 - \sigma_{rx}^2} \tag{13.9}$$

Using Eq. (13.4), we can determine σ_f/L i.e., fiber dispersion per unit length. Therefore, maximum link length under dispersion limited condition will be $L \cdot \sigma_{al}/\sigma_f$. The flow chart for link design is given in Fig. 13.1.

Example 13.1

Design an optical fiber link for a length 2.9 km and data rate 90 Mbit/s. The required BER is 10^{-9}. Available components are:

(a)   Laser at 850 nm; rms spectral width 1 nm; optical output power 0 dBm; transmitter rise time 2 ns,

(b)   LED at 850 nm; rms spectral width 50 nm; optical output power -13 dBm; transmitter rise time 15 ns,

(c)   Fiber GI type; $n_1 = 1.43$; $\Delta = 0.015$; $\alpha = 4$ dB/km; $D_{mat} = 70$ ps/(nm·km) at 850 nm; unit factory length 1 km,

(d) PIN-photodiode receiver of sensitivity presuming typical parameters is described by

$$P_r = 200 \cdot 10^{-17}\,\text{Ws} \cdot R \left[1 + \sqrt{1 + 1.42 \cdot 10^{-8}\,\text{s} \cdot R + \frac{1.468 \cdot 10^9\,\text{s}^{-1}}{R}} \right] \qquad (13.10)$$

where R is the data rate.

(e) APD receiver with sensitivity better than PIN receiver sensitivity by 11 dB at the same data rate and BER. Take receiver rise time with either PIN or APD to be 1 ns.

Procedure:

With the given information, procedural steps involved in the link design are as follows:

(i) Keeping in view the availability of components, region of operation is 800 nm - 900 nm.

(ii) Following Section 4.7, different coupling losses can be determined. Let these losses be $L_{sf} \approx 2$ dB, $L_{ff} \approx 0.5$ dB and $L_{fd} \approx 0.2$ dB.

(iii) Let us consider the simple and economical LED-GI-PIN combination. The sensitivity of the receiver at $R = 90$ Mbit/s from Eq. (13.10) will be -30.3 dBm. The power margin P_m from Eq. (13.1) is

$$P_m = (-13 + 30.3 - 2 - 1 - 11.6 - 0.2)\ \text{dB} = 2.5\ \text{dB} \qquad (13.11)$$

Since the power margin is less than 4 dB, this combination will not meet the power budget requirement.

(iv) Now either the source or detector has to be upgraded. Let us replace the PIN-photodiode receiver by APD receiver. Sensitivity of the receiver will be -41.3 dBm. The power margin is

$$P_m = (-13 + 41.3 - 2 - 1 - 11.6 - 0.2)\ \text{dB} = 13.5\ \text{dB} \qquad (13.12)$$

This combination satisfies the power budget requirement. Let us now make the time budget. The modal and material dispersion rms pulse widths can be determined from Eqs.(13.6) and (13.7) respectively. Total system rise time from Eq. (13.2) with a safety margin of 10% is

$$\sigma_{sys} = 1.1\sqrt{(7.5)^2 + (0.09)^2 + (10.15)^2 + (0.5)^2}\ \text{ns} = 13.89\ \text{ns} \qquad (13.13)$$

Allowable rms pulse width ($= 1/4R$) is 2.78 ns. Since σ_{sys} is more than the allowable rms value, time budget is not satisfied.

(v) Now let us take Laser-GI-PIN combination. The power margin is

$$P_m = (0+30.3-2-1-11.6-0.2) \text{ dB} = 15.5 \text{ dB} \tag{13.14}$$

Thus, power budget requirement has been satisfied. The system rise time is

$$\sigma_{sys} = 1.1\sqrt{(1)^2+(0.09)^2+(0.2)^2+(0.5)^2} \text{ ns} = 1.25 \text{ ns} \tag{13.15}$$

As σ_{sys} is less than the allowable rms value, time budget is also satisfied. Hence, this combination will meet the given system requirements.

(vi) Maximum link length for the above combination under attenuation limited condition can be determined by using

$$L_a(\text{max}) = \frac{P_t-(P_r-11\,\text{dB})-L_{sf}-L_{fd}-P_m}{\alpha+L_{ff}/L_o} \tag{13.16}$$

In the above equation, P_t and P_r are in dBm; L_{sf}, L_{ff}, L_{fd} and P_m in dB; L_0 in km and α in dB/km. Substituting the values of various parameters, we get

$$L_a(\text{max}) = \frac{-P_r+4.8}{4.5} \text{ km} \tag{13.17}$$

P_r is computed using Eq. (13.10) for different R and then $L_a(\text{max})$ is determined from the above equation. The results are shown in Fig. 13.3 below. The maximum link length under dispersion limited condition for the above combination is given by

$$L_d(\text{max}) = \frac{\sqrt{\left(\frac{1}{4R}\right)^2-\sigma_{tx}^2-\sigma_{rx}^2}}{1.1\sqrt{\left(\frac{n_1\Delta^2}{20\sqrt{3}c}\right)^2+\left(|D_{mat}|\,\sigma_\lambda\right)^2}} \tag{13.18}$$

In the above equation, factor 1.1 in the denominator represents a safety margin of 10% in the fiber dispersion as P_m in the calculation of $L_a(\text{max})$. When values of various parameters are substituted in the above equation, it gives

$$L_d(\max) = \frac{\sqrt{\left(\frac{1}{4R}\right)^2 - 1.25}}{0.084} \text{ km} \tag{13.19}$$

where R is in Gbit/s. Numerical results computed from the above equation are also shown in Fig. 13.3.

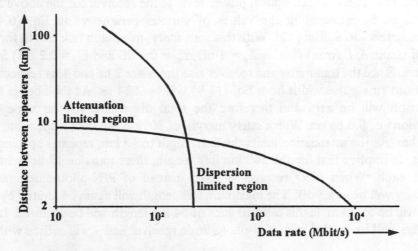

Fig. 13.3: Variations of maximum distance between repeaters with data rate under attenuation and dispersion limited conditions

It is seen from the figure that at a data rates of 10 Mbit/s, channel is attenuation limited and the maximum repeater spacing is 9 km. At higher data rate, say 200 Mbit/s, channel becomes dispersion limited and the maximum repeater spacing is 6.6 km.

Example 13.2

An optical communication link is to be designed for a data rate 100 Mbit/s, BER 10^{-9} and link length 540 km. The available components are:

(a) Laser at 1300 nm; rms spectral width 3 nm; optical output power -5 dBm; transmitter rise time 2 ns.

(b) Fiber SI-SM type, $n_1 = 1.43$; $\Delta = 0.0037$; $\alpha = 0.6$ dB/km; $D_{mat} = 3.2$ ps/(nm·km) at 1300 nm, $L_0 = 3$ km.

(c) For signal detection either a PIN-photodiode or an APD receiver can be used. The PIN has the same specifications as APD ($M < 100$, $F(M) \approx M^{0.5}$). $C_d = 1$ pF; $C_s = 1$ pF;

$C_a = 2$ pF; $R_F = 6.5$ kΩ, $R_{eq} \gg R_F$; $T_0 = 300$ K; $I_{dark} = 15$ nA; $\{i_a^2\} = 0.6 \times 10^{-24}$ A^2/Hz; $\{v_a^2\} = 0.04 \times 10^{-24}$ V^2/Hz; $R_0 = 0.8$ A/W (for PIN) and receiver rise time 1 ns.

Procedure:

As the link length is quite large, one has to use repeaters in-between. As discussed in Section 13.1, the BER of each link must be better than the overall system BER, say 10^{-10}. For this BER, an average S/N ratio of about 16 dB is required [12]. Let us now consider Laser-GI-PIN combination. The required optical power level at the receiver for the above S/N ratio can be determined by substituting the values of various parameters in Eq. (6.48a). It will be approximately -36.8 dBm [12]. With this sensitivity, maximum link length for the attenuation limited channel, L_a(max) (taking $L_{sf} = 1$ dB, $L_{ff} = 0.1$ dB and $L_{fd} = 0.2$ dB) from Eq. (13.16) is 42 km. Since the transmitter and receiver rise times are 2 ns and 1 ns respectively, allowable maximum rms pulse width from Eq. (13.9) will be 2.24 ns. As the fiber is SM type, modal dispersion will be zero and therefore the total dispersion will be same as the material dispersion i.e., 9.6 ps/km. With a safety margin of 10% in the fiber dispersion, L_d(max) will be 212.1 km. As the attenuation limits the link length to 42 km, repeater spacing can be at most 42 km. It implies that for the 540 km link length, there must be 13 links of about 41.5 km length each. When APD receiver is used instead of PIN-photodiode receiver, receiver sensitivity will be -45.3 dB. The maximum link length will again be limited by the attenuation and it will be 55.4 km. In this case, 10 links of 54 km length will be required. It implies that 12 repeaters will be needed with PIN-photodiode receiver and 9 will suffice with APD receiver.

13.5   REFERENCES

[1]   Agrawal, G. P.: Fiber Optic Communication System. John Wiley & Sons Inc., 1992.
[2]   Elion, G. R.; Elion, E. A.: Fiber Optics in Communication Systems, Electro-optics Series/2, Marcel Dekker Inc., 1978.
[3]   Gower, J.: Optical Communication Systems. Second Edition, Prentice-Hall, 1993.
[4]   Jain, V. K.; Gupta, H. M.: Optical fiber communication link design. J. Opt. Commun. 6(1985)2, 58-66.
[5]   Jain, V. K.: An approach to optical fiber link design. Students' Journal IETE, 29 (1988)2, 35-41.
[6]   Keiser, G.: Optical Fiber Communications. Second Edition, McGraw-Hill, 1991.
[7]   Kleekamp, C.; Metcalf, B.: Designer's Guide to Fiber Optics. Cahners, 1978.
[8]   Personick, S. D.: Optical Fiber Transmission System. Plenum Press, 1981.
[9]   Powers, J. P.: An Introduction to Fiber Optic System. Aksen Associates Inc., 1993.
[10]   Senior, J. M.: Optical Fiber Communication-Principles and Practice. Second Edition, Prentice-Hall, 1992.
[11]   Sharma, A. B.; Halme, S. J.; Butosov, M. M.: Optical Fiber Systems and Their Components. Springer-Verlag, 1981.
[12]   William, B. Jones, J.: Optical Fiber Communication Systems. Holt Rinehart and Winston Inc., 1988.

14 OPTICAL NETWORKS

This Chapter first presents an introduction to the principle of network topologies and network design valid for both electrical and optical networks (Section 14.1). Some fundamental technical and non-technical backgrounds of optical networks are discussed in Section 14.2. The following two Sections are dedicated to fiber-supported networks (Section 14.3) which include optical as well as electronic components and all-optical networks (Section 14.4) where all network functions are exclusively realized by optical components including the switching function as well as the access. Thus, Section 14.3 primarily considers use of fiber optics in traditional and upgraded networks such as local area networks, transport networks and access networks whereas Section 14.4 is focussed on high-capacity end-to-end optical networks. Finally, Section 14.5 highlights some important aspects of digital broadband fiber networks and information-superhighways as an important precondition for high-quality multimedia communication.

14.1 PRINCIPLE OF NETWORKS

Looking back to the history of telecommunication networks, which started with Alexander Graham Bell in the middle of the 19th century, the network in its early stages conveyed only telephone information in an analog format, with signals being multiplexed in the frequency domain. Modem technology was used to facilitate the transmission of low-speed computer data. Subsequently, the development of semiconductor device technology enabled the realization of digital processing. The use of digital communication, where the signals are electrically multiplexed in the time domain, reduced the number of restrictions imposed upon the transmitted information. This progress continues and led to the development of ATM-based multiplexing.

A network shared by a number of users should economically convey any kind of information that the users require. This is the ultimate goal of telecommunication liberations and optical technology which is expected to play a major part in its realization. In all areas of the telecommunication network, including the communication-line network, transmission and switching system, and terminal equipment, technology has moved from analog to digital, from copper to fiber optics and, in particularly, in transport networks, from hardware- to software-based equipment. The optical communications is now progressing from point-to-point links to optical networks. Irrespective of type of network such as local area network (LAN) or subscriber loop, an important question is: Which technology meets the network requirements most economically? This means:

> The fundamental goal of designing networks is to provide bandwidth at low cost to the end user.

14.1.1   NETWORK TOPOLOGIES

Although there is a trend to integrate networks which has become possible by digitalization, various other types of network are still existent. In order to classify these networks, parameters such as area, transmission scheme (digital or analog), bandwidth (narrow- or broadband), physical layer, service and network function can be used. Table 14.1 presents some typical examples.

Table 14.1: Classification of networks

Classification parameter	Network (examples)
Area	Local area network (LAN) Metropolitan area network (MAN) Widespread area network (WAN) National and international network Global network
Transmission scheme	Digital network Analog network
Bandwidth	Broadband network Narrowband network
Physical layer (medium)	Fiber network Copper network Satellite network
Services	Distribution services network (e.g., TV) Interactive services network (e.g., POTS, ISDN) Data services network Multimedia services network
Function	Transport, trunk or core network Feeder network Broadcast network Access network (e.g., subscriber loop) Overlay network

Each network has both physical and logical (functional) characteristics. The logical characteristics are defined by the communication mechanism among the network elements. The physical characteristics are defined by geographical locations of transmission, switching, processing and storage elements.

The topology of a particular network e.g., a LAN or a public telephone network, is defined as the type of connectivity between stations such as workstations, personal computers (PCs) or telephones. Thus the network topology defines how the stations, normally referred as nodes, are geometrically arranged. The choice of a certain network topology is primarily determined either by its services which have to be transported and to be switched (e.g., interactive multimedia services) or its function (e.g., broadcasting of TV channels). The major topologies are

- Bus topology,
- Star topology,
- Ring topology,
- Tree topology and
- Mesh topology.

Each of these network topologies has its own advantages in terms of reliability, flexibility, expandability and performance characteristics. As an example, Fig. 14.1 shows the three basic LAN topologies which are the bus, star and ring configuration.

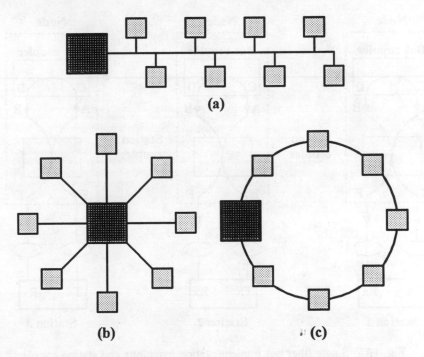

Fig. 14.1: Typical network topologies (a) bus, (b) star and (c) ring topology

The principle of networks based on the bus, star, ring, tree and mesh topologies are discussed in the following Subsections in more detail.

14.1.1.1 BUS TOPOLOGY

The *bus topology* consists of a common shared bus line (Fig. 14.1a). Each individual station is directly connected to the bus and signals travel in both directions on the bus from the point of origin until being dissipated at terminations. Any station transmission is detected by all other stations so a bus is especially suitable for broadcast operation. Bus topologies are normally used to connect PCs or workstations in a LAN or to distribute multiple video channels within a city. In the latter application, normally one station acts as the transmitter (the hub station) while the other stations act as receivers. But in LAN, each user has random access to the network.

In fiber optic LANs abbreviated as FOLANs, bus topology was initially considered. In the nineties, bus topologies were more and more replaced by ring topologies. Disadvantage of the bus topology is that signal loss in dB increases linearly with the number of taps i.e., users. Three commonly used structures for this topology are described below.

(i)   SINGLE FIBER BUS

This is a direct fiber optic adaptation of an electrical bus configuration. Each node consists of bus and station couplers as shown in Fig. 14.2.

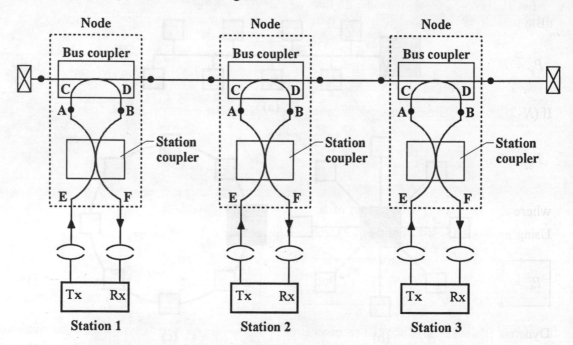

Fig. 14.2: Single fiber bus implementation using bus and station couplers

Split ratio of the bus coupler can be varied resulting into different amount of power tapped from/launched onto the bus by a station. For any given number of nodes, there is an optimum value of split ratio which will maximize the available power budget/system gain [54]. Station coupler is a 3 dB coupler which splits the transmitter power equally in both the output ports A and B (Chapter eight). These ports couple power onto the bus in both the directions through the bus coupler. Similarly, an incoming signal from either direction of the bus is coupled to either port A or B and is then passed onto the receiver through the station coupler. Since the power division in ports A and B is for power launching in opposite directions on the bus, it is natural to have a 3 dB station coupler.

Let P_t be the transmitted power by a node in dBm, the power launched on the fiber bus in dBm will be

$$P_{b1} = P_t - \left(L_{cn} + L_{sc} + L_{sp} + L_{bc} + L_{sp}\right) \text{ dB} \tag{14.1}$$

Here, L_{cn} and L_{sp} are the connector and splice losses typically in the order of 0.5 dB and 0.1 dB respectively; L_{sc} the station coupler loss consisting of excess loss of about 0.5 dB and power splitting loss of 3 dB and L_{bc} the bus coupler loss which consists of about 0.5 dB excess loss and $-10 \log_{10}(k)$ dB loss due to power splitting where k is the coupling ratio (see Section 8.2 or Section 9.4.3).

Let P_{b2} be the power available on the bus in dBm. The power received P_r at the receiver in dBm will be

$$P_r = P_{b2} - \left(L_{sp} + L_{bc} + L_{sp} + L_{sc} + L_{cn}\right) \text{ dB} \tag{14.2}$$

If $(N-2)$ intermediate nodes exist between transmitting and receiving nodes, then

$$P_{b2} = P_{b1} - (N-2) \left(L_{sp} + \tilde{L}_{bc} + L_{sp}\right) \text{ dB} \tag{14.3}$$

where $\tilde{L}_{bc}$ consists of excess loss of about 0.5 dB and transmission loss of $-10 \log_{10}(1-k)$ dB. Using Eqs. (14.1), (14.2) and (14.3), P_r in terms of P_t and N will be

$$P_r = P_t - \left[9.4 + 0.7(N-2) - 20\log_{10}(k) - 10(N-2)\log_{10}(1-k)\right] \text{ dB} \tag{14.4}$$

Dynamic range required for the receiver on this bus can be determined by computing minimum and maximum values of P_r at any receiver. Minimum P_r is obtained when transmitter and receiver are located at the extreme ends i.e., N in Eq. (14.4) is equal to the maximum number of nodes supported by the bus. Maximum P_r occurs when the transmitting node is an immediate neighbour of the receiver and can be obtained from Eq. (14.4) by taking $N=2$.

(ii) DUAL FIBER BUS

In dual fiber bus structure, all transmitters are connected to one segment of the bus and receivers to the other. These segments are interconnected at one end as shown in Fig. 14.3. This is also known as folded bus [42]. In this structure, station couplers are basically Y-couplers as described in Chapter eight.

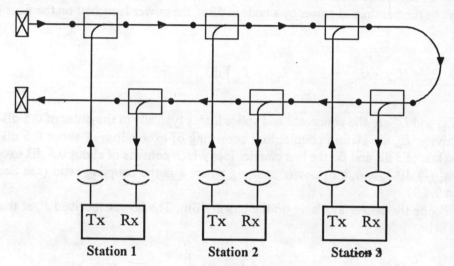

Fig. 14.3: Folded bus implementation using Y-couplers

For a bus of N stations, the received power level P_r will be

$$P_r = P_t - \left[2(L_{cn}+L_{sp}+L_{yc}) + f(N)(L_{sp}+\tilde{L}_{yc}+L_{sp})\right] \text{ dB} \tag{14.5}$$

where $\tilde{L}_{yc}$ is the loss associated with the Y-coupler, which consists of power transmission loss $-10\log_{10}(1-k)$ dB and excess loss (≈ 0.5 dB). L_{yc} is the loss of the Y-coupler, but it consists of split loss $-10\log_{10}(k)$ dB and excess loss (≈ 0.5 dB). The term $f(N)$ represents the number of intermediate couplers between receiver and transmitter [54]. Assuming the same values of L_{cn} and L_{sp} losses as in single fiber bus, above equation gives

$$P_r = P_t - \left[2.2 - 20\log_{10}(k) + f(N)\left(0.7 - 10\log_{10}(1-k)\right)\right] \text{ dB} \tag{14.6}$$

Dynamic range required by the optical receiver can be determined by computing the minimum and maximum values of P_r received by the extreme right station. Minimum P_r is received when the transmitter is on the extreme left and is given by

$$P_r(\text{min}) = P_t - \left[2.2 - 20\log_{10}(k) + (N-1)(0.7 - 10\log_{10}(1-k))\right] \text{dB} \qquad (14.7)$$

P_r will be maximum when a station received power from its own transmitter i.e., $N = 0$ in Eq. (14.6). Thus

$$P_r(\text{max}) = P_t - \left[2.2 - 20\log_{10}(k)\right] \text{dB} \qquad (14.8)$$

(iii) DUAL DISJOINT FIBER BUS, DDFB

In DDFB structure, there are two disjoint segments each permitting flow of signals in opposite directions. Station coupler is not required and each station needs two transmitters and two receivers as shown in Fig. 14.4.

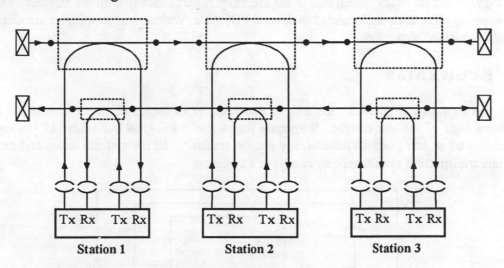

Fig. 14.4: Dual disjoint fiber bus implementation using station couplers

Loss analysis of this bus structure is quite similar to that of a single fiber bus except for a few minor differences. In this configuration, as there is no station coupler, terms involving split loss and excess loss of this coupler and one splice loss will not be there. This results in a saving of 3.6 dB each while launching on and tapping off the bus. The modified equation will be

$$P_r = P_t - \left[2.2 + 0.7(N-2) - 20\log_{10}(k) - 10(N-2)\log_{10}(1-k)\right] \text{dB} \qquad (14.9)$$

The passive linear bus structures discussed above can support only a few stations with the available sources and detectors. Assuming a power margin of 40 dB, single fiber bus and folded fiber bus can support only eight and six stations respectively and dual disjoint fiber bus supports around 12 stations [54].

14.1.1.2 STAR TOPOLOGY

In the *star topology*, each station is connected to a central node or a hub through point-to-point links (Fig. 14.1b). The hub, which can be an active or a passive device, directs the flow of traffic, e.g., multimedia traffic, to all other stations. An active hub allows to control the network such as routing of messages between central node to particular outlying nodes or between outlying nodes. In a star network with a passive central node, e.g., a passive optical splitter, the incoming signals are equally shared among all the stations. In this case, the star topology becomes a broadcasting system. Sometimes several passive stars are interconnected to accommodate large number of users or to reduce the fiber clusters at the coupler. It will lead to a multistar topology. In star topology, signal loss in dB increases logarithmically with the number of nodes. Star topology has been implemented in several FOLANs. Various star topologies are discussed in the following Subsections.

(i)   PASSIVE STAR

In this topology (Fig. 14.5), transmitter and receiver of each station are connected to the opposite ends of the star coupler. It appears that a star is a logical bus where all the taps are concentrated at one point. Whenever any station transmits, all the stations listen and any two stations transmitting simultaneously results in a collision.

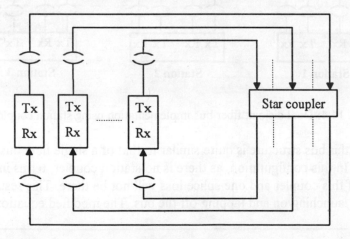

Fig. 14.5: Passive star LAN

In the star coupler, power received in any output port P_r in dBm is given by

$$P_r = P_t - L_{st}$$

(14.10)

where P_t is the transmitted power in dBm and L_{st} is loss due to star coupler in dB. It consists of power split loss $10 \log_{10}(N)$ dB for N×N star coupler and excess loss which may be about 2 dB. Therefore,

$$P_r = P_t - 10\log_{10}(N) - 2 \, \text{dB} \qquad (14.11)$$

It is seen from the above equation that the received power level in dBm varies as $\log_{10}(N)$, whereas in bus topology it varies linearly with N (Eqs. (14.4), (14.6) and (14.9)). It means that the number of nodes N places more stringent requirement on the received power level in the bus topology than in the star topology. Another very important advantage of the star topology is that all the receivers get the same amount of power and hence the requirement of dynamic range is minimal in this topology. Of course, there may be some variations due to imperfect star coupler, difference in the path lengths or connector losses etc..

Disadvantages of star topology are: (i) no scope of direct expandability i.e., N×N star can support at most N nodes. Some expandability is possible using multistars, but these have their own limitations, (ii) clustering and presence of long lengths of cables from the star to each node and (iii) unreliable collision detection. The non-ideality of the star coupler, difference in the path lengths or source power variations may make the two signals differ from each other by more than 5 dB - 6 dB. In that case, stronger signal would mask the weaker signal and the collision would not be reliably detected. In order to avoid these problems, several networks based on active star, directional star, multistar have been proposed.

(ii) ACTIVE STAR

Active star implementation involves the individual detection of each signal (Fig. 14.6).

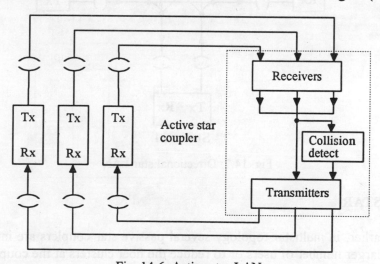

Fig. 14.6: Active star LAN

When two or more signals are present at the receiver outputs, a collision is detected. These signals are then combined and the composite signal is transmitted to all the stations. Since the optical signal is regenerated at the star coupler and a transmitter is dedicated to one or a few outgoing lines, there is no power splitting as in the case of passive star coupler. This improves the available power margin and as a consequence more number of nodes can be supported by the active star coupler. This, of course, is achieved at the cost of reliability.

(iii) DIRECTIONAL STAR

In directional star, several small couplers are joined together to perform the function of power splitting and combining. It is so constructed that any node can receive transmission of all other nodes except its own. It means that the reception of any signal by a transmitting station is a collision condition. A LAN can be implemented by cascading these directional couplers (Fig. 14.7). However, the high excess loss of directional couplers puts a severe limitation on the number of nodes that can be interconnected which in turn limits the size of network.

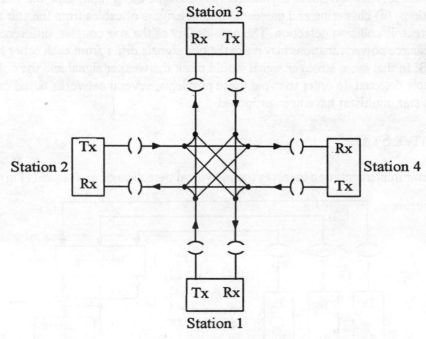

Fig. 14.7: Directional star LAN

(iv) MULTISTAR

As stated earlier, in multistar topology several passive star couplers are interconnected to accommodate larger number of users or to reduce the fiber clusters at the coupler. One way of connecting them is by merely joining each star coupler by another (Fig. 14.8). This is referred as multistar network. With multistar, any signal transmitted by a station will be received by all the stations on the bus, but the received power level will depend on whether transmitting and receiving stations are connected on the same star coupler or on the different couplers. Depending on the size of network and other constraints like path difference, power splitting at the star coupler etc., difference in the received power levels may be quite high. Another problem associated with the multistar is that a station may receive the same signal two or more times via different couplers. This along with the dynamic range problem complicate the receiver design in the network. It is not found feasible to connect more than four or five stars of size 8×8 (or 16×16 with higher receiver complexity) if multiple loop backs are not to create serious problems [54].

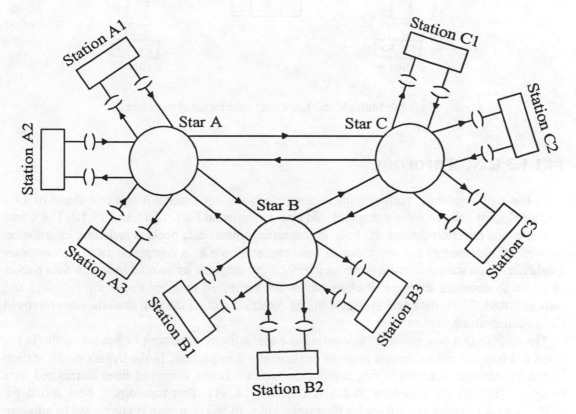

Fig. 14.8: Multiple star LAN

The problems in multistar network can be overcome by using a repeater between any two star couplers, which permits signals to be transmitted only in one direction at a time (Fig. 14.9). These selectively unidirectional repeaters regenerate the signals as well as perform the echo suppression. As a consequence, signals received at stations connected to any star coupler have the same order of magnitudes irrespective of the transmitter location. Similarly, echoes never reach any of the transmitting transceiver avoiding the problem of loop backs. Price paid for the advantages is in terms of reduction in reliability due to the presence of active components in the repeaters.

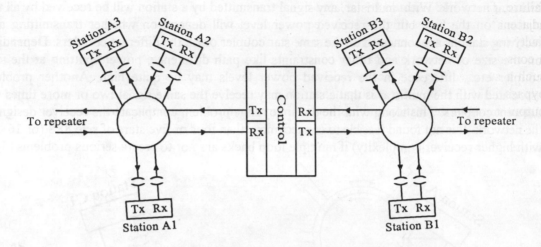

Fig. 14.9: Multiple star LAN with unidirectional repeaters

14.1.1.3 RING TOPOLOGY

In the *ring topology*, each stations is connected to one preceding it and one ahead of it by point-to-point links to form a single closed ring as shown in Figs. 14.1c and 14.10. The signal travels in one direction through this path. In this arrangement, data packets including information as well as address bits are sent from an upstream station e.g., a computer, to its downstream neighbour. Each station contains an active interface to recognize its own address in a data packet in order to accept a message. Packets which are addressed to other nodes are repeated and retransmitted. Thus, data packets travel in one direction around the ring until they are removed by the transmitting station.

The stations in a ring network can operate in either a listen, a transmit or bypass mode. In the listen and transmit modes, station receives or transmits data packets. In the bypass mode, station is not operational. Currently, ring topologies such as Token Ring and fiber distributed data interface (FDDI) are important in LANs (Section 14.3.1). This topology is best suited for FOLAN as it requires point-to-point fiber optic links. In this, a station is connected to adjacent stations through in- and outbound links. The interface at each node is an active device. It converts the incoming optical signal to an electrical signal. If the signal is destined to a particular station, it is removed otherwise it is remodulated onto the outgoing optical carrier using an electrical-to-optical converter. In this process, each node introduces some delay.

In ring topology, node or link failure may result in making the network non-operational. In order to prevent the ring breakdown due to link failure, one simple remedy is to provide an additional backup ring either in the same direction (dual ring structure) or in the reverse direction (counter rotating ring). In FOLANs, it can be provided either by using dual fiber or by using wavelength division multiplexing (WDM) on the same fiber (Section 14.3.1). In case of node failure, faulty node should be isolated or bypassed. This is done by using loop back at node adjacent to faulty nodes (Fig. 14.10b). Second method is by using an optical switch to bypass the faulty node. At present, such switches are rather slow (switching time typically 5 ns to 10 ns) and a considerable amount of data may be lost during the switching time. Further, insertion loss of such a switch is high (typically 1 dB to 2 dB) [16]. The third method is by using a high loss bypass fiber [1]. The signal output of bypass fiber is 20 dB to 30 dB lower than the normal output signal of the station and has negligible interfering effect on the latter signal.

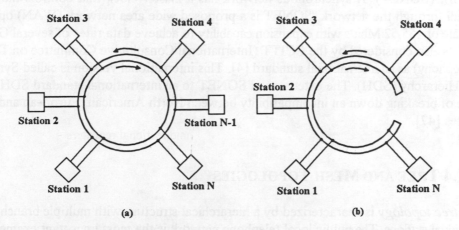

Fig. 14.10: (a) Two counter-propagating rings in FDDI and (b) formation of single ring when both the rings fail at some point

When a node fails, adjacent node no longer gets the normal signal. It then haunts for the low level signal which is acquired, amplified and processed. Fourth method is to use the bypass links. These bypass links connect alternate nodes (Fig. 14.11). A node normally locks onto the incoming primary (normal) links. When this signal is absent, it switches to the secondary (bypass) links. Each node transmits data on both primary and secondary links. This method requires extra fiber and some routing decisions.

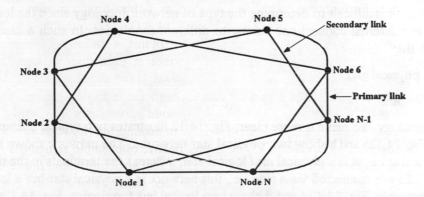

Fig. 14.11: Primary and secondary links for bypassing a faulty node

Fiber optic ring networks have been operating at data rates varying from 4 Mbit/s to 200 Gbit/s [16]. The American National Standard Institute (ANSI) has produced a FOLAN standard at 100 Mbit/s. This standard is known as Fiber Distributed Data Interface (FDDI) [8, 52]. The FDDI uses ring topology and is an extension of IEEE 802.5 (token passing ring) protocol. A variant of FDDI, called FDDI-II has been proposed for switched traffic such as private branch exchange (PBX).

The other fiber optic network that has been proposed by ANSI is Synchronous Optical NETwork (SONET). A synchronous network has a master clock that controls the timing of events all through the network. SONET is a proposed wide area network (WAN) operating at a base rate of 155.52 Mbit/s with expansion capability to achieve data rates of several Gbit/s. The SONET is also considered by the CCITT (International Consultative Committee on Telegraphy and Telephony) as an international standard [4]. This international version is called Synchronous Digital Hierarchy (SDH). The extension of SONET to an international standard SDH holds the promise of breaking down an incompatibility between North American, European and Japanese data rates [47].

14.1.1.4 TREE AND MESH TOPOLOGIES

The *tree topology* is characterized by a hierarchical structure with multiple branches leading to individual stations. The public local telephone network is the most important example of such a configuration. This format makes the tree topology of potential interest for multimedia communications (Section 14.5).

The *mesh topology* is among the most complex configuration. In this topology, stations may be directly linked with most other stations. It can be compared to ring topology with partial tree or star topology pattern.

14.1.1.5 PHYSICAL AND LOGICAL TOPOLOGIES

Sometimes, it is difficult to determine the type of network topology since the location of user terminals and internal connections belong to different topologies. In such a case, we have to distinguish the

- physical and
- logical

network topology. To make it more clear, Fig. 14.12 illustrates two typical examples. As a first example, Fig. 14.12a and b show two physical star networks. The network shown in Fig. 14.12a is actually a star i.e., it is a physical *and* logical star, whereas the terminals in the network given in Fig. 14.12b are connected via a ring i.e., this network is a physical star but a logical ring. As a second example, Fig. 14.12c and d show two logical bus topologies. Fig. 14.12c represents a physical bus, whereas Fig. 14.12d is a physical star. Network topologies can also be mixed which is very common in practical network design. Typical examples of mixed topologies in the LAN are shown in Fig. 14.13. To avoid or at least to reduce blocking which occurs in networks with more traffic (e.g., in the public telephony network), route diversity is normally applied to the network which is illustrated in Fig. 14.14.

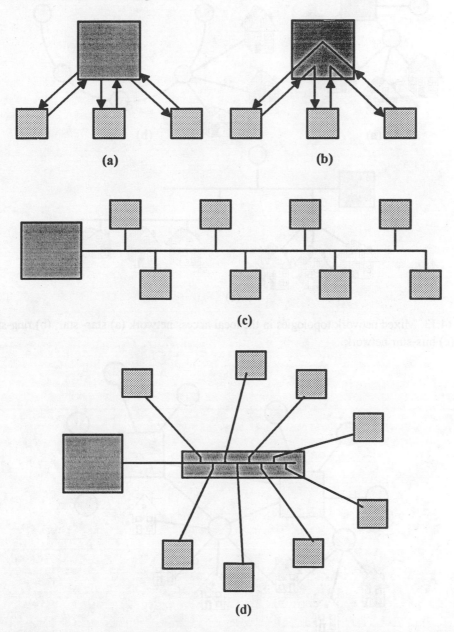

Fig. 14.12: Physical and logical network topologies (a) physical and logical star, (b) physical star and logical ring, (c) physical and logical bus and (d) physical star and logical bus

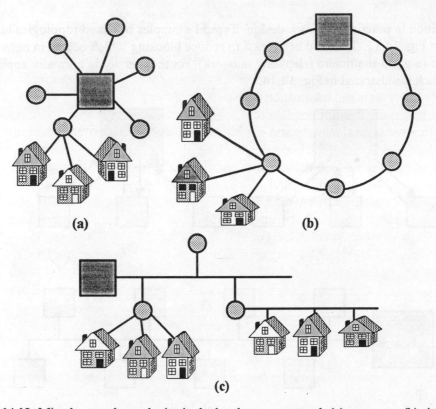

Fig. 14.13: Mixed network topologies in the local access network (a) star- star, (b) ring-star and (c) bus-star network

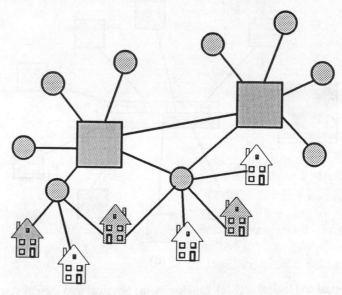

Fig. 14.14: Route diversity in a star-star topology

14.1.2   NETWORK DESIGN

Network design is primarily performed by employing appropriate software tools. Input parameters required are, for example, number and location of users (i.e., nodes), number and location of crossconnects and information about the expected traffic. As this book is focussed on optical communication, it is not possible to discuss network design in detail. We have to restrict our interest to some general remarks and questions. One question of prime importance is: Which network topology will match the users demands most economically in future?

A typical telephony local loop network average only 2 km in reach. The need for thousands of local switches and hundred of central offices for a typical nation has led to the creation of hierarchical telecommunication networks where the traffic is incredibly thin in the local loop and incredibly thick at the core and the international level. Thus, these telephony networks are characterized by a rather complex hierarchical structure including a large number of switching centres. On the other hand, LANs employed for computer communications are normally based on simple bus or ring topologies such as Token Ring or FDDI (Section 14.3.1). These networks operate without any central switching node and without a complex hierarchical structure. With converging information markets (telecommunication, computing and consumer electronics), both types of networks i.e., global telephony network and LAN will also converge. However, independent of convergence, the principle functions of the network to

- route and

- switch

the information to be transmitted remains the same. It becomes clear by the networks mentioned above that there is always a trade-off between the routing and switching function.

> To optimize the economics of a network, there is a traditional balance between switching and transmission costs.

In principle and in the extreme, a local or even national network could be designed with a single switching node site rather than about 10,000 local exchanges for a typical nation at present. If we exploit, for example, the inherent reach capability of optical fiber and extend the local loop to about 40 km, we then see complete national network with less than 100 switches possible. At this level, there is no need to create a hierarchical structure. Wavelength routing using linear and non-linear properties offers the prospect of non-blocked switchless communication for up to 20 million customers (Section 14.4). In this case, the network has become purely a local loop [10]. While designing future networks, the following simple structure will most likely be of prime importance (Table 14.2).

Table 14.2: Structure of future optical networks

Services
ATM
SDH
WDM

Advanced networks have to support transmission and switching of a various types of services such a data services (e.g., computer links, ISDN and B-ISDN services, telebanking), voice (e.g., digital telephone service, digital audio broadcasting), video (e.g., video-on-demand, TV broadcasting) and interactive multimedia services (e.g., teleeducation, telemedicine). Since all these services operate at different bit rates in the range of some kbit/s to some Mbit/s, a flexible transmission technique is required.

The most promising switching and transport technique for carrying interactive multimedia traffic is the asynchronous transfer mode (ATM). It will soon penetrate widely into public networks to carry voice, pictures, videos as well as data. In the trunk and feeder networks, a fixed bit rate based on synchronous digital hierarchy (SDH) will most likely be used to transport the ATM traffic (Table 14.3).

Table 14.3: Synchronous transport modes (STM) of synchronous digital hierarchy (SDH)

Level	Bit rate
STM-1	155.52 Mbit/s
STM-4	622.08 Mbit/s
STM-16	2488.32 Mbit/s
STM-64	9953.28 Mbit/s

In modern LANs the traditional plesiochronous digital hierarchy (PDH) ring network between the local exchange (LEX) and the service access point (SAP) will be replaced by a SDH feeder ring network [36, 37, 41]. This variable and intelligent synchronous optical ring network will carry digital narrowband services (2 Mbit/s) as well as broadband services (155 Mbit/s). It will use add-drop multiplexers and subscriber accesses with flexible multiplexers to select the information.

In the distribution networks, asymmetric and high bit rate digital subscriber lines (ADSL and HDSL) equipment are installed in the short run to provide new services such as 2 Mbit/s asymmetric switched broadband, service-on-demand (SoD), fast Internet access and 2 Mbit/s symmetric switched broadband (Section 14.3.3.2). In addition, a coaxial cable network is installed for cable TV (CATV) distribution. In the long run, ADSL and HDSL will be replaced by the migration of the fiber to the buildings and the homes. In order to increase transmission capacity in the trunk networks and to interconnect the large communication centres, an optical overlay network based on *wavelength or optical frequency division multiplexing* (WDM or OFDM) can be employed. WDM and OFDM are excellent and powerful technologies to enhance flexibility in the access networks and to upgrade existing networks.

14.2 BACKGROUND OF OPTICAL NETWORKS

The explosive advancement of light wave technology over the past decade has revolutionized long distance communications, whereas it is still unclear what role photonics technology will play in short-hop applications such as LANs and telephone subscriber loops [18]. Research on photonics switching has been motivated by the assumption that the advantages of fiber optic transmission would spread to distribution networks, intermediate crossconnects and switching nodes as well. In the early nineties, it was predicted that fiber would reach out and touch every home, ultimately replacing coaxial cables and twisted pair copper wires for telephone and cable television communications.

Highly motivated by the impressive results of fiber optic technology in the eighties and nineties, various scenarios predicted a telecommunication infrastructure based on all-optical networks including optical frequency division multiplexing (OFDM), information super-highways, coherent receivers with high-quality tunable local lasers, all-photonics switching in the trunk network as well as in the local loop and finally fiber-to-the-home (FTTH) with ATM-based digital broadband accesses which offer interactive multimedia services and a large number of interactive high definition TV (HDTV) channels. However, it has not yet been fulfilled due to mainly the economic reasons. The high cost is not in fiber, but in the optoelectronics that go with it. Thus, fiber local loops have stopped at the curb or an even-more distant node.

The introduction of high-quality, interactive multimedia services requires high bit rate accesses and system upgradation to much larger capacities e.g., from the basic hierarchical level STM-1 to STM-4 or STM-16. No doubt, FTTH with single-mode fiber represents the best technical solution for broadband access. However, high cost-barriers make it difficult to realize a global implementation of FTTH. In addition, various other important parameters limit the pace of realizing FTTH as explained below. On the other hand, there is a need for advanced networks and there is a continuously increasing demand of higher transmission capacity, new services (in particularly interactive services) and multimedia communication [56]. Introduction of video-on-demand (VoD) is also influencing the demand for transmission.

High replacement costs for users normally require continued use of existent copper infrastructure. Thus, systems and strategies for transport and access networks depend heavily on the installed infrastructure and the range of services permitted by regulation. In particular, the existing access network i.e., the last kilometre or the last mile to the customer premises, represents the most critical bottleneck in terms of bandwidth and is the most sensitive cost component of any upgradation of local loop infrastructure. Therefore, many telecommunication network operators explore in the development of less expensive techniques which allow a first upgradation of the existent copper infrastructure and offer the user a first generation broadband access. However, in some areas where a large-scale redevelopment takes place (e.g., east Germany after the reunification in 1990) or in developing countries where little installed infrastructure is existent, other alternatives are available. In such areas, the environmental constrains on the last kilometre are considerably relaxed and FTTH or at least fiber-to-the-building (FTTB) represents the best approach.

Upgraded networks will act as a bridge between existent narrowband and future broadband telecommunication networks. However, it is often not clear which technology and which network configuration will match the unknown users demands most efficiently (Fig. 14.15).The decision on when and in particularly how to upgrade the network troubles the network operators all over the world.

Fig. 14.15: Bridging current and future telecommunication infrastructure

The development of electronic broadband upgrading methods or the design of fiber-based or all-optical networks requires a detailed consideration of the following important parameters:

- technical aspects e.g., bandwidth demands,

- economic aspects,

- market aspects e.g., converging market,

- social and human aspects and

- political aspects such as liberalization of network operators.

As far as this book is concerned, the first parameter (i.e., the technical aspects) is of prime importance. Non-technical aspects, however, become more and more important when realizing a network. In most cases, these non-technical aspects are even more important than the technical aspects. For this reason, some of the non-technical aspects are briefly discussed below.

Most of the devices required to realize fiber-based networks are either developed or even commercially available. Thus, the *technical aspect* seems to be less important as compared to the others. The great challenge of technical research and development is to realize low-cost devices (in particularly optoelectronic devices) and to design high-quality networks on a low-cost level i.e., to provide cheap bandwidth to the end users. Another important task is to highlight alternatives. Certainly, fiber optic transmission systems exhibit a nearly unlimited bandwidth, but intense research work on data compression methods (e.g., MPEG-2) have resulted in a significant reduction of bandwidth demands. Applying MPEG-2, for example, bandwidth demands can be reduced by a factor of ten to hundred. Thus, installed copper links are sufficient for a number of medium- and even broadband applications.

Economic aspects are primarily responsible for the delay in realizing FTTH and other highly-sophisticated techniques such as coherent detection. Although FTTH is the best technical solution among all access scenarios, at present it is the least economical alternative because of the translation of optical signal to electrical signals is too costly on house-to-house basis. In realizing broadband networks and broadband accesses, network designers must always keep in mind the following: The user is not interested in fiber or copper, in analog or digital; he only would like to communicate with an acceptable quality on a low-cost level. Of course, he is interested in new services too, but he is not willing to pay much more than for the traditional services. It is expected that by the year 2010, more than 60% of households will use CATV, 37% satellite receivers, 80% personal computers (PCs), 60% multimedia PCs and about 60% digital TV [23].

Another important aspect to be considered is the trend in *converging information markets* such as (i) telecommunication companies (Telcos) and network operators, (ii) computer industries and (iii) consumer electronic industries and audio-visual broadcasting operators as shown in Fig. 14.16.

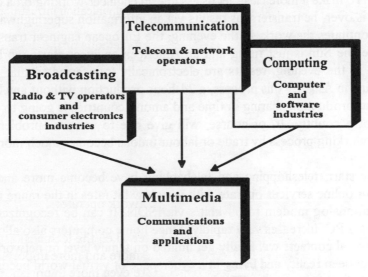

Fig. 14.16: Converging markets

The convergence of Telcos and consumer electronic yields, for example, new types of video services such as video-on-demand (VoD). Telcos and computing offer services such as data-based email, telebanking, home shopping and teleeducation also known as online university. Convergence of all three markets finally yields broadband interactive multimedia services. Thus, from the technical and consumer point of view the trend in converging markets can be very advantageous. On the other hand, this trend is going to produce some problems in regulation and standardization. New standardization committees consisting of members from all three industrial segments are needed e.g., DAVIC [48]. There are also expert groups within the international telecommunications union (ITU) who are working on standardization of multimedia networks, but ITU only represents the telecommunication part [26].

Social and human aspects, more and more become an important parameter when designing new telecommunication infrastructure systems. Undoubtly, we are on the way to a global communication society and during the next decades communication technology will most likely change human live much more than any other technology. Some of these changes have already been recognized. For example,

- Production sharing (e.g., hardware production in South America, Software production in Asia and Marketing in Europe),
- Teleshopping, telelearning, home banking, home office and
- TV around the clock.

Instead of a shared spatial production among different countries, production can also be shared globally in time. To make it more clear, let us assume an engineer working on a computer in Asia. When day time is over, he transfers his results via an information superhighway to an European engineer who continues the work. In the evening, the European engineer transfers his results to America where the Sun is just rising and an American engineer starts working on the same project. Again in the evening, results are electronically transferred to that colleague who is working and living in Asia. By this process, a 24-hour production without any break is obtained. It is evident that production sharing in time and among countries is going to change the social infrastructure of a country. It, of course, will give rise to various problems. For example, controlling the working process by trade or labour unions becomes much more complicated.

After a slow start, teleshopping and telebanking have become more and more attractive although current online services operate with very low bit rates in the range from 1.2 kbit/s to 56 kbit/s via an analog modem to 64 kbit/s via ISDN. It can be recognized that number of households using a PC increases very rapidly. Since home computers also allow to do business work at home, social contacts will finally get limited on family level or network communication level. The gap between reality and living in a cyberspace or virtual world becomes smaller. This will get strengthen further by viewing TV shows around the clock. Already, many young persons spend more time in front of TV screen than in school. An increased crime rate is observed, because they frequently watch low quality movies based on crimes. These examples illustrate that advantages, drawbacks and inherent problems of a global communication society are closely interrelated. More than in the last decades, diversification of value is required when new technologies are planned to be introduced. Consequences of new technologies must be estimated as far as possible. Besides technological aspects, engineers should take into account this important aspect also i.e., they have to carry both the technical *and* social responsibilities. In principle, technical changes can be performed very rapidly. Social changes which finally determine the pace of technical changes are, however, relatively slow processes.

The global networks interconnect different countries and cultures. Therefore, each global network has also its *political aspect*. In this book, it is not possible and also not required to discuss the political components of worldwide networks. However, one important aspect should be mentioned which directly influences the technical aspect of the network. In many countries of the world, telecommunication market is or will be liberalized in the near future. Thus, monopoly of the public Telcos will come to an end. Telecommunication market will be opened to everyone and private telecommunication companies will start to provide and operate their own networks. Since these companies normally will come from different segments of the industry (energy supply companies, consumer electronic companies, newspaper and video publishing companies et. al.), different techniques will likely to be applied by the different operators. This will give rise to the problems of regulation and standardization.

The aspects and problems described above sometimes yield a curious situation. Since it is not really known what are actually the users demands, how these demands will change in future and which network will match these demands most economically. Industrial companies and network operators are restrained in realizing new networks. Since every reconfiguration of a public network involves enormous costs, there is always a fear of investment in the wrong network. As a result, the required network configuration will be created most likely in the following steps:

(i)   Upgrade the existent copper infrastructure by electrical means such as

   • ADSL (asymmetrical digital subscriber line) and/or

   • HDSL (high bit rate digital subscriber line) and/or

   • alternative techniques like microwave and satellite subscriber lines and/or

   • data compression by MPEG-2

(ii)   Upgrade the existent copper infrastructure by fiber optics such as

   • FTTC (fiber-to-the-curb) and/or

   • HFC (hybrid fiber coaxial) and/or

   • HDSL or VHDSL (very high bit rate digital subscriber line) supported by fiber

(iii)   FTTB (fiber-to-the building) and FTTH (fiber-to-the-home)

It becomes clear from above that there is a continuous progress in fiber optics irrespective of the various problems existent at present and most likely in the near future. In particular, the following statement is valid.

> Optical systems are progressing from
> point-to-point links to optical networks!

The aim of the following Sections is to show in step-by-step how advanced fiber optic technology is going to change the telecommunication infrastructure. In contrast to this Section wherein technical as well as non-technical problems have been discussed, the following Sections are focussed on technical aspects only.

14.3   FIBER-SUPPORTED NETWORKS

This Section is focussed on networks which use fiber optics to increase the network capacity in terms of bit rate, flexibility and transparency. In contrast to all-optical networks which are exclusively based on optical components, fiber-supported networks use electronic as well as optical components. Three types of networks are discussed in the following Sections: LAN in Section 14.3.1, transport network in Section 14.3.2 and public access network also known as the local subscriber loop in Section 14.3.3.

14.3.1   LOCAL AREA NETWORKS

Now-a-days, all important companies have currently installed their own packet switched LAN such as Ethernet, Token Ring or FDDI (fiber distributed data interface) to interconnect their PCs or workstations. Normally, these networks are very convenient for conventional traffic. As available bandwidth is shared between different users, their efficiency is, however, seriously reduced when traffic increases. The bit rate is limited to 10 Mbit/s with Ethernet, 16 Mbit/s with Token Ring or 100 Mbit/s with FDDI or fast Ethernet. These networks, however, cannot match future data rate requirements due to the need for large data file or high definition moving pictures broadcasting.

In this Section, principle of traditional (conventional) LAN architectures is briefly summarized first (Subsection 14.3.1.1) as a basis for upgrading fiber-supported LANs (Subsection 14.3.1.2) and future high-sophisticated architectures employing ATM and WDM (Subsection 14.3.1.3). It should be noted that various expertise and experiences in designing advanced fiber optic LANs referred as FOLANs can be used in the field of designing local subscriber loops (Section 14.3.3), where an upgradation is also very much required in many industrial nations.

14.3.1.1 CONVENTIONAL LANS

Conventional LANs provide a medium-speed, low cost communication system over a limited distance, usually linking PCs, workstations and servers in a building or campus. LANs are ideal for supporting traditional business applications using textual user interfaces. The overall transmission speed of conventional LANs varies from 4 Mbit/s at the low end to 100 Mbit/s at the current high end, depending on the system and the technology used.

There are in principle two methods for a LAN device (workstation or PC) to gain access to the network's resources (bandwidth): Carrier Sense Multiple Access with Collision Detection (CSMA/CD) and Token Passing. To form a larger network, LANs are frequently interconnected with bridges and routers.

(i) ETHERNET

Ethernet, developed at XEROX/DEC/INTEL and now a standard known as IEEE-802.3, is a LAN technology with a bus or tree topology. Every station on the network is tied onto a common bus and can hear the traffic from every other station as described in Section 14.1. The different Ethernet systems available are classified by a common symbolic structure. Table 14.4 gives some examples.

Table 14.4: Examples of conventional Ethernet systems

Example	Bit rate in Mbit/s	Transmission scheme (baseband or passband)	Maximum distance (x 100 m) between stations or applied physical layer
(i)	10	Baseband	5
(ii)	10	Baseband	T
(iii)	10	Baseband	F

In the examples (ii) and (iii), physical layer "T" and "F" represent twisted pair and fiber cable. All other Ethernet systems such as 10Base5 (also referred as Thick-Ethernet) and 10Base2 (Thin-Ethernet) use coaxial cable as the physical layer. Ethernet standard 10Base5 is available since 1972, whereas fiber based Ethernet standard 10BaseF is only available since 1992. The use of fiber allows to link repeaters or hubs even when distance is large. Therefore, 10BaseF is used in particular to interconnect locally separated LANs and to form larger networks. For this application, passive (10BaseFP) or active (10BaseFA) optical star couplers are employed. As an example, Fig. 14.17 shows a LAN based on 10Base2.

Now-a-days, Ethernet consists of 500 m bus segments with coaxial, twisted pair or fiber cable. Different segments can be connected via an amplifier (repeater) creating a tree topology. With 10BaseT Ethernet, the physical layout becomes a star with a wiring cabinet (hub) at its centre, but the logical topology of the network remains a bus. In the eighties and nineties, 10BaseT Ethernet has become a runaway best-seller simple to install and easy to operate [38]. Ethernet systems typically support 10 Mbit/s and fast Ethernet 100 Mbit/s. Ethernet with bit rates higher than 100 Mbit/s e.g., Gbit/s-Ethernet, are under development.

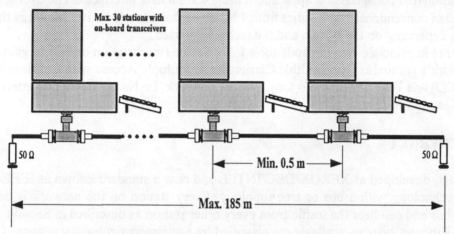

Fig. 14.17: LAN with Ethernet 10Base2

The characteristic features of traditional Ethernet are:

- 10 Mbit/s transmission bit rate (data rate shared among the stations on Ethernet),
- CSMA/CD (Carrier Sense Multiple Access with Collision Detection),
- Bus or tree topology,
- Coaxial, twisted pair or fiber,
- 500 m maximum bus length,
- 50 m maximum bus-station length,
- 2.5 m minimum and 2.8 km maximum station distance,
- 100 stations maximum per bus segment and
- 1024 stations total.

(ii) RING TOPOLOGY

Each station in a ring topology can switch from listen mode to the transmit mode (Section 14.1). A number of methods for switching from listen mode to transmit mode have been designed and implemented. Two popular ones are IEEE 802.5 Token Ring and the fiber distributed data interface (FDDI). Token Rings, which were the brain child of IBM Corporation and developed in the IEEE 802.5 committee, rely on a special control frame, called a token. It is passed around a ring from one station to another (Fig. 14.18). Only the station in possession of the token has the right to transmit. Token Ring was initially specified for a data rate of 4 Mbit/s and has been extended to 16 Mbit/s at the end of eighties. Current Token Ring networks allow to support 255 stations, however Token Ring networks can be cascaded by means of a ring hub (Fig. 14.19).

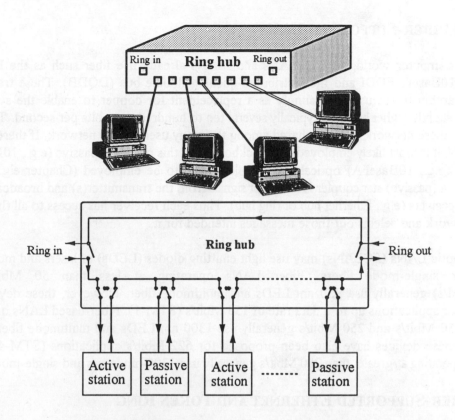

Fig. 14.18: LAN with Token Ring configuration and ring hub

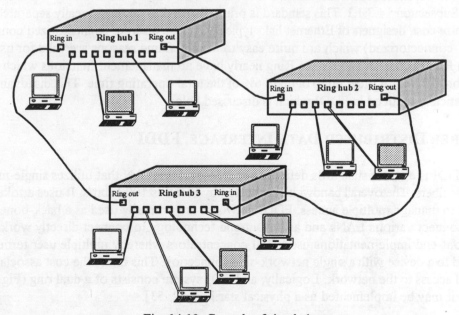

Fig. 14.19: Cascade of ring hubs

14.3.1.2 FIBER-SUPPORTED LANs

In the computer world, several LANs and MANs already use fiber such as the Ethernet standard 10BaseF, FDDI and the distributed-queuing double bus (DQDB). These traditional network architectures use fiber simply as a replacement for copper to enable the system to operate at slightly higher bit rates, typically several ten to hundred megabits per second. The total capacity in these networks is time-shared among the many users of the network. If there is fiber in the LAN, it is most likely employed as a backbone. For this purpose, passive (e.g., 10BaseFP) or active (e.g., 10BaseFA) optical star couplers have to be employed (Chapter eight). The function of a (passive) star coupler is to accept signals from the transmitter(s) and broadcast them to all the receivers (e.g., Ethernet hub or ring hub). Thus each receiver has access to all the traffic in the network and selects out those messages intended for it.

Fiber optic LANs (FOLANs) may use light emitting diodes (LEDs) or lasers and multimode fibers or single-mode fibers. Fiber LANs operating at less than 50 Mbit/s (or 50 Mbaud/s) generally use 850 nm LEDs and multimode fiber. However, these devices are available for applications up to a data rate of 155 Mbit/s (STM-1). Fiber-based LANs operating between 50 Mbit/s and 250 Mbit/s generally use 1300 nm LEDs and multimode fiber. In the nineties, these devices have also been proposed for 622 Mbit/s applications (STM-4). LAN systems operating at greater than 250 Mbit/s generally use 1300 nm lasers and single-mode fiber.

(i) FIBER-SUPPORTED ETHERNET AND TOKEN RING

The fiber-supported Ethernet standard 10BaseF has already been briefly described in the previous Subsection 14.3.1.1. This standard is primarily used to connect locally separated LANs. To minimize cost, designers of Ethernet links typically choose discrete connectorized components (e.g., ST-connectorized) which are quite easy to handle. Same reasons are valid for using fiber in Token Ring. Ethernet and Token Ring nearly have no management functions which result in network breaks (out of order time) of up to 6% of the total operating time. This disadvantage has been drastically reduced in FDDI which is discussed below.

(ii) FIBER DISTRIBUTED DATA INTERFACE, FDDI

The FDDI is a set of standards defining a shared-medium LAN that utilizes single-mode and multimode fibers. The overall bandwidth supported by FDDI is 100 Mbit/s. It uses a token-based discipline to manage multiple access. FDDI networks are primarily used as a back-bone technology to connect various LANs and as a front-end technology to connect directly workstations. Some front-end implementations use FDDI concentrators whereby multiple user terminals are connected to a device with a single network-side connection. This limits the cost associated with individual access to the network. Logically, an FDDI system consists of a dual ring (Fig. 14.20), although it may be implemented as a physical star [27, 52, 53].

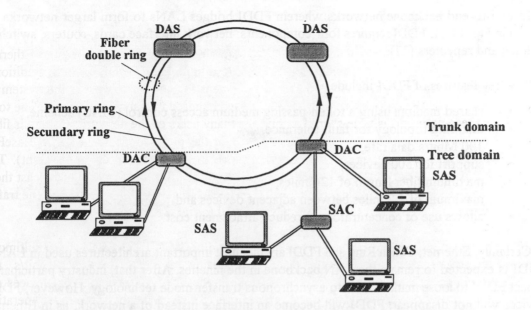

Fig. 14.20: Fiber distributed data interface (DAS: dual attached station; SAS: single attached station; DAC: dual attached concentrator; SAC: single attached concentrator)

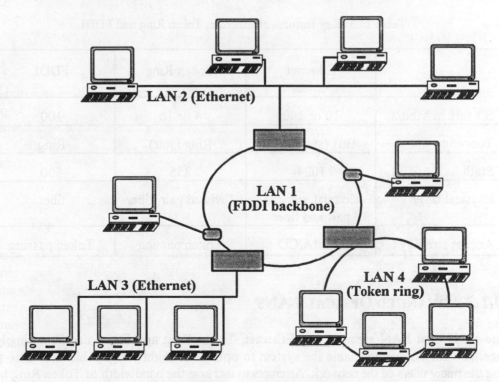

Fig. 14.21: LAN with FDDI backbone

In end-to-end backbone networks wherein FDDI bridges LANs to form larger networks as shown in Fig. 14.21, FDDI requires four components: network interface cards, routers, switches or hubs and repeaters [17].

The key features of FDDI include

- shared medium using a token-passing medium access control (MAC) scheme,
- dual ring topology for fault tolerance,
- 100 Mbit/s data rate,
- support for 500 devices,
- maximum fiber length of 124 miles,
- maximum of 1.2 miles between adjacent devices and
- allows use of concentrators to reduce attachment cost.

Certainly, Ethernet, Token Ring and FDDI are the most important architectures used in LANs. FDDI is expected to remain the LAN backbone in the nineties. After that, industry participants expect FDDI to loose market share to asynchronous transfer mode technology. However, FDDI devices will not disappear. FDDI will become an interface instead of a network, as in Ethernet and Token Ring. The following table summarizes and compares these traditional LAN architectures.

Table 14.5: Key features of Ethernet, Token Ring and FDDI

	Ethernet	Token Ring	FDDI
Bit rate in Mbit/s	10 or 100	4 or 16	100
Topology	Bus (star, tree)	Ring (star)	Ring
Stations	1024	255	500
Physical layer	Coaxial, twisted pair and fiber	Twisted pair, fiber	fiber
Access strategy	CSMA/CD	Token passing	Token passing

14.3.1.3 ADVANCED OPTICAL LANs

The traditional LAN architectures Ethernet, Token Ring and FDDI use fiber simply as a replacement for copper to enable the system to operate at slightly higher bit rates time-shared among the many users of the network. Attempts to increase the bandwidth of Token Ring beyond 16 Mbit/s have been unsuccessful as Token Ring users turn to FDDI or to ATM (Subsection i below) for upgradation. However, ATM seems to become the standard of LANs in the next century; whereas FDDI seems to become only a supporting interface instead of a network.

The traditional networks cannot match future data rate requirements such as high definition moving pictures or large data file transmission. To match these requirements, a high-speed, low-noise backbone network is required. Fiber optic transmission systems employing 1300 nm or 1550 nm lasers, single-mode fibers and WDM (Subsection ii below) are in principle suited to operate at bit rates up to some Tbit/s. Although photonics technology is capable of supporting many Tbit/s of throughput, the digital electronic components at the nodes of the light wave network can drastically limit total throughput to less than 1 Gbit/s. This problem is severe with the popular ring architecture such as FDDI topology. Each node of the ring path must process all the network traffic, so the traffic through any node and hence through the entire network, is limited to the electronic processing capability of a single node i.e., roughly 1 Gbit/s or less. Thus:

> The traditional LAN architectures Ethernet, Token Ring and FDDI cannot be extended to Tbit/s capacities because of a basic *electronic bottleneck*.

Among numerous approaches to opening up this electronic bottleneck, the following two are very promising. The first, multihop employs a new network architecture to achieve high capacity with existing devices; the second, wavelength division multiple access (WDMA), with new devices in a relatively conventional architecture [18].

The association of a WDM optical network with a packet switched network appears to be the best solution to resolve the issues mentioned above. The packet switched network transmits usual traffic at a limited bit rate. The WDM network transmits large amounts of information between nodes at a very high bit rate. Information for the WDM network management is inserted in the normal traffic on the packet switched network. Thus, we add the flexibility of packet switching to the high capacity of circuit switched optical WDM networks.

(i) ATM-BASED LANS

The asynchronous transport mode (ATM) is based on cell transmission. Each cell has a fixed length of 53 bytes, where 5 bytes are used as header and 48 bytes as payload. The ATM cell header contains routing information for moving the cell from one node to the next, whereas the payload contains the information to be transmitted. As shown in Table 14.6, ATM combines features of the synchronous transfer mode (STM) and the packet transfer mode (PTM). A lot of interest has been shown in ATM technology due to its flexibility and its ability to support interactive multimedia broadband services [3]. In future, range of services will be very diverse in terms of bit rate, network occupancy (continuous or burst), connection duration and frequency. In this heterogeneous environment, ATM will represent a key multiplexing and routing technique.

Table 14.6: Comparison of STM, ATM and PTM

STM	ATM	PTM
Line-switching	Packet-switching	Packet-switching
Connection-based	Connection-based	Connectionless
Circuit	Virtual circuit	Virtual Circuit

As far as network operation is concerned, the possibility of transporting and routing different bit rates is the most important advantage of using ATM.

> Variable bit rate traffic can
> easily be handled with ATM.

In addition, certain levels of quality of services (QoS) can be maintained. ATM offers much greater capacity than existing shared medium LANs. The information given above justifies the introduction of ATM into LAN on the basis of its ability to handle multimedia traffic. Currently, ATM switches are exclusively realized by electronics. In future, these switches can become an electronic bottleneck. To solve this problem, optical ATM switches are needed [34, 43].

(ii) WDM-BASED LANs

In the nineties, most optical communication systems like point-to-point connections or networks only use a single optical carrier wave to transmit the whole information. By applying this standard technique, fiber bandwidth is only inefficiently used. It may be noted that in the 1550 nm region, an optical fiber provides a spectral bandwidth of about 15000 GHz! In WDM networks, a large number of equally spaced optical carriers are used. Each optical carrier wave can be modulated by a high bit rate signal. With, for example, ten optical carriers, overall bit rate transmitted through the fiber is ten times higher than the bit rate transmitted through a single carrier system. Thus:

> The introduction of WDM technologies greatly
> enhances fiber transmission capacity.

With WDM, full potential of optical fibers can be explored and network transmission bandwidth will be increased. Moreover, WDM offers interesting benefits with respect to LANs [49]:

- Increased flexibility: Without the introduction of new optical fibers in the network, new path connections can be realized. Changes in the logical network structure can easily be performed. Addition of new wavelengths allows to respond to an unforeseen traffic load or to specific requirements such as creating a high bit rate leased line or video transmission. Thus, WDM gives the network architect more freedom.

- Optical platform and transparency: Transmission technologies are evolving from PDH to SDH and ATM. The optical path layer imposes no restriction on the electrical path transmission mode carried by optical paths; even analog signals are allowed. Thus, the optical path layer provides a platform into which different transmission mode networks can be included. Different types of signals may be superimposed on the same medium due to the optical components transparency to the formats of the signals to be transmitted. It results in multiple service capability.

- Enhanced crossconnect node processing: When WDM is used, the electrical processing bottleneck can be at least partly eliminated through the introduction of wavelength routing of the paths at the crossconnect nodes. This technique is referred to as optical routing because wavelength defines the route of a given signal through the network. Optical routing enhances the throughput of crossconnect nodes.

In WDM routed networks, at each node optical signals are either routed through to a subsequent stage or dropped out of the network to an optoelectronic interface by using an add-drop multiplexer (ADM). This multiplexer is a key component in WDM networks. It can select any desired channel from a set of wavelength multiplexed channel (i.e., dropping one channel) or insert a new channel (i.e., adding a channel). By controlling the optical channel routing, any logical network can be set up across a given physical structure. The resulting network may be treated as an optical network WDM layer or an optical overlay network.

14.3.2 TRANSPORT NETWORK

In transport network, such as telephone interoffice or intercity links and transoceanic links, fiber is exclusively used in the last two decades by many telecom operators. Single-mode fiber is the preferred transmission medium for long-distance communications at present, whereas in the eighties the graded-index fiber was exclusively used. Most terrestrial long-distance fiber transmission systems operate at line rates up to 2.5 Gbit/s and have a loss limited range of 30 km to 50 km between electronic regenerators. They usually operate at 1.3 μm wavelength to take advantage of the zero dispersion at this wavelength. In the eighties, more than 90% of the fiber in the world's telecommunications network had a dispersion zero at 1.3 μm.

During the nineties, 1.55 μm has become the favourite wavelength since fiber loss is minimum at 1.55 μm. Now-a-days, fiber loss no longer imposes a serious limitation on high-capacity light wave transmission systems due to the advent of highly reliable and highly efficient Erbium-doped fiber amplifiers (EDFAs) operating in the 1.55 μm wavelength region (Chapter seven). The most important obstacles to be overcome in order to enhance single wavelength system capacity are fiber dispersion, fiber non-linearities and electronic "bottlenecks" at very high speed. Since a huge quantity of non-dispersion-shifted fiber referred as standard single-mode fiber (SSMF) has already been installed world-wide for transmission at bit rates up to 2.5 Gbit/s, investigations of means to upgrade existing systems to 10 Gbit/s have become an important task.

> Laser chirp and fiber dispersion is the basic limitation of transmission distance in high-speed transport networks.

In order to upgrade the transport network, an optical network layer is technically feasible and very advantageous. This high-capacity backbone optical fiber network is based on wavelength division multiplexing (WDM), which uses different optical carriers to route the traffic while reducing demand on electronics [6, 15, 20, 44, 55]. Thus WDM represents an extension of the present network hierarchy by adding an optical layer. Actually, two stages of WDM evolution are anticipated, with transmission-only structures appearing first and networks with optical routing and switching appearing later.

> Adding WDM channels upgrades capacity of the transport network!

Optical layers employing WDM also increase flexibility and improve scalability of the network. They are upgradable and support national information infrastructure (NII) and global information infrastructure (GII). By employing optical routing, WDM networks become more advantageous in terms of transparency which is the enabling factor towards all-optical networks [14]. In wavelength routing transmission, path is determined by wavelength and access point. Electronic switches are required only in the access nodes located in the electronic layer. The optical layer consists of two important elements, optical crossconnects (OXC) and optical add-drop multiplexers (OADM) as shown in Figs. 14.22 and 14.23. These elements should be seen as a complement to the ATM and SDH technologies, which will provide the transport mechanism for many years. It becomes clear that WDM allows rapid enhancement of embedded fiber capacity and eliminates costly electronic switching through wavelength selective add-drop multiplexing.

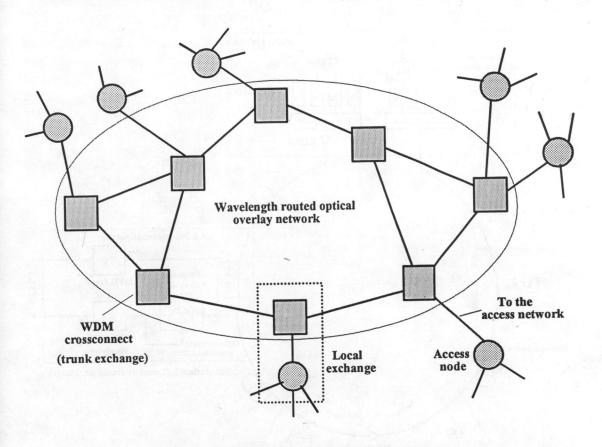

Fig. 14.22: Network with WDM-based optical layer

An OXC should be used when reconfiguration of an optical network is required. The basic function of the OXC is to crossconnect wavelength channels separately between a number of incoming and outgoing fiber lines (Section 14.4). Thereby, a new type of blocking can occur: wavelength blocking. An OADM, however, has more specialized application and can be much smaller in size compared to an OXC. An OADM will preferably be used in an optical WDM ring (or bus) as shown in Fig. 14.23. In this case, the ring is defined and there is no requirement to reconfigure the network topology. This means that the basic functionality is to add and drop one or more wavelength channels from the ring. In some cases, there is a need to be able to select which wavelength channel is to add or to drop.

In order to select a certain wavelength channel, either a tunable direct detection receiver or a coherent receiver can be employed. Because of complexity and cost of coherent receivers, tunable direct detection receiver are primarily considered now-a-days. By using tunable filters, a 32-channel hybridly-integrated tunable receiver has been realized [59]. The channels are located in the wavelength range of 1537 nm to 1569 nm with a separation of about 1 nm. The receiver is tuned digitally and the tuning time is approximately 10 ns.

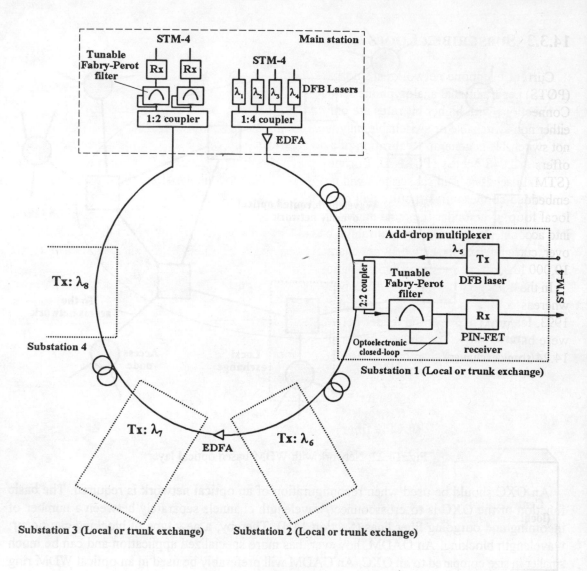

Fig. 14.23: WDM ring network

At present, fiber optic technology is more and more used to transport traffic also from the local exchange (LEX) to the distribution point nearby to the subscriber. Also this part of the network, frequently referred as the feeder network, is actually a transport network rather than an access network. In a typical national network, the extension of the feeder network is about 1700 m in average, whereas the distribution network which connects the subscriber to the distribution point has an average extension of about 300 m. Although fiber optics is used in feeder networks for some years, only about 0.14% of the feeder network is based on fiber links in the nineties. This figure clearly illustrates that copper is the favourite transport medium in the feeder network at least during the next decade.

14.3.3 SUBSCRIBER LOOPS

Current telephone networks are normally based on switchable plane old telephone services (POTS) i.e., traditional analog phone services and digital 64 kbit/s services (i.e, ISDN services). Connections with higher bit rates are only available as permanent links (leased lines) which are either non-switchable or switchable only within a small closed group of local users. But they are not switchable in general. Realization of a national- or even worldwide switched network which offers a 2.048 Mbit/s (PCM-30, Europe), a 1.544 Mbit/s (T1, US) or even a 155.52 Mbit/s (STM-1) access for all subscribers will not be available most likely within the next decade. The embedded copper infrastructure will certainly continue to make up a significant portion of the local loop, even under the most aggressive fiber deployment scenario. This fact must be taken into account while considering ultrahigh bit rate accesses via fiber-to-the-home (FTTH). Moreover, current telephone networks are characterized by a hierarchical architecture including about 10,000 local exchanges for a typical nation.

In the trunk networks, fiber is exclusively used for about one decade in most industrial nations, whereas copper is still the favourite medium in the subscriber loop at present. By the end of 1993, less than 0.2% of the feeder networks and less than 0.01% of the distribution networks were based on fiber transmission links in a typical industrial country [66]. As an example, Fig. 14.24 shows a typical telephony subscriber loop.

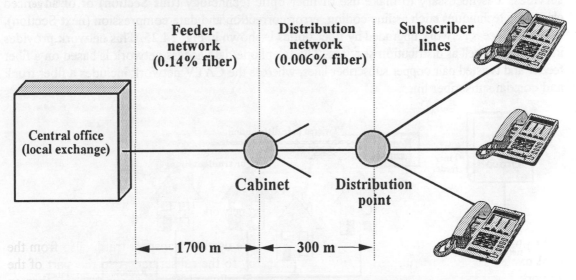

Fig. 14.24: Typical traditional telephony access network

Indeed, there are two different service categories: (i) the telephone and data services (e.g., POTS, ISDN) which come via the telephone networks and (ii) the video distribution services (e.g., TV) via cable TV (CATV) networks. The latter is basically unidirectional, requiring huge capacity and its selection is optional. The distribution or broadcast service uses the frequency division multiplexing (FDM) scheme which is quite different from the baseband pulse code modulation (PCM) scheme used in telecommunication systems. In most residential areas, there are already installed networks of twisted pair copper cables for phone and data services and separate coaxial cables for TV distribution maintained by the network operators. Traditionally, telephone networks are operated by postal, telephone and telegraph ministries (PTTs) or telephone companies (Telcos) and CATV networks by normally independent CATV operators. The CATV companies are expected to transform themselves into public telecommunication operators, offering the full range of multimedia and communication services. In developing new business opportunities, CATV operators depart from their traditional entertainment business (i.e., video distribution) and move, via sophisticated video services (pay-per-view, video-on-demand) to fully interactive services. This new business requires a transformation of the one-way broadband non-switched CATV network into a two-way switched network. Most CATV operators upgrade their networks using hybrid fiber coaxial (HFC) technology wherein fiber nodes serve about 100 - 1500 households. In future, fiber nodes will penetrate deeper into the network to service level of dozens of subscribers.

Telcos in turn search for opportunities in multimedia and video services to extend their business i.e., Telcos take the opposite direction. Coming from their traditional POTS in a narrowband two-way switched network, they upgrade the network so that it can carry more sophisticated information and transactional services in combination with broadband video services. To use installed telephone twisted pair cable network and CATV network for new services, it is necessary to make use of fiber optic technology (this Section) or of advanced electronic techniques such as line coding, error correction and data compression (next Section). A typical access network upgraded by fiber optics is shown in Fig. 14.25. This network provides interactive as well as distribution services. The telephone/data access network is based on a fiber feeder and twisted pair copper subscriber lines, whereas the CATV network includes a fiber trunk and coaxial subscriber lines.

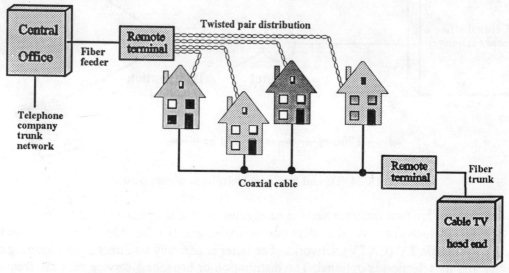

Fig. 14.25: Fiber-supported telephony and cable TV network

Telephone and CATV network operators face great strategic challenges in upgrading the physical layer of the very cost sensitive access network due to both selection of transmission medium and system technology. Optical fiber cable, coaxial cable and twisted pair copper cable are the most important wireline transmission media, while there is a seemingly exhaustive choice of alternative system technologies. Relevant technical and economical options are fiber-to-the-curb (FTTC) with enhanced copper architecture, fiber-to-the-building (FTTB), and HFC systems. However, in optical access networks or optical subscriber loops,

> the vision is to provide ATM-based interactive broadband multimedia services to the end users by means of fiber-to-the-home (FTTH).

Fiber-to-the-desktop (FTTD) or fiber-to-the-terminal (FTTT) are also under consideration. However, fundamental problems in realizing cost effective FTTH accesses have led to moving from the FTTH vision in the eighties to more cost effective solutions such as FTTC or HFC in the nineties. In addition, intense developments in the field of digital signal processing, coding, detection and estimation as well as VLSI electronic circuits have resulted in the utilization of twisted pair local subscriber loops as broadband transmission links. Thus, these new technologies have increased the potential transmission capacity of copper pairs, originally meant for POTS to the point where certain leading-edge broadband services can be provided to customers well in advance of direct fiber access. Meanwhile, high-bit rate digital subscriber loops (HDSL) have increased the transmission bandwidth of copper wire pairs to more than 2 Mbit/s as compared to the largely existing non-loaded copper loops (Section 14.4). Similarly, asymmetric digital subscriber line (ADSL) permits the use of twisted pair copper cables for some telephone and video services [24]. With the decline in costs of optical modules this will, however, change in the long run and FTTH becomes realistic.

In conclusion, there are multiple alternate approaches available to offer broadband access to the subscriber at present, in the near and far future:

- Asymmetric digital subscriber loop (ADSL),
- High bit rate digital subscriber loop (HDSL),
- Very high bit rate digital subscriber loop (VHDSL or VDSL),
- Microwave access,
- Satellite access (e.g., cellular television),
- Hybrid fiber-coaxial access (HFC),
- Fiber-to-the-curb (FTTC),
- Fiber-to-the-building (FTTB),
- Fiber-to-the-home (FTTH) and
- Fiber-to-the-terminal (FTTT) or fiber-to-the-desktop (FTTD).

At present, it is still unknown which of above approaches will match cost and delivery of new services (e.g., VoD, multimedia services) most efficiently [45]. As described in Section 14.2, decision is influenced by various technical as well as non-technical parameters such as available techniques, costs, liberalization of telecommunication companies (Telcos) and process of converging markets (telecommunication, audio-visual consumer electronics and computer industries). The latter one additionally yields problems in regulation and standardization.

In most areas, wherein an embedded copper infrastructure is already existent, the reconfiguration of the network will most likely be performed in three steps: First, an upgradation of the existing copper subscriber loops by electrical means (Section 14.4), second, fiber-to-the-curb, fiber-to-the-building, hybrid fiber/coaxial systems or fiber-wireless mix and third, fiber-to-the-home. The architectures mentioned in the second step are similar, at least upto their termination i.e., they use fiber from a central office or hub to each node where the fiber is linked to copper twisted pair, coaxial cable or a radio antenna. These architectures are commonly referred as fiber-in-the-loop (FITL) systems.

In the following Section, state of the art FITL networks are discussed in more detail.

14.3.3.1 FIBER-IN-THE-LOOP

As mentioned above, access networks will have to support a variety of narrowband and broadband services. Started in the 1980s, telephone companies (Telcos) began deploying fiber-based digital loop carriers, a technology in which fiber extends from the central office to remote terminals [29, 32]. In this, the signals are converted from optical into electrical waves and demultiplexed onto individual copper lines, bundled together as trunks for connecting to the home. Like the Telcos, CATV operators use optical fiber to decrease operating costs. A bonus is enhanced signal quality. Just a few hundred metres of coaxial cable attenuate RF signals noticeably, so that a typical all-coaxial plant must boost them with a dozen or more amplifiers spaced 300 m to 700 m apart. These amplifiers are not only expensive to maintain but also introduce noise into the system [7].

The strategic aspects of how far the fiber should be deployed toward the customer (including the corresponding sizes of the optical nodes), in order to decrease the copper loop length and hence to increase the capacity available per customer, require comprehensive analysis including required quality of service (QoS) as well as the overall financial budget. QoS of a network is mainly determined by average delay, delay variations, cell loss for ATM and error rate. Financial budget depends on discount system, operation, maintenance, powering and life-cycle costs.

There are at least two alternative approaches for FITL deployment: fiber-to-the-building (FTTB) or fiber-to-the-home (FTTH) and fiber-to-the-curb (FTTC). In the first approach, the optical fiber is installed from the central office (CO) or local exchange (LEX) up to the customer premises equipment (CPE). In the second approach, the last drop of the network is still copper-based twisted pair for POTS or narrowband ISDN and coaxial for CATV. Regarding FTTH, the stumbling block is the optoelectronics which is too expensive for an individual subscriber to bear.

As a compromise, FTTC was developed which takes away fiber from the home and out to the curb, so that the costs could be spread over a group of 4 to 48 users [7]. In FTTC architecture, fiber runs from a remote terminal (itself connected to the central office by fiber) to a curbside pedestal. At the pedestal, the optical signal is converted into an electrical one and then demultiplexed for delivery to individual homes over twisted pair. These optical networks typically serve eight homes. FTTC is a proven technology employed, for example, in North America and Europe, in particular Germany [2, 62, 64]. However, existing phone FTTC can still be regarded as "virtual twisted pair over fiber".

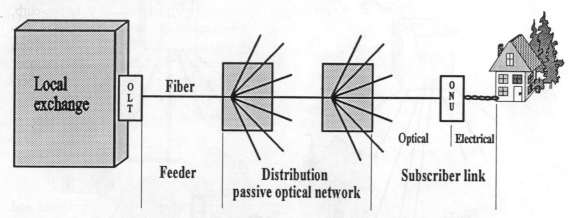

Fig. 14.26: Architecture of a fiber-supported telephony optical network

First generation FITL systems for the most time are based on the delivery of POTS, ISDN and, in general, n·64 kbit/s based services between an optical line termination (OLT) and one or several optical network units (ONU) through a passive optical network (PON) based on passive optical star couplers as shown in Fig. 14.26. At present, optical line termination and optical network unit are not commonly referred by same terms and same abbreviations. Depending on company and area, the following synonyms are also used for these devices:

- Optical line termination (OLT), office or optical interface unit (OIU), central unit (CU)
- Optical network unit (ONU), subscriber interface unit (o), distribution unit (DU) and optical network termination (ONT)

FITL includes FTTC, FTTB, FTTH, FTTT sometimes referred as FTTx. It also includes hybrid fiber/coaxial (HFC) systems, fiber-wireless mix and certain combinations of these. Some of the most important FTTx systems are illustrated in Fig. 14.27. In FTTx, the ONU is placed closer and closer to the customer when shifting from FTTC, FTTB, FTTH to FTTT.

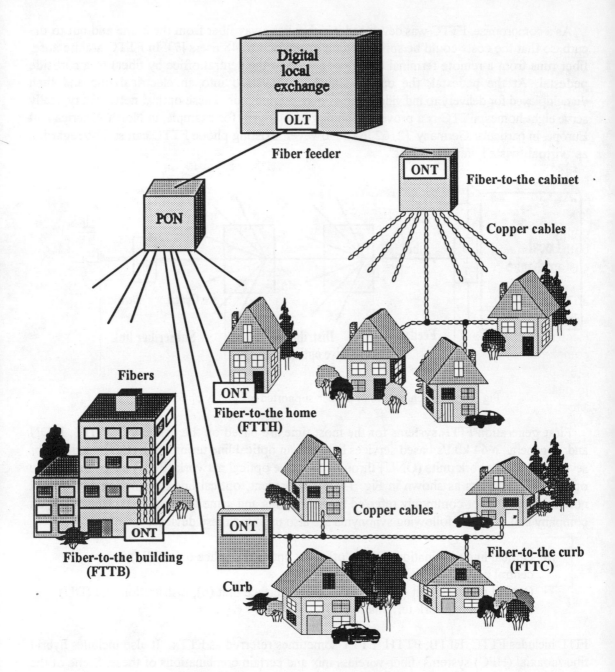

Fig. 14.27: Classification of FTTx systems

In realizing FITL networks, the overall costs and the cost per access or per household are the most important parameters. For a few telephone lines, the fiber loop systems are as cost effective as metallic loop systems. However, if the total cost for telephone lines to all customers in a fixed distribution area is considered, then in many cases fiber loop systems are more cost-effective than metallic loop systems. For instance, if many apartment buildings are included in one area where each building requiring only one ONU, the cost savings achieved from the apartments compensates for the higher costs incurred by the single houses in the same area. In other words, fixed distribution areas are characterized by the average number of telephone circuits per building. Therefore, high priority for fiber deployment is given to areas of large average circuit numbers per building [36, 37].

(i)   FIBER-TO-THE-CURB

As shown in Fig. 14.27 above, FTTC is based upon an optical line terminator (OLT) in the local exchange, a passive optical network (PON), an optical network unit (ONU) at the curb and copper cable subscriber lines. At present, each PON carries signals at 20 Mbit/s - 30 Mbit/s in both directions and serves about 1000 telephone subscribers. An ONU typically serves 8 to 64 customers.

The combination of OLT, PON and ONU i.e., the optical network between local exchange and subscribers is called the optical distribution network (ODN). An ODN does not include any opto-electrical or electro-optical conversion. Instead of a PON, the ODN may also include an active component such as an optical amplifier. In this case, the ODN is called an active ODN, in the other case a passive ODN. The ODN can be topologically structured as point-to-point or point-to-multipoint. The first one is treated as a special case of the second one with a splitting ratio 1:1. Most important optical distribution networks are passive double star and bus configuration as shown in Figs. 14.28 and 14.29 [41].

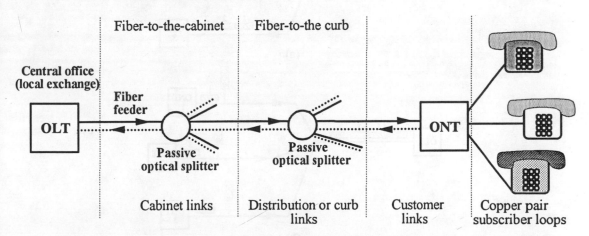

Fig. 14.28: Passive optical distribution network (ODN) in a double star configuration

In most telephony FITL networks, the TDM/TDMA technology is used to transmit and to share the interactive service signals between the OLT at the local exchange point and the connection ONUs at the user site (Fig. 14.30).

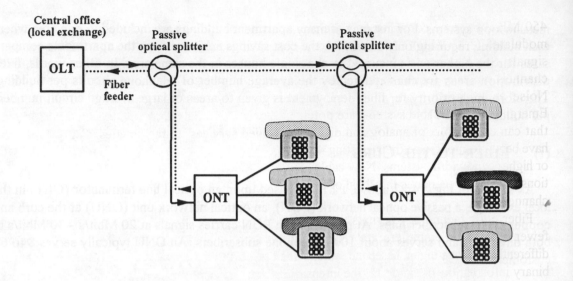

Fig. 14.29: Passive optical distribution network (ODN) in a bus configuration

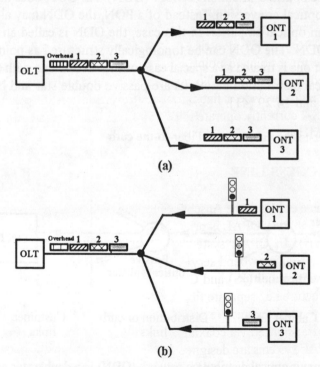

Fig. 14.30: TDM and TDMA (a) OLT to ONU (TDM), (b) ONU to OLT (TDMA)

Fiber in the access network should not only provide interactive services (IS) such as POTS and ISDN-based services, but also distribution services (DS) such as CATV. In a traditional CATV network, coaxial cables used normally operate in the frequency range of 50 MHz to

450 MHz. In some countries only the range up to 300 MHz is used. In this range, amplitude-modulated (AM-VSB) PAL signals are transmitted together with FM radio and digital audio signals. Since higher frequencies attenuate faster than the lower frequencies, adding more TV channels typically requires additional coaxial amplifiers in the traditional coaxial cable system. Noise and distortion are added and picture quality deteriorates along the amplifier chain. Emerging coaxial cable systems are potentially have 1000 MHz (1 GHz) of analog bandwidth that can carry a mix of analog and digitally encoded services. In the nineties, CATV operators have begun to modernize their existing 450 MHz systems into 550 MHz to 750 MHz, 860 MHz or higher bandwidth systems. New cable system components are often rated for 1 GHz applications. However, economic threshold of coaxial systems is at 450 MHz which yields about 60 TV channels [30].

Fiber optic technology permits the use of greater RF spectrum (i.e., more channels) because fewer active RF amplifiers are required. Fiber optics in existing CATV networks operates differently than in a digital baseband system. Instead of turning a laser on and off to represent the binary information "0" and "1", the intensity (or the frequency) of a highly linear laser is modulated with the broadband RF signal. This permits the signal (e.g., 80 channels, 550 MHz) to be sent via fiber optics to a point close to the subscriber in the system where it is converted back into an electrical RF signal and distributed to the homes.

Lithium niobate (LiNbO$_3$) integrated-optical modulators are used in 1320 nm CATV transmitters. External modulators which give rise to chirpless modulation can double the maximum transmission distance for a point-to-point link over that offered by transmitters using directly modulated distributed-feedback laser. To meet the CATV transmission requirements, RF and optical components must provide a flat frequency response over the band of interest. Many installations in the USA currently operate with 550 MHz, while European frequency plans will generally cover 860 MHz.

(ii) OPTICAL ACCESS LINE

The optical access line or "Optische Anschlußleitung (OPAL)" is the German version of Fiber-in-the-loop [19, 50, 57, 58, 64]. OPAL is primarily based on fiber-to-the curb (FTTC) and fiber-to-the building (FTTB). In OPAL systems, a considerable number of CATV channels are implemented in addition to narrowband services. The principle of an OPAL network is shown in Fig. 14.31. Interactive services (IS) and CATV are transmitted on parallel PONs that share a common cable infrastructure i.e., separate fibers are used for IS and CATV. An optical network unit (ONU) is used to transmit interactive services via copper pairs, while broadband distribution services (DS) will be transmitted via coaxial cables.

The established OPAL systems are designed for an overall bit rate of approximately 20 Mbit/s to 50 Mbit/s and a maximum distance of about 10 km between the OLT and the ONU. The interactive system consists of an OLT set up at the location of a local switching centre. Up to four PONs can be connected to one OLT. Thereby, each PON interface backs up the operation of at least 32 ONUs.

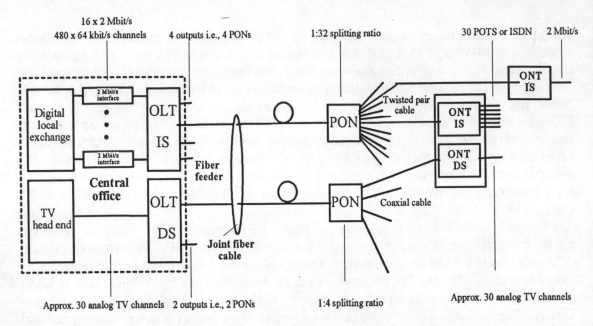

Fig. 14.31: Principle of an optical access line (OPAL)

OPAL systems have been tested in seven field projects (OPAL 1 to OPAL 7) in the west part of Germany in 1990 to 1992. After the reunification of Germany, OPAL systems have been widely used to realize advanced fiber-based networks in the eastern part of Germany (OPAL '93, OPAL '94, and OPAL '95). At present, fiber optics provides more than one million homes in the new federal states with existing interactive narrowband services and cable TV.

(iii) FIBER-TO-THE-BUILDING

Similar to FTTC, fiber is installed from the local exchange via the feeder network to a service access point (node). A passive optical distribution network (e.g., a ATM-PON) connects the buildings to this node. Instead of electronic upgradation techniques such as ADSL and HDSL (Subsection 14.3.3.2), FTTB allows to provide new services like 2 Mbit/s asymmetric and symmetric switched broadband.

(iv) HYBRID FIBER AND COAXIAL NETWORK

To provide video broadcasting, FTTC requires an overlay of fiber and coaxial cables, where the coaxial cabling normally also carries required power to the optical units. Another option is digital broadband FTTC which requires advanced electronic upgradation techniques as mentioned above and in addition to the ONU, each subscriber must have a digital set-top box to convert signal back to analog. A compromise to FTTC is a full service integration on hybrid networks taking advantage of cable bandwidth split.

Hybrid fiber-coaxial (HFC) networks normally have the following bandwidth allocation:

Return band (voice, data):	5 MHz - 30 MHz,
Guard band:	30 MHz - 54 MHz,
Forward band (TV, video, voice, data):	54 MHz - 750 MHz.

A typical HFC topology is shown in Fig. 14.32 [46].

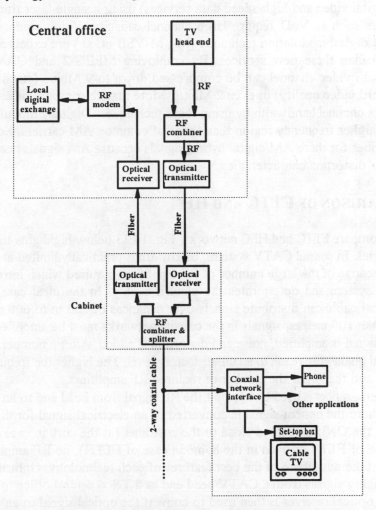

Fig. 14.32: Hybrid fiber and coaxial local loop

In the traditional telephone network, each home has a separate copper loop. In a hybrid network serving about 30,000 subscribers, approximately 300 km of coaxial cabling is required, instead of the approximately 188,000 km of copper cable needed in a twisted pair network. It should be remembered that the public switched telephone network is a logical star. It excels in point-to-point applications. In contrast, the CATV network is a logical bus which is designed for broadcast video i.e., it is point-to-multipoint network. The interest of the phone companies in a hybrid fiber and coaxial network, partially come from the need to deliver services beyond POTS such as video and high-speed data. HFC networks have received widespread support as a broadband access platform for the delivery of both entertainment video and two-way telecommunication services. A hybrid system with 750 MHz bandwidth can provide about 70 AM-VSB broadcast TV channels and more than 40 256-QAM digital channels. Each QAM channel occupies 6 MHz bandwidth in the 500 MHz - 750 MHz range and operates at 40 Mbit/s rate (for compressed digital video and high-speed data services) using a single-laser transmitter.

New services such as VoD require large channel capacity. Video compression technique (MPEG-2) and digital modulation technique (M-, M-VSB et. al.) are expected to open a new approach to realize these new services. By employing MPEG-2 and QAM (or QPSK) a 270 Mbit/s studio video channel can be compressed down to 9 Mbit/s (standard TV quality), 6 Mbit/s (standard video quality) or even 2 Mbit/s. More than four programs can be transmitted within a 6 MHz channel bandwidth by using these techniques. Digitally modulated carriers are located in the higher frequency region than those of common AM carriers. Some systems use more than one fiber for these AM/digital hybrid signals because AM signals need extremely low noise and better distortion characteristics.

(v) COMPARISON OF FTTC AND HFC

In order to compare FTTC and HFC networks, Fig. 14.33 below highlights the characteristics of both topologies. In coaxial CATV systems, bandwidth is practically limited at about 450 MHz to 750 MHz because of the large number of RF amplifiers required which introduce additional noise into the system and deteriorates the picture quality. In the ideal case, 1 GHz is also possible. Coaxial cable can distribute signals over distances of 300 m to 600 m. For distances much greater than 100 metres, signals in the coaxial networks must be amplified and equalized. Every time the signal is amplified, noise and distortion is added. After a number of amplification stages, the signal quality can reach to an unsatisfactory level. The higher the frequency, the higher the attenuation and the larger the number of required RF amplifiers.

In HFC systems, a fiber is used to transmit the RF signal from head end to an optical network unit (ONU), where the optical signal is converted to an electrical signal for distribution to the customers. As the ONU is located close to the customer (at the curb in case of FTTC, in the building in case of FTTB or even in the home in case of FTTH), no RF amplifier is required. HFC networks take advantage of the best features of each technology. Optical fiber is used to transmit high quality signals from a CATV head end or a Telco central office to a point close to the homes. An optical receiver is then used to convert the optical signal to an electrical signal. The signal is then distributed within the neighbourhood using "short range" coaxial tree and branch design. The coaxial network also collects return path signals from neighbourhoods which are transmitted back to the head end or central office via a return path optical link. The historical CATV one-way tree and branch coaxial network with dozens of cascaded amplifiers has been transformed through the capability of linear fiber optics to a low-cost, interactive, fiber/coaxial network with higher reliability, bandwidth and performance. The key to these advances has been the development of linear DFB laser transmitters that are capable of modulating multiple AM-VSB video signals onto optical signals. High performance linear DFB lasers are now manufactured in large quantities at low cost. They, together with digital compression, have become the enabling technology of the so-called "information superhighway" which enables the transmission of interactive broadband multimedia services.

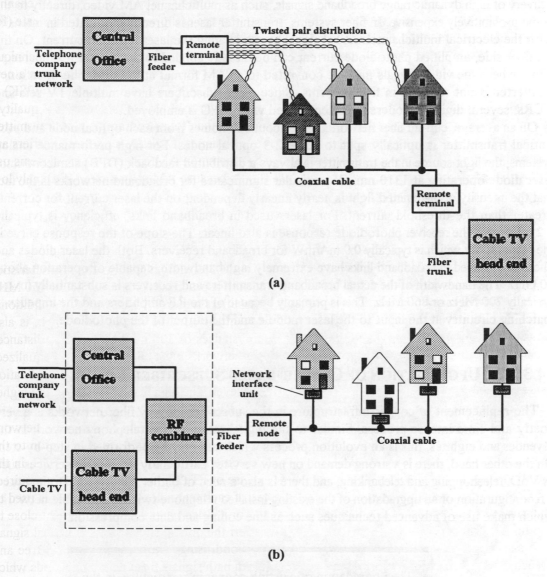

Fig. 14.33: Comparison of (a) FTTC and (b) HFC Topologies

In fiber systems, some advantages are there as compared to coaxial and HFC systems. Most important are very low attenuation, high bandwidth and immunity from electro-magnetic interferences. However, some disadvantages of fiber systems are also exist. Fiber optics techno-logy is sometimes less preferable than standard electronics. In particular, the cost of large dynamic range optoelectronic transmitters and receivers is significantly higher than their electronic counterparts. The costs for passive optical components, such as couplers and connectors are also much higher than the electronic versions. These costs make fiber optic systems for the delivery of high dynamic range broadband signals, such as multichannel AM video, directly to the home prohibitively expensive. In fiber systems, transmitter laser is directly modulated in intensity with the electrical multichannel CATV signal by modulating the laser injection current. On the receiver side, amplified photodiode current can be directly applied to the TV set. Compared to HFC, where the video signals must be converted to an AM format at the subscriber location, a converter is not required in fiber systems. Since most subscribers have multiple TV sets and VCRs, several digital decoders are often needed when HFC is employed.

On an average, current fiber networks serve about 500 homes from each optical node and each optical transmitter is optically split to serve 2-3 optical nodes. For high performance analog systems, the light source in the transmitter is always a distributed feedback (DFB) semiconductor laser diode operating at 1310 nm. Of particular significance for broadband networks is the fact that the intensity of the emitted light is nearly linearly dependent on the laser current for currents greater than the threshold current. For lasers used in broadband links, efficiency is typically 0.2 mW/mA. The receiver photodiode response is also linear. The slope of the response curve is the responsivity which is typically 0.9 mA/mW for broadband receivers. Both the laser diodes and photodiodes used in broadband links have extremely high bandwidth, capable of operation above 10 GHz. The bandwidth of the actual broadband transmitters and receivers is substantially lower, typically 700 MHz or 860 MHz. This is primarily because of the RF amplifiers and the impedance matching circuitry at the input to the laser module and the output of the photodiode.

14.3.3.2 UPGRADATION OF COPPER-BASED SUBSCRIBER LOOPS

The replacement of copper infrastructure in the subscriber loop by fiber networks is a very costly and time intense process. Similar to the digitalization of the telephony network in the seventies and eighties, this is an evolution process which can only be performed in step-by-step. On the other hand, there is a strong demand on new services particularly interactive services such as VoD, teleshopping and telebanking, and there is also a need of higher bandwidth. Both require a reconfiguration or an upgradation of the existing installed telephone twisted pair cable networks which make use of advanced techniques such as line coding and data compression.

> The upgradation of existing copper infrastructure in the subscriber loop bridges current analog telephony and future digital broadband applications.

Intense developments in the past resulted in several upgradation techniques such as

- Asymmetric Digital Subscriber Line (ADSL),
- High-bit rate Digital Subscriber Line (HDSL) and
- Very-High-bit rate Digital Subscriber Line (VHDSL or VDSL).

All these XDSL techniques have the common goal of transmitting digital data at relatively high bit rates over the existing installed twisted pair subscriber lines of the local telephony access network. By delivering high bit rates, these techniques should give the customers a feeling of the advantages of (multimedia) high bit rate interactive and distribution services. Thus, all upgradation solutions act as a mean of testing and introducing new services as the prelude to widespread optical fiber deployment in the local loop. XDSL will help new services to come up. As soon as these services are established, fiber architectures are more and more required.

Twisted pair copper cables exhibit a variety of different defects and impairments which have to be compensated by the techniques mentioned above. Most critical is the extreme non-linear loss characteristic and the high value of loss itself. In the frequency range of about 100 kHz to 400 kHz, loss ranges from 50 dB to 70 dB. Thus all upgrading methods require modern state-of-the-art digital signal processing such as digital adaptive equalization, signal estimation and digital multilevel modulation. Employing commercial twisted pair copper cables with 0.4 mm core diameter, a 2 Mbit/s signal can be transmitted over a distance of about 4 km and a 6 Mbit/s signal over a distance of about 1.7 km.

Upgrading methods combined with data compression techniques such as JPEG, MPEG-1 and MPEG-2 enables to transmit digital videos over existing copper pairs with acceptable quality. As shown in Fig. 14.34, digital data compression yields a bit rate reduction up to a factor of 50. However, this factor is again reduced by a factor of about two with error correcting codes.

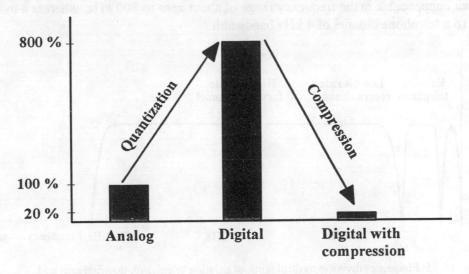

Fig. 14.34: Benefit of digital data compression techniques

In the following, the principle of upgradation techniques will briefly be discussed. A more detailed description of these techniques is given e.g., in [11, 66].

(i) ASYMMETRIC DIGITAL SUBSCRIBER LINE

The asymmetric digital subscriber line (ADSL) is a technique which provides down-stream transmission channels of up to 1.544 Mbit/s or 6.144 Mbit/s (USA) and 2.048 Mbit/s (Europe) over the existing installed twisted pair subscriber lines [11, 66]. By combining ADSL with digital video compression techniques, colour TV signal can be delivered without disturbing the telephony service already provided on the same line. In addition to the high-bit-rate downstream channel, a low bit rate return channel of some kbit/s enables users to select and control downstream information. With such a potential, it is not surprising that ADSL is attracting a large interest, especially in the short term. The ADSL will be of great significance to private customer.

Realization of ADSL systems requires highly sophisticated electronic devices, including digital adaptive equalizers to compensate the extreme non-linear loss characteristic of a typical twisted pair copper cable and high-quality filters to separate existing telephony services, high bit rate downstream channel and low bit rate upstream (return) channel. ADSL is based on frequency division multiplexing (FDM), where downstream and upstream channels are moved to higher frequencies by means of modulation. The telephony service remains in the original baseband frequency domain (Fig. 14.35). It becomes clear from this figure that the principle of ADSL is similar to HFC (compare Subsection 14.3.3.1). The latter one is based on fiber and coaxial cable, whereas the first one is based on twisted pair copper cable. In ADSL systems, FDM is realized by modulation schemes such as well known quadrature amplitude modulation (QAM), carrierless amplitude/phase (CAP) modulation or discrete multi-tone (DMT) modulation [11]. Thus, an ADSL device operates similar to a conventional modem. An ADSL device is matched to a twisted pair copper cable in the frequency range of about zero to 800 kHz, whereas a modem is matched to a telephone channel of 4 kHz bandwidth.

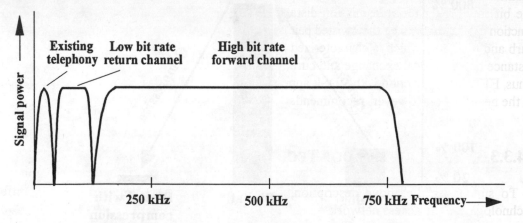

Fig. 14.35: Frequency division multiplexing of existing telephony, downstream and upstream data service

(ii) HIGH BIT RATE DIGITAL SUBSCRIBER LINE

In contrast to ADSL, high bit rate digital subscriber line (HDSL) provides a symmetric data transmission in both direction i.e., from local exchange to customer and from customer to local exchange. Instead of one twisted pair copper cable as used in ADSL systems, HDSL normally operates with two copper pairs.

The HDSL provides bit rates up to 1.544 Mbit/s or 6.144 Mbit/s (USA) and 2.048 Mbit/s (Europe). A HDSL system is based on two transceivers, one is placed in the local exchange (LEX), other at the subscriber side. The transceivers are connected via two copper pairs i.e., local exchange transmitter is connected with the subscriber receiver and subscriber transmitter is connected with the local exchange receiver in a duplex connection. Similar to ADSL, HDSL requires some highly sophisticated electronics such as a frame synchronization, clock generation and recovery, bit scrambling, cyclic redundancy check (CRC) and some other control functions.

With an universal 2.048 Mbit/s subscriber line, HDSL can be employed for a number of old as well as some new services. Since HDSL provides a symmetric bidirectional data transmission, this technique is, in particular, advantageous for business users, whereas ADSL is especially useful for private subscribers who are first of all interested in unidirectional distribution services such as TV and VoD. First recommendations have been drawn up in 1992 by ANSI (1.544 Mbit/s) and in 1994 by ETSI (2.048 Mbit/s).

(iii) VERY HIGH BIT RATE DIGITAL SUBSCRIBER LINE

A subscriber line with bit rates of much more than 2.048 Mbit/s is called very high bit rate digital subscriber line (VHDSL or VDSL). Typical bit rates are 34 Mbit/s and 140 Mbit/s (Europe), 45 Mbit/s (USA), and 155 Mbit/s (SDH). Due to high bit rate, bidirectional data transmission can only be performed by employing two separated copper pairs. Moreover, higher the bit rate, lower the transmission distance. Therefore, HDSL is useful, in particular, in the junction cable network i.e., the twisted pair copper network between the distribution point at the curb and the home which is characterized by a relatively short line length. To bridge the longer distance between local exchange and curb, coaxial or fiber cable (i.e., FTTC) can be employed. Thus, FTTC in conjunction with VDSL appears quite suitable for switchable broadband accesses in the near future. However, recommendations on this do not exist yet.

14.3.3.3 SUBSCRIBER LOOP TECHNIQUES: A COMPARISON

To summarize the above description, following table gives a brief overview of different technologies in the access network.

Table 14.7: Techniques in the subscriber loop

Technique	Bit rate	Physical layer	Remark
ADSL	1.5 to 2.0 Mbit/s	One twisted pair copper cable	Primarily for private users
HDSL	2.0 Mbit/s	Two twisted pair copper cables	Primarily for business users
VHDSL	> 10 Mbit/s		
FTTC	nx64 kbit/s (ISDN) Option: cable TV	Fiber, twisted pair cable or coaxial cable	Suitable architecture for clustered growth application
HFC	nx64 kbit/s (ISDN) analog and digital TV	Fiber and coaxial cable	Suitable technology for immediate implementation
RITL	about 2 Mbit/s	Satellite, Microwave	RITL: Radio-in-the-loop (wireless) Quick installation

Neither ADSL nor HDSL can cope with more than a couple of digital video channels over relatively short distances of about 2 km. This is one of the main reasons Telcos pursue upgradation scenarios that combine telephone and interactive narrowband services with "switched digital video" instead of video broadcasting in an FTTC network topology. The question is where to put the switching functionality for channel selection: in the curb, in the host digital terminal or higher in the network? In the coming years, centralized broadband switching will be lodged in the digital switching system, but in the long run the switching function will migrate down the network to the curb and ultimately into the home (FTTH).

In future, broadband-ISDN (B-ISDN) based on ATM and SDH will most likely support interactive multimedia traffic (Section 14.5). Since optical fibers represent the best physical layer to transport this high-speed traffic, the deployment of FITL systems such as FTTC, FTTH and HFC is most important [60].

14.4 ALL-OPTICAL NETWORKS

During the past two decades, great progress has been made in optical fiber communication technologies toward higher capacity and longer repeater spans. To accommodate the upcoming broadband ISDN (B-ISDN) era where vast amounts of information including data, voice and pictures will be provided to subscribers through optical fiber cables, high-speed technologies that can handle more than 100 Gbit/s must be developed not only for transmission lines, but also for transmission nodes. Key techniques of an all-optical network also referred as all-photonics network are (i) high-speed optical pulse generation and modulation, (ii) all-optical multiplexing and demultiplexing, (iii) optical timing extraction, (iv) all-optical repeater and regenerator and (v) soliton transmission. The applications of an all-optical network range from trunk networks to LANs and subscriber loops. All-optically multiplexed networks may include both optical TDM (OTDM) as well as optical FDM (OFDM) or WDM.

As mentioned in the previous Sections, current switches are exclusively based on electronic designs supporting a maximum bit rate of about 10 Gbit/s. Research in electronic components and materials may slightly increase performance of electronic circuits in future [12, 51]. Even handling of bit rates up to 40 Gbit/s seems to be possible. Fiber optics, however, offer the possibility to transmit bit rates up to some Tbit/s which opens the door towards ultrafast information superhighways. In traditional systems,

> electronic switches will become
> *network bottlenecks* !

To overcome this problem, all-optical networks including fiber-based transmission, photonics (i.e., optical) switching and optical broadband access are required [22]. Predictions that local networks such as subscriber loops may eventually provide broadband service via fiber to 100 percent of potential subscribers is driving continued research on all-photonics information infrastructure. However, realization of an all-optical network including trunk networks, local area networks and subscriber loops is an evolution process performed in step-by-step during the next decades. As a first step, photonics technology will primarily be employed to support and upgrade existing networks and network components. In this step, photonics technology will be used to

- realize optical overlay networks based on WDM,

- upgrade switching systems by optical interconnects and

- support crossconnect offices by function sharing.

Wavelength or optical frequency division multiplexing (WDM or OFDM) is an excellent and powerful technology to upgrade the traditional networks. WDM can be employed either to increase transmission capacity in the trunk networks, to enhance flexibility in access networks or both. The deployment of WDM-based multiwavelength transmission systems will also have, in turn, a significant impact on the prospects of photonics switching systems. The transmission link is, of course, just a chain in a larger communication network. In crossconnect offices, fibers come from different routes. Some traffic needs to be dropped in a particular office, but most traffic needs to be routed on to its final destination while other traffic is added to the network. In current networks, sorting, dropping and routing of transmission signals are performed electronically by detecting and demultiplexing all the broadband optical signals to lower rate (typically 50 Mbit/s) to perform these functions. This process always occurs despite the fact that much of the traffic is destined to be routed through the office. It should be noted that in WDM or OFDM, coherent systems can still play a role in future even if the sensitivity gain of these systems over direct detection systems has been largely offset by the use of optical amplifiers (Chapter seven). Frequency multiplexing may be used as a distribution or an access technique.

Now-a-days, photonics is established as the supreme information (cable) transport medium whilst electronics is exclusively used in information processing.

Currently, switching is exclusively based on electronic processing.

However, by using *optical interconnects*, photonics technology has started to enter the electronic domain of present switching systems. Fiber-based interconnections increase speed of switching and routing and will play a significant role in the hardware of switching offices in the near future (Fig. 14.36). Thus, optical interconnects can be regarded as a pilot-technique bringing photonics technology into the crossconnect offices [33].

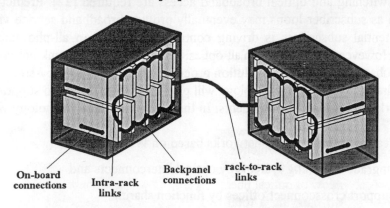

On-board connections   Intra-rack links   Backpanel connections   rack-to-rack links

Fig. 14.36: Optical interconnects in crossconnect offices

As mentioned above, in a communication network most traffic is to be routed through the office instead of being dropped. To perform both functions of a switching office, an appropriate technologies can be employed. Since the routing function does not require any complex signal processing, photonics technology is, in particular, well suited to perform this routing task. By using photonics, the throughput of a switching office can be significant. Because of the high functionality of electronics, the switching and dropping functions will most likely be performed by electronic components in the next two decades. In Fig. 14.37, it is shown how throughput of a switching office can be upgraded by *function sharing* i.e., photonics for routing and electronics for switching and dropping.

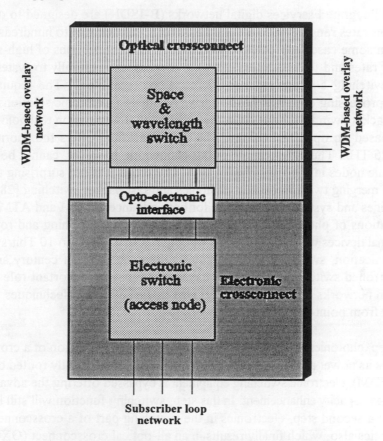

Fig. 14.37: Function sharing in a crossconnect office

All-optical networking can pave the way for a simpler, less hierarchical networks. The trend of reduction of hierarchical levels, which is already noticeable in the existing structure, will only be possible by an extensive use of optics for both transmission and routing. It is conceivable that by using the low-loss window of optical fibers (1.2 μm - 1.6 μm wavelength), we could construct multiple access networks carrying a total traffic of perhaps $5 \cdot 10^4$ Gbit/s which is 50 Tbit/s, roughly four orders of magnitude greater than the traffic flow through a high-capacity electronic telephone switch. At present, of course, we are far from realizing this high potential [18]. But certainly, photonics is moving from the classical transmission function to the more sophisticated network functions such as routing and switching. More and more functions performed by electronics will be substituted by optics in the future. The speed of this movement primarily depends on subscribers demands, power of alternate techniques and preparation of network operators to take risks.

14.4.1 PHOTONIC SWITCHING

Broadband integrated services digital networks (B-ISDN) are designed to offer a variety of services with bit rates ranging from several kbit/s (e.g., teleactions) to hundreds of Mbit/s (e.g., HDTV) and in some cases approaching Gbit/s (e.g., interconnections of high-speed LANs). A multiplicity of rates and the burstiness of traffic sources lead naturally to systems based on the fast packet switching concept realized by the ATM technology. The requirements of data buffering and processing of packet headers have resulted in electronic solutions for most of the existing fast packet switching nodes. On the other hand, transmission technology in broadband networks is based on optic technology. The optical medium offers the enormous bandwidth approaching 25 THz. The full potential of this bandwidth, however, cannot be used in systems having electronic nodes in the transmission path. Therefore, it is not surprising that a significant effort is put on merging two approaches: photonics and fast packet switching [28, 35]. Photonics switching devices and systems are required for both advanced WDM and ATM applications.

For applications of photonics to support high-throughput switching and routing, novel all-optical functional devices will be required. A throughput of 1 Tbit/s to 10 Tbit/s will be required in telecommunication, switching and computer systems in the 21st century and beyond [61]. Optically controlled switching devices are expected to play an important role in high capacity communication networks based on TDM, OTDM as well as WDM techniques. It is required to bring the fiber from point-to-point into network [5].

As a first step, photonics will be used to perform the routing function of a crossconnect office which improves its power and throughput. With an end-to-end optically routed connection (e.g., by means of WDM), electronic switching equipment is bypassed offering the advantage of bit rate transparency and capacity enhancement. In this step, switching function will still be performed by electronics. In a second step, electronics in the switching part of a crossconnect office can be replaced by optics also, which finally results in an all-optical crossconnect (OXC).

The functionality of an all-optical crossconnects would require that each wavelength from a set of wavelengths on a given input fiber can be connected to any desired output fiber. Photonics crossconnects must be able to spatially route individual wavelength channels onto the desired output fiber. Thus, we have to distinguish space-division switches composed of a wavelength independent crosspoint with wavelength multiplexers and demultiplexers, wavelength-selective crosspoints and full non-blocking crossconnects with wavelength-converting elements to avoid potential collisions of two identical wavelengths on the same output fiber. As an example, Fig. 14.38 shows an all-photonics 8x8 crossconnect. The function of this OXC is splitted in space switching i.e., fiber-to-fiber switching and wavelength switching. It becomes clear that an OXC based on space and wavelength switching are logically same as traditional crossconnects based on space and time-slot switching which are installed in current digital PCM-based crossconnect offices. Wavelengths (time-slots) are routed and switched separately.

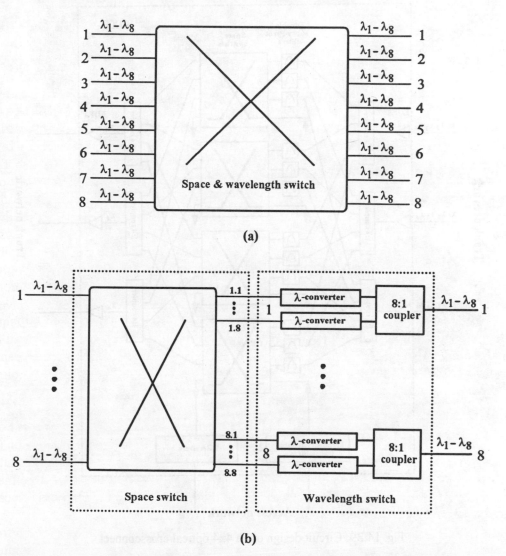

Fig. 14.38: (a) All-photonics 8x8 crossconnect splitted in (b) space and wavelength switching

An OXC, for example, can be realized by using star couplers, tunable Fabry-Perot filters, optical space switches and optical amplifiers to avoid losses inside the crossconnect. As an example, Fig. 14.39 shows circuit design of a 4x4 optical crossconnect, which allows to switch and drop network traffic. To avoid wavelength blocking, wavelength converters can be employed (not included in Fig. 14.39). Wavelength converters are the most critical components in an all-photonics crossconnect since they drastically increase complexity and costs. Therefore, when designing all-optical networks with all-photonics switching, wavelength conversion should be avoided wherever possible.

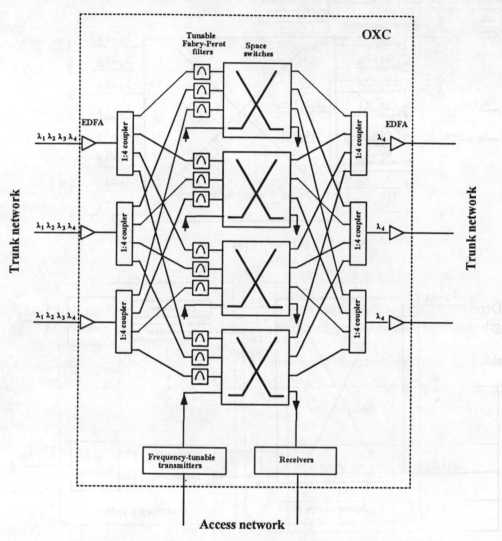

Fig. 14.39: Circuit design of an 4x4 optical crossconnect

In all-optical networks, the number of wavelengths employed in the network is an optimizable system parameter as shown Fig. 14.40. In one extreme, only one wavelength is used and no wavelength conversion is needed but flexibility is low (Fig. 14.40a). In the other extreme, when number of wavelengths is high (Fig. 14.40d), many wavelength converters are required and cost increases.

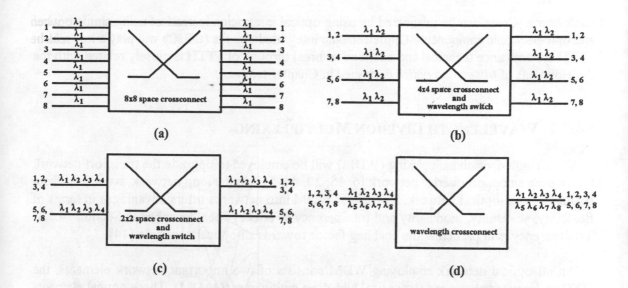

Fig. 14.40: All-photonics 8x8 crossconnects (a) 8x8 space and zero wavelength switch, (b) 4x4 space and 2-wavelength switch, (c) 2x2 space and 4-wavelength switch and (d) zero space and 8-wavelength switch

Depending on input and output lines to be switched either only a space switch, only a wavelength switch or both switching functions can be employed as shown in Table 14.7.

Table 14.7: Switching operations required in 4x4 space & 2-wavelength crossconnect of Fig. 14.40b

Input line	Output line	Switching operation	
		Space switching	Wavelength switching
1	3	yes	no
1	4	yes	yes
1	2	no	yes

In end-to-end digital connections, optics is used only for link connections whereas all functionality is in the electronics. It must be remembered that most traffic in a local or trunk exchange is going through the exchange and must not be switched. In end-to-end optically routed connections, electronics is bypassed.

By employing fiber-based transmission and all-photonics crossconnect, the network becomes all-optical except the subscriber loop where optics normally stop at the last mile to the home. The realization of an all-optical access at home is less a question of technique but more a question of economics. Low-cost optical transceivers are primarily required to realize an optical access at each home. Costs can be minimized by using optical integration instead of using single optical and optoelectronic components. Optoelectronic integrated circuits (OEIC) will play a key role in FTTH systems since technical and economical breakthrough of FTTH is closely related with the development of *integrated optics* described in Chapter twelve.

14.4.2 WAVELENGTH DIVISION MULTIPLEXING

Wavelength division multiplexing (WDM) will be employed to upgrade the transport network by realizing an optical overlay network [6, 15, 20, 44, 55]. In the loop network, we are far from realizing an all-optical network. The use of WDM into networks offers advantages in terms of flexibility, scalability, modularity and transparency as already been described in Section 14.3.1. Transparency is in particular the enabling factor towards all-optical networks [14].

An all-optical network employing WDM consists of two important network elements, the OXC as described above and the optical add-drop multiplexer (OADM). These optical elements are complement to the present ATM and SDH technologies, which will provide the transport mechanism for many years. An OXC should be used when reconfiguration of an optical network is required. As explained above, the basic function of a OXC is to crossconnect wavelength channels separately between a number of incoming and outgoing fiber lines. A new type of blocking can occur: wavelength blocking. An OADM, however, has more specialized application and can be much reduced in size compared to an OXC. An OADM will preferably be used in optical WDM rings (or busses) as shown in Fig. 14.23. In this case, the ring is defined and there is no requirement to reconfigure the network topology. This means that the basic functionality is to add and to drop one or more wavelength channels from the ring. In some cases, there is a need to be able to select which wavelength channel is to add or to drop.

14.5 TRENDS IN OPTICAL COMMUNICATION NETWORKS

In this Section, three of the most important applications of advanced optical communication networks are discussed. First is the interactive broadband digital network also referred as B-ISDN (Subsection 14.5.1), second the information superhighway (Subsection 14.5.2) and third the transmission of interactive multimedia services (Subsection 14.5.3).

14.5.1 BROADBAND DIGITAL NETWORK

The current telephony networks (public networks) are based on POTS and/or narrowband integrated digital services (N-ISDN services) with a bit rate of 64 kbit/s. These services are globally switchable. Thus, these networks can be regarded as interactive multipoint-to-multipoint networks. Each subscriber at any location in the world can have a connection to each other subscriber at any other location in the world by simply dialling the individual telephone number. Bit rates higher than 64 kbit/s, such as the primary rate of 2.048 Mbit/s, are normally available on leased lines only which are in general non-switchable. Thus, these lines can be regarded as fixed point-to-point connections. Because of the relatively low bit rate in present public networks, bit rates higher than the primary rate of 2.048 Mbit/s are already referred as broadband:

> "broadband, ... a service or system requiring transmission channels capable of supporting rates greater than the primary rate" (CCITT, now ITU).

This definition has, however, to be changed in future. An improved definition of what is broadband could be the following:

> "broadband, ... a service or system requiring transmission channels capable of supporting rates equal to the STM-1 rate"

which is 155.52 Mbit/s. In the far future, even STM-4 rate (622.08 Mbit/s) or STM-16 rate (2488.32 Mbit/s) seems to be possible to define the term broadband.

In the future mass market for broadband access and connectivity, the key challenges will be associated with the physical design of large broadband switching systems having aggregate throughout three orders of magnitude higher than current largest switches. If one considers a large switching office with 100,000 terminating lines, each producing a demand for 64 kbit/s connectivity when active (at peak usage times 10% of the lines are active), then one obtains a peak aggregate throughput of $100,000 \cdot 0.1 \cdot 64$ kbit/s $= 0.64$ Gbit/s. If we increase the line rate to 155 Mbit/s and keep everything else the same, then we need an aggregate throughput capability of 1.5 Tbit/s! Thus, the broadening of the bandwidth of the telecommunication network brings with it the need for faster, higher-capacity switching systems.

In a broadband digital network such as B-ISDN, packet transport will likely to be an important component of the total traffic carried and switched in the network. Different amounts of information at varying data rates and in different formats have to be transported and routed. In this environment of multiple-bit-rate data streams with potentially burst traffic, packet switching can be the appropriate switching technology. Thus, an important ingredient of the future broadband networks will be very-high-capacity switches for the public network. Currently, switches (packet or non-packet) are exclusively based on electronic designs, but the advent of multi-Gbit/s optic transmission systems may cause these to become *network bottlenecks*. In future, there will be a tradeoff of transmission and switching. Emerging technologies such as SDH rings that can be used to reconfigure the connectivity of networks at the path level and facility level make it feasible to build networks with a smaller number of larger switching systems. Declining transmission costs and powerful data base technologies will continue to push the trend toward fewer numbers of larger switches and increased utilization of network reconfiguration at the path and facilities levels. Thus:

> The number of switches in future networks may be smaller than in the past.

Certainly, broadband local access is one of the most challenging elements in rebuilding the global information infrastructure and involves numerous technical and economic tradeoffs. The increasing sophistication of user equipment makes it necessary to take the user equipment (intelligent terminal, PBXs, LANs, computers etc.) into consideration when designing telecommunication network services.

> Broadband networks are capable of supporting integrated interactive multimedia services including voice, data, image and full-motion video.

Most important are switchable wide-area networks for B-ISDN access. A typical broadband network configuration is shown in Fig. 14.41. A broadband network supporting subscriber line bit rates of 155.52 Mbit/s or more requires highly sophisticated techniques such as

- Optical overlay network,
- Wavelength division multiplexing (WDM),
- Fiber-in-the-loop (FITL) such as fiber-to-the-home (FTTH),
- Asynchronous transfer mode (ATM, most likely photonics ATM) and
- Passive optical networks (PON).

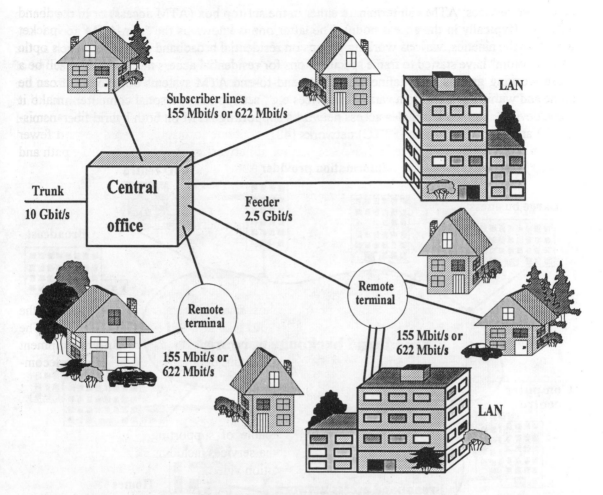

Fig. 14.41: Broadband-ISDN (B-ISDN) distribution network

ATM is a very promising switching and transport technique for carrying multimedia traffic in the B-ISDN. It will soon penetrate widely into public networks to carry voice, pictures as well as data. The throughput of ATM networks will grow rapidly as the provision of fiber optic subscriber loops stimulates the demand for broadband multimedia services. To meet this increasing demand, higher-throughput ATM switches will be needed.

Combining the high functionality of electronics with the high bandwidth of photonics is an important issue in the construction of high-throughput ATM switches. Important steps towards advanced ATM are: (i) electronic ATM switches, (ii) electronic ATM switches with optical interconnects, (iii) photonics ATM switches and (iv) photonics ATM switches using a frequency-routing-type time-division interconnection network. The broadband ATM-based FITL system which uses passive optical network (PON) technology is referred to as the ATM-based PON (APON). For the use of ATM and B-ISDN protocols, two scenarios are considered for interactive services: ATM can terminate either in the set-top box (ATM access) or in the access network (typically in the access node). The latter one is known as the "non-ATM access network". In the nineties, various working groups on residential broadband networking such as the "ATM Forum" have started to frame specifications for residential access networks. The scope of these working groups is to define a complete end-to-end ATM systems both to and from the home and within the home, to a variety of devices e.g., set-top box, personal computer and other home devices. The work includes access networks supporting ATM on both hybrid fiber-coaxial (HFC) and fiber-to-the-curb (FTTC) networks [48].

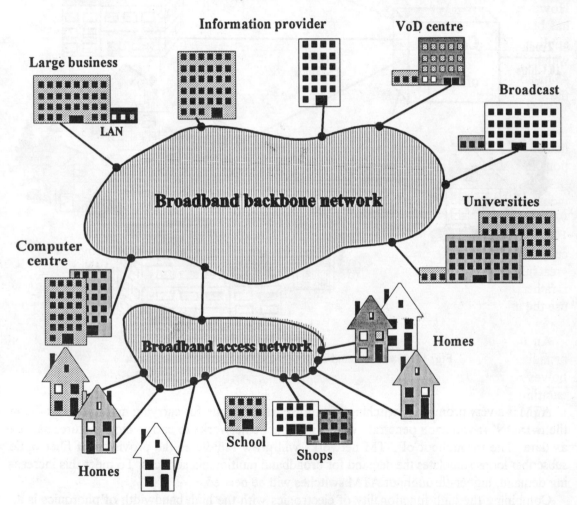

Fig. 14.42: Proposal of a broadband network

A proposal of a broadband network is shown in Fig. 14.42. This network contains an ATM backbone network (transport or trunk network) including broadband transport and switching with a throughput of 2.5 Gbit/s (STM-16), 10 Gbit/s (STM-64) or more. In this network, numerous information providers, VoD centres, computer centres, hospitals, universities are connected via servers. In addition, LANs and access networks (loop networks) with broadband access for the users (home, school, shop) are linked to this backbone network.

14.5.2 INFORMATION SUPERHIGHWAY

Scanning the daily press, technical journals and coloured prospects of telecommunication industry, words such as information superhighway or data highway can often be recognized. However, behind these well known and frequently used words, the idea is still rather vague. It has been associated with definitions that span from simple transmission pipes to the extensive application of ATM for switched interactive broadband services. The ever expanding Internet with its global reach is often regarded as the prototype superhighway. From this perspective, the 45 million km of single-mode fiber installed around the world also provides the infrastructure for the superhighway.

Certainly, the dominant force on the scene is Internet. It evolved directly from government-funded development of packet switching and internetworking technologies for defence applications in the 1970s (minicomputers) and their application to more general university research in the 1980s (LANs, PC workstations), with most of the Internet now being provided by private industry. The expansion of Internet connectivity and ease of access resulting from World Wide Web (WWW) technology has provided a major stimulant to commercial uses of the Internet in 1990s and to information infrastructure more generally. Of course, Internet can be regarded as current information superhighway since a very large amount of information is provided to very great number of users. On the other hand, transmission bit rate and overall network throughput is rather low which soon will become a network bottleneck. More and more Internet consumers use the late night hours to scan (surf) through the network.

An upgraded version of Internet referred as the fast Internet, should be able to provide broadband access offering interactive high-quality multimedia services. In the longer term, however, photonics networking using ultrafast optical processing offers some of the most exciting prospects. This technology is sure to lead to experimental networks capable of multiplexing and routing data packets from a 100 Gbit/s digital highway. Similar to motorcar highways (Fig. 14.43),

> information superhighways should be able to transport as much traffic as possible.

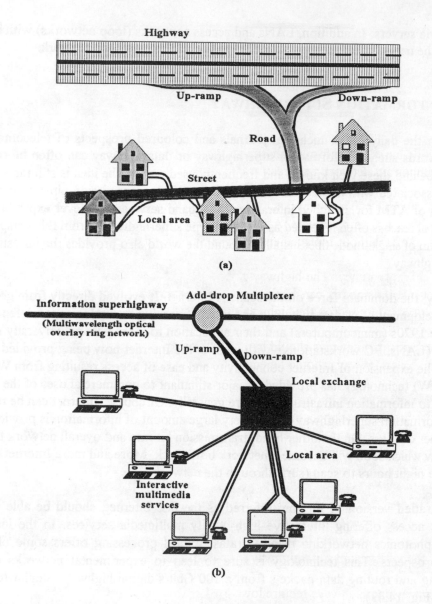

Fig. 14.43: Similarities of (a) motor highway and (b) information superhighway

By using this simple analogy of information highway and motorcar highway, an information super-highway can be defined either in terms of bit rate (speed of motorcars) or services (type of vehicles). The information superhighway should support all types of networks and all types of services including conventional POTS services and telefaxing, videophony and e-mail services, digital TV or high definition television (HDTV) broadcasting, video-on-demand (VoD), computer games, teleeducation, teletransactions (home shopping, home banking), access to libraries and others [9]. The overall bit rate of an information superhighway should be as high as possible e.g., 2.5 Gbit/s, 10 Gbit/s or even more. Thus, the following definition can be given:

> The information superhighway is a network which supports high bit rates and all types of services.

A network based on information superhighways should be at least characterized by the following important network features:

(i) Flexibility: The highway allows to respond to an unforeseen traffic load or to specific requirements. A future capacity demand can be quickly accommodated without replacement or procurement of another network or system.

(ii) Transparency: The highway provides a platform into which different transmission modes can be included. Thus, different types of signals may be superimposed on the highway. It results in multiple service capability.

Information superhighways should be used to support broadband networks. For this reason, information superhighways should be employed as

- network backbones,
- interoffice connections,
- intercity or interregion links and
- transoceanic links.

As a high-speed backbone network, an information superhighway connects other networks through junctions which are known as gateways. The preferred topology of the information superhighway is a ring. Consumers are connected to the ring by up- and down-ramps either directly (business consumers) or via one or more switching centres (private users). As a broadband pipe, the information superhighway is an ideal medium to transport interactive multimedia signals.

Information superhighways require low-loss transmission links which are able to carry ultrahigh bit rates.

> Since optical fibers are able to operate up to some Tbit/s, fiber optic is a key technology of information superhighways.

However, current switches are exclusively based on electronic designs supporting a maximum bit rate of about 10 Gbit/s only. Electronic switches will become *network bottlenecks*! Consumers and network operators are still uncertain as how to find an up- and down-ramp to the information superhighway, the so called "last mile" that will connect them to hundreds of entertainment sources and interactive capabilities.

Since the existing networks represent a very big investment, upgrading techniques must be applied in the near future. As explained in Section 14.3.3, several architectures have been proposed and some are currently being tested in field trials. One of the most popular architectures, due to its relatively low incremental cost per customer, is hybrid fiber coaxial (HFC). It consists of an optical fiber feed from its network source to a coaxial cable tied to a few hundred homes. It combines the strength of a high capacity, low loss fiber backbone with the low cost of coaxial in the last mile to the customer. Another architecture being tried is fiber-to-the-curb (FTTC). It offers significant technical advantages for switched digital services, but has a higher cost per subscriber than HFC. Other promising architectures include 28 GHz wireless and direct satellite, each having its own distinct strengths and weaknesses. Another very promising architecture at least in the short run is the asymmetric digital subscriber line (ADSL) as described in Section 14.3.3.

14.5.3 INTERACTIVE MULTIMEDIA SERVICES

At present multimedia is one of the most popular words. Without doubt, multimedia is one of the key technologies influencing how people will use computers over the next decades.

> Multimedia is characterized by the simultaneous use of text, voice, pictures, videos and data.

Thus, *multimedia service* is a service in which the interchanged information consists of more than one type i.e., a multimedia information may include text, data, voice, graphics and video [39]. In stand-alone multimedia desktop systems, the different types of media are usually taken from a CD-ROM drive. In multimedia networks, however, various multimedia services may be received from different sources. In the extreme case, these sources may be located at different places on the Earth. Thus, multimedia services may require multiple connections, instead of only one connection in traditional telecommunication services (e.g., POTS, ISDN). However, by means of hypertext, the user sees a single call which is similar to Internet applications.

First of all, multimedia services require powerful multimedia-configured personal computers (PCs) with powerful audio and video cards, scanner and a CD-ROM drive. First CD-ROM drivers for PC applications appeared in 1984 providing storage for 600 Mbytes to 680 Mbytes of information including data, text, voice, pictures and short video sequences. At the end of this century, in industrial countries about 40% of all households will use a PC and about 25% a multimedia-configured PC. However, most multimedia-configured PCs will be used as stand-alone desktop systems, whereas networked multimedia systems are still in the minority. Communication technology is critical in enabling multimedia to migrate from dedicated desktop applications to more efficient multimedia network applications. The key problem is the amount of bandwidth needed to transmit multimedia data streams. Nevertheless,

> multimedia is moving from desktop applications to network applications.

Multimedia network and multimedia desktop systems will offer a nearly unlimited number of applications for business as well as private users. A brief sample of typical multimedia applications is given in the following list [40]:

- telemedicine,
- teletraining (e.g., imaginary medical operations, self learning),
- teleeducation (remote or distance learning, on-line university, virtual colleges),
- video conferences with documents and graphics (virtual meeting rooms),
- video libraries,
- health care,
- remote library,
- digital encyclopaedia and atlas,
- virtual museums,
- virtual reality or cyberspace (telepresence, "step into the TV"),
- virtual trips
- computerized story-telling,
- multimedia magazines, newspapers, technical publications, and journals,
- animation and video presentations of products and
- highly interactive video games.

In general, a variation of multimedia applications can be used simultaneously. For example, a user's PC with multiple windows may have an online video window with a friend who has just phoned (the voice is given on the loudspeakers), while another window may display multimedia e-mail including text, pictures and video sequences. A third window may show some CD-ROM-based information, a fourth window the daily electronic news and a fifth window a TV movie. The user can decide to display one of the PC windows on a high-resolution large-scale TV screen and a pair of Hifi loudspeakers (Fig. 14.44). By this example, it becomes clear again that computer, telecommunication and video (TV) markets are converging as already explained in Section 14.2.

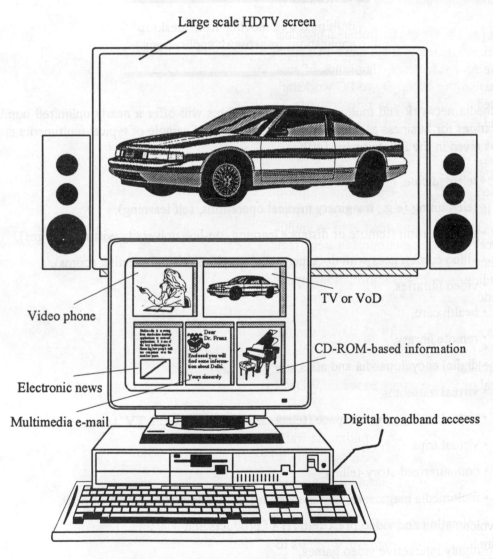

Fig. 14.44: Typical multimedia workstation

As discussed above, multimedia systems can be splitted into

• desktop multimedia systems and

• multimedia network systems also referred as online multimedia.

Desktop multimedia applications are primarily restricted on CD-ROM applications and interactions can only be performed between the PC and the CD-ROM. Multimedia network applications which can be of interactive and distribution type are nearly unlimited. In contrast to desktop systems, interactions can be performed worldwide.

At present, we are far from real broadband networked multimedia applications since most PCs are primarily used as stand-alone PCs supporting business applications such as word processing. In the past few years, thousands of small and large companies installed departmental LANs increasing the effectivity of PC workstations. However, users of departmental LANs are again isolated from users in other departments. Therefore, companies have started to realize enterprise-wide LAN interconnections [39].

A very similar evolution is expected for connectivity in relation to multimedia in the next few years. Since multimedia applications normally require large bandwidth, traditional LANs can only be used for a very limited number of low-bit rate multimedia services (data, voice and graphics). The limitations of standard LANs are primarily related to the limited effective bandwidth per user. Therefore, the next step in multimedia evolution is to develop advanced high-speed networks that span metropolitan areas, regions, nations and even continents. Highly sophisticated multimedia networks require advanced communication technologies such as fiber-based networks including fiber-based information superhighways. In the ideal case, these networks will also include optical routing and switching.

As a scenario, Fig. 14.45 shows a multimedia ring network supporting switchable 155 Mbit/s (STM-1). It becomes clear from this figure that the whole network is only based on three hierarchical levels i.e., two rings and a star. This network can easily be upgraded by realizing an optical overlay network based on WDM.

> Multimedia traffic is characterized by a
> wide range of bit rates.

The voice services only require 64 kbit/s (i.e., ISDN bit rate), whereas compressed video services require bit rates in the range of 2 Mbit/s to 9 Mbit/s. The latter bit rate yields a quality similar to analog PAL TV. Therefore, a flexible transfer technology is needed. Without any doubt, the asynchronous transfer mode (ATM) represents the most advantageous transport and multiplexing technology for interactive multimedia services. The cell structure of ATM allows to transmit and to multiplex low as well as high bit rate multimedia services. In the trunk network, ATM cells will be carried by using a high-speed synchronous network based on SDH (synchronous digital hierarchy).

The combination of ATM, SDH and WDM represents a high-sophisticated technology for broadband digital networks with interactive switchable multimedia access.

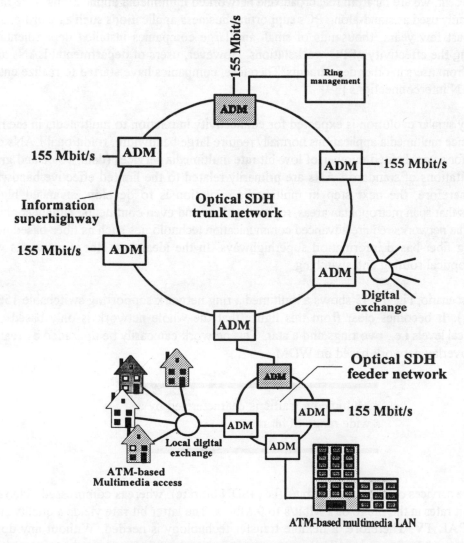

Fig. 14.45: Scenario of a broadband digital network for interactive multimedia access

All-optical networks with fiber-based high-speed multiwavelength information superhighways, optical crossconnects of large throughput and optical add-drop multiplexers represent the ideal physical layer to realize high-speed switchable interactive multimedia accesses for the end users. However, such networks will not be available during the next two decades. As explained in the previous Chapters, various upgrading methods will be employed to enhance the efficiency of the existing network. Since the most critical problem of current networks is the limited bandwidth, data compression techniques will play a key role during the next few years. Upgradation techniques such as ADSL and HDSL and data compression will act as a bridge from existing to future all-optical network.

A vision of a broadband digital fiber network is shown in Fig. 14.45 above [31]. This network is again primarily based on two rings as discussed above: one serves as an optical SDH trunk or transport network and the other as an optical SDH feeder network. Private as well as business users are connected to the feeder network via optical add-drop multiplexers (ADMs). Each user has got an ATM-based multimedia access which allows to transmit and receive data signals with different bit rates varying from e.g., 64 kbit/s to maximum 155.52 Mbit/s. As this network is based on WDM, where different wavelengths are added and dropped within the ADMs, it is well suited to accommodate the expected growth in traffic from video, multimedia and a variety of data communication services.

When user interfaces are realized optically, the above network becomes an *all-optical broadband network* where the data do not undergo any optical-to-electrical conversion within the network as it is routed to its destination.

14.6 REFERENCES

[1]   Alabanese, A.: Fall safe nodes for light guide digital networks. Bell System Tech. J. 61(1982)2, 247-257.

[2]   Albuquerque, A. A. de; et. al.: Field trials for fiber access in the EC. IEEE Commun. Magazine, 32(1994)2, 40-48.

[3]   Armbrüster, H.: The flexibility of ATM: supporting future multimedia and mobile communications. IEEE Commun. Magazine 33(1995)2, 76-84.

[4]   Ballart, R.; Ching, Y.: Sonet-Now it's the standard optical network. IEEE Commun. Magazine 27(1989)3, 8-15.

[5]   Böttle, D.; et. al.: Schritte in Richtung vollständig optischer Netze. Elektrisches Nachrichtenwesen, 3. Quartal 1994.

[6]   Brackett, C. A.: A scalable multiwavelength multihop optical network: a proposal for research on all-optical networks. IEEE J. LT-11(1993), 736-752.

[7]   Brunet, C. J. : Hybridizing the local loop. IEEE Spectrum, 31(1994)6, 28-32.

[8]   Burr, W.: The FDDI optical data link. IEEE Commun. Magazine 24(1986)5, 8-23.

[9]   Carter, D. J. : Services for super highways. ECOC (1994), 933-936.

[10]   Cochrane, P.; et. al.: The hidden benefits of optical transparency. IEEE Commun. Magazine, 32(1994)10, 90-97.

[11]   Cole, N. G.: Asymmetric digital subscriber line technology - a basic overview. BT Technol. J. 12(1994)1.

[12]   Deyhimy, I.: Gallium Arsenide joins the giants. IEEE Spectrum, 32(1995)2, 33-40.

[13]   Driel van, C-J. L.; Grinsven van, P. A. M.; Pronk, V.; Snijders, W. A. M.: The (r)evolution of access networks for the information superhighway. Communications Magazine, 35(1997)6, 104-112.

[14]   Fioretti, A.; Masetti, F.; Sotom, M.: Transparent routing: the enabling factor towards all-optical networking. ECOC (1994), 503-509.

[15]   Goodman, M. S.: Multiwavelength networks and new approaches to packet switching. IEEE Commun. Magazine, 27(1989)10, 27-35.

[16]   Gupta, H. M.: Fiber Optic Local Area Networks. Technical Report No. CS-TR-2130, Department of Computer Science, University of Maryland, USA, September 1988.

[17]   Haber, L.: FDDI finally gains backbone network prestige, but new technologies contend. Lightwave, 12(1995)1, 52.

[18]   Henry, P. S.: High-capacity lightwave local area networks. IEEE Commun. Magazine, 27(1989)10, 20-26.

[19]   Hiergeist, F.: Fiber is on its way to conquering German access networks. Photonics Spectra, (1995)2, 94-96.

[20]   Hill, G. R.: Wavelength domain optical network techniques. Proceedings of IEEE 17(1989)1, 121-132.

[21]   Hill, G. R.; et. al.: A transport network layer based on optical network elements. IEEE J. LT-11(1993)5/6, 667-680.

[22]   Hinterlong, S. J.: When electrons can't keep up it's time to switch to photonics. Photonics Spectra, 33(1995)2, 88-91.

[23]   Hopkins, M.; et. al.: A multi-faceted approach to forecasting broadband demand and traffic. IEEE Commun. Magazine, 33(1995)2, 36-42.

[24]   Hsing, T. R.; Chen, C.-T.; Bellisio, J. A.: Video communications and services in the copper loop. IEEE Commun. Magazine, 31(1993)1, 62-68.

[25]   Ims, L. A.; Stordahl, K.; Olsen, B. T.: Risk analysis of residential broadband upgrade in a competitive and changing market. Communications Magazine, 35(1997)6, 96-103.

[26]   Irmer, T.: Shaping future telecommunications: the challenge of global standardization. IEEE Commun. Magazine, 32(1994)1, 20-28.

[27]   Jain, R.: FDDI: current issues and future plans. IEEE Commun. Magazine, 31(1993)9, 98-105.

[28]   Jajaszczyk, A.; Monftah, H. T.: Photonics fast packet switching. IEEE Commun. Magazine, 31(1993)2, 58-65.

[29]   Kashima, N.: Optical Transmission for the Subscriber Loop. Artech House Publishers, 1993.

[30]   Kyees, P. J.; et. al.: ADSL: a new twisted-pair access to the information highway. IEEE Commun. Magazine, 33(1995)4, 52-59.

[31]   Landegem, Van T.; Prycker, De. M.; Brande, Van den F.: Eine Vision des Netzes der Zukunft. Elektrisches Nachrichtenwesen, 3. Quartal 1994.

[32]   Lin, Y.-K. M.; et. al.: Fiber-based local access network architectures. IEEE Commun. Magazine, 27(1989)10, 64-73.

[33]   Lösch, K.: Potential and implementation of optical technologies for high bit rate short distance interconnections. ECOC (1995), 165-171.

[34]   Masetti, F.; et. al.: ATMOS (ATM optical switching): results and conclusion of the RACE R2039 project. ECOC (1995), 645-652.

[35]   Midwinter, J. E.: Photonic switching - how and why ! ECOC (1994), 29-35.

[36]   Miki, T.: Toward the service-rich era. IEEE Commun. Magazine, 32(1994)2, 34-39.

[37]   Miki, T.: Optical network development towards multimedia information era. ECOC (1995), 11-18.

[38]   Miller, A.: From here to ATM. IEEE Spectrum, 31(1994)6, p. 20-24.

[39]   Minoli, D.; Keinath, R.: Distributed Multimedia through Broadband Communications Services. Artech House, 1993.

[40]   Mizusawa, J.-I.: Multimedia services - a Japanese perspective. ECOC (1995), 509-515.

[41]   Mochida, Y.: Technologies for local-access fibering. IEEE Commun. Magazine, 32(1994)2, 64-73.

[42]   Nassehi, M. M.; Tobagi, F. A.: Fiber optic configurations for local area networks. IEEE J. SAC-3(1985)11, 941-949.

[43]   Nishio, M.; et. al.: A new architecture of photonic ATM switches. IEEE Commun. Magazine, 31(1993)4, 62-68.

[44]   O'Mahony, M. J.: The potential of multiwavelength transmission. ECOC (1994), 907-913.

[45]   Oliver, R.: The role of different architecture alternatives in the access network. ECOC (1995), 493-500.

[46]   Paff, A.: Hybrid fiber/coax in the telecommunication infrastructure. IEEE Commun. Magazine, 33(1995)4, 40-45.

[47]   Powers, J. P.: An Introduction to Fiber Optic System. Aksen Associates Inc., 1993.

[48]   Prycker, M. de.: Standards: a success factor for multimedia services. ECOC (1995), 501-508.

[49]   Ramaswami, R.: Multiwavelength lightwave networks for computer communication. IEEE Commun. Magazine, 31(1993)2, 78-88.

[50]   Rocks, M.; Flor, E.: Experiences with FITL programmes in Germany and future strategies. ECOC (1994), 159-167.

[51]   Roper, S.: Future fiber-optic networks depend on advances in semiconductor technology. Lightwave, 11(1994)12, 52-56.

[52]   Ross, F. E.: FDDI - a tutorial. IEEE Commun. Magazine 24(1986)5, 10-17.

[53]   Ross, F. E.: An overview of FDDI. IEEE J. SAC-7(1989)9, 1043-1051.

[54]   Sanghi, R. K.; Jhunjhunwala, A.: Implementation considerations in fiber optic bus networks. IETE Technical Review, 6(1989)2, 97-110.

[55]   Sato, K.: Network evolution with optical paths. ECOC (1994), 919-926.

[56]   Stordahl, K.; Murphy, E.: Forecasting long-term demand for services in the residential market. IEEE Commun. Magazine, 33(1995)2, 44-49.

[57]   Tenzer, G. : Glasfaser bis zum Haus - Fibre to the Home. German Telecom, 1991.

[58]   Tenzer, G. : Challenges in telecommunication markets. German Telecom, 1991.

[59]   Tong, F.; et. al. A 32-channel hybridly-integrated tuneable receiver. ECOC (1995), 203-206.

[60]   Verbiest, W.; et. al.: FITL and B-ISDN: a marriage with a future. IEEE Commun. Magazine, 31(1993)6, 60-66.

[61]   Wada, O.: Progress in optoelectronic device technologies for optical interconnection and photonic switching. ECOC (1995), 159-163.

[62]   Warzanskyj, W.; Ferrero, U.: Access network evolution in Europe: a view from EURESCOM. ECOC (1994), 135-142.

[63]   Wasen, O. J.; et. al.: Forecasting broadband demand between geographic areas. IEEE Commun. Magazine, 33(1995)2, 50-57.

[64]   Weippert, W.: The Evolution of the access network in Germany. IEEE Commun. Magazine, 32(1994)2, 50-55.

[65]   Weiss, S. A. : Fiber and coax: a marriage of convenience. Photonics Spectra, (1994)2, 78-82.

[66]   Wellhausen, H.-W., Heuser, S.: Effiziente Nutzung vorhandener Kupfer-Ortsanschlußleitungsnetze. Der Fernmeldeingenieur (1993)8/9.

15   OPTICAL SPACE COMMUNICATIONS

Optical free-space communication systems for intersatellite links and deep space missions show significant advantages in comparison to alternate microwave systems. This Chapter describes the various aspects of optical space communications. The Chapter is organized as follows. Section 15.1 gives a brief introduction and Section 15.2 discusses the applications of optical space communication systems. Their comparison with microwave communication systems is made in Section 15.3. Characteristic features of coherent optical space communication systems are explained in Section 15.4. Drawbacks and realization problems of such systems vis-a-vis coherent optical fiber communication systems are given in Section 15.5. The system description and design aspects viz., free-space propagation formula, optical transmitter, background noise, pointing, acquisition and tracking and receiver are discussed in Section 15.6.

15.1   INTRODUCTION

The complexity of space communication systems is growing rapidly with the increase in the transmission requirements. Most of these systems are exclusively based on microwave links. The optical communication systems become more and more attractive as the interest in high-capacity and long-distance space links grows. Advances in laser communication system architectures and optical components technology make such high capacity links feasible. Nevertheless, the choice of an optical source (semiconductor laser diode, solid-state laser or gas laser) as well as the choice of a proper modulation/demodulation technique are still an open question.

The two basically different receiver techniques for optical space communication systems are: direct detection and coherent (heterodyne or homodyne) detection. In direct detection systems, only light intensity modulation (IM) is practical, whereas in heterodyne and homodyne systems many modulation schemes well-known from radio frequency (RF) transmission can be used; examples include ASK, FSK, PSK, DPSK etc. As is well-known, these schemes have significant advantages in regard to the receiver sensitivity, bit rate and transmission distance which, for example, can reach millions of km in space.

A compromise in sensitivity gain and realization demands is represented by the DPSK heterodyne and phase-diversity receivers. The sensitivity gain of both these receivers is only 3 dB less than the maximum sensitivity gain that can be achieved using a coherent optical homodyne receiver. In contrast to an optical PSK homodyne receiver, these receivers do not require an optical phase-locked loop (OPLL) which is an advantage in system realization.

15.2   APPLICATIONS OF OPTICAL SPACE COMMUNICATIONS

The potential use of optical free-space communication systems can be divided into three main applications as shown in Fig. 15.1: (i) interorbit links (IOLs), (ii) intersatellite links(ISLs) and (iii) deep space missions (DSMs).

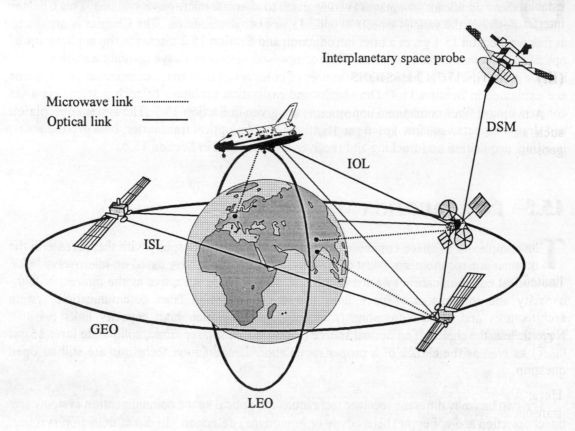

Fig. 15.1: Applications of optical space communications

(i)   INTERORBIT LINKS

These links are of great interest for data transmission from a low Earth orbit (LEO) vehicle, such as Earth resource satellites, manned spaced stations, polar platforms of Hermes to a geostationary Earth orbit (GEO) vehicle, usually a data relay satellite (DRS). The return link (LEO-GEO) requires high data rates in the order of 500 Mbit/s, while the forward link (GEO-LEO) only demands telemetry data rates of about 25 Mbit/s.

(ii) INTERSATELLITE LINKS

A second important application, especially for commercial voice, television and data transmission is an intersatellite link between two geostationary telecommunication satellites. This link prevents double hopping arising from the use of an additional third Earth station in establishing a link between two participants living on opposite sides of the Earth. Moreover, interferences with terrestrial microwave links will be avoided.

(iii) DEEP SPACE MISSIONS

A third application of prime importance is the high capacity data transmission from planets such as Mars (78 million km from Earth) or Saturn (1278 million km from Earth) to a geostationary Earth satellite.

15.3 COMPARISON WITH MICROWAVE SYSTEMS

The promise of an optical communication system is mainly based on the strongly improved gain and reduction in the beam width of the antenna. The antenna gain in dB for a diffraction limited beam at short optical wavelengths is given by

$$G_a = 20\log_{10}\left(\frac{\pi D_a}{\lambda}\right) \tag{15.1}$$

Here, λ represents the wavelength and D_a the antenna (telescope) diameter. The half-power transmitter beam width full angle in radian is given by [24]

$$\psi_a = 1.03\left(\frac{\lambda}{D_a}\right) \tag{15.2}$$

It may be noted that Eqs. (15.1) and (15.2) are valid for microwave as well as optical antennas.

Most of the differences between optical and microwave space links are due to the difference in wavelengths. The much shorter optical wavelength results in a very small beam width and a very small spot size. As shown in Fig. 15.2, spot size of an optical beam transmitted from planet Mars is only about 10 percent of the Earth's diameter, whereas in case of microwave beam it is about 100 times. The small spot size of an optical beam results in substantial increase in the received optical power and therefore improvement in the receiver performance.

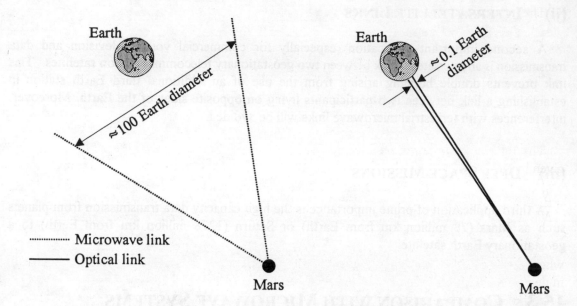

Fig. 15.2: Spot size at the Earth for microwave beam and optical beam transmitted from Mars

Block diagrams of microwave and optical communication systems with typical values of transmitting power and antenna gains are shown in Fig. 15.3.

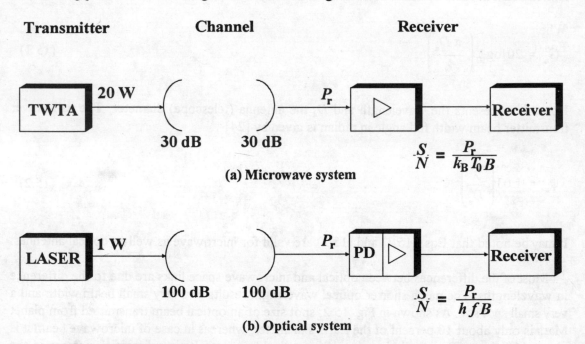

Fig. 15.3: Block diagrams of (a) microwave and (b) optical communication systems

The much higher antenna gain achieved at the optical frequency allows to

- decrease the antenna (telescope) diameter (for example $D = 10$ cm),
- increase the data rate and
- increase the transmission distance which can, for example, reach millions of km in space.

The intermediate frequency (IF) signal-to-noise (S/N) ratio for the microwave receiver is given by

$$\frac{S}{N} = \frac{P_r}{k_B T_0 B} \tag{15.3}$$

where P_r represents the received signal power, k_B the Boltzmann's constant, T_0 the temperature in degree Kelvin and $B = 1/T$ (T is the bit duration) the Nyquist bandwidth of the IF signal. The corresponding expression for the coherent (heterodyne) optical receiver is given by [22]

$$\frac{S}{N} = \frac{P_r}{hfB} \tag{15.4}$$

When practical bandwidth is taken to be two times of the Nyquist bandwidth, denominators in Eqs. (15.3) and (15.4) will get multiplied by a factor of two.

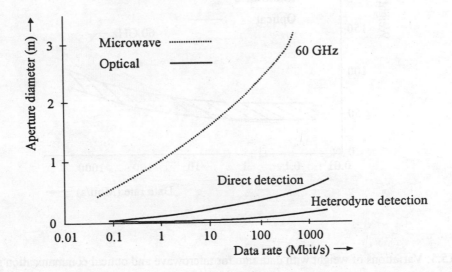

Fig. 15.4: Variations of antenna aperture diameter with data rate for microwave and optical systems

A comparison of Eqs. (15.3) with (15.4) reveals that the thermal noise ($k_B T_0$) in microwave systems corresponds to the quantum noise (hf) in optical systems. This analogy is not surprising. It is well-known from thermodynamics that $k_B T_0$ and hf are the two limiting forms of the noise psd of an oscillator at frequency f and in thermal equilibrium at temperature T_0. At the microwave frequencies, $hf \ll k_B T_0$. As a consequence, thermal noise becomes a limiting factor, but it may not be at the optical frequencies [13, 22, 25]. In optical space communications, thermal and other sources of noise like background noise, optical preamplifier noise etc. may become comparable with the quantum noise. In that case, an additional parameter F_i called noise figure (NF) of the receiver will come in the denominator of Eq. (15.4). It varies from 1 to ∞. Lesser is the NF, better is the receiver performance. Variations of antenna aperture diameter and terminal weight with data rate for microwave and optical systems are shown in Figs. 15.4 and 15.5 respectively [5].

It is infer from these figures that the use of optical technology in space communications offers a tremendous reduction in antenna diameter and weight as compared to microwave technology. This is quite promising for satellite communications particularly when the data rate is high. The high data rates is essentially required to obtain a better resolution and a large frame rate of image sequences. For Earth observation satellite data rate of 1 Gbit/s is desired, whereas from the planet Mars more than 10 Mbit/s is the goal (compared to a few kbit/s with the existing RF systems).

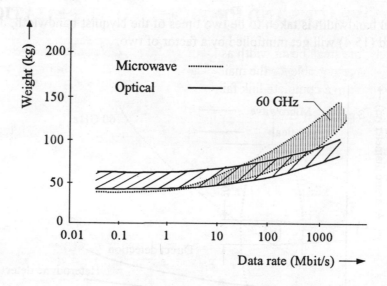

Fig. 15.5: Variations of weight with data rate for microwave and optical communication systems

15.4   COHERENT OPTICAL SPACE COMMUNICATIONS

The main advantage of using a coherent optical space communication system lies in the receiver sensitivity gain of more than 20 dB. It may be mentioned that a sensitivity improvement of 6 dB doubles the transmission distance (for example 30 million km with a direct detection system will become more than 60 million km with a coherent system). This substantial improvement in transmission distance is particularly attractive for the long-range space links in DSMs. For the same transmission distance, highly sensitive coherent optical receivers require much less transmitter power than an optical system with a direct detection receiver. And, therefore, the power consumption of satellite becomes very efficient. Moreover, the influence of background noise (for example from the Sun, Moon, Earth, Venus etc.) is reduced in coherent systems [18]. The main advantages of using coherent systems in space are the following:

- High capacity long-range optical space links are practicable,
- ASK, FSK, DPSK and PSK modulation techniques (binary and multilevel) with and without coding scheme are feasible,
- less sensitivity to background noise and
- better frequency selectivity permits deployment of high density multichannel systems.

15.5   DRAWBACKS AND PROBLEMS OF REALIZATION

Unfortunately, the small beam width as the main advantage of optical space communication systems is also responsible for the main disadvantage. As shown in Fig. 15.6, small satellite vibrations may result in a complete link failure.

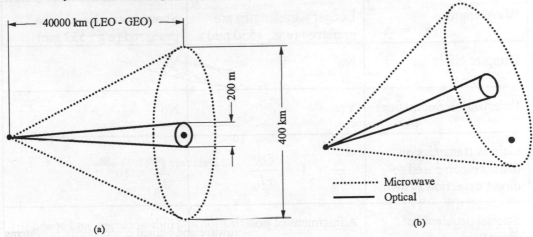

Fig. 15.6: (a) Main advantage and (b) drawback of optical space communication systems

The very small beam width complicates pointing, acquisition and tracking (PAT). Therefore, in addition to the communication subsystem, high-accuracy and speed PAT subsystems are required to reduce the influence of satellite vibrations. The main problem in the realization of optical space communication systems is the non-availability of optical components at the wavelengths used in space. In optical fiber communication systems, higher wavelengths are preferred (for example 1500 nm), whereas in optical space systems lower wavelengths (for example 1064 nm, 532 nm or visible) are preferred. It is worth mentioning that the fiber attenuation loss (due to material effects) decreases at longer wavelengths, whereas the free-space beam spread loss due to geometrical effects decreases at shorter wavelengths (refer to Eq. (15.8) derived latter).

Optical components viz., optical phase modulators, optical isolators, couplers etc. are available at wavelengths typically used in optical fiber applications such as 850 nm, 1300 nm or 1500 nm. In contrast to this, no such components are available at the Nd:YAG laser wavelength of 1064 nm or at its double the frequency i.e., 532 nm wavelength (except high-cost single productions). The Doppler frequency shift ($\approx \pm 10$ GHz) as a result of the relative motion of two linked satellites is another problem in realizing a coherent optical space communication system. This problem can be tackled by including a tuneable local laser with a tuneable frequency span of 20 GHz and an automatic frequency control (AFC) in the receiver circuit. In fiber optics, Doppler effect is non-existent. In the following table, main differences between coherent optical fiber and coherent optical space communications have been summarized.

Table 15.1: Comparison of coherent optical fiber and space communication systems
* L_d is the distance in km for direct detection system, GR the improvement in receiver sensitivity in dB (see Chapter nine) and α the attenuation of fiber in dB/km.

	Coherent optical fiber communication systems	Coherent optical space communication systems
Wavelength	Longer wavelengths are preferred (e.g., 1550 nm)	Shorter wavelengths are preferred (e.g., 532 nm)
Doppler effect	No	Yes
Polarization fluctuations	Yes	No
Gain in transmission distance compared to direct detection systems*	$\dfrac{L}{L_d} = 1 + \dfrac{GR}{\alpha L_d}$	$\dfrac{L}{L_d} = 10^{\frac{GR}{20\text{ dB}}}$
Special problems of realization	Adjustment of polarization	High-accuracy and speed PAT subsystems and reduction of Doppler effect

15.6   SYSTEM DESCRIPTION AND DESIGN

A physical model of laser communication system is shown in Fig. 15.7. The relationship between transmitted and received optical power is described by the range equation. This equation provides a characterization of propagation in the communication channel, free-space propagation loss, transmitting and receiving antenna gains.

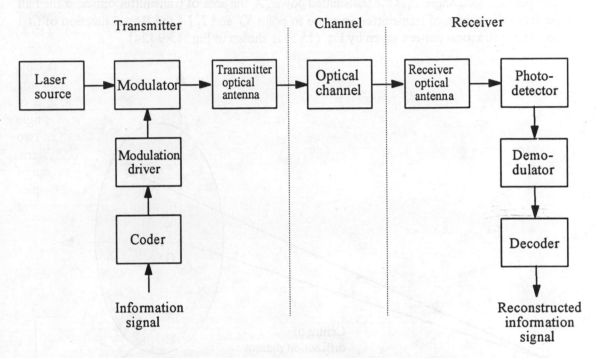

Fig. 15.7: Physical model of optical communication system

15.6.1   FREE-SPACE PROPAGATION FORMULA

When the source radiating diameter D_s is too small, size of its image in the receiver plane is determined by diffraction at the transmission aperture. The diffraction pattern produced by a uniformly illuminated circular aperture of diameter D_t consists of a set of concentric rings. The intensity per unit solid angle centred at point 'Q' in the receiver plane (Fig. 15.8) can be expressed as [24].

$$I(\alpha) = \left[\frac{2J_1(\pi D_t \alpha/\lambda)}{\pi D_t \alpha/\lambda}\right]^2 I(0) \tag{15.5a}$$

where

$$I(0) = \pi D_t^2 \frac{P_t}{4\lambda^2} \tag{15.5b}$$

The parameter $I(0)$ (also equals to $P_t A_t / \lambda^2$) represents the intensity at the centre of diffraction pattern per unit solid angle. P_t is the transmitted power, A_t the area of transmitter optics, α the half angle from the centre of transmitter aperture to point 'Q' and J_1 [.] the Bessel function of first order. The diffraction pattern given by Eq. (15.5) is shown in Fig. 15.9 [24].

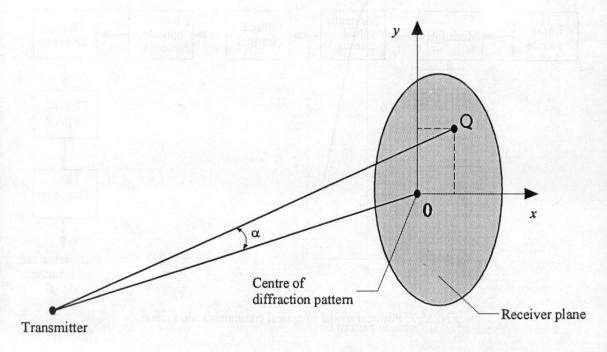

Fig. 15.8: Geometry of transmitter antenna diffraction pattern in the receiver plane

The power incident upon the receiver plane, $P(\theta)$ about the centre of diffraction pattern assuming medium transmissivity factor to be unity is given by

$$P(\theta) = I(0) \int_0^{2\pi} \int_0^{\theta} \left[\frac{2J_1(\pi D_t \alpha / \lambda)}{\pi D_t \alpha / \lambda} \right]^2 \alpha \, d\alpha \, d\phi \tag{15.6}$$

If the receiver optical antenna of diameter D_r is located at a distance L from the transmitter and centred on the optic axis, diffraction angle θ is approximately $D_r/2L$. For a large L, power density in the receiver plane is nearly constant at a maximum value of $I(0)$. Therefore, the maximum received power level from Eq. (15.6) is given by

$$P_r = I(0) \int_0^{2\pi} \int_0^{D_r/2L} \alpha \, d\alpha \, d\phi = \frac{\pi^2 P_t D_t^2 D_r^2}{16 L^2 \lambda^2} \tag{15.7}$$

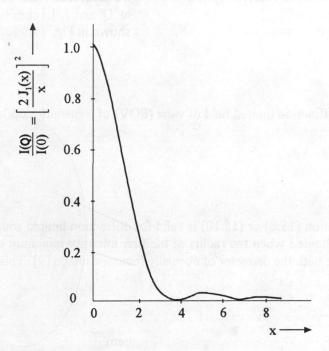

Fig. 15.9: Diffraction pattern of uniformly illuminated circular aperture

The equation above can be put in the form

$$P_r = P_t G_t \left(\frac{\lambda}{4\pi L} \right)^2 G_r \tag{15.8}$$

It is referred as range equation. Here, $(\lambda/4\pi L)^2$ represents the free-space loss, G_t and G_r are the gains of the transmitting and receiving antenna/optics respectively. These are given by

$$G_t = 4\pi \frac{A_t}{\lambda^2} \tag{15.9a}$$

and

$$G_r = 4\pi \frac{A_r}{\lambda^2} \tag{15.9b}$$

where A_r is the area of the receiver optics. The range equation can also be written as

$$P_r = P_t \frac{A_r}{\Omega_{dt} L^2} \tag{15.10}$$

where Ω_{dt} is the diffraction limited field of view (FOV) of transmitter optics. It is given by

$$\Omega_{dt} = \frac{\lambda^2}{A_t} \tag{15.11}$$

The range equation (15.8) or (15.10) is valid for diffraction limited sources. A source is said to be diffraction limited when the radius of the first intensity minimum or dark ring becomes comparable in size with the diameter of normally focussed image [9]. This will happen if

$$D_s < \frac{1.22 f_c \lambda}{D_t} \tag{15.12}$$

where f_c is the focal length of the transmitter optics. When a source satisfies Eq. (15.12), P_r can be determined by using either Eq. (15.8) or Eq. (15.10) [9]. Otherwise P_r will be reduced by a factor Ω_{dt}/Ω_t where Ω_t is the transmitter optics FOV [3].

The P_r may also get reduced due to obscuration caused by the secondary mirrors or field stops of the telescopes (see Figs. 15.11 and 15.16), poor quality of the optics etc. And also due to inaccuracy in the alignment of the optical transmitter and receiver i.e., the pointing error. For a fixed pointing error $\pm\psi_e$, optical beam width has to be increased to compensate the pointing error. It will cause the reduction of power by a factor of $(1+2\psi_e/\psi_a)^2$ [6]. Generally, ψ_e is a random variable. It is considered to be Gaussian distributed with zero mean and variance σ_e^2. Setting $6\sigma_e \le \psi_a$ gives the probability of $\psi_e > \psi_a$ less than 0.01 [24]. When σ_e^2 is known, one can find out the required ψ_a and hence D_a.

As shown in Fig. 15.7, space optical communication systems have four components: optical transmitter, channel, receiver and PAT subsystem involving both transmitter and receiver (not shown explicitly). These components have been briefly discussed in the following Sections.

15.6.2   OPTICAL TRANSMITTER

The optical transmitter consists of coder, modulator, optical source and transmitting antenna. Depending upon the type of source, modulation may be impressed during the generation of light or after the light has been generated. The transmitting antenna is a lens system that focuses the optical beam for transmission through the optical channel.

(i)   OPERATING WAVELENGTHS AND SOURCES

Use of shorter wavelengths in optical space communications give rise to smaller spot size. It produces improvement in the receiver performance/increase in the transmission distance. While selecting a wavelength, in addition to above, many other factors like availability of source, output power level, life time etc. are also to be taken into account.

Three basic laser technologies are existing for space communications. These are based on: (i) CO_2 gas laser operating in the infrared region at 10 μm, (ii) Laser diode pumped Nd:YAG laser operating at 1.064 μm and can be frequency doubled and (iii) Semiconductor laser diode operating in the region 0.78 μm-1.55 μm. During the last decade, CO_2 laser communication links appeared most attractive because of their higher electrical-to-optical conversion efficiency and longer life time. The problems with this laser are its long wavelength and less reliability by virtue of a gas laser. The rapid development of laser diode technology since 1980 has considerably changed this scenario. Now the major light sources are semiconductor laser devices (lasers and laser arrays) and Nd:YAG lasers.

Semiconductor lasers because of their high reliability, efficiency, small size and weight are excellent light sources in optical space communications. Their major drawback is that a single device typically cannot radiate the required power (few hundred milliwatt to a watt) in a stable single-mode operation. This problem can be overcome by combining the power of several semiconductors lasers. The objective is to increase the source brightness (define as the power divided by product of source area and solid angle into which light is emitted), rather than the power. The reason for this is that higher source brightness and not just more source power, translates into higher intensity at the receiver [3].

The source brightness can be increased by combining the power of several semiconductor lasers either incoherently or coherently. The incoherent methods employ lasers that emit light at different polarizations or different wavelengths (a combination of two is also possible). The power of lasers is combined by using a polarization beam combiner or wavelength selective elements like filters or wavelengths dispersive elements like gratings. The incoherent method of power combining is conceptually a simpler approach, but the entire transmitter head becomes a sophisticated optical system that includes, in addition to the laser themselves, many other optical components. Further, increase in the brightness level is not much.

The coherent methods are usually much more complicated to implement. The potential payoff is mainly an increased source brightness due to phase-locking of the lasers (similar to microwave phased antenna arrays). The three basic methods of coherent power combining are: mutual coupling, injection locking and coherent amplification. In mutual coupling method, no laser in the array has a privileged status. There is some coupling among the lasers, which under certain conditions results in their synchronization. The amount of coupling depends on the ratio of separation $(\Delta \lambda_i)$ between the intrinsic wavelengths of oscillation of the interacting lasers and the centre wavelength (λ) of the laser transition linewidth [3]. It becomes more difficult to phase-lock the array as the number of elements increases [14]. In injection locking, there is single master laser oscillator. Portions of its emitted radiation are coupled simultaneously into all the other lasers in the array forcing them to oscillate at its frequency. In coherent amplifications, light generated in the master laser is split and fed simultaneously into gain elements, where a travelling wave amplification is employed. If the output of the master laser is coherent over its near-field pattern, outputs of the amplifiers are automatically phase-locked. It has the advantage of avoiding the generally stringent conditions required for establishing phase-locking [3].

In both the injection locking and coherent amplification, failure of master oscillator leads to the failure of the entire source. It may not be so critical in the mutual phase-locking. However, failure of a single device may alter the modal shape of the entire array and make it unusable. The mutual phase-locking is more amenable than the other methods to monolithic implementation.

In the optical space communications, when the background noise or receiver noise or both are high, receiver performance is severely degraded. This can often be improved by using pulse laser i.e., higher amplitude and narrower pulses, while still maintaining the same average power level. The minimum pulse duration is determined by the detector bandwidth and the maximum power level by the largest peak power that the laser source can generate. Due to peak power limitations, semiconductor lasers in low duty cycle operation will emit much less average power than their continuous wave (CW) rating. This problem can be overcome by employing a different type of source. Among all the possible alternatives, Nd:YAG laser seems to be the most attractive. It is a solid-state laser and can be operated in both CW and pulse modes. The most efficient pumping band for Nd:YAG laser is 0.81 μm. It is relatively easy to fabricate AlGaAs laser sources which emit at the pump wavelength. Even when Nd:YAG laser operates in a pulse mode (employing Q-switching, cavity dumping or mode locking), pump is continuously operated. Thus, this laser may be considered as a highly efficient DC-to-pulse converter to the light that is (efficiently) generated by the semiconductor laser arrays. It may be mentioned that beam quality requirements from the semiconductor laser arrays are less stringent when they are used as pump sources and not directly as the light sources for communications. Pulse position modulation (PPM) is one of the modulation schemes which requires the use of a pulse laser. In this modulation scheme, laser pulse is delayed into one of the "m" possible time slots during each frame interval. Therefore, correct decision of a single pulse conveys $\log_2(m)$ bits of information.

Theoretically, spectral linewidth of a semiconductor diode pumped Nd:YAG laser is less than 1 Hz. For a typical semiconductor laser diode, it is in the order of 1 MHz. Further, in Nd:YAG laser, excess intensity noise is negligibly small. Therefore, it provides improvement in the receiver performance for single-detector configuration [16]. Its main drawback is extremely low (below 0.5 percent) electrical-to-optical conversion efficiency. This drawback has now been overcome to a great extent. These lasers with an efficiency of 8.5% to 17% have been developed and demonstrated [20].

(ii) TRANSMITTING ANTENNA

The transmitting antenna should always be designed as close to the diffraction limit as possible since this results in the smallest size. As mentioned earlier, narrowness of beam width is limited by the pointing error. The above two requirements put a limitation on the antenna diameter to be 10 centimetres to several metres depending on the wavelength of operation.

The transmitting antenna can be based either on refractive optics or reflective optics. Both types of antennas are shown in Fig. 15.10 and Fig. 15.11 respectively [24]. For small size apertures, lenses are practical for forming the transmitter antenna. When the aperture size is several centimetres in diameter, reflective type is preferred because of lower cost and weight.

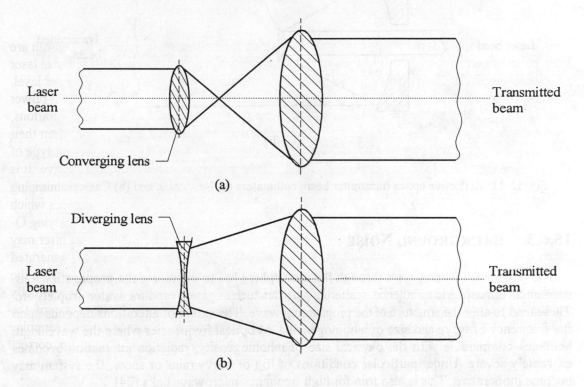

Fig. 15.10: Refractive optics transmitter beam collimators with (a) converging and (b) diverging lens

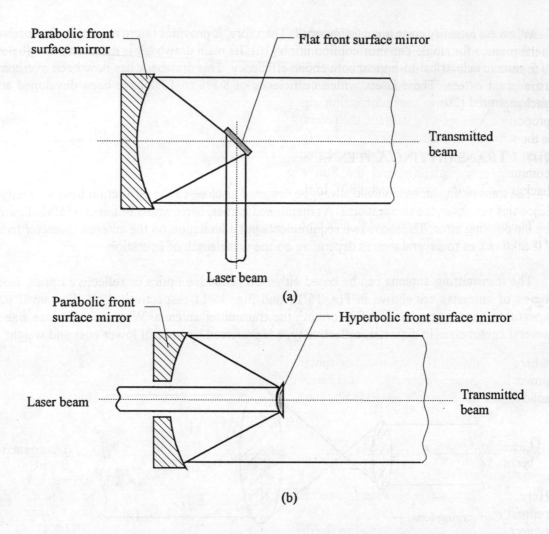

Fig. 15.11: Reflective optics transmitter beam collimators (a) Newtonian and (b) Cassegrainian

15.6.3   BACKGROUND NOISE

The laser beam when passes through the atmosphere has to overcome the propagation loss, encounters atmospheric turbulence, scattering particles such as gases, vapours, water droplets etc. These tend to alter the structure of the propagating wave. The extent of alterations depends upon the frequency of wave and size of inhomogeneity. At optical frequencies where the wavelength becomes comparable with the physical size of inhomogeneity, radiation interaction becomes extremely severe. Under particular conditions of fog or heavy rains or snow, the system may become inoperative. This is also true for high frequency microwave links [24].

When the propagation medium is free-space as in IOLs, ISLs or DSMs, principal channel effect is the propagation loss. The transmission is free of distortion and the objective of system design is to overcome loss by the appropriate choice of source, transmitter and receiver antennas or by increasing the receiver sensitivity or their combinations. Of course, optical channel will introduce background noise in the information signal. This background noise when present in large proportion, can severely degrade the receiver performance and may lead to link failure. The Sun is the strongest background noise source (power spectral density per mode at $\lambda = 1.064$ µm has been reported to be $1.9 \cdot 10^{-20}$ Watt/Hz·mode) [18]. If the receiver FOV is quite large and the communication satellites and the Sun are in line, interruption of communication due to background noise is quite likely [30]. In the following, background noise has been discussed.

The background noise power P_b pick up by the receiver in a bandwidth B around the frequency f is given by [6]

$$
P_b = \begin{cases} N(f)BA_r\Omega_r & \text{if } \Omega_r \in \Omega_s \\ N(f)BA_r\Omega_s & \text{if } \Omega_s \in \Omega_r \end{cases}
\tag{15.13}
$$

where $N(f)$ is the spectral radiance function for the single polarization state. It represents the power radiated at frequency f in a unit bandwidth into a unit solid angle per unit source area. The solid angle Ω_s is the angle subtended by the source when viewed from the detector. It is given by

$$
\Omega_s = \frac{A_s}{L^2}
\tag{15.14}
$$

Here, L is the distance between source and receiver and A_s the source area. The parameter, Ω_r represents the FOV of the receiver. It is the solid angle within which all arriving plane waves project their diffraction pattern onto the detector. By standard geometric analysis in Fig. 15.12, it is given by

$$
\Omega_r \approx \frac{A_d}{f_c^2}
\tag{15.15}
$$

Here, A_d is the detector area and f_c the focal length of the lens at the receiver. Eq. (15.13) for $\Omega_r \in \Omega_s$ is used when the angular extent of source is larger than the receiver FOV. Such sources include sky, sunlit Earth etc. When the angular extent of the source is smaller than the receiver FOV, noise power is determined by using Eq. (15.13) for $\Omega_s \in \Omega_r$. It may be the case with sources like stars, planets, Sun etc.

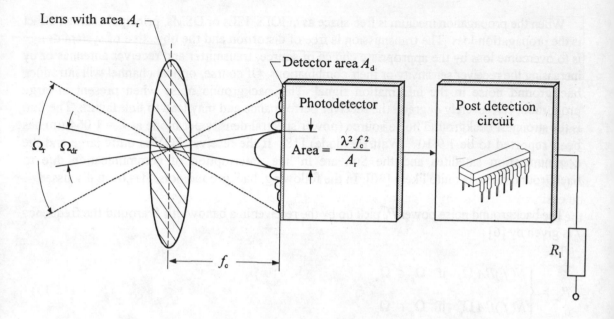

Fig. 15.12: Schematic diagram for relationship between Ω_r and Ω_{dr}

When $\Omega_r \in \Omega_s$, number of background spatial modes received by the receiver is

$$m_t = \frac{\Omega_r}{\Omega_{dr}} \tag{15.16a}$$

Otherwise (i.e., when $\Omega_s \in \Omega_r$), number of modes is given by

$$m_t = \frac{\Omega_s}{\Omega_{dr}} \tag{15.16b}$$

Here, Ω_{dr} represents the diffraction limited FOV of the receiver. It is the minimum FOV consisting of set of arrival angles whose pattern superimpose on the detector surface and are indistinguishable in terms of direction of arrival. As shown in Fig. 15.12, it is given by

$$\Omega_{dr} \simeq \frac{(\lambda f_c)^2/A_r}{f_c^2} = \frac{\lambda^2}{A_r} \tag{15.17}$$

Using Eqs. (15.13) and (15.16), background noise psd $S_n = P_b/(B \cdot m_t)$ per spatial mode for both the types of sources (i.e., when $\Omega_r \in \Omega_s$ or $\Omega_s \in \Omega_r$) becomes

$$S_n = N(f)\lambda^2 \tag{15.18}$$

This psd will get multiplied by a factor of two if the other polarization states of the background noise are considered [18]. The Eq. (15.18) is used to determine S_n at wavelengths greater than 3 μm. For shorter wavelengths, radiation due to Sun reflectance has also to be considered along with the self-emission. To determine S_n in these instances, one may resort to spectral radiance, $H(f) = N(f)\Omega_s$ having the units of power per unit bandwidth per unit area [6]. This function can be easily measured without the explicit knowledge of Ω_s. Data on $H(f)$ are readily available in the literature [24].

The background noise power at the receiver is given by

$$P_b = \begin{cases} H(f)BA_r\Omega_r/\Omega_s & \text{if } \Omega_r \in \Omega_s \\ H(f)BA_r & \text{if } \Omega_s \in \Omega_r \end{cases} \tag{15.19}$$

As before, noise psd for spatial mode in both the cases will be same and is given by

$$S_n = H(f)\frac{\lambda^2}{\Omega_s} \tag{15.20}$$

Using the relationship

$$H(f) = H(\lambda)\frac{\lambda^2}{c} \tag{15.21}$$

Eq. (15.20) can be written as

$$S_n = \frac{H(\lambda)\lambda^4}{c\,\Omega_s} \tag{15.22}$$

Eqs. (15.18) and (15.22) are used to determine S_n at different wavelengths and for various sources. Some of these values are given in the following table [18].

Table 15.2: Power spectral density per mode (in W/Hz) for various celestial bodies at selected wavelengths

	0.85 μm	1.06 μm	1.3 μm	10.6 μm
Earth	$2.6 \cdot 10^{-26}$	$4.0 \cdot 10^{-26}$	$5.5 \cdot 10^{-26}$	$2.0 \cdot 10^{-22}$
Moon	$1.8 \cdot 10^{-26}$	$2.7 \cdot 10^{-26}$	$3.6 \cdot 10^{-26}$	$6.5 \cdot 10^{-22}$
Sun	$1.3 \cdot 10^{-20}$	$1.9 \cdot 10^{-20}$	$2.6 \cdot 10^{-20}$	$7.0 \cdot 10^{-20}$
Venus	$2.1 \cdot 10^{-25}$	$3.1 \cdot 10^{-25}$	$4.1 \cdot 10^{-25}$	$3.1 \cdot 10^{-22}$

For determining the receiver performance, time modes are also to be considered along with the spatial modes. For the description of time modes, let the wave velocity is c and there is no dispersion. Then in time t, receiver measures the field which had previously occupied a path length $L_p = ct$. This field can be described by expanding it into a series of orthogonal modes using spatial Fourier series. For this expansion, the q th mode varies with distance and time according to the exponential factor

$$\exp\left[jq\frac{2\pi}{L_p}(z-ct) \right] \qquad (9.23)$$

where q is an integer. It is seen from the above equation that the frequency of the q th mode is qc/L_p. The condition for orthogonality of the modes is that different values of q differ by an integer. This means that the frequency separation between adjacent modes is $\Delta f = c/L_p$. In a receiver of bandwidth B, there will be $B/\Delta f = BL_p/c$ orthogonal modes. Since L_p is equal to ct, receiver measures the state of excitation of Bt such modes. Therefore, the rate of arrival of independent spatial modes at the receiver is B [8].

15.6.4   POINTING, ACQUISITION AND TRACKING

In order to establish an optical space link, transmitted field in addition to having to overcome the path loss must also be properly aimed towards the receiver. Likewise, the receiver must determine the direction of arrival of the transmitted beam. The operation of aiming a transmitter in the proper direction is referred as pointing. The receiver operation of determining the direction of arrival of an impinging beam is called spatial acquisition. The subsequent operation of maintaining the pointing and acquisition throughout the communication time period is called spatial tracking. Thus, the PAT function is to keep aligned the opposite terminals before and throughout the communication. Further, PAT has to compensate the motion of opposite terminals during the round trip time of light by using a point-ahead angle. In a coherent receiver, it has to take care of Doppler frequency shift or perform the frequency pointing, acquisition and tracking also.

The pointing function of PAT subsystem is basically an open-loop system. The inherent errors of the open-loop pointing are typically orders of magnitude higher than the divergence of transmitter beam, so that the tracking function cannot yet lock onto the received beam. The acquisition system performs the function of reducing the pointing error to allow the tracking system to take over. The required accuracy makes a closed-loop operation mandatory. The tracking function operates throughout the communication to keep the antenna aligned within the required sub-microradian accuracies. The tracking is a closed-loop process. Generally, the communication beam of opposite terminal is used by the tracking subsystem to obtain the pointing information for its own antenna.

To understand the pointing, acquisition and tracking operations, consider a link to be established between station 1 and station 2. An important step in establishing the link involves the process of spatial acquisition i.e., aiming the receiving antenna in direction of the arriving field. The acquisition subsystem can use either an expanded transmitter laser beam as a beacon or a separate beacon [2]. In the latter case, optical beacon is an unmodulated laser beam at a wavelength different from the transmitter. It is located at or near the receiver. Alternatively, in a duplex system, each transmitter can serve as a beacon to the other terminal.

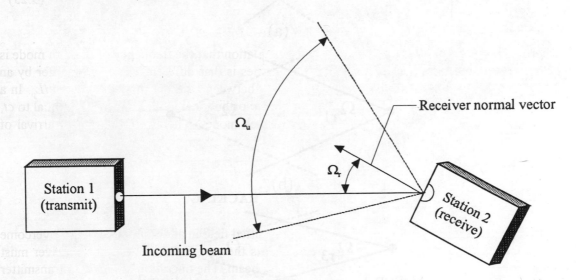

Fig. 15.13: One-way beam acquisition sequence

The acquisition can be accomplished by one-way or two-way procedures. Let us consider expanded laser beam as a beacon. As shown in Fig. 15.13, a single transmitter or solar-illuminated planets as in deep space communication at one point transmit to a single receiver located at another point. If the pointing is satisfactory, receiver knows the transmitter direction within some uncertainly solid angle Ω_u defined from the receiver location. The receiver will then align the antenna normal to the direction of arriving field within some pre-assigned resolution solid angle Ω_r. In two-way acquisition procedure (refer to Fig. 15.14), station 1 transmits a beam wide enough (solid angle Ω_{u1}) to cover its pointing error. The station 2 will search its uncertainly FOV Ω_{u2} to acquire. After successful acquisition, station 2 transmits to the station 1 with beam width Ω_{r2} using the arrival direction obtained from the acquisition. The station 1 can now acquire the return beam with its desired resolution Ω_{r1}. For still narrower beams, the procedure can be repeated again. The link is now established with the desired resolution and the communication can begin.

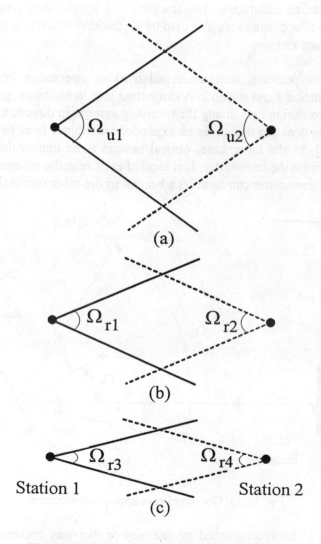

Fig. 15.14: Two-way beam acquisition sequence

When a separate beacon is used for acquisition (refer to Fig. 15.15), station 2 after the acquisition of transmitted beam sends a beacon signal towards the station 1. The station 1 first receives the beacon and then transmits its modulated laser beam towards the beacon direction of arrival. When the beams are extremely narrow, the station 2 may have moved out of the station 1 beam width during the round trip transmission time. In that case, station 1 must point-ahead its transmission from the measured beacon arrival direction.

For two-way acquisition, both the stations 1 and 2 must contain a transmitting laser and an optical receiver. The modulated laser beam serves as a beacon for the return direction. Usually, separate wavelengths are used for the optical beam in each direction. For point-ahead correction, command control must adjust transmit direction relative to receive direction. Such an adjustment requires accurate satellite altitude control [7].

Although acquisition and tracking systems are distinct from the communication system, the two are interrelated in defining system performance. As clear from above, there are basically three link types: acquisition, tracking and communication which operate at the same time. Performance criteria for the above three links are quite different. For acquisition, criteria are typically the acquisition time, false alarm rate (FAR), probability of detection (p_d) and if a multiple detections scheme is used, number of detections needed. For the tracking link, key consideration is the angle error induced by the receiver circuitry. This angle error is commonly referred as noise effective angle (NEA). For the communication link, key parameters are: required data rate, probability of error p_e or bit error rate (BER) and probability of burst error (PBE) [1]. The bit error occurs due to the presence of various internal and external sources of noise in the optical receiver. It is normally determined by the average/peak signal power and rms noise power during a bit interval. The burst error occurs when the signal irradiances at the distant receiver falls below the level required to maintain average BER. When considered from the pointing and tracking point of view, burst error occurs when the instantaneous mispoint loss exceeds the value included in the link power budget.

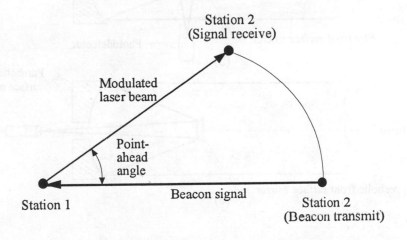

Fig. 15.15: Point-ahead model for optical pointing

15.6.5   OPTICAL RECEIVER

The optical receiver basically consists of a receiving optics (antenna) followed by a demodulator/detector. The latter may be based on incoherent or coherent techniques depending upon the type of modulation scheme used. The various receiver components are described below.

(i)   RECEIVING ANTENNA

The receiving antenna can be a reflective or refractive type collimator. It will gather the optical energy over an aperture of diameter D_r and produce a collimated beam of diameter somewhat smaller than the photodetector area to allow for pointing inaccuracy of the receiver. The photodetector will be placed at the focal point of antenna (Fig. 15.16) [24].

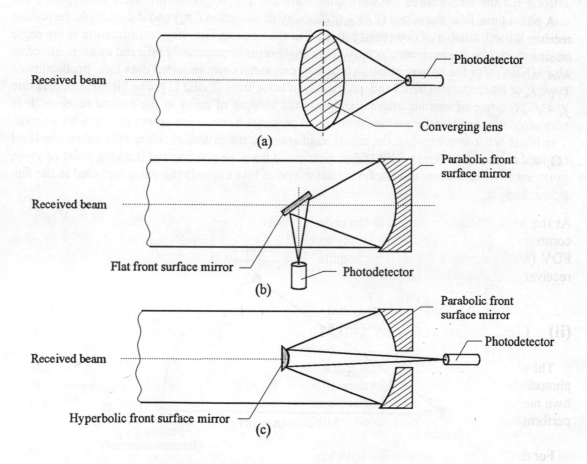

Fig. 15.16: Receiver optical antennas (a) converging lens, (b) Newtonian and (c) Cassegrainian

In a direct detection receiver, diameter of the receiving antenna should be as large as possible to gather the maximum amount of signal energy. Increasing the antenna diameter not only increases the average signal level, but also decreases the ripple in it. In a coherent receiver (both homodyne and heterodyne), there are different constraints. As the aperture diameter increases, effect of intensity fluctuations decreases, but contribution to the variance from phase fluctuations increases. Thus, there is some compromise value of receiver antenna diameter which minimizes the signal variance [24].

Irrespective of whether it is a direct detection or a coherent receiver, amount of optical power focussed onto the detector is equal to the amount of power incident over the focussing lens. Thus, the photodetected power can be equivalently computed by determining received power over the lens area instead of the focussed power over the detector area. The former is obtained directly from the range equation. It may be mentioned that it does not explicitly involve the actual photodetecting area [6].

A photodetector will respond to all radiation focussed on its photoemissive surface. Thus, the receiver FOV given by Eq. (15.15) will depend on detector area A_d and focal length f_c rather than on size A_r of receiving lens. This is in contrast to a RF system wherein receiver FOV depends on λ^2/A_r when diffraction limited. Thus, the optical system FOV can be adjusted independently of A_r. Typically, the focal length of receiver lens is approximately taken as square root of A_r (i.e., $f_c \approx \sqrt{A_r}$). With this, the receiver FOV becomes

$$\Omega_r \cong \frac{A_d}{f_c^2} \approx \frac{A_d}{A_r} = \left(\frac{A_d}{\lambda^2}\right)\left(\frac{\lambda^2}{A_r}\right) \qquad (15.24)$$

At the optical frequency, A_d is in the order of millimetres and λ in the order of microns. As a consequence, (A_d/λ^2) is greater than unity and Ω_r is many times larger than the diffraction limited FOV (λ^2/A_r) for the same A_r. This is quite important in assessing optical visibility and optical receiver noise [6].

(ii) OPTICAL DETECTORS AND MINIMUM REQUIRED POWER EVALUATION

The photodetectors used in optical space communications include PIN-photodiode, avalanche photodiode (APD), photomultiplier tube (PMT) and photon counter. These detectors have their own merits and demerits. As a consequence, an optical receiver with a particular detector may perform better than with other detectors under certain conditions.

For direct detection intersatellite links based on Nd:YAG laser, detector used is either a PIN-photodiode or an APD. For applications requiring gain, silicon APD is generally preferred because of its internal gain. The gain of an APD is limited to about 50-300 because it is a smaller and lighter device. Its excess noise factor increases with the increase in the average gain, M and is usually in the range of 2-5 for the above gain values. When the gain is not required, a PIN-photodiode would be preferred because of its higher quantum efficiency ($\eta \sim 0.8$). At double the frequency of Nd:YAG laser (i.e., $\lambda = 532$ nm), PMT is generally preferred. Its main advantages are very high gain and low excess noise factor. Typical value of gain lies between $10^4 - 10^6$ and excess noise factor between 1 and 2. However, aside from the physical limitation of size and reliability of high voltage components (5 kV or higher), PMTs have lower quantum efficiencies. This is the result of limited materials and deposition techniques for the photoactive surfaces and the inability to collect efficiently all generated photoelectrons. In addition to this, each photoelectron follows a slightly different trajectory as it passes from photocathode to dynodes and finally to anode. Thus, there is a dispersion of arrival times for the photoelectrons. It leads to reduction in the receiver bandwidth. Usually, higher the gain, greater is the reduction in the bandwidth [15]. In coherent communication systems, PIN-photodiode is generally used.

Irrespective of whether the receiver is of direct detection (PIN/APD/PMT) or coherent detection (homodyne/heterodyne) type, background noise sources like Sun, sunlit Earth, Venus etc. may enter in its FOV. Since photodetector in the optical receiver acts as a mixer for the incident optical fields, it will produce signal-background, background-background beat noise components along with the desired signal component. In a coherent receiver, in addition to above noise components, there will be a LO-background beat noise component too. These components degrade the signal-to-noise (S/N) ratio [18].

In space communications, strength of the received signal may become considerably low due to increase in the transmission distance. In such cases, use of a semiconductor optical amplifier (OA) as preamplifier in the receiver may be of great help. When the amplifier output is detected by a photodetector, it will produce signal-ASE and ASE-ASE beat noise components. In a heterodyne receiver, there will be an additional LO-ASE beat noise component. In a TWA, ASE-ASE beat noise component is spread over a wide continuum instead of being concentrated in discrete modes [17].

In the following, performance evaluation of commonly used coherent and incoherent receivers in the background noise environment has been discussed. The receiver structures considered are: (a) coherent PSK homodyne receiver, (b) coherent FSK heterodyne receiver, (c) direct detection (PIN + OA) receiver for OOK, (d) direct detection (APD) receiver for OOK, (e) direct detection (PMT) receiver for OOK, (f) direct detection (APD) receiver for m-PPM and (g) direct detection (PMT) receiver for m-PPM. The general expressions for the output S/N ratio in terms of received optical power level P_r and other receiver parameters are given. These expressions are used to obtain the variations of p_e with P_r in the graphical form for a given data rate R. The minimum required P_r for achieving a given p_e at a given R can easily be determined with the help of these curves. This minimum P_r is subsequently used in the preparation of link power budget.

(a) COHERENT PSK HOMODYNE RECEIVER

A general schematic diagram of an optical communication receiver (valid for different configurations after the deletion of some blocks) is shown in Fig. 15.17.

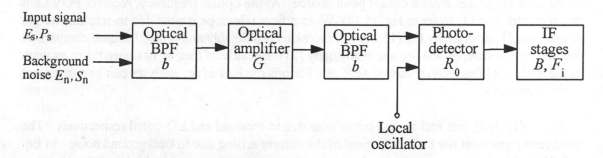

Fig. 15.17: Schematic diagram of optical communication receiver

For PSK homodyne receiver, optical amplifier, following optical BPF and IF stages blocks will not be there. The photodetector current

$$i(t) \propto |E_r(t) + E_n(t) + E_l(t)|^2 \qquad (15.25)$$

where E_r, E_n and E_l are the electric fields due to signal, background noise and local oscillator (LO) respectively. Following [11, 18, 21], various signal and noise components from Eq. (15.25) under the condition that the optical filter bandwidth b is much greater than $B = R = 1/T$ (where T is the bit duration), are given by

Signal Current: $\quad i_s(t) = R_0 2\sqrt{P_r P_l} \cos(\theta(t)) \qquad (15.26a)$

DC Current: $\quad I_{ch} = R_0\left[P_r + P_l + m_t S_n b\right] \qquad (15.26b)$

RMS noise current: $\overline{i_{nh}^2} = eI_{ch}2B + R_0^2 S_n\left[2P_r + 2P_l + m_t S_n b\right]2B$

$$\qquad (15.26c)$$

$$+ \frac{2k_B T_0}{R_l}F_i 2B$$

where $\theta(t) = \theta_l(t) - \theta_r(t)$. The parameters P_r and $\theta_r(t)$ represent the power and phase of received optical signal respectively. Similarly, P_l and $\theta_l(t)$ represent the above quantities for LO signal. In binary PSK signalling scheme, transmitter is considered to send a light pulse with a phase of either zero or π radians. The bandwidth occupied by PSK signal is taken as $B=1/T$. The $R_0 = e\eta/hf$ represents the responsivity of the photodetector, where η is the quantum efficiency, e the electron charge, f the optical frequency of received signal ($f = \omega/2\pi$) and h the Planck's constant. The R_l represents the photodetector load resistance and F_i the NF of the following stages.

In the derivation of Eq. (15.26), it has been considered that the received data field is in single mode as it originates from a distant point source. At the optical frequency, receiver FOV Ω_r is much greater than Ω_{dr} (refer to Fig. 15.12). When a fiber telescope is used due to size and weight constraints, Ω_{dr} as given by Eq. (15.17) will increase considerably and it may become comparable with Ω_r. In that case, m_t will become nearly unity [21]. Because of this, m_t has been taken as unity in the analysis of this receiver as well as others. For other values of m_t, analyses can be carried out in the same way.

In Eq. (15.26b), first and second terms arise due to received and LO signal respectively. The third term represents the DC component of the current arising due to background noise. In Eq. (15.26c), first term represents the shot noise component. The second, third and fourth terms represent the contribution due to signal-background, LO-background and background-background beat noise respectively. The last term represents the thermal noise contribution of the photodetector load resistance and following stages.

The peak output S/N ratio for binary PSK signalling scheme from Eq. (15.26) is given by

$$\frac{S}{N} = \frac{\overline{i_s^2(t)}}{\overline{i_{nh}^2}} \qquad (15.27)$$

Substituting $i_s(t)$ and $\overline{i_{nh}^2}$ from Eq. (15.26a) and Eq. (15.26c) respectively and after further simplification, above equation becomes

$$\frac{S}{N} = \frac{2P_r}{hfBF_h} \qquad (15.28)$$

where

$$F_h = \frac{1}{\eta}\left[1 + \frac{P_r}{P_l} + \frac{S_n b}{P_l}\right] + 2\frac{S_n}{hf}\left[1 + \frac{P_r}{2P_l} + \frac{S_n b}{2P_l}\right] + \frac{k_B T_0}{hf}\frac{F_i}{L_m} \qquad (15.29a)$$

and

$$L_m = \frac{1}{2}R_0^2 P_l R_l \qquad (15.29b)$$

In a balanced homodyne receiver, signal-background and background-background components will be cancelled out [29]. Therefore, F_h from Eq. (15.29a) becomes

$$F_h = \frac{1}{\eta}\left[1 + \frac{P_r}{P_1} + \frac{S_n b}{P_1}\right] + 2\frac{S_n}{hf} + \frac{k_B T_0}{hf}\frac{F_i}{L_m} \qquad (15.30)$$

The p_e is, therefore, given by

$$p_e = \frac{1}{2}\,\text{erfc}\left(\sqrt{\frac{S}{N}}\right) \qquad (15.31)$$

Variations of p_e with P_r computed from the above equation for $\eta = 0.7$, $\lambda = 1064$ nm, $b = 5$ nm, $n_{sp} = 1.0$, $F_i = 2$ dB, $T_0 = 300$ K, $R_1 = 100\ \Omega$, $S_n = 1.9 \cdot 10^{-20}$ Watt/Hz·mode, $R = 560$ Mbit/s and $P_1 = 10$ dBm are shown in Fig. 15.18.

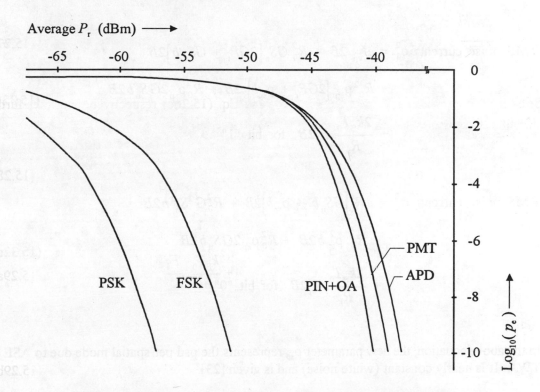

Fig. 15.18: Variations of $\log_{10}(p_e)$ with average P_r for different receiver configurations

(b) COHERENT FSK HETERODYNE RECEIVER

Analysis of this receiver structure can be carried out in the same way as for PSK homodyne receiver. The error curve i.e., the variations of p_e with P_r for binary orthogonal FSK signalling scheme remain same as in the earlier case except that it will shift towards right by 6 dB (refer to Fig. 15.18).

(c) DIRECT DETECTION (PIN + OA) RECEIVER FOR OOK

Block diagram of this receiver structure remains same as in Fig. 15.17 except that LO and IF stages blocks will not be there. In this case, various signal and noise components of the photodetector current under the same condition as in homodyne PSK receiver i.e., $b \gg B$ are given by

Signal current: $\quad\quad I_{sp} = R_0 GP_r$ $\hspace{4cm}$ (15.32a)

DC current: $\quad\quad\quad I_{cp} = R_0 \left[GP_r + GS_n b + \rho_{sp} b \right]$ $\hspace{2.5cm}$ (15.32b)

RMS noise current: $\sigma_1^2 = eI_{cp} 2B + R_0^2 GS_n \left[2GP_r + GS_n b \right] 2B$

$$+ R_0^2 \rho_{sp} \left[2GP_r + \rho_{sp} b \right] 2B + R_0^2 \rho_{sp} 2GS_n b 2B$$

$$+ \frac{2k_B T_0}{R_1} F_i 2B \quad \text{for bit "1"}$$
$\hspace{10cm}$ (15.32c)

RMS noise current: $\sigma_0^2 = eR_0 \left[GS_n b. + \rho_{sp} b \right] 2B + R_0^2 G^2 S_n^2 b 2B$

$$+ R_0^2 \rho_{sp}^2 b 2B + R_0^2 \rho_{sp} 2GS_n b 2B$$

$$+ \frac{2k_B T_0}{R_1} F_i 2B \quad \text{for bit "0"}$$
$\hspace{10cm}$ (15.32d)

In the above equation, the new parameter ρ_{sp} represents the psd per spatial mode due to ASE for TWA. It is nearly constant (white noise) and is given [23]

$$\rho_{sp}(f) = (G - 1)n_{sp} hf$$
$\hspace{9cm}$ (15.33)

where n_{sp} and G are the population inversion parameter and gain of the optical amplifier respectively.

For this signalling scheme, probability of errors for bit "1" and "0" are given by

$$P_{e1} = \frac{1}{2} \text{erfc} \left[\frac{R_0 G P_r - D}{\sqrt{2}\sigma_1} \right]$$

(15.34)

and

$$P_{e0} = \frac{1}{2} \text{erfc} \left[\frac{D}{\sqrt{2}\sigma_0} \right]$$

(15.35)

where D is the threshold level. If an optimum threshold level which equalizes p_{e1} and p_{e0} is used, then average p_e from Eqs. (15.34) and (15.35) will be

$$P_e = P_{e1} = P_{e0} = \frac{1}{2} \text{erfc} \left(\sqrt{\frac{S}{2N}} \right)$$

(15.36)

Here, S/N represents the signal-to-noise ratio and is given by

$$\frac{S}{N} = \frac{R_0^2 \, G^2 \, P_r^2}{\left(\sigma_1 + \sigma_0\right)^2}$$

(15.37)

The Eq. (15.32c) after some mathematical simplifications can be written as

$$\frac{\sigma_1^2}{2 \eta e R_0 G^2 P_r B} = F_{p1}$$

(15.38a)

where

$$F_{p1} = \frac{1}{\eta G}\left[1 + \frac{S_n b}{P_r} + \frac{\tilde{n}_{sp} hfb}{P_r}\right] + 2\left[\tilde{n}_{sp} + \frac{S_n}{hf}\right] \cdot \left[1 + \frac{S_n b}{2P_r} + \frac{\tilde{n}_{sp} hfb}{2P_r}\right]$$

$$+ \frac{k_B T_0}{hf}\frac{F_i}{G^2 \tilde{L}_m} \tag{15.38b}$$

$$\tilde{n}_{sp} = (1 - 1/G)n_{sp} \tag{15.38c}$$

and

$$\tilde{L}_m = \frac{1}{2} R_0^2 P_r R_1 \tag{15.38d}$$

Here, the sign "~" is used to identify direct detection. Similarly, Eq. (15.32d) can also be expressed in the following form:

$$\frac{\sigma_0^2}{2\eta e R_0 G^2 P_r B} = F_{p0} \tag{15.39a}$$

where

$$F_{p0} = \frac{1}{\eta G}\left[\frac{S_n b}{P_r} + \frac{\tilde{n}_{sp} hfb}{P_r}\right] + 2\left[\tilde{n}_{sp} + \frac{S_n}{hf}\right]$$

$$\cdot \left[\frac{S_n b}{2P_r} + \frac{\tilde{n}_{sp} hfb}{2P_r}\right] + \frac{k_B T_0}{hf}\frac{F_i}{G^2 \tilde{L}_m} \tag{15.39b}$$

From Eqs. (15.38a) and (15.39a)

$$(\sigma_1 + \sigma_0)^2 = 2\eta e R_0 G^2 P_r B\left(\sqrt{F_{p1}} + \sqrt{F_{p0}}\right)^2 \tag{15.40}$$

Substitution of Eq. (15.40) in Eq. (15.37) and further simplification gives

$$\frac{S}{N} = \frac{P_r}{2hfB\left(\sqrt{F_{p1}} + \sqrt{F_{p0}}\right)^2} \tag{15.41}$$

Variations of p_e with P_r obtained from Eqs. (15.36) and (15.41) at $\lambda = 850$ nm and for the same values of the other parameters as in PSK homodyne receiver are shown in Fig. 15.18.

(d) DIRECT DETECTION (APD) RECEIVER FOR OOK

Schematic diagram of this receiver configuration will remain same as in Fig. 15.17 except that optical amplifier, following BPF, LO and IF stages blocks will not be there. Further, PIN-photodiode will be replaced by APD. In this case, various signal and noise components under the condition $b \gg B$ are given by

Signal current: $\qquad I_{sa} = R_0 M P_r \tag{15.42a}$

DC current: $\qquad I_{ca} = M^2 F R_0 \left[P_r + S_n b \right] \tag{15.42b}$

RMS noise current: $\sigma_1^2 = eI_{ca}2B + M^2 R_0^2 S_n \left[2P_r + S_n b \right] 2B$

$$+ \frac{2k_B T_0}{R_1} F_i 2B \quad \text{for bit "1"} \tag{15.42c}$$

RMS noise current: $\sigma_0^2 = eM^2 F R_0 S_n b 2B + M^2 R_0^2 S_n^2 b 2B$

$$+ \frac{2k_B T_0}{R_1} F_i 2B \quad \text{for bit "0"} \tag{15.42d}$$

where M and $F = M^x$ are the multiplication and excess noise factors of the APD respectively. The optimum value of M which maximizes S/N ratio is given by [4]

$$M_{opt}^{x+2} = \frac{4k_B T_0 F_i / R_1}{xe R_0 (P_r + S_n b)}$$

(15.43)

Following the same approach as in direct detection (PIN + OA) receiver for OOK, p_e can be obtained by using Eqs. (15.36) and (15.41) with F_{p1} and F_{p0} replaced by F_{a1} and F_{a0} respectively. The F_{a1} and F_{a0} are given by

$$F_{a1} = \frac{F}{\eta}\left[1 + \frac{S_n b}{P_r}\right] + 2\frac{S_n}{hf}\left[1 + \frac{S_n b}{2P_r}\right] + \frac{k_B T_0}{hf}\frac{F_i}{M_{opt}^2 \tilde{L}_m}$$

(15.44a)

and

$$F_{a0} = \frac{F}{\eta}\frac{S_n b}{P_r} + 2\frac{S_n}{hf}\frac{S_n b}{2P_r} + \frac{k_B T_0}{hf}\frac{F_i}{M_{opt}^2 \tilde{L}_m}$$

(15.44b)

Variations of p_e with P_r for the same parameters as in direct detection (PIN + OA) receiver for OOK and $x = 0.5$ are shown in Fig. 15.18.

(e) DIRECT DETECTION (PMT) RECEIVER FOR OOK

In contrast to APD receiver, M_{opt} and F are almost independent of the received optical signal power level in PMT receiver. Further, M_{opt} is very high ($\approx 10^5$) and F is quite low (≈ 1.5). Therefore, the variations of p_e with P_r for this receiver can be obtained as for the APD receiver for OOK by taking $M = 10^5$, $F = 1.5$, $\lambda = 532$ nm, $\eta = 0.3$ and keeping the other parameters same. It is shown in Fig. 15.18.

(f) DIRECT DETECTION (APD) RECEIVER FOR m-PPM

In the m-PPM signalling scheme, let each word convey n bits of information. It means

$$m = 2^n$$

(15.45)

For the same R as in OOK (also in binary PSK and FSK), the frame interval T_R (Fig. 15.19) is given by

$$T_R = n \cdot T$$

(15.46)

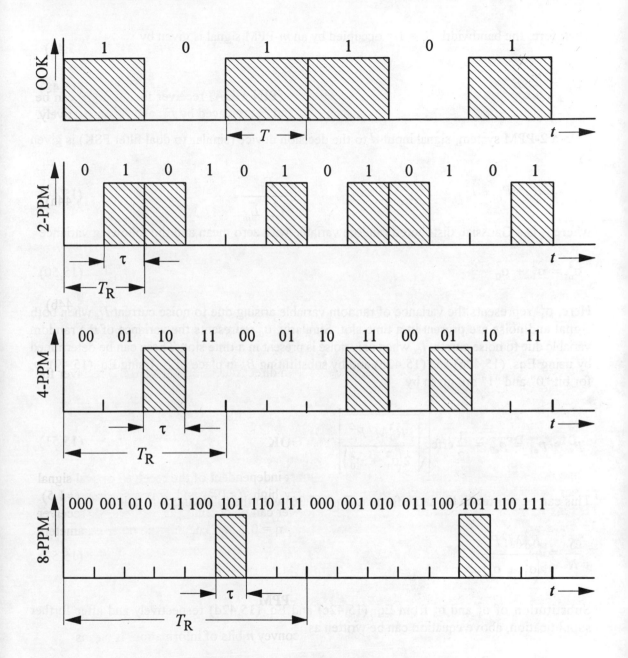

Fig. 15.19: OOK and m-PPM signals (m = 2, 4 and 8)

The slot interval τ from Eq. (15.46) will be

$$\tau = \frac{n \cdot T}{m} \qquad (15.47)$$

Therefore, the bandwidth $B_P = 1/\tau$ occupied by an m-PPM signal is given by

$$B_P = \frac{m}{n} B \tag{15.48}$$

In a 2-PPM system, signal input d to the decision device (similar to dual filter FSK) is given by

$$d = \pm R_0 M P_r + I_n \tag{15.49}$$

where I_n is a Gaussian distributed random variable with zero mean and the following variance

$$\sigma_n^2 = \sigma_1^2 + \sigma_0^2 \tag{15.50}$$

Here, σ_1^2 represents the variance of random variable arising due to noise current I_1, when both signal and noise are present in a time slot. Similarly, σ_0^2 represents the variance of the random variable due to noise current I_0, when only noise is present in a time slot. These can be determined by using Eqs. (15.42c) and (15.42d) and by substituting B_P in place of B. Using Eq. (15.49), p_e for bit "0" and "1" are given by

$$P_e = P_{e1} = P_{e0} = \frac{1}{2} \operatorname{erfc} \left[\sqrt{ \frac{R_0^2 M^2 P_r^2}{2 \left(\sigma_1^2 + \sigma_0^2 \right)} } \right] \tag{15.51}$$

This can be written in the same form as Eq. (15.36) with the following S/N ratio

$$\frac{S}{N} = \frac{R_0^2 M^2 P_r^2}{\left(\sigma_1^2 + \sigma_0^2 \right)} \tag{15.52}$$

Substitution of σ_1^2 and σ_0^2 from Eq. (15.42c) and Eq. (15.42d) respectively and after further simplification, above equation can be written as

$$\frac{S}{N} = \frac{P_r}{2hfB_P \left(F_{a1} + F_{a0} \right)} \tag{15.53}$$

In an m-PPM system, decision regarding the presence of pulse in a time slot is made on the basis of $(m-1)$ comparisons. Following the same approach as for an orthogonal m-FSK signalling scheme, probability p_{ew} of wrongly decoding a word is given by

$$p_{ew} = 1 - \int_{-\infty}^{\infty} \left[\int_{-\infty}^{R_0 MP_r + I_1} p(I_0) \, dI_0 \right]^{m-1} p(I_1) \, dI_1 \qquad (15.54)$$

where $p(I_1)$ and $p(I_0)$ are the probability density functions of I_1 and I_0 respectively. These are considered to be Gaussian distributed with variance σ_1^2 and σ_0^2 respectively. Substitution of $p(I_1)$ and $p(I_0)$ in the above equation and further simplification gives

$$p_{ew} = 1 - \frac{1}{\sqrt{2\pi\sigma_1^2}} \int_{-\infty}^{\infty} \left[1 - \frac{1}{2} \text{erfc} \left(\frac{R_0 MP_r + I_1}{\sqrt{2}\sigma_0} \right) \right]^{m-1} \cdot \exp \left[-\frac{I_1^2}{2\sigma_1^2} \right] dI_1 \qquad (15.55)$$

This p_{ew} can be used to obtain the upper bound on the probability of error p_e. Following [26], it is given by

$$p_e \leq \frac{m/2}{(m-1)} p_{ew} \qquad (15.56)$$

If the quantity inside the second square bracket in Eq. (15.55) is very small («1), then Eq. (15.55) can be approximated as [10]

$$p_{ew} \approx \frac{(m-1)}{2} \cdot \text{erfc} \left[\sqrt{\frac{R_0^2 M^2 P_r^2}{2\left(\sigma_1^2 + \sigma_0^2\right)}} \right] \qquad (15.57)$$

For a 2-PPM system, Eq. (15.57) reduces to Eq. (15.51) which is an exact expression for p_e. For an m-PPM system ($m > 2$) and $p_e < 10^{-4}$, Eq. (15.57) is a good approximation of p_{ew} given in Eq. (15.55). Variations of p_e with average P_r for different m at $\lambda = 1064$ nm and for the same values of the other parameters as in direct detection (PIN + OA) receiver and $x = 0.5$ obtained from Eq. (15.57) are shown in Fig. 15.20.

(g) DIRECT DETECTION (PMT) RECEIVER FOR m-PPM

The error curves for this receiver can be obtained as for direct detection (APD) receiver for m-PPM by taking $\lambda = 532$ nm, $M = 10^5$, $F = 1.5$, $\eta = 0.3$ and keeping the other receiver parameters same. It is shown in Fig. 15.21.

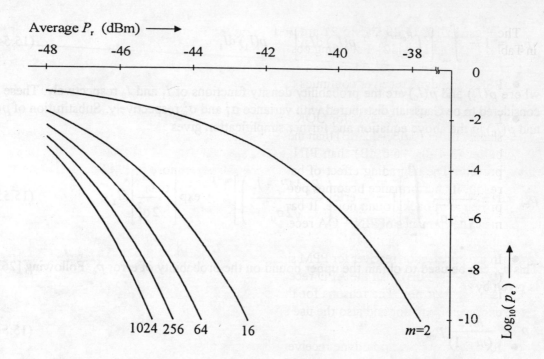

Fig. 15.20: Variations of $\log_{10}(p_e)$ with average P_r for direct detection (APD) receiver for m-PPM

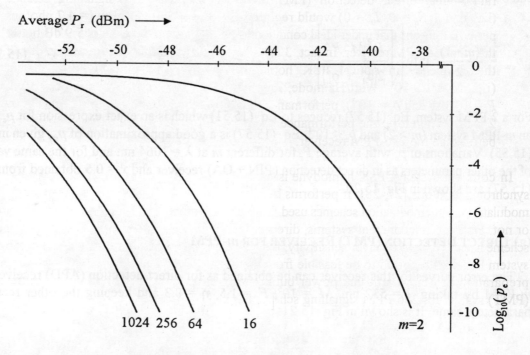

Fig. 15.21: Variations of $\log_{10}(p_e)$ with average P_r for direct detection (PMT) receiver for m-PPM

The Figs. 15.18, 15.20 and 15.21 are used to present some of the results in the tabular form in Tables 15.3 and 15.4 [12]. Following observations are made from the above figures and tables.

- PSK homodyne receiver continues to perform better than the other receivers.

- In direct detection receiver for OOK signalling scheme, PIN + OA receiver performs slightly better (0 dB - 2.2 dB) than the APD receiver. Further, PMT receiver performs better (3.4 dB - 6.6 dB) than PIN + OA receiver when background noise is not present. The degrading effect of background noise is more in PMT receiver. As a result, its performance becomes poorer (0.5 dB - 1 dB) than PIN + OA receiver in presence of background noise. It basically means that in background noise environment, performance of PIN + OA receiver is better than the APD and PMT receivers.

- In direct detection receiver for PPM signalling scheme, PMT receiver performs better (0.1 dB - 5.2 dB) than the APD receiver. This is true whether the background noise is present or not. The reasons for this better performance are higher gain and low excess noise factor and also the use of double the frequency.

- Ideal optical PSK homodyne receiver (i.e., $P_1 \rightarrow \infty$, $\eta \rightarrow 1$, $S_n \rightarrow 0$, $T_0 \rightarrow 0$) at a data rate of 560 Mbit/s would require -66.0 dBm optical power (= 2.4 photons per bit) to achieve an error rate of 10^{-3} (typical for Mars-Earth link). At the same data and error rates, ideal direct detection (PMT) receiver for 256-PPM signalling scheme (i.e., $\eta \rightarrow 1$, $S_n \rightarrow 0$, $T_0 \rightarrow 0$) would require power level of -60.1 dBm (= 4.7 photons per bit). It means that under ideal conditions, PSK homodyne receiver is 5.9 dB better than the 256-PPM receiver. In fact, 3 dB advantage arises only due to the use of half the frequency in optical PSK homodyne receiver. In practical environment (i.e., $S_n = 1.9 \cdot 10^{-20}$ Watt/Hz·mode, $T_0 = 300$ K, $\eta = 0.7$ for PSK and 0.3 for PPM, $P_1 = 10$ mW and $p_e = 10^{-3}$), performance of the PSK homodyne receiver degrades by 2.6 dB. Therefore, the performance of PSK homodyne receiver becomes 10.9 dB better than the 256-PPM receiver.

In coherent systems, PSK homodyne is the best system, in particularly in combination with synchronization bits [27, 28]. It performs better than the incoherent systems irrespective of the modulation and demodulation schemes used. This is true whether the background noise is present or not. Further, in incoherent systems, direct detection (PMT) system for 1024-PPM signalling scheme has the lowest receiver sensitivity, but its practical realization is difficult. The 256-PPM system has been reported to be feasible from the practical point of view [19]. Therefore, in the preparation of Mars-Earth link power budget in the following Section, only direct detection (PMT) system for 256-PPM signalling scheme has been considered.

Table 15.3: Minimum required average power P_r in dBm at a data rate of 560 Mbit/s with and without background noise

* It will be slightly different at different λ, but its order of magnitudes remains same.
£ For FSK heterodyne, required P_r will be 6 dB higher than in PSK homodyne.
$ Computation of P_r is a good approximation of exact expression for $p_e < 10^{-4}$.

p_e	S_n^* Watt/ Hz·mode	PSK Homodyne£ λ=1064 nm η=0.7	PIN+OA for OOK λ=850 nm η=0.7	APD for OOK λ= 850 nm η=0.7	PMT for OOK λ= 532 nm η=0.3	APD for 256-PPM$ λ=1064 nm η=0.7	PMT for 256-PPM$ λ=532 nm η=0.3
10^{-9}	0.0	-58.1	-40.2	-38.6	-43.6	-45.2	-49.1
	$1.9 \cdot 10^{-20}$	-57.6	-40.0	-37.8	-39.0	-44.8	-48.8
10^{-6}	0.0	-60.1	-41.4	-40.3	-46.2	-46.6	-50.8
	$1.9 \cdot 10^{-20}$	-59.6	-41.1	-39.3	-40.2	-46.0	-50.2
10^{-3}	0.0	-63.9	-43.4	-43.4	-50.0	-48.6	-53.1
	$1.9 \cdot 10^{-20}$	-63.4	-43.1	-41.8	-42.6	-47.7	-52.5

Table 15.4: Minimum required power P_r in dBm and photons count at a data rate of 560 Mbit/s and probability of error $p_e = 10^{-3}$.

* At 1064 nm wavelength.
$ At 532 nm wavelength.

System	Optical power P_r in dBm	Number of photons per bit
PSK* (Practical)	-63.4	4.3
256-PPM$ (Practical)	-52.5	27.1
PSK* (Ideal)	-66.0	2.4
256-PPM$ (Ideal)	-60.1	4.7

15.6.6 LINK BUDGET

As an example, the following table presents the power budget for Mars-Earth link in the tabular form.

Table 15.5: Power budget for Mars-Earth link at data rates of 10 Mbit/s and 1 Mbit/s for $p_e = 10^{-3}$

System parameter	Data rate = 10 Mbit/s		Data rate = 1 Mbit/s	
	PSK homodyne λ = 1064 nm	256-PPM direct detection λ = 532 nm	PSK homodyne λ = 1064 nm	256-PPM direct detection λ = 532 nm
Minimum P_r	-80.9 dBm	-65.0 dBm	-90.9 dBm	-70.5 dBm
G_r	149.4 dB	155.4 dB	149.4 dBm	155.4 dB
Space loss	360.7 dB	354.7 dB	360.7 dB	354.7 dB
G_t	117.4 dB	123.4 dB	117.4 dB	123.4 dB
Misalignment loss	6.1 dB	6.1 dB	6.1 dB	6.1 dB
Atmospheric loss	10.0 dB	10.0 dB	10.0 dB	10.0 dB
Safety margin	5.0 dB	5.0 dB	5.0 dB	5.0 dB
Required average P_t	34.1 dBm	32 dBm	24.1 dBm	26.5 dBm
	2.6 W	1.6 W	0.3 W	0.4 W
Peak P_t	2.6 W	405.7 W	0.3 W	114.4 W

In the above table, diameter of transmitting antenna on Mars is presumed to be 25 cm. Similarly, let the diameter of receiving antenna on Earth be 10 m. Using Eqs. (15.9a) and (15.9b), gains of the transmitting and receiving antennas at 532 nm are 123.4 dB and 155.4 dB respectively. At this wavelength, free-space loss i.e., $(\lambda/4\pi L)^2$ considering Mars-Earth distance to be 78 million km is 254.7 dB. Substitution of these values of G_t, free-space loss and G_r in Eq. (15.8) gives

$$P_r(\text{dBm}) = P_t(\text{dBm}) - 75.9 \text{ dB} \tag{15.58}$$

If misalignment loss of 6.1 dB, atmospheric loss of 10.0 dB and safety margin of 5.0 dB are taken into account, then the required P_t at 532 nm from Eq. (15.58) will be

$$P_t(\text{dBm}) = P_r(\text{dBm}) + 97.0 \text{ dB} \tag{15.59}$$

At 1064 nm wavelength and for the same diameters of the transmitting and receiving antennas, gain of each antenna will be reduced by 6 dB as compared to 532 nm wavelength. At this new wavelength, space loss will increase by 6 dB. The required transmitted power in this case, considering the same additional losses and safety margin, is given by the equation

$$P_t(\text{dBm}) = P_r(\text{dBm}) + 115.0 \text{ dB} \tag{15.60}$$

For PSK homodyne receiver, minimum required average P_r at $\lambda = 1064$ nm, $R=1$ Mbit/s and $p_e = 10^{-3}$ from Eq. (15.31) is -90.9 dBm. Using Eq. (15.60), required average P_t will be 24.1 dBm. It means an average/peak power level of 0.3 W. This result, however, can further be improved by employing convolutional coding as proofed in [28].

For 256-PPM direct detection receiver at 532 nm and for the same data and error rates, minimum required average P_r from Eqs. (15.56) and (15.57) is -70.5 dBm. Therefore, required average P_t from Eq. (15.59) is 26.5 dBm. It corresponds to an average power level of approximately 0.4 W and a peak power level of 114.4 W. With the current source technology, it is possible to generate this power level in a pulse mode. Thus, it is feasible to establish the above link using either of the system.

At $R = 10$ Mbit/s and for the same p_e and λ, required peak power level for PSK homodyne system is 2.6 W. It is difficult to generate this peak power level in a continuous mode. The 256-PPM direct detection system at the above data rate will require a peak power level of 405.7 W. Though this power level is much higher than the power level required by the PSK homodyne system, it is practically possible to generate such a peak power level in a pulse mode.

15.7   REFERENCES

[1]   Begley, D. L.; Kobylinski, R. A.; Ross, M.: Solid-state laser cross-link systems and technology. International J. Satellite Commun. (1988)6, 94-105.

[2]   Boutemy, J.: Use of CCD arrays for optical link acquisition and tracking. Proc. SPIE on Optical Systems for Space Applications, Vol. 810(1987), 215-222.

[3]   Chinlon, L.: Optoelectronics Technology and Lightwave Communication Systems. Van Nostrand Reinhold Company Inc., 1989.

[4]   Franz, J.: Optische Übertragungssysteme mit Überlagerungsempfang. Springer-Verlag, 1988.

[5]   Franz, J.: Optical space communication. Lecture notes at IIT Delhi, 1993.

[6]   Gagliardi, R. M.; Karp, S.: Optical Communications. John Wiley & Sons Inc., 1976.

[7]   Gagliardi, R. M.: Satellite Communications. Van Nostrand Reinhold Inc., 1984.

[8]   Gordon, J. P.: Quantum effects in communication systems. IRE Proc., 50(1962)9, 1898-1908.

[9]   Gower, J.: Optical Communication Systems. Second Edition, Prentice-Hall, 1993.

[10]   Jain, V. K.: Study on Space Optical Communication System. Technical Report, DLR Institute for Communication Technology, Germany, 1991.

[11]   Jain, V. K.: Optical preamplifier in noncoherent space communications. J. Opt. Commun. 14(1993)2, 75-88.

[12]   Jain, V. K.: Effect of background noise in space optical communication systems. AEÜ 47(1993)2, 98-107.

[13]   Joindet, M.: Heterodyne receiving techniques in microwave and optics: comparison of some concepts. Workshop on Digital Communication. Tirrenia, 1985, 179-191.

[14]   Katz, J.: Phase Locking of Semiconductor Injection Lasers. TDA Progress Report 42-66, Jet Propulsion Laboratory, Pasadena, CA, 1981, 101-114.

[15]   Katzman, M.: Laser Satellite Communication. Prentice-Hall, 1987.

[16]   Kobylinski, R. A.: Coherent Nd:YAG laser communications. Proc. SPIE on Optical Technologies for Space Communication Systems, Vol.756(1987), 117-121.

[17]   Kobayashi, S.: Semiconductor optical amplifiers. IEEE Spectrum, 21(1984)5, 26-32.

[18]   Leeb, W. R.: Degradation of signal-to-noise ratio in optical space data links due to background illuminations. Applied optics, 28(1989)15, 3443-3449.

[19]   Lesh, J. R.: Optical communication research to demonstrate 2.5 bits/detected photon. Commun. Magazine, 1982, 35-37.

[20]   Lesh, J. R.; Deutsch L. J.; Weber, W. J.: Plan for the development and demonstration of optical communications for deep space. Proc. SPIE on Optical Space Communication II, Vol.1552 (1991), 27-35.

[21]   Letterer, R.; Krichbaumer, W.; Wallmeroth, K.: Signal-to-noise ratio considerations for an analog direct detection receiver with integrated optical amplifier. ESA Contract Number 7660/88/JS, 1989.

[22]   Oliver, B. M.: Thermal and quantum noise. Proc. IEEE, 53(1965)5, 436-454.

[23]   Olsson, N. A.: Lightwave systems with optical amplifiers. IEEE J. LT-7(1989)7, 1071-1082.

[24]   Pratt, W. K.: Laser Communication Systems. John Wiley & Sons Inc., 1969.

[25]   Sobel, H.: The application of microwave techniques in lightwave systems. IEEE J. LT-5(1987)3, 293-299.

[26]   Taub, H.; Schilling, D. L.: Principles of Communication Systems. Second Edition, McGraw-Hill, 1986.

[27]   Wandernoth, B.: 1064 nm, 565 Mbit/s PSK transmission experiment with homodyne receiver using synchronization bits. Electron. Lett. 27(1991)19, 1692-1693.

[28]   Wandernoth, B.: 5 photon/bit low complexity 2 Mbit/s PSK transmission breadboard experiment with homodyne receiver applying synchronization bits and convolutional coding. ECOC (1994), 59-62.

[29]   Yamashita, S.; Okoshi, T.: Analysis of Optical Fiber Communication System with Optical Amplifiers. Tech. Rep. IEICE Japan, No. OCS-90-20, 1990.

[30]   Zhu, Z.: Principles and design of bidirectional optical ISL using GaAlAs laser at λ=0.85 μm and direct detection PPM. International J. Satellite Commun. (1988)6, 81-90.

APPENDIX

A.1   TRANS-ATLANTIC-TRANSMISSION LINKS

In the following table, characteristic features of *trans-Atlantic transmission* (TAT) links are presented. Capacity of all the systems is given in terms of number of voice channels or bit rate. It may be mentioned that (i) TAT does not allow TV transmission and (ii) TAT, HAW (Hawaii cable) and TPC (Trans-Pacific cable) are equivalent e.g., TAT-8 is equivalent to HAW-4 as well as to TPC-3.

Table A1: Characteristic features of TAT links

Year	Name	Medium	Wavelength in μm	Number of voice channels/ bit rate
1956	TAT-1	Coaxial cable	-	84
1959	TAT-2	Coaxial cable	-	84
1963	TAT-3	Coaxial cable	-	128
1965	TAT-4	Coaxial cable	-	128
1970	TAT-5	Coaxial cable	-	720
1976	TAT-6	Coaxial cable	-	4,000
1983	TAT-7	Coaxial cable	-	4,000
1988	TAT-8	Fiber	1.30	40,000
1991	TAT-9	Fiber	1.55	40,000
1992	TAT-10	Fiber	1.55	40,000
1993	TAT-11	Fiber	1.55	40,000
1997	TAT-12	Fiber with optical amplifiers	1.55	60,000 (5 Gbit/s)
1997	TAT-13	Fiber with optical amplifiers	1.55	60,000 (5 Gbit/s)
1999	TAT-13 upgrade	Fiber with WDM	1.55	3·5Gbit/s
2001	TAT-14	Fiber with WDM	1.55	>50·STM1

A.2   GRIN-ROD LENS

The graded index (GRIN)-rod lens consists of a cylindrical glass rod typically 0.5 mm to 2 mm in diameter which exhibits a parabolic RI profile with a maxima at the centre. Light propagation through the lens is determined by the lens dimension and wavelength of light since the RI is wavelength dependent. The GRIN-rod lens can produce a collimated output beam with a divergent angle α between 1° and 5° from a light source situated on or near to the opposite lens face as shown in Fig. A.1.

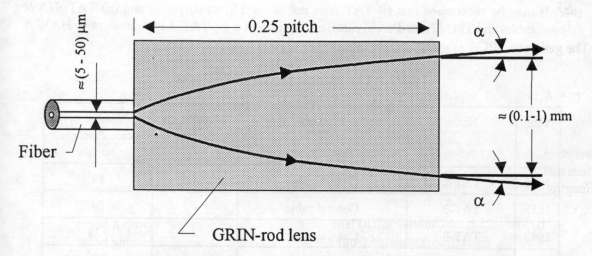

Fig. A.1: Formation of a collimated beam from a GRIN-rod lens

There are number of factors which can cause divergence of the collimated beam. These include errors in the lens cut length, the finite size of the fiber core and chromatic aberration. Conversely, GRIN-rod lens can be used to focus an incoming light beam on to a small area located at the centre of opposite lens face.

The ray propagation through the GRIN-rod lens is approximately described by the following ray equation

$$\frac{d^2r}{dz^2} = \frac{1}{n}\frac{dn}{dr} \tag{A.1}$$

where r is the radial co-ordinate, z the distance along the optical axis and n the RI at a point. Further, the RI at a distance r from the optical axis in a graded-index medium ($g = 2$) from Eq.(4.42) is approximately given by

$$n(r) \approx n_1\left(1 - \frac{Ar^2}{2}\right) \tag{A.2}$$

where n_1 is the RI on the optical axis and A the positive constant to be decided by the designer. Substitution of Eq. (A.1) in Eq. (A.2) and with an approximation, we get

$$\frac{d^2r}{dz^2} \approx -Ar \tag{A.3}$$

The general solution of the above equation following [1] is given by

$$r = K_1\cos(\sqrt{A}z) + K_2\sin(\sqrt{A}z) \tag{A.4}$$

where K_1 and K_2 are constant. It means that the input rays follow a sinusoidal path through the lens medium. Its period is called one full pitch. These lenses are manufactured with several pitch lengths. Three major pitch lengths are shown in Fig. A.2 and their descriptions are as follows.

(i) The 0.25 pitch (quarter pitch) lens produces perfectly collimated output beam when the input beam emanated from a point source or focuses an incoming beam to a point at the centre of the opposite lens face (Fig. A.2a). The focal point of such a lens is coincident with the lens face. It thus provides an efficient direct butt connection to optical fiber [2].

(ii) The 0.23 pitch lens is designed such that its focal point lies outside the lens when a collimated beam is projected on the opposite lens face. It is often used to convert the divergent beam from a fiber or laser diode into a collimated beam (Fig. A.2b) [3].

(iii) The 0.29 pitch lens is designed such that both focal points lie just outside the lens end faces. It is often used to convert a diverging beam from a laser diode into a converging beam (Fig. A.2c). It is quite useful for coupling laser diode output into an optical fiber or alternatively for coupling the output from an optical fiber into a detector [2].

The majority of GRIN-rod lenses have diameters in the range 0.5 mm and 2 mm and may be employed with either SM or MM (SI or GI) fiber.

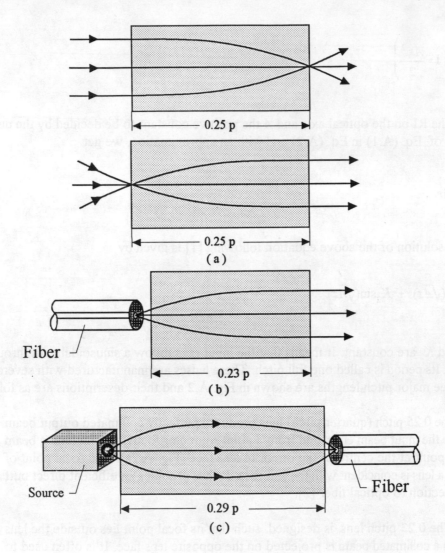

Fig. A.2: Operation of (a) the quarter pitch, (b) the 0.23 pitch and
(c) the 0.29 pitch GRIN-rod lenses

REFERENCES

[1] Miller, S. E.: Light propagation in generalized lenslike media. Bell Syst. Tech. J., 44(1965), 2017-2064.

[2] Senior, J. M.: Optical Fiber Communication-Principles and Practice. Second Edition, Prentice-Hall, 1992.

[3] Sono, K.: Graded index rod lenses. Laser Focus, 17(1981), 70-74.

INDEX